SINGLE-DISH RADIO ASTRONOMY: TECHNIQUES AND APPLICATIONS

A SERIES OF BOOKS ON RECENT DEVELOPMENTS IN
ASTRONOMY AND ASTROPHYSICS

Publisher

THE ASTRONOMICAL SOCIETY OF THE PACIFIC
390 Ashton Avenue, San Francisco, California, USA 94112-1722
Phone: (415) 337-1100   E-Mail: catalog@astrosociety.org
Fax: (415) 337-5205   Web Site: www.astrosociety.org

ASP CONFERENCE SERIES - EDITORIAL STAFF
Managing Editor: D. H. McNamara   LaTeX-Computer Consultant: T. J. Mahoney
Associate Managing Editor: J. W. Moody   Production Manager: Enid L. Livingston
Production Assistant: Andrea Weaver

PO Box 24453, Room 211 - KMB, Brigham Young University, Provo, Utah, 84602-4463
Phone: (801) 422-2111   Fax: (801) 378-4049   E-Mail: pasp@byu.edu

ASP CONFERENCE SERIES PUBLICATION COMMITTEE:
Alexei V. Filippenko   Geoffrey Marcy
Ray Norris   Donald Terndrup
Frank X. Timmes   C. Megan Urry

A listing of all of the ASP Conference Series Volumes and IAU Volumes
published by The ASP may be found at the back of this volume

ASTRONOMICAL SOCIETY OF THE PACIFIC
CONFERENCE SERIES

Volume 278

SINGLE-DISH RADIO ASTRONOMY: TECHNIQUES AND APPLICATIONS

Proceedings of the NAIC-NRAO Summer School held at
National Astronomy and Ionosphere Center, Arecibo Observatory,
Arecibo, Puerto Rico, USA
10-15 June 2001

Edited by

Snežana Stanimirović
National Astronomy and Ionosphere Center, Arecibo Observatory, Arecibo, Puerto Rico, USA

Daniel R. Altschuler
National Astronomy and Ionosphere Center, Arecibo Observatory, Arecibo, Puerto Rico, USA

Paul F. Goldsmith
National Astronomy and Ionosphere Center, Department of Astronomy, Cornell University, Ithaca, New York, USA

and

Christopher J. Salter
National Astronomy and Ionosphere Center, Arecibo Observatory, Arecibo, Puerto Rico, USA

© 2002 by Astronomical Society of the Pacific. All Rights Reserved

No part of the material protected by this copyright notice may be reproduced or utilized in any form or by any means – graphic, electronic, or mechanical including photocopying, taping, recording or by any information storage and retrieval system, without written permission from the publisher.

Library of Congress Cataloging in Publication Data
Main entry under title

Card Number: 2002111600
ISBN: 1-58381-120-6

ASP Conference Series - First Edition

Printed in United States of America by Sheridan Books, Chelsea, Michigan

Contents

Preface ix

Participants xi

Conference Photograph xvi

The National Astronomy and Ionosphere Center's (NAIC) Arecibo
  Observatory in Puerto Rico 1
  D. R. Altschuler

Part 1. Fundamentals of Single-Dish Radio Astronomy

Why Single-Dish? 27
  D. T. Emerson

Radio Telescopes and Measurements at Radio Wavelengths 45
  P. F. Goldsmith

Measurement in Radio Astronomy 81
  D. B. Campbell

The Receiver System – cm Regime 91
  R. D. Norrod

The Receiver System – mm Regime 99
  J. M. Payne

Back-ends 113
    J. R. Fisher

Spectrometry and Autocorrelation 123
    J. Hagen

A Heuristic Introduction to Radioastronomical Polarization 131
    C. Heiles

Part 2. Single-Dish Observing Disciplines and Associated Techniques

Continuum 1: General Aspects 155
    J. J. Condon

Continuum 2: Specific Applications 173
    C. J. Salter

The Rudiments of Spectral Line Radio Astronomy 187
    H. S. Liszt

Spectral Line Advanced Topics 209
    J. M. Dickey

Pulsar Observations I. – Propagation Effects, Searching, Distance
    Estimates, Scintillations and VLBI 227
    J. M. Cordes

Pulsar Observations II. – Coherent Dedispersion, Polarimetry, and Timing   251
    I. H. Stairs

Planetary Radar Astronomy 271
    G. J. Black

Part 3. Calibration and Data Reduction

Single-Dish Calibration Techniques at Radio Wavelengths 293
    K. O'Neil

Millimeter Wave Calibration Techniques	313
P. R. Jewell	
Reduction and Analysis Techniques	329
R. J. Maddalena	
Single-Dish Analysis within AIPS++	353
J. P. McMullin, R. W. Garwood, T. Cornwell, A. Kemball, and J. Braatz	

Part 4. Special Topics

Short-Spacings Correction from the Single-Dish Perspective	375
S. Stanimirović	
Stray Radiation: Causes, Curses, and Cures	397
F. J. Lockman	
The Effects of the Atmosphere	413
L. Olmi	
RFI and How to Deal with It	433
J. R. Fisher	
Spectrum Management	447
T. Ghosh	
Focal Plane Arrays	453
J. M. Payne	
Bolometers for Submillimeter and Millimeter Astronomy	463
W. Holland, W. Duncan, and M. Griffin	
Single-Dish Radio Telescopes of the World	493
C. J. Salter	

Part 5. Poster Papers

First *VLBI* Observations with Arecibo in an International *S2* Ad-hoc Array	507
I. Molotov, A. Chuprikov, S. Likhachev, C. J. Salter, T. Ghosh, F. Ghigo, and S. Dougherty	

Exploring the Sensitivity Limit: 21-cm Observations of Low Surface
  Brightness Galaxies with the Arecibo Telescope 511
    K. O'Neil, J. Eder, G. Bothun, and J. Schombert

Simulating the Performance of Large Format Sub-mm Focal-plane Arrays 515
    E. Chapin and D. H. Hughes

On the Life and Death of OH/IR Stars 519
    B. M. Lewis

Arecibo Spectral Baselines in the Presence of Continuum Emission . . . 521
    T. Ghosh and C. J. Salter

Project Phoenix: SETI Observations from 1200 to 1750 MHz with the
  Upgraded Arecibo Telescope 525
    P. R. Backus and the Project Phoenix Team

Epilogue 529
    L. Solanch

Index 531

Single–Dish Radio Astronomy: Techniques and Applications
ASP Conference Series, Vol. 278, 2002
S. Stanimirović, D. R. Altschuler, P. F. Goldsmith, and C. J. Salter

Preface

This volume contains the lectures and poster papers presented at the first NAIC-NRAO School on Single Dish Radio Astronomy, which took place at the Arecibo Observatory in June 2001. The justification for this School was a combination of several factors. The first was the availability of both the Upgraded Arecibo 305-m telescope and the new Robert C. Byrd Green Bank Telescope (GBT) – both major new tools for astronomers in the 21st century. The second was the feeling that 'would be' users should benefit from a set of lectures whose breadth ranged from radio astronomy basics to in-depth discussions of some of the special techniques that have been developed over the years, but which are rarely described in print. Finally, we hoped that by including "hands on" projects, we would give participants a real feeling for radio astronomical observations at centimeter and millimeter wavelengths.

We believe that the goals of the organizers were fulfilled thanks to the efforts of the organizing committees, the lecturers, the participants, and the excellent support from the staff of the whole Arecibo Observatory. This School was the first event to use the new "Learning Center" of the Arecibo Observatory Visitor and Education Facility, and construction personnel were working overtime finishing things even the day before the start of the school. The feedback from the participants has been positive, and we certainly hope that this will be the first in an on-going series of such schools. New groups of students, as well as professional astronomers with astronomical backgrounds at different wavelengths, will have an on-going need to be introduced to the basics of centimeter and millimeter wavelength radio astronomy, as well as to be brought up to speed in respect of the latest developments. With the new instruments of NAIC[1] and NRAO[2] (plus others around the world), this type of school will serve a purpose for years to come.

The organization for this School was carried out by two groups working together. The Scientific Organizing Committee included staff from NAIC and NRAO. Its members, Daniel Altschuler, Darrel Emerson, Paul Goldsmith, Phil Jewell, Ron Maddalena, Chris Salter, and Snežana Stanimirović, assembled the lecture program, and invited lecturers to participate. The Local Organizing Committee for this School was responsible for the logistics at Arecibo, and included José Alonso, Daniel Altschuler, Ramesh Bhat, Rey Medina, Karen O'Neil, Chris Salter, and Snežana Stanimirović. The success of the School was

---

[1] The Arecibo Observatory is part of the National Astronomy and Ionosphere Center, operated by Cornell University under a cooperative agreement with the National Science Foundation.

[2] The National Radio Astronomy Observatory is a facility of the National Science Foundation operated under cooperative agreement by Associated Universities, Inc.

dependent on the efforts of all these individuals, as well as other NAIC and NRAO staff. At Arecibo, we are particularly grateful to the Visitor Center and VSQ staff for their support of the activities and to Edith Alvarez, secretary to the school, who worked diligently to resolve the day-to-day problems that inevitably arise in an endeavor such as this. We are indebted to the NASA Puerto Rico Space Grant for their support of the school. We would also like to acknowledge support for all aspects of this School from the National Science Foundation through both the National Radio Astronomy Observatory and the National Astronomy and Ionosphere Center.

A subgroup of the organizing committees metamorphosed into an editorial committee with responsibility for editing the contributions of the lecturers and assembling their material into the required form for publication as this volume. The editors would especially like to thank Hector Hernandez Arzola for his careful reading of the final manuscript with 'fresh and alert' eyes. The editors are also grateful to members of the Arecibo astronomy and electronics departments for their valuable assistance.

Our hope is that these lectures will prove to be useful for astronomers of varied backgrounds and experience who become involved with radio astronomy at different levels.

The editors:

Snežana Stanimirović
Daniel Altschuler
Paul Goldsmith
Chris Salter

Participants

ALONSO JOSÉ, National Astronomy and Ionosphere Center, Arecibo Observatory, Arecibo, Puerto Rico, USA ⟨alonso@naic.edu⟩

ALTSCHULER DANIEL, National Astronomy and Ionosphere Center, Arecibo Observatory, Arecibo, Puerto Rico, USA ⟨daniel@naic.edu⟩

ARAYA ESTEBAN, University of Puerto Rico, San Juan, Puerto Rico, USA ⟨earaya@astro4.cnnet.clu.edu⟩

ASGEKAR ASHISH, Raman Research Institute, Bangalore, India ⟨ashish@rri.res.in⟩

BHAT RAMESH, National Astronomy and Ionosphere Center, Arecibo Observatory, Arecibo, Puerto Rico, USA ⟨rbhat@naic.edu⟩

BLACK GREGORY, National Radio Astronomy Observatory, Green Bank, West Virginia, USA ⟨gblack@nrao.edu⟩

CABANELA JUAN, Saint Cloud State University, Saint Cloud, Minnesota, USA ⟨juan@lua.stcloudstate.edu⟩

CABRERA IVELISSE, University of Puerto Rico, Mayaguez, Puerto Rico, USA ⟨icabrera@coqui.net⟩

CAMPBELL DONALD, National Astronomy and Ionosphere Center, Cornell University, Ithaca, New York, USA ⟨campbell@astrosun.astro.cornell.edu⟩

CASTRO EDGAR, National Astronomy and Ionosphere Center, Arecibo Observatory, Arecibo, Puerto Rico, USA ⟨ecastro@naic.edu⟩

CATINELLA BARBARA, Cornell University, Ithaca, New York, USA ⟨barbara@astro.cornell.edu⟩

CERSÓSIMO HOMERO, University of Puerto Rico, Humacao, Puerto Rico, USA ⟨homero_ac@hotmail.com⟩

CHANDLER ADAM, Caltech, Pasadena, California, USA ⟨amc@srl.caltech.edu⟩

CHAPIN EDWARD, I.N.A.O.E., Puebla, Pue., Mexico ⟨echapin@inaoep.mx⟩

CHUNG SUNMI, Wesleyan University, Middletown, Connecticut, USA ⟨schung@mail.wesleyan.edu⟩

COLES WILLIAM, University of California, San Diego, California, USA ⟨bcoles@ucsd.edu⟩

CONDON JIM, National Radio Astronomy Observatory, Charlottesville, Virginia, USA ⟨jcondon@nrao.edu⟩

CORDES JIM, Cornell University, Ithaca, New York, USA ⟨cordes@astrosun.astro.cornell.edu⟩

DAUGHTERY DANIEL, University of Alabama, Huntsville, Alabama, USA ⟨sdaugher@cs.uah.edu⟩

DEREMER LINDSAY, Wellesley College, Wellesley, Massachusetts, USA ⟨lderemer@wellesley.edu⟩

DEL RÍO M. SOLEDAD, I.N.A.O.E., Puebla, Pue., Mexico ⟨sole@inaoep.mx⟩

DICKEY JOHN, University of Minnesota, Minneapolis, Minnesota, USA ⟨john@astro.umn.edu⟩

EHLEROVA SONA, Astronomical Inst., Acad. Sciences, Prague, Czech Republic ⟨sona@ig.cas.cz⟩

EMERSON DARREL, National Radio Astronomy Observatory, Tucson, Arizona, USA ⟨demerson@nrao.edu⟩

EYDENBERG MICHAEL, New Mexico Institute of Mining and Technology, Socorro, New Mexico, USA ⟨mseyden@nmt.edu⟩

FIGUEROA NATALIA, University of Puerto Rico, Mayaguez, Puerto Rico, USA ⟨nati-rum@yahoo.com⟩

FISHER RICK, National Radio Astronomy Observatory, Green Bank, West Virginia, USA ⟨rfisher@nrao.edu⟩

FORD JOHN, NRAO, Green Bank, West Virginia, USA ⟨jford@nrao.edu⟩

FREIRE PAULO, National Astronomy and Ionosphere Center, Arecibo Observatory, Arecibo, Puerto Rico, USA ⟨pfreire@naic.edu⟩

GALLO LUIGI, San Diego State University, La Mesa, California, USA ⟨lgallo@mintaka.sdsu.edu⟩

GANESAN RAJAGOPALAN, National Astronomy and Ionosphere Center, Arecibo Observatory, Arecibo, Puerto Rico, USA ⟨ganesh@naic.edu⟩

GHOSH TAPASI, National Astronomy and Ionosphere Center, Arecibo Observatory, Arecibo, Puerto Rico, USA ⟨tghosh@naic.edu⟩

GOLDSMITH PAUL, National Astronomy and Ionosphere Center, Cornell University, Ithaca, New York, USA ⟨pfg@astrosun.astro.cornell.edu⟩

GOMEZ MERCEDES, Observatorio Astro. de Cordoba, Cordoba, Argentina ⟨mercedes@oac.uncor.edu⟩

GREVE TOMAS, University College London, London, UK ⟨trg@star.ucl.ac.uk⟩

GROSSI MARCO, University of Wales, Cardiff, UK ⟨marco.grossi@astro.cf.ac.uk⟩

GUTIERREZ JORDI, Universitat Politécnica de Catalunya, Barcelona, Spain ⟨jordi@fa.upc.es⟩

HAGEN JON, National Astronomy and Ionosphere Center, Arecibo Observatory, Arecibo, Puerto Rico, USA ⟨hagen@naic.edu⟩

HEALY KEVIN, Arizona State University, Tempe, Arizona, USA ⟨kevin.healy@asu.edu⟩

HEILES CARL, University of California, Berkeley, California, USA ⟨cheiles@astro.berkeley.edu⟩

HERNANDEZ-TOLEDO HECTOR, Instituto de Astronomia, UNAM, Mexico D.F., Mexico ⟨ hector@astroscu.unam.mx ⟩

HOBBS GEORGE, Jodrell Bank Observatory, University of Manchester, Macclesfield, Cheshire, UK ⟨ ghobbs@jb.man.ac.uk ⟩

HOFNER PETER, University of Puerto Rico, Rio Piedras, Puerto Rico, USA ⟨ hofner@naic.edu ⟩

HOLLAND WAYNE, UK-Astronomy Technology Centre, Royal Observatory, Edinburgh, UK ⟨ wsh@roe.ac.uk ⟩

JACOBY BRYAN, Caltech, Pasadena, California, USA ⟨ baj@astro.caltech.edu ⟩

JEWELL PHIL, National Radio Astronomy Observatory, Green Bank, West Virginia, USA ⟨ pjewell@nrao.edu ⟩

KAPLAN DAVID, Caltech, Pasadena, California, USA ⟨ dlk@astro.caltech.edu ⟩

KEARNS KRISTIN, University of Wisconsin, Superior, Wisconsin, USA ⟨ kkearns@uwsuper.edu ⟩

KOPON DEREK, Cornell University, Ithaca, New York, USA ⟨ dak29@cornell.edu ⟩

KRCO MARKO, Colgate University, Hamilton, New York, USA ⟨ mkrco@mail.colgate.edu ⟩

LAI SHIH-PING, University of Illinois, Urbana, Illinois, USA ⟨ slai@astro.uiuc.edu ⟩

LANE WENDY, Naval Research Lab, Washington DC, USA ⟨ lane@rsd.nrl.navy.mil ⟩

LEBRÓN SANTOS MAYRA, Max Planck Institute für Radioastronomie-Bonn, Bonn, Germany ⟨ mlebron@mpifr-bonn.mpg.de ⟩

LEWIS DION, University of Tasmania, Hobart, Tasmania, Australia ⟨ Dion.Lewis@atnf.csiro.au ⟩

LEWIS MURRAY, National Astronomy and Ionosphere Center, Arecibo Observatory, Arecibo, Puerto Rico, USA ⟨ blewis@naic.edu ⟩

LISTER MATTHEW, National Radio Astronomy Observatory, Charlottesville, Virginia, USA ⟨ mlister@nrao.edu ⟩

LISZT HARVEY, National Radio Astronomy Observatory, Charlottesville, Virginia, USA ⟨ hliszt@nrao.edu ⟩

LOCKMAN FELIX, National Radio Astronomy Observatory, Green Bank, West Virginia, USA ⟨ jlockman@nrao.edu ⟩

LÓPEZ-CRUZ OMAR, Universidad de Guanajuato, Guanajuato, Mexico ⟨ omar@astro.ugto.mx ⟩

MADDALENA RON, National Radio Astronomy Observatory, Green Bank, West Virginia, USA ⟨ rmaddale@nrao.edu ⟩

MASTERS KAREN, Cornell University, Ithaca, New York, USA ⟨ masters@astrosun.astro.cornell.edu ⟩

MCMULLIN JOSEPH, National Radio Astronomy Observatory, Charlottesville, Virginia, USA ⟨ jmcmulli@nrao.edu ⟩

MURGIA MATTEO, Istituto di Radioastronomia CNR, Bologna, Italy
⟨murgia@ira.bo.cnr.it⟩

NEAKRASE JENNIFER, Arizona State University, Tempe, Arizona, USA
⟨Jennifer.Neakrase@asu.edu⟩

NICOLLS MICHAEL, Cornell University, Ithaca, New York, USA
⟨mjn25@cornell.edu⟩

NORROD ROGER, National Radio Astronomy Observatory, Green Bank, West Virginia, USA ⟨rnorrod@nrao.edu⟩

NOWAKOWSKI LESZEK, University of Puerto Rico, Mayaguez, Puerto Rico, USA ⟨leszekan@coqui.net⟩

OLMI LUCA, University of Puerto Rico, Rio Piedras, Puerto Rico, USA
⟨olmi@fcrao1.astro.umass.edu⟩

O'NEIL KAREN, National Astronomy and Ionosphere Center, Arecibo Observatory, Arecibo, Puerto Rico, USA ⟨oneil@naic.edu⟩

PANTOJA CARMEN, University of Puerto Rico, Rio Piedras, Puerto Rico, USA
⟨cpantoja@astro2.cnnet.clu.edu⟩

PAYNE JEFFREY, University of Western Sydney Nepean, Sydney, Australia
⟨4snvoa2@gte.net⟩

PAYNE JOHN, National Radio Astronomy Observatory, Tucson, Arizona, USA
⟨jpayne@nrao.edu⟩

PHILLIPS VAL, University of Colorado, Denver, Colorado, USA
⟨phillips_val@hotmail.com⟩

RIVERA-CASTILLO NEFTALIS, University of Puerto Rico, Arecibo, Puerto Rico, USA ⟨sotero@coqui.net⟩

RUIZ ABRAHAM, University of Puerto Rico, Humacao, Puerto Rico, USA
⟨a_ruiz@cuhac.upr.clu.edu⟩

SALTER CHRIS, National Astronomy and Ionosphere Center, Arecibo Observatory, Arecibo, Puerto Rico, USA ⟨csalter@naic.edu⟩

SANDSTROM KARIN, Harvard-Smithsonian Center for Astrophysics, Cambridge, Massachusetts, USA ⟨ksandstr@fas.harvard.edu⟩

SANTOS ESTHER, University of Puerto Rico, Mayaguez, Puerto Rico, USA
⟨gremypocha@yahoo.com⟩

SEWILO MARTA, University of Wisconsin, Madison, Wisconsin, USA
⟨sewilo@astro.wisc.edu⟩

SOLANCH LARRY, University of Georgia, Athens, Georgia, USA
⟨solanch@hal.physast.uga.edu⟩

SPEKKENS KRISTINE, Cornell University, Ithaca, New York, USA
⟨spekkens@astro.cornell.edu⟩

SPRINGOB CHRIS, Cornell University, Ithaca, New York, USA
⟨springob@astro.cornell.edu⟩

STAIRS INGRID, National Radio Astronomy Observatory, Green Bank, West Virginia, USA ⟨istairs@nrao.edu⟩

STANIMIROVIĆ SNEŽANA, National Astronomy and Ionosphere Center, Arecibo Observatory, Arecibo, Puerto Rico, USA ⟨ sstanimi@naic.edu ⟩

STENNES MICHAEL, NRAO, Green Bank, West Virginia, USA ⟨ mstennes@nrao.edu ⟩

UMANA GRAZIA, Istituto di Radioastronomia CNR, Noto, Italy ⟨ umana@ira.noto.cnr.it ⟩

VARGAS CARLOS, University of Puerto Rico, Mayaguez, Puerto Rico, USA ⟨ cvargas@naic.edu ⟩

VENKATARAMAN ARUN, National Astronomy and Ionosphere Center, Arecibo Observatory, Arecibo, Puerto Rico, USA ⟨ arun@naic.edu ⟩

VEKINIS PETER, University of Arizona, Steward Observatory, Tucson, Arizona, USA ⟨ peter@t12.org ⟩

WATSON CHRISTER, University of Wisconsin, Madison, Wisconsin, USA ⟨ watson@astro.wisc.edu ⟩

WRAY LISA, National Astronomy and Ionosphere Center, Arecibo Observatory, Arecibo, Puerto Rico, USA ⟨ lwray@naic.edu ⟩

YOUNG OWL ROLAINE, University of California LA, Los Angeles, California, USA ⟨ rolaine@astro.ucla.edu ⟩

CONFERENCE PHOTO

CONFERENCE PHOTO KEY

1. Juan Cabanela
2. Sona Ehlerova
3. Ron Maddalena
4. Daniel Altschuler
5. Shih-ping Lai
6. Edith Alvarez
7. Karen O'Neal
8. Wendy Lane
9. Ashish Asgekar
10. Jo Ann Eder
11. Homero Cersósimo
12. Matthew Lister
13. Karen Sandstrom
14. Sun Mi Chung
15. Ivelisse Cabrera
16. Marco Grossi
17. Snezana Stanimirovic
18. Paulo Freire
19. Matteo Murgia
20. Ramesh Bhat
21. Lisa Wray
22. Jordi Gutierrez
23. Dion Lewis
24. Edward Chapin
25. Mike Nicols
26. John Dickey
27. Marta Sewilo
28. Grazia Umana

1. Alexandra Santos
2. Natalie Figueroa
3. Val Phillips
4. Carlos Vargas
5. Adam Chandler
6. Brian Jacoby
7. Jim Cordes
8. Edgar Castro
9. Hector Hernandez
10. Mike Stennes
11. Karen Masters
12. Jennifer Neakrase
13. Daniel Altschuler
14. Paul Goldsmith
15. Jon Hagen
16. George Hobbs
17. John Ford
18. Roger Norrod
19. David Kaplan
20. Daniel Dougherty
21. Kristine Spekkens
22. Chris Springob
23. Kevin Healy
24. Marta Sewilo

1. Daniel Altschuler
2. Lindsay De Remer
3. Esteban Araya
4. Larry Solanch
5. Omar Lopez Cruz
6. Peter Vekinis
7. Jeffrey Payne
8. Soledad del Rio
09. Barbara Catinella
10. Kristin Kearns
11. Mercedes Gomez
12. Phil Jewel
13. Leshek Novakovski
14. Ingrid Stairs
15. Jim Condon
16. Mayra Lebron
17. Tapasi Ghosh
18. Abraham Ruiz
19. Christer Watson
20. Derek Kopon
21. Tomas Greve
22. Carmen A. Pantoja
23. Neftali Rivera
24. Rolaine Young Owl
25. Luigi Gallo
26. Joseph Mc Mullin
27. R. Ganesan
28. Harvey Liszt
29. Rick Fisher
30. Darrel Emerson
31. Chris Salter

The National Astronomy and Ionosphere Center's (NAIC) Arecibo Observatory in Puerto Rico

Daniel R. Altschuler

National Astronomy and Ionosphere Center, Arecibo Observatory, HC 3 Box 53995, Arecibo, Puerto Rico 00612, USA

Abstract. This paper reviews the history of the Arecibo Observatory, its genesis, construction, and the two upgrades through which the remarkable 305-m telescope continues to contribute significant results in many areas of astronomy and atmospheric physics.

1. Introduction

The Arecibo Observatory is part of the National Astronomy and Ionosphere Center (NAIC), a national research center operated by Cornell University under a cooperative agreement with the National Science Foundation (NSF). The Observatory operates on a continuous basis, 24 hours a day every day, providing observing time, electronic, computer, travel and logistic support to visiting scientists. The Observatory is located on the Caribbean island of Puerto Rico, about 10 km south of the town of Arecibo, which is located on the north coast of the island (Figure 1).

All research results are published in the scientific literature which is publicly available. As the site of the world's largest single-dish radio telescope, the Observatory is recognized as one of the most important national centers for research in radio astronomy, planetary studies (via radar and passive observation) and space and atmospheric science. Use of the Arecibo Observatory is available on an equal, competitive basis to scientists from throughout the world. Observing time is granted on the basis of the most promising research as ascertained by a panel of independent referees who review all proposals sent to the Observatory by interested scientists.

In addition, the Observatory hosts an Optical Laboratory with a variety of instrumentation used for the passive study of terrestrial airglow. A lidar (Light Detection and Ranging) facility is used to measure the neutral winds, composition, and temperature of the middle atmosphere. This instrumentation complements the incoherent scatter radar used to study the Earth's ionosphere, giving Arecibo a unique capability for aeronomic research.

The Observatory had its origins in an idea of Professor William E. Gordon, then of Cornell University, who was interested in the study of the Ionosphere. During the fifties, Gordon's research led him to the idea of radar back scatter studies of the ionosphere. His persistence culminated in the construction of the Arecibo Observatory, which began in the Summer of 1960. Three years later,

Figure 1. The island of Puerto Rico is about 100 by 40 miles in size. You can just see a white dot about one quarter from the west shore (left) and one quarter from the north shore (top). This is the Arecibo Reflector. (NASA)

Figure 2. A panoramic view of the Arecibo Observatory in the Karst of northern Puerto Rico (Courtesy Stephane Aubin, Ciel et Espace).

the Arecibo Ionospheric Observatory (AIO) was in operation under the direction of Gordon, the formal opening ceremony taking place on November 1, 1963.

In 1960, the science of radio astronomy was still in its infancy. The ubiquitous 21-cm line emitted by neutral hydrogen (HI) atoms in the interstellar space of the Milky Way had been detected in 1951 with a special horn antenna built at Harvard's Lyman Laboratory by Edward Purcell and his graduate student Harold Ewen (Ewen & Purcell 1951), but studies of neutral hydrogen in galaxies was still in its infancy. The first extragalactic HI detection was obtained in 1953 when Frank Kerr and collaborators detected the line from the Magellanic Clouds (Kerr, Hindman, & Robinson 1954), and by 1960 HI in just a few tens of galaxies had been detected. The problem was the extreme weakness of the signals from distant sources which could therefore only be detected with very large antennas and sensitive receivers.

Also in 1960, the investigations which lead to the discovery of Quasars by Maarten Schmidt in 1963 had just begun, and only a few powerful radio galaxies had by then been identified (Schmidt 1963).

It would be seven years before pulsars were discovered in November 1967 by Jocelyn Bell and Anthony Hewish (Hewish et. al. 1968) using an antenna they had built in Cambridge, England.

Between April and July of 1960, an 85-foot radio telescope of the NRAO located at Green Bank, West Virginia, pointed at the stars Tau Ceti and Epsilon Eridani, some eleven light years away. This was project Ozma, the first search for signals from another technology, conducted by Frank Drake, later to become Director of the Arecibo Observatory (from 1966 to 1968).

Radar astronomy had gotten a few feeble echoes from the Moon as early as 1946, but the planets remained elusive targets. On March 10, 1961, echoes from Venus were finally obtained by the JPL Goldstone radars. Two antennas of 26 meter diameter, one used to transmit a radio wave using a modest (by today standards) 10 kilowatt transmitter and the other to receive the echo, were used. After a round trip of 113 million km, the weak echo was detected by the second antenna after about six minutes of travel time. This work paved the way for a greatly improved value of the Astronomical Unit (Muhleman, Holdrige, & Block 1962).

Over forty long years have elapsed since then and the Arecibo telescope, still the largest telescope on Earth, has become a household word. To the scientific community it is known for its many contributions to science, one of which earned the Nobel Prize in Physics in 1993, and to the public it is known from uncountable media exposures, including two major films (Golden Eye and Contact), as well as for the many (false) stories about how at the Arecibo Observatory we regularly communicate with "Them".

Those who see the Arecibo radio telescope for the first time are astounded by the enormity of its reflecting surface, or radio mirror. This huge "dish" is 305 m (1,000 feet) in diameter, 167 feet deep, and covers an area of about twenty acres. Suspended 450 feet above the reflector is the 900-ton platform supporting the antennas and receivers which can be positioned with millimeter precision to point at any direction in the "Arecibo" sky, a forty degree cone of visibility about the local zenith (between -1 and 38 degrees of declination). Just below the triangular frame of the upper platform is a circular track on which the azimuth arm turns. The azimuth arm is a bow-shaped structure, 304 feet (93 m) long. Fixed to the curved part of the arm is a second track, on one side of which a carriage house moves and on the other side moves the Gregorian dome (not named for a pope or for music but in honor of James Gregory, one of the foremost mathematicians of the seventeenth century, being the first professor of mathematics at the University of Edinburgh). This can be positioned anywhere up to twenty degrees from the center of the arm.

Similar in design to a bridge, the platform hangs in midair on eighteen cables, which are strung from three reinforced concrete towers. One tower is 365 feet high and the other two are 265 feet high. All three tower tops are at the same elevation. Each tower is back stayed to large concrete ground anchors with seven 3.25-inch diameter steel bridge cables (Figure 2).

2. Gestation

The headlines of the New York Times for October 5, 1957 read:

"Soviet Fires Earth Satellite Into Space; It Is Circling the Globe at 18,000 M.P.H.; Sphere Tracked in 4 Crossings Over U.S."

The Soviets had launched Sputnik 1, (Satellite 1), the first man-made satellite, an aluminum sphere with a diameter of twenty-two inches and a weight of 184 pounds (83 kg), on October 4, 1957. It circled the globe 500 miles above the surface every 96 minutes and sent a beeping sound that both disturbed and fascinated the world. It was disturbing because it meant that the Soviets had the capability to send missiles toward the US. It was fascinating since it was the first step in our conquest of space. It represented a sensational goal that temporarily set the score as USSR – 1, USA – 0 in the long and frightening game called the Cold War.

But, four months later, on January 31, 1958, a US Army team lead by Wernher von Braun, and composed of experts who had worked on the German war efforts to build rockets (the infamous V2), and had post-war emigrated to the US, launched Explorer 1 from Cape Canaveral in Florida. The small 30 pound

(14 kg) satellite carried a Geiger-Mueller counter built by James Van Allen, a physicist at the University of Iowa, that could detect high energy particles. Explorer 1 discovered what later came to be known as the Van Allen Radiation Belts (Van Allen, McIlwain, & Ludwig 1959).

One of the consequences of Sputnik was the creation of the National Aeronautics and Space Administration (NASA). In July 1958, Congress passed the National Aeronautics and Space Act (commonly called the "Space Act"), which created NASA as of October 1, 1958. On February 7, 1958, the Advanced Research Projects Agency (ARPA) was established by the Department of Defense (DoD). ARPA was responsible "for the direction or performance of such advanced projects in the field of research and development as the Secretary of Defense shall, from time to time, designate by individual project or by category". In 1969 ARPA created the ARPAnet to research the transfer of data between computers across systems, thus initiating the predecessor of the Internet.

The introduction to the December 1958 engineering report No. 3 of the School of Electrical Engineering of Cornell University authored by W. E. Gordon, H. G. Booker, & B. Nichols entitled "Design study of a radar to explore the earth's ionosphere and surrounding space" reads as follows (Gordon, Booker, & Nichols 1958a):

"The discovery that free electrons in the Earth's ionosphere incoherently scatter signals that are weak but detectable with a powerful radar makes possible the exploration of the upper atmosphere and surrounding space by radar. The fact that the radar components, while sensitive, are all within the state of the art means that the exploration can begin as soon as the radar is assembled.

The radar for the first time will measure directly the electron density and electron temperature as functions of height and time through the ionosphere not only in the recognized layers but also between and above them. The formation and disappearance of these layers, their structure and the diurnal and seasonal changes in them will be observed. These observations should contribute substantially to our understanding of the ionosphere and its effect on radar waves.

In addition to exploration of the ionosphere, the radar has the following exciting capabilities: a) the observation of transient streams of charged particles traveling through space near the earth, b) the search for the existence of a ring current around the earth, c) radar observation of the planets Venus and Mars and an improved measurement of the astronomical unit of distance, d) the possibility of radar observation of the Sun and its irregular atmosphere, e) the sensitivity to observe heretofore undetected radio stars in a limited region of the sky."

The ionosphere begins at a height of about 60 km, a region where solar ultraviolet radiation and X-rays ionize the gas. Gordon wanted to use a radar to study the ionosphere. Most of the radar energy would pass through the ionosphere and be lost into space, but a tiny fraction would be scattered in all directions by the electrons and a small part of that would be returned back to Earth (backscattered) where a very sensitive antenna could detect and study it, providing information about the ionosphere. Thus the fundamental scientific

Figure 3. In this photo taken in 1960, the beginning of earth movement can be seen on the site of the future Arecibo Observatory.

mission for the AIO as seen by Gordon was to determine the electron density and temperature of the ionosphere as a function of height and time.

Being "within the state of the art" meant a reflector with a diameter of 1,000 feet (305 meters), a transmitter with one megawatt of power operating at a frequency of 430 MHz (70-cm wavelength), and the best receivers then available. The frequency was low enough that the surface of the reflector needed only to be within about 3 cm of perfect to be an efficient reflector. Nevertheless this was still a major engineering challenge for such a large surface.

The minimum size of the reflector was determined by Gordon in a paper published in the Proceedings of the IRE (Institute of Radio Engineers) in 1958 (Gordon 1958b). This is the same journal which in 1922 published a paper by Guglielmo Marconi, in which the concept of Radar was spelled out (Marconi 1922), and where in 1933 Karl Jansky published his results that mark the birth of radio astronomy (Jansky 1933).

Gordon's paper also stated: "Streams of charged particles originating in outer space and flowing near the Earth may or may not be detected, depending on their range from the radar and electron density..." Perhaps this did not escape those at ARPA who had to decide on funding. In the wake of Sputnik it offered a possible way to detect these satellites which would leave a transient trail of ionization as they moved in their orbit at a height of some 500 miles.

Originally, a fixed parabolic reflector was envisioned, pointing in a fixed direction with a 500 foot tower to hold equipment at the focus. Such a design would have had a very limited use for other potential areas of research – plane-

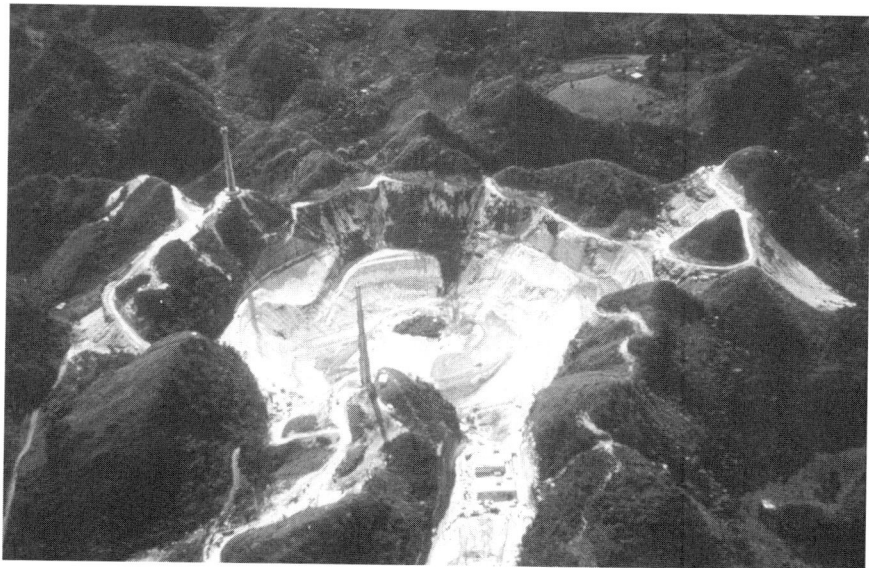

Figure 4. On this photo of 1962, two towers are almost completed, as is the operations building and long stairs to the Visiting Scientists Quarters.

tary studies and radio astronomy – which require the ability to point at different positions in the sky, and to track those positions for an extended period.

It was Ward Low, an official at ARPA and a secret hero in the story of Arecibo, who thought that this was a serious limitation and who suggested to Gordon that he should get in touch with the Air Force Cambridge Research Laboratory (AFCRL), in Boston, Massachusetts. There, a group headed by Phil Blacksmith was working on spherical reflectors, while another group was engaged in an intense study of the upper atmosphere and its effects on radio wave propagation. This effort was related to Signal Intelligence, the attempt to obtain information from intercepted radio waves. Cornell University proposed the project to ARPA in the summer of 1958 and a contract was signed between the AFCRL and the University in November of 1959.

3. Construction

There were four requirements for the site. It had to be in the tropics so that all objects in the solar system would pass overhead. This was an important consideration for planetary radar work since the telescope could not look down to the horizon. If a natural bowl with roughly the size of the reflector could be found it would save significantly on construction costs as otherwise it would be necessary to dig a gigantic hole or build a tall suspension for the reflector. The

Figure 5. The triangular platform was raised with pulleys attached to auxiliary cables strung between the three towers.

site should also have a reasonable climate, be of easy access, and in a politically stable country (Figure 3).

Donald J. Belcher, a professor of Civil Engineering at Cornell and an expert in mapping and aerial photography, was asked to locate an appropriate hole in the ground. In passing, in 1953, Professor Belcher had been involved in locating Brasilia, the new capital of Brasil. After considering such places as Hawaii, Mexico, Cuba and some smaller islands in the Caribbean, Puerto Rico was selected. Belcher studied the Karst topography of northern Puerto Rico and located three possible sites, one in the municipality of Florida, one in that of San Sebastian and the third in Barrio Esperanza, Arecibo. The latter was chosen after an on-site inspection. Not only did it fulfill all the above criteria, it also helped that at the time a faculty member of the engineering department of the University of Puerto Rico at Mayaguez, Braulio Dueño, (now an emeritus professor), was pursuing a doctorate degree at Cornell. There he met and worked with Gordon.

Despite the general shape of the sinkhole within which the telescope was built, nature had to be improved upon by excavating about 270,000 cubic yards of soil, some of it rock requiring blasting. The contractors for this also had to place 200,000 cubic yards of compacted fill to shape the excavation.

Each of the three concrete towers, two of them 265 feet high and the third 365 feet high (111 m), was poured in a slip form that was raised at a rate of about 9 inches/hour so that new concrete was exposed about five hours after it was poured. A total of 9,100 cubic yards of concrete was utilized, the equivalent

Figure 6. On March 3, 1963, the azimuth arm was on its way to the platform.

of 1,000 standard cement trucks. A concrete production plant was installed on site to make this possible.

A 265 feet tower took about 375 hours to pour, ($\sim$ 16 days of continuous pouring). Each tower sits on a reinforced concrete footing 36 x 38 ft by 15 ft deep. The anchors contain a total of 4,650 cubic yards of concrete. The towers are labeled T4, T8, and T12 following the numbers on the face of a watch, T12 being the one due north.

The feed platform, suspended 150 m above the ground, had a total weight of 550 tons. (It later was increased to about 900 tons after the Gregorian upgrade.) It is suspended from four three-inch diameter cables that run from each platform corner to the top of the corresponding tower. Five three-and-a-quarter inch diameter cables run from the top of the towers to the concrete anchors resting on the ground. (The reason for the different size and number of the cables is related to the different angle at which they carry the loads).

Visitors to the Observatory often wonder how the platform was raised in the first place. In fact, it was assembled at the center of the bowl. A cable spider was strung between the towers which supported a block and tackle to raise the platform. The blocks were made of 100 tons of concrete embedded in the ground.

Figure 7. June 13, 1963, the mesh of the reflector is placed over the supporting cables.

The platform lift at the end of October 1962 proceeded at about 50 feet per hour during which a delicate balance was kept between the backstays and the spider and main cables so that tower deflection stayed within the permissible limits of ±2 inches. This was the most critical operation of the entire project and took three days. After the lift, the three inch diameter main cables, which had been hanging from the three corners of the platform, were attached to the tops of the towers. Next, the ring girder, holding the azimuth track and the feed arm were raised to the platform (Figures 4, 5 and 6).

The contractor for the reflector proceeded with construction after the platform was in place. The reflector consisted of 16 x 19 gauge standard inch square mesh soldered at every joint and held by a network of cables suspended from the concrete rim of the reflector. Three hundred and eighteen three-eighths inch cables run east-west, and these are stabilized by 10 (increased to 39 after the first upgrade) one and one quarter inch cables running north-south. These are tied to concrete blocks on the ground to provide the required circular profile (since a free hanging cable will have the form of a catenary – not a circle).

The reflector is a portion of a sphere of radius 870 feet (265 m), has a surface area of 18 acres, and had a total weight of 207 tons (Figure 7).

Figure 8. August 14, 1963, the 430-MHz line feed is lifted.

Figure 9.   An aerial view of the site in 1969.

Finally on August 14, 1963, the first 96-foot long line feed operating at 430 MHz was lifted through the central opening of the reflector and attached to the receiver and transmitter (Kavanagh 1963; Gordon 1964), see Figures 8 and 9.

As illustrated in Figure 10, a spherical reflector does not have a focal point along a principal axis, as is the case for a parabolic reflector. In fact, it has no principal axis since you can view the surface along any radius and it looks the same. Waves incident on a sphere will be concentrated along a focal line which lies along a radius parallel to the direction of incidence. Therefore, if a device can be built to collect the waves in an efficient manner, by placing the device along radii at different angles it becomes possible to point the telescope at different directions in the sky without moving the reflector.

It is, however, not trivial to build an efficient device to collect the waves - a line feed - and in fact the first one used, was quite inefficient. It was a 96-foot (29 meter) long tapered aluminum device with a square cross section and an arrangement of slots through which radiation would pass (this original line feed now graces the entrance of the Observatory's Learning Center). In order for the waves arriving at the line feed at different heights to add coherently, the cross section of the line feed along its length and the design and placement of hundreds of slots through which radiation enters (or exits in the case of transmission) must be controlled precisely. Line feeds also have an inherent limitation caused by dispersion (the phase of the wave being a function of frequency) so that a line feed will operate optimally at one central frequency for which it was designed and increasingly poorly as the frequency diverges from this central frequency. This limits the range of frequencies over which a line feed can operate (the

The NAIC Arecibo Observatory in Puerto Rico 13

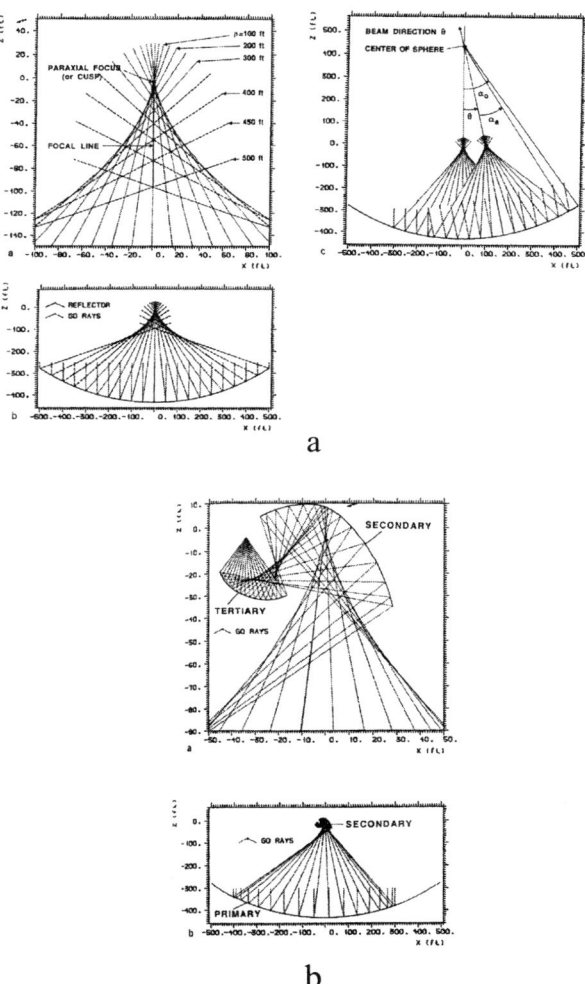

Figure 10. Ray optics for the Arecibo spherical reflector (Kildal 1989). a. Top left. Rays from the rim of the reflector (500 ft radius) will intersect 96 feet below the paraxial focus - the point at which rays from near the center intersect. Bottom. Overall focusing of rays from the 1000 ft aperture. Top Right. Steering of the beam with a 700 ft illumination pattern. b. Top. Dual reflector feed geometry. Bottom. Complete Gregorian system.

bandwidth - which for Arecibo line feeds is between 10 and 40 MHz) which in many cases limited the sensitivity of the telescope and forced the use of several different line feeds to obtain the desired measurements.

Soon after its inauguration, on April 7 1964, the first radar contact with Mercury was achieved, under the leadership of Gordon H. Pettengill, then associate director of the Observatory. The Arecibo radar then made its first surprising discovery: the rotation rate of Mercury was not 88 days, as previously thought, but 59 days (Pettengill & Dyce 1965). An early experiment, under the guise of a study of "lunar temperature", was able to intercept and study the characteristics of a soviet radar operating on the Arctic coast. The giant reflector captured these signals reflected from the Moon and, after a careful study of the changing geometry over time, was able to determine the source of the signal (Gerson 1984).

On October 1, 1969, the National Science Foundation (NSF) took over the facility from the DoD and the Observatory was made a national research center. In September 1971 the AIO became the National Astronomy and Ionosphere Center (NAIC).

4. The First Upgrade

The desire to be able to operate at higher frequencies, both for radio astronomy and for planetary radar, was behind the idea that it was worthwhile to upgrade the surface of the reflector. It was concluded that this would be possible after it was found that the suspended platform was much more stable (something needed for work at centimeter wavelengths) than had originally been thought. The proof of this stability came in August 28, 1966, as Hurricane Inez roared across Puerto Rico subjecting the platform to 70 miles per hour winds. The platform moved barely half an inch (LaLonde 1974).

A high frequency S-band (2380 MHz, 12.6 cm) radar with half a megawatt of power was proposed to greatly enhance the planetary radar capability. Whereas the 430 MHz radar had been able to study the Moon and the terrestrial planets (Mercury, Venus and Mars), the S-band radar would provide much higher resolution on these bodies and be able to study objects at the distance of Jupiter and Saturn. The surface of Venus, eternally shrouded under a thick cloudy atmosphere would become "visible".

A scientific panel, convened by the NSF in July of 1967, and chaired by Robert Dicke at Princeton University concluded that upgrading the reflector of the Arecibo telescope was the best and most economical way to obtain a radio telescope of unprecedented capabilities. However, the NSF felt that until a successful line feed could be constructed, investing in a multimillion dollar upgrade of the reflector was not warranted. It was not until early in 1972 that the difficult problem of designing a more efficient line feed was solved by Alan Love and Merle LaLonde. Merle had acted as chief engineer during the AIO construction. (The ultimate resolution of the limited bandwidth problem, as we shall see later, was part of the second Arecibo upgrade). Once this problem was resolved, the NSF made funds available; this after the Dicke committee expressed in 1969 that, "Resurfacing the Arecibo telescope should be carried out immediately."

Figure 11. In this view of the first upgrade, a large fraction of the reflector had already been replaced.

To work at S-band, several changes were needed, including the new transmitter and a new receiver and, most importantly, the replacement of the reflector by one of much higher precision. As a general rule, the surface of a reflector will need to be constructed with a precision of one twentieth the wavelength at which it will be used. So for a reflector to operate efficiently at 430 MHz (70 cm), such as the original surface, it was sufficient to have the surface not deviate from a perfect sphere by more than about 3 cm, quite a feat over a diameter of 300 meters, but feasible. To have it work efficiently at the wavelength of an S-band radar meant a precision of better than 6 millimeters, a tremendous challenge. The new reflector was expected to achieve a surface accuracy of 3 mm RMS. The old mesh was replaced by 38,778 precisely shaped aluminum panels each 40 by 80 inches in size. A special factory was built on site to fabricate the panels out of perforated aluminum sheets. The holes in the panels allow 44 percent of available sunlight to pass through the reflector allowing vegetation to grow beneath and in this way to control erosion. The total weight of aluminum on the reflector is 692,000 pounds (350 tons) and without the holes it would be much higher. The holes also set an absolute upper limit to the operating frequency since at some point the holes will affect the surface reflectivity adversely. The panel frames were made of aluminum belt. A total of 227 miles of this material was used, enough to build a railing all around the island of Puerto Rico (Figures 11 and 12).

Figure 12. The last panel, number 38,778 is placed on the reflector in November 1973.

The upgraded telescope was dedicated on 15-16 November 1974. As part of the ceremonies the famous Arecibo Message was sent by the new radar. The three-minute-short message was beamed toward the globular cluster M13.

Over the next twenty years, the study of the "21-cm line" of neutral hydrogen, made possible by this new surface, became an important area of Arecibo research. Also in 1974, a professor and his graduate student from the University of Massachusetts arrived at Arecibo to pursue a search for pulsars. The search was initiated during the time when the reflector was being upgraded, since this work did not interfere with continued observations at 430 MHz. The results of studying one of the new pulsars discovered, the binary pulsar PSR1913+16, led to the confirmation of the existence of gravitational waves as predicted by Einstein's theory of gravitation. (Taylor, Fowler, & McCulloch 1979). The 1993 Nobel Prize in Physics was awarded to Joseph Taylor and Russell Hulse for their work at Arecibo.

5. The Second Upgrade

The second and major upgrade to the telescope was completed in 1997 and provides a geometrical optics correction for the spherical aberration of the primary reflector, allowing receivers to be fed by standard horns instead of by line feeds. Several important changes to the telescope were introduced (Goldsmith 1996).

A 16-meter high stainless steel mesh screen surrounding the one kilometer perimeter of the reflector shields the feeds from radiation coming from the ground

Figure 13. The telescope as viewed through the ground screen.

as the illumination pattern spills over. This was completed in August 1993 (Figure 13).

A pair of large sub reflectors (a 22-meter diameter secondary and a 9-meter tertiary reflector) are enclosed in the Gregorian dome, 30 meters in diameter, suspended from the feed arm. The dome was assembled at the center of the main reflector and then lifted to the platform on May 16, 1996. The secondary and tertiary reflectors correct the spherical aberration of the primary and bring radio waves to a point focus where a set of receivers mounted on a rotating floor can be positioned. This eliminates the above mentioned limitations of the line feeds (Figures 14, 15 and 16).

The new S-band transmitter, located in a special room inside the dome, doubles the power of the previous one (to 1 MW). A new 3.3 MW turbine was installed on site to provide power for this radar. Together with the other improvements, this enhances radar sensitivity by a factor of more than ten in some cases. The capability of transmitting from the Gregorian at 430 MHz allowed operation of the atmospheric radar in a dual beam (simultaneously from the line feed and the Gregorian) mode providing new capabilities to measure ionospheric winds.

To support the additional weight of the Gregorian dome considerable work was needed to stiffen the platform and support the 50% increase in the suspended weight. Two additional main and backstay cables with new anchors were installed at each tower. A further system of three pairs of vertical cables

Figure 14. On May 16, 1996, the new Gregorian dome was lifted to the platform.

runs from each corner of the platform to large concrete blocks under the reflector. They are attached to giant jacks so as to adjust the height of each corner with millimeter precision.

A total of 26 new electric motors were installed to control the various new drive systems. These motors drive the azimuth, the Gregorian dome, and the carriage house to any position with millimeter precision. The tertiary reflector can also be moved to improve focusing and pointing, receivers are moved into focus on a rotating floor and the dynamical tie downs are activated as needed to maintain platform position. A complete new system of receivers and associated electronics is located in the dome and covers the frequency range from 300 MHz to 10 GHz. Signals are sent to the control room via optical fibers. New computers were installed and software developed for data acquisition and to perform monitor and control tasks. Special purpose instrumentation, such as a new spectrometer, a new radar decoder and new pulsar signal processors were designed and built at the Observatory.

To reach the required efficiency at the highest frequency range of operation (5 – 10 GHz), the surface of the reflector has been surveyed by photogrammetry with a precision of one millimeter, after which each of the almost 40,000 panels was adjusted to reach an overall RMS of better than 2 mm. A new VLBA4 system has enabled the telescope to participate in VLBI studies of weak sources, and in the future a seven-feed L-band system, now under construction, will further enhance the capabilities of the Arecibo telescope. This system will allow large-scale surveys of the sky with unprecedented sensitivity, in particu-

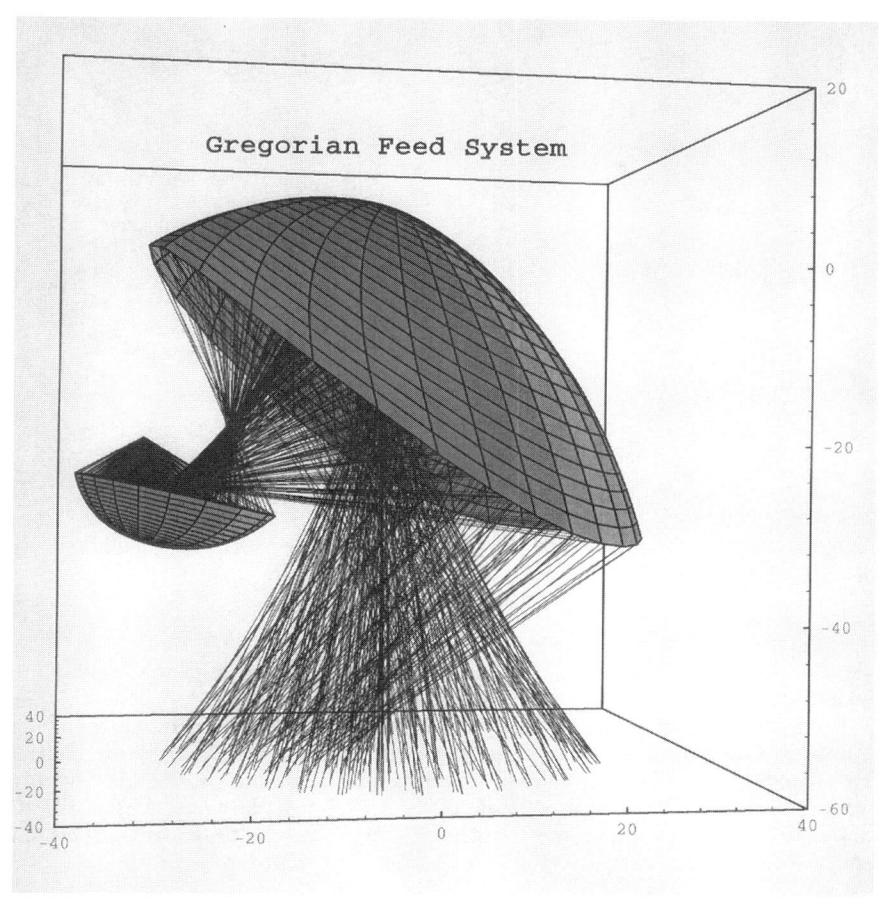

Figure 15. Two sub reflectors correct for the spherical aberration of the primary as shown. Compare with Figure 10(b).

Figure 16. A view inside the Gregorian showing part of the secondary reflector, the tertiary (bottom) and the feed tower.

lar searching for new pulsars, studying the neutral hydrogen in the Galaxy and searching for intergalactic hydrogen clouds.

In preparation for the upgrading work, during the exchange of some structural components, the azimuth arm had to be parked for several weeks. During that time a search for new pulsars at high galactic latitudes was undertaken by Alex Wolszczan (Wolszczan & Frail 1992). As a result, he found PSR1257+12, a pulsar with a planetary system, the first extra-solar one known, albeit not associated with an ordinary star.

The capabilities of the new S-band radar are astounding. As an example, the Near Earth Asteroid 1999 JM 8 was observed at a distance of about 9 million km from Earth. Images of this 3-km sized object with a resolution of 15 meters were obtained (Figure 17). With its radar vision, Arecibo is the leading facility for studies of the properties of planets, comets and asteroids.

The Arecibo telescope has scrutinized our atmosphere, from a few kilometers to a few thousand kilometers, where it smoothly connects with interplanetary space. In our Galaxy it detects the faint pulses emitted hundreds of times per second from millisecond pulsars. Also, from the farthest reaches of the Universe quasars and galaxies emit radio waves which arrive at earth hundreds of millions of years later as signals so weak that they can only be detected by a giant eye such as this one.

The giant size of the reflector is what makes the Arecibo telescope so special to scientists. It is the largest curved focusing antenna on the planet, which means

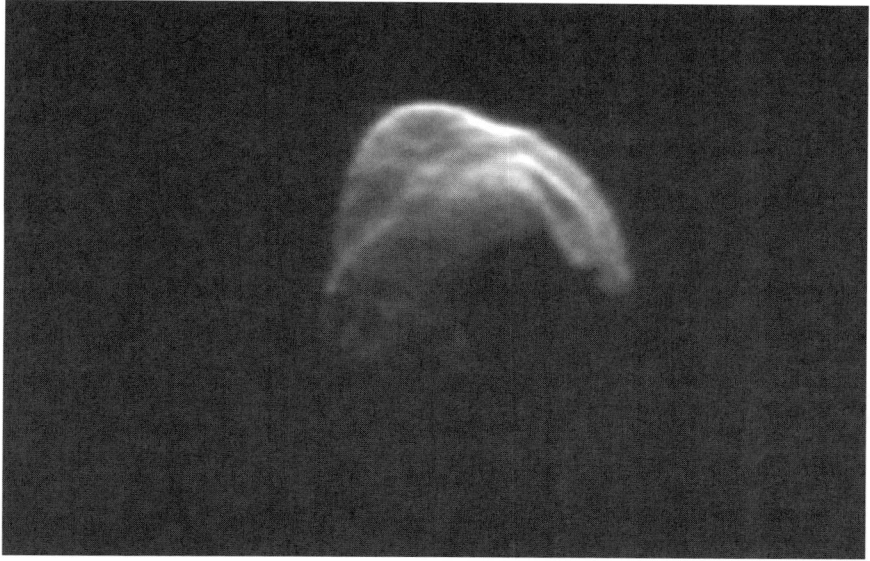

Figure 17. Asteroid 1999 JM8 is an irregular shaped object about 3 km in size. Its closest approach to Earth occurred in August of 1999 at a distance of 8.5 million km. Small impact craters can be seen in this Delay-Doppler image (Benner et. al. 2001, in preparation) (see http://echo.jpl.nasa.gov/~lance/1999JM8/press_release.html).

that it is the world's most sensitive radio telescope. Other telescopes may require several hours observing a given radio source to collect enough energy for analysis whereas at Arecibo this may require just a few minutes of observation.

6. Education and Outreach

Finally, let me also point out that beginning in 1997, with the inauguration of the "Angel Ramos Foundation" Visitor Center, the role of the Observatory in public outreach and education was greatly enhanced. It is the only facility of its kind serving the general public and the public and private schools of Puerto Rico, and we believe sets an example for other initiatives at other research centers throughout the Nation. Pride in the Observatory, and an effective fund raising campaign, caused local Puerto Rican organizations, in particular the Angel Ramos Foundation, a philanthropic organization that contributes to improving the educational, cultural and civic conditions of Puerto Rican society, to contribute the funds necessary for the construction of the Center. The NSF provided the funds for the exhibits (Figure 18).

Prior to 1997, 35,000 visitors per year came to the Observatory to view the telescope and listen to a rudimentary 10 minute audio tape. With the Visitor Center, this number has increased to 125,000, about one third being of school

Figure 18. The "Angel Ramos Foundation" Visitor Center.

age. Visitors enjoy a modern facility with a professionally prepared exhibit program offering a wide ranging menu. A special theater presentation "A Day in the Life of the Arecibo Observatory" plus various outdoor exhibits such as a scale model of the solar system, are part of the offering. And then comes the main dish – the unique view of the giant dish.

The opening in 2001 of the Conference Center, (a project supported in part by the Angel Ramos Foundation), a separate building housing a conference room and service areas, further enhances our activities, providing a venue for the science-teacher workshops which take place and have already hosted 200 teachers, plus scientific workshops such as this "Single-Dish School" (Figure 19).

In 2001, the Arecibo Telescope was declared an Electrical Engineering Milestone by the Institute of Electrical and Electronic Engineers (IEEE) and an Historic Mechanical Engineering Landmark by the American Society of Mechanical Engineers (ASME) in recognition of its significance.

About 140 persons are employed by the Observatory providing everything from food to software support for the operation. A scientific staff of about 16 divides their time between scientific research and assisting visiting scientists. Engineers, computer experts, and technicians design and build new instrumentation and keep it in operation. A large maintenance staff keeps the telescope and associated instrumentation, as well as the site, in optimal condition. A staff of telescope operators support observing twenty-four hour per day. Here I wish to acknowledge the collective work of all these dedicated people, of which I know first hand, and without which nothing would be possible.

References

Ewen, H. I., & Purcell, E. M. 1951, Nature, 168, 356

Gerson, N. C. 1984, Studies in Intelligence, 28, 2 (declassified NSA document K-TSR-04-98)

Goldsmith, P. F. 1996, IEEE Potentials, August-September, 38

Gordon, W. E., Booker, H. G., & Nichols, B. 1958a, Engineering report of the School of Electrical Engineering of Cornell University, 3

Gordon, W. E. 1958b, Proc. IRE, 58, 1824

Gordon, W. E. 1964, Science, 146, 26

Hewish, A. S., Bell, J., Pilckington, J. D. H., Scott, P. F., & Collins, R. A. 1968, Nature, 217, 709

Jansky, K. 1933, Proc. IRE, 21, 1387

Kavanagh, T. 1963, Engineering News Record, January 10 1963

Kerr, F. J., Hindman, J. V., & Robinson, B. J. 1954, Aust. J. Phys., 7, 297

Kildal, P. 1989, Radio Science, 24, 601

LaLonde, L. M. 1974, Science, 186, 213

Marconi, G. 1922, Proc. IRE, 10, 215

Muhleman, D. O., Holdrige, D. B., & Block, N. 1962, AJ, 67, 191

Pettengill, G. H., & Dyce, R. B. 1965, Nature, 206, 1240

Schmidt, M. 1963, Nature, 197, 1040

Taylor, J. H., Fowler, L. A., & McCulloch, P. M. 1979, Nature, 277, 437

Van Allen, J. A., McIlwain, C. E., & Ludwig, G. H. 1959, J. Geophys Res., 64, 271

Wolszczan A., & Frail D. 1992, Nature, 255, 145

Figure 19. The Conference Center and the original line feed, now a monument gracing its front.

Part 1
Fundamentals of Single-Dish Radio Astronomy

NRAO trio, Phil Jewell, Harvey Liszt and Darrel Emerson, discussing who put the rum in the coffee.

Single–Dish Radio Astronomy: Techniques and Applications
ASP Conference Series, Vol. 278, 2002
S. Stanimirović, D. R. Altschuler, P. F. Goldsmith, and C. J. Salter

Why Single-Dish?

Darrel Emerson

National Radio Astronomy Observatory, 949 N. Cherry Avenue, Campus Bldg. 65, Tucson, Arizona 85721-0655, USA

Abstract. Single-dish radio telescopes and interferometers/aperture synthesis telescopes are complementary. For some tasks - such as obtaining an accurate measurement of large-scale structure in a region of the sky - the single-dish may be the only option. For other tasks, for example where higher resolution is required, then an interferometer may be the only choice. For some tasks - such as where the critical parameter may be total collecting area, no matter how it is configured - then either single-dish or interferometric instruments may be suitable.

The fundamental difference between single-dish and interferometric instruments is in the range of angular scale sizes to which they are sensitive. However, purely practical points, such as the relative complexity of interferometer instrumentation with inevitably lower flexibility and the relative difficulty in upgrading instrumentation, and in many cases a steeper learning curve for user software, sometimes make a single-dish the "telescope of choice" for a given experiment. On the other hand, the difficulty in obtaining sufficient amplitude stability in single-dish instrumentation sometimes makes an interferometric instrument, which relies on cross-correlation of the signal rather than auto-correlation, a better choice.

For a future generation of telescopes, in particular in the mm-wave and submm-wave region, perhaps most observing plans will require a combination of single-dish and interferometric data. Design of the ALMA telescope is nearly complete; this will consist of about 64 individual 12-meter antennas, operating from 30 GHz up to 950 GHz. The current design specifies that all 64 antennas will sometimes be operated as independent single-dish instruments, observing the same or different sources. This is in addition to the interferometric mode where all 64 antennas operate together as a multi-element interferometer.

With a future generation of radio telescopes, the distinction between purely single-dish instruments and purely interferometer telescopes may become blurred.

1. Introduction

As is well known, for some observations a single-dish telescope is superior to an interferometer, and vice versa. Some observations require a combination of single-dish and interferometric data; with the new generation of both single-

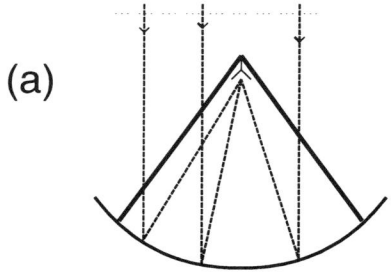

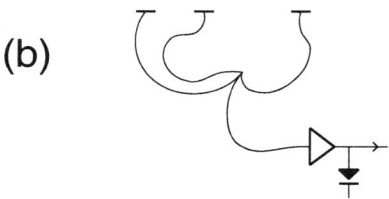

Figure 1. A parabolic dish and a phased array of dipoles. The dish brings the signals together by making the free-space pathlengths equal, while the phased array makes the lengths of cables equal. They are equivalent.

dishes and interferometric instruments operating into the mm and sub-mm parts of the spectrum, this is becoming more and more important. This paper outlines some of the pros and cons of interferometers versus single-dish telescopes; some of the factors are fundamental, but many are purely practical, relating to the relative difficulties of building satisfactory instrumentation. This article compares some pros and cons of single-dish and interferometric telescopes, and finishes with mention of a new generation of telescope that performs both functions equally effectively.

2. Why Single-Dish?

To answer the question "Why Single-Dish?" you have to look at what the alternatives might be. The obvious alternative in many cases might be an aperture synthesis array such as the VLA in New Mexico, but for purpose of illustration we will also consider a hypothetical phased array of smaller elements.

Figure 1(a) shows a parabolic single-dish looking at a fixed source in the sky, with an incoming wavefront being brought to a single focal point. At the focal

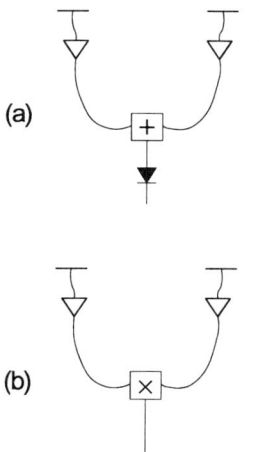

Figure 2. Two elements of an array could be combined as in (a) by adding the signals in phase and then detecting, or as in (b) by multiplying the signals. (a) is in effect what happens in a single-dish, where the "elements" are different parts of the antenna surface - for example, the left half and the right half of the dish.

point, there is a feed to receive the radiation reflected from the dish, and a radio astronomy receiver - basically an amplifier followed by a detector. Details of the radio astronomy receiver are given in other lectures, but the two components of the receiver relevant in this discussion are:

1. a sensitive amplifier, which inevitably will add its own background noise to the incoming signal, and

2. the detector. Usually we are interested in measuring the power coming from a certain direction in the sky, so the detector ideally has a square-law characteristic, converting the amplified voltage received from the antenna into something proportional to power. The detector then incorporates some averaging or integration, so that we end up recording the average power coming into the antenna from the chosen direction.

The parabolic shape has the property that all rays entering the dish from a parallel wavefront are brought to the focus with equal paths - in other words with equal time delay, so that all the signals across the wavefront arrive at the focus in phase with each other. To choose signals from a different direction, the easiest way, at least conceptually, is to tilt the antenna. For small changes of position, instead of moving the whole antenna surface, it can be equivalent just to move the focal point a little.

Figure 1(b) shows part of a phased array of dipoles. The 3 incoming rays shown schematically in Figure 1(a) are here captured by 3 independent dipoles.

The signal from each dipole is brought by a cable to a common point, where the voltages are added before being amplified and detected in the receiver. If the source to be observed is directly overhead, and if the lengths of each cable from the dipole to the receiver are all identical, then all the individual signals will arrive exactly in phase at the receiver before being amplified and detected. If we want to make the phased array of dipoles look at a different point in the sky, in principle we could pick up the array and tilt it, just as with the parabolic dish. However, we have probably built the phased array because we wanted to avoid the engineering difficulties of tilting a large steel structure; the smarter way of shifting the beam is to make the cables to each dipole of progressively different lengths, such that the dipoles closer to the source have a longer cable to travel through, and those dipoles further away from the radio source have correspondingly shorter cables. If we get all the lengths exactly right, then signals from the desired point in the sky will all arrive at the receiver in phase, and those from the original point, directly overhead, will not. We have tilted the beam, replacing mechanical difficulties by electronic complications. Another thing we would probably choose to do with this phased array: the cables from dipoles to central receiver will have loss. It would be better to amplify the signal at each dipole, before sending it down the cable; this is shown schematically in Figure 2(a). The amplifiers must all be very closely matched in phase and gain response, as well as hopefully adding the minimum of extra background noise to the wanted signal. This phased array has become simpler mechanically, but more complicated electronically, than the simple "single-dish" antenna with its single receiver at the focal point. However, if the amplifiers are of equal sensitivity, and if the area of the single-dish is equal to the area filled by the field of dipoles, the total sensitivities and angular resolution will be nearly identical.

The true aperture synthesis array takes this one stage further, as illustrated in Figure 2(b). Taking the same field of dipoles, the signals are brought together but are paired up. Rather than, as shown in Figure 2(a), all signals being added together in phase and then detected, pairs of signals are brought to the same location and multiplied together. With N individual dipoles or antenna elements, there will now be $N(N-1)/2$ (complex) multiplications. The output of each multiplier is then averaged (integrated) before being recorded in a computer. Without going into the details, each multiplication yields one Fourier term of a description of the sky distribution within the field of view of each antenna element. The computer combines the products with different permutations of phase offset added to the products. In the simplest case, this could just be via a Fourier transform. Ignoring the many details involved in this, the main point is that the electronic phase shift using the variable length cables of Figure 1(b) has been replaced by software. Overall the electronics may not be simpler - we have after all $N(N-1)/2$ multipliers rather than just one detector. The real gain is that in software, long after the observations, we can steer the beam over the sky - possibly to several million discrete points, in the case of some VLA maps. It is the ultimate multi-beam system.

The aperture synthesis technique is so powerful, that it sounds too good to be true. Software replacing an enormous amount of hardware to give a million-pixel multi-beam system. Murphy's law says there has to be a catch. And there is.

3. Some Very Simple Algebra

This applies to Figure 2 above, but remember that the telescope of Figure 1(a) is in principle no different from that of Figure 1(b). In the parabolic reflector, small pieces of signal from different parts of the reflector are brought to one point, in phase, by free-space propagation from the dish surface to the focus. With the phased array, small signals picked up by individual dipoles are brought to the focus along pieces of cable. That's really the only difference.

For simplicity, I will consider a phased array of just 2 dipoles, and compare the response of the systems shown in Figures 2(a) and 2(b). Equivalently, we can consider the signal "a" as coming from the left half of the parabolic dish surface, and "b" from the right half. One dipole receives a signal voltage "a" and the other dipole a voltage "b". The voltages **a** and **b** are each the vector sum of radiation coming from every part of the sky within the field of view of each element; in some directions the received signals will contribute voltages to **a** and **b** that will be in phase, in other directions components of the voltages will be in anti-phase. Each dipole has its own low-noise amplifier, but the amplifier superposes on the signal at each dipole a noise voltage **A** and a noise voltage **B**, respectively. **A** and **B** are assumed equal in magnitude, but completely uncorrelated with each other. So, overall the amplified voltage from the first dipole at the detector circuit is $(\mathbf{A} + \mathbf{a})$, and from the second dipole $(\mathbf{B} + \mathbf{b})$. For simplicity, the amplifiers are assumed to have unity gain.

For the phased array example shown in Figure 2(a), the voltage coming out of the square-law detector will then be proportional to:

$$[(\mathbf{A} + \mathbf{a}) + (\mathbf{B} + \mathbf{b})]^2$$

or:

$$\mathbf{A}^2 + \mathbf{B}^2 + \mathbf{a}^2 + \mathbf{b}^2 + 2(\mathbf{A} \cdot \mathbf{a} + \mathbf{A} \cdot \mathbf{b} + \mathbf{B} \cdot \mathbf{a} + \mathbf{B} \cdot \mathbf{b} + \mathbf{A} \cdot \mathbf{B} + \mathbf{a} \cdot \mathbf{b}).$$

This squared voltage is averaged, or integrated. Now, if two uncorrelated quantities are multiplied together, the time-averaged product tends to zero. There will be statistical fluctuations, which are very important and which define the ultimate signal-to-noise ratio, but they will be neglected in this discussion. In the above expressions, $\mathbf{A} \cdot \mathbf{a}$, $\mathbf{A} \cdot \mathbf{b}$, $\mathbf{B} \cdot \mathbf{a}$, $\mathbf{B} \cdot \mathbf{b}$ and $\mathbf{A} \cdot \mathbf{B}$ represent products of uncorrelated quantities, so their time averaged value will tend to zero. $\mathbf{A}^2$ and $\mathbf{B}^2$ are proportional to the receiver amplifier noise power. "a^2" and "b^2" are the summed power of all the astronomical signals received by the two elements. "$\mathbf{a} \cdot \mathbf{b}$" is the vector product of the astronomical signals. So, in this phased array case, apart from statistical noise we are left with just:

$$\mathbf{A}^2 + \mathbf{B}^2 + \mathbf{a}^2 + \mathbf{b}^2 + 2\mathbf{a} \cdot \mathbf{b}. \tag{1}$$

The first two terms represent of course just the sums of the noise powers in the two receivers. The next two terms, $\mathbf{a}^2 + \mathbf{b}^2$, represent the sums of the signal powers received in the two sub-elements of the array. The final term, $2\mathbf{a} \cdot \mathbf{b}$ is a signal voltage containing, amongst other things, relative positional information of the signals received in the two sub-elements. (The factor two in $2\mathbf{a} \cdot \mathbf{b}$ can be regarded just as a calibration factor and is not important to this discussion.)

Remember that the voltages **a** and **b** have both an amplitude and a phase, and that the phase will in general be different in the two subelements of the array; **a** and **b** can be anything between completely in phase, and completely out of phase, so their time-averaged product can be either positive or negative, or even zero. It is this term that gives the phased array potentially higher angular resolution than either of the two elements alone.

Now consider the correlation interferometer case, as in Figure 2(b). The two signals, including amplifier noise, are multiplied rather than being added and squared. So, we have

$$(\mathbf{A} + \mathbf{a}) \cdot (\mathbf{B} + \mathbf{b})$$

or

$$\mathbf{A} \cdot \mathbf{B} + \mathbf{A} \cdot \mathbf{b} + \mathbf{a} \cdot \mathbf{B} + \mathbf{a} \cdot \mathbf{b}.$$

Now all terms except the last are products of uncorrelated voltages, so their time average will tend to zero. We are left with just

$$\mathbf{a} \cdot \mathbf{b} \qquad (2)$$

the same as the last term of the total power mode phased array. The good thing about this is that, apart from the inevitable statistical noise, we have made ourselves immune to changes in the noise power in the amplifiers, **A** and **B**. Whereas in equation (1) a slight change the magnitude of **A** or **B** would likely swamp the much smaller signals **a** and **b**, in equation (2) we would not even detect such an instrumental change. This is very good.

However, remember that **a** and **b** have an amplitude and a relative phase, so the time averaged magnitude of $\mathbf{a} \cdot \mathbf{b}$ inevitably has a somewhat arbitrary sign and amplitude. The individual voltages, **a** and **b**, are in general made up of signals from many different sources at different positions within the field of view of each sub-element. For example, suppose there are just 2 sources "s" and "t" in the field, putting a signal voltage $\mathbf{a} = \mathbf{s_1} + \mathbf{t_1}$ into the first sub-element, and a signal voltage $\mathbf{b} = \mathbf{s_2} + \mathbf{t_2}$ into the second sub-element of the array. The product $\mathbf{a} \cdot \mathbf{b}$ now becomes:

$$(\mathbf{s_1} + \mathbf{t_1}) \cdot (\mathbf{s_2} + \mathbf{t_2})$$

or

$$\mathbf{s_1} \cdot \mathbf{s_2} + \mathbf{s_1} \cdot \mathbf{t_2} + \mathbf{s_2} \cdot \mathbf{t_1} + \mathbf{t_1} \cdot \mathbf{t_2}.$$

Remembering that **s** and **t** are uncorrelated, the time averaged result is:

$$\mathbf{s_1} \cdot \mathbf{s_2} + \mathbf{t_1} \cdot \mathbf{t_2}.$$

Depending on the relative phases of $\mathbf{s_1}$, $\mathbf{s_2}$ and $\mathbf{t_1}$, $\mathbf{t_2}$ at the two sub-elements, the product $\mathbf{s_1} \cdot \mathbf{s_2}$ may well be negative, while the product $\mathbf{t_1} \cdot \mathbf{t_2}$ might be positive. There is an infinite number of possible combinations of relative amplitudes and phases that can give:

$$(\mathbf{s_1} \cdot \mathbf{s_2} + \mathbf{t_1} \cdot \mathbf{t_2}) = 0.$$

In other words, no matter how strong these two sources in the beam are, the multiplying interferometer may be totally unaware of their presence. Actually it is worse than this. In any situation with many individual sources, if for every source where the signals s_1, s_2 arrive in phase at the two sub-elements, there is another source whose equal signals t_1, t_2 arrive out of phase at the two elements, then we will be blissfully unaware of the presence of the sources. This happens every time two sources are separated by $n\lambda/(2D)$, with n any odd integer, λ the observing wavelength and D the interferometer spacing. This is exactly what happens with a very extended source. If every in-phase position is balanced by an out-of-phase component of equal amplitude, we see nothing. Interferometers are totally blind to very extended sources.

A more general way of looking at this, which is covered in more detail in other lectures, is in terms of the relative spatial frequency responses. Figure 3 shows (solid line) the spatial frequency response of a single 12-meter dish. By "spatial frequency" we just mean the reciprocal of particular scale sizes. For a given wavelength, spatial frequency is proportional to antenna baseline, so the horizontal axis in Figure 3 is just labelled in meters. At zero spatial frequency, it is sensitive to the $\mathbf{a}^2 + \mathbf{b}^2$ ... terms, i.e. the sum of all the powers of all the sources within its field of view. With increasing spatial frequencies, the sensitivity is progressively worse. Beyond its spatial frequency cutoff, a given single-dish could not distinguish between a series of closely spaced ridges on the sky, and a continuous, smooth distribution - although it would detect the total flux coming from that part of the sky. The dashed line in Figure 3 shows the spatial frequency response of a 2-element correlation interferometer using a pair of 12-meter dishes, separated center-to-center by 15 meters. The interferometer has full sensitivity to ridges on the sky that are "resonant" with its particular spacing in wavelengths, but has NO sensitivity to very large scale structure - the $\mathbf{a}^2 + \mathbf{b}^2$ ... terms are just not seen. The remaining dashed lines in Figure 3 show the response of an example aperture synthesis instrument, using 12-meter dishes, arranged to give antenna spacings of 15, 30, 45 .. meters.

So where does that leave us? Single-dishes are very good at looking at large features and measuring the total amount of energy being received from a given patch of sky. Correlation interferometers, which make up aperture synthesis arrays such as the VLA, are very good at looking at fine scale structure, but are (for example) totally blind to the 3-K background, (though not necessarily to its fluctuations!).

Sometimes, you do not need both at once. You may be studying the fine detail on a small object, and your science may not require "the big picture". In other circumstances, you may be interested in total flux from extended regions of the sky, and not be concerned with fine scale details within your field of view. But sometimes, you may need both single-dish and interferometric data - if you want an accurate measurement of large scale structure, total flux within a region, and at the same time you want to study fine detail within your object, then you may require both interferometric and single-dish observations. This is often a very difficult task, for practical reasons - data formats, telescope scheduling, calibration and so on - but sometimes it is essential.

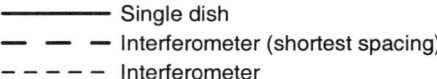

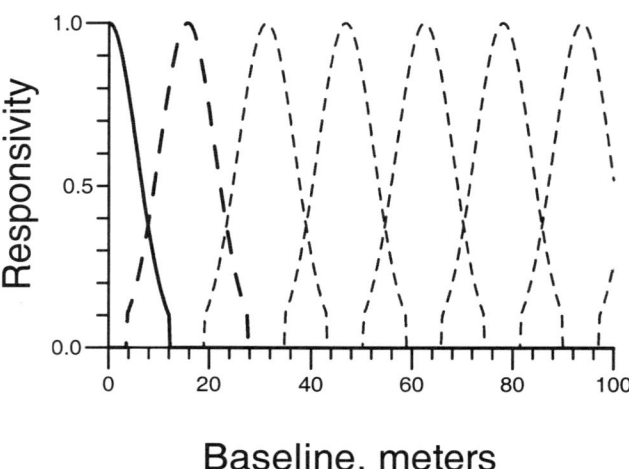

Figure 3. An example of the responsivity of a single-dish telescope, and of an interferometer using identical antennas, as a function of spatial frequency or scale size. Antennas 12 meters in diameter are assumed, with a minimum antenna spacing of about 15 meters. Note that the single-dish theoretically has a finite response extending from baselines of zero to 12 meters. (Scale size on the sky, in radians, is given by M/D, with M the observing wavelength and D the baseline. So, at a given wavelength, spatial frequency is just proportional to baseline, which is marked as the horizontal axis of this figure.) In practice, as illustrated in Figures 7 and 8, single-dish observations will often have baselines and zero levels adjusted, which is equivalent to filtering out the largest scale size of structure. With an interferometer consisting of 12-meter antennas separated by 15 meters, the smallest separation of the antenna surfaces is 3 meters. Theoretically, such an interferometer would have finite sensitivity to such scale sizes, but in practice other factors, such as pointing errors, dish surface errors and the antenna feed illumination pattern limit the range of valid spatial frequencies. The bold, dashed line in Figure 3 shows an example interferometer response, assuming two 12-meter antennas separated center-to-center by 15 meters. Dashed curves are also given to show the relative spatial frequency sensitivity of interferometers pairs, still using 12-meter dishes, with spacings 30, 45, 60 ... meters.

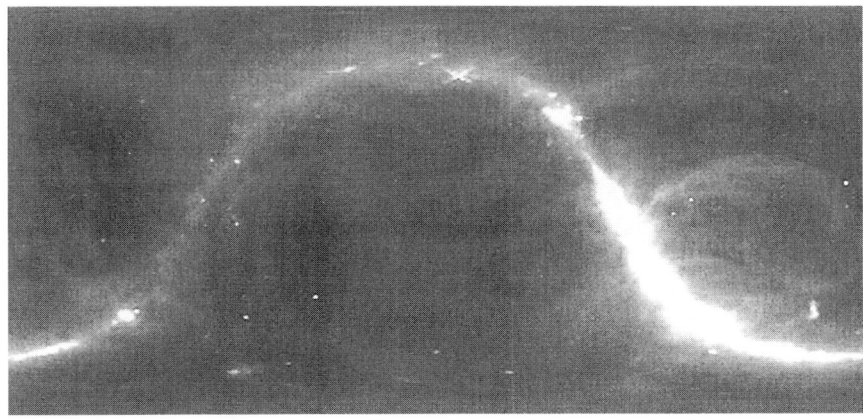

Figure 4. A map of the entire sky, made with single-dish telescopes at a frequency of 408 MHz. The plot is in a rectangular RA-Dec coordinate system, with RA= 0, Dec= 0 at the center of the field. This map is complete in all scale sizes from about 1 up to 360°, and includes also, for example, the uniform contribution of the 3-K cosmic background (Haslam et al. 1982).

The moral of this story is that astronomers should not consider themselves "single-dish astronomers" or "aperture synthesis astronomers". You need to pick the right tool for the job. In some cases, you will need both tools.

4. An Example

Figure 4 depicts an image of the entire sky as measured at 408 MHz (Haslam et al. 1982), plotted in rectangular RA-Dec coordinates. The data are complete on all scale sizes from zero to the limit of the telescope resolution - the 3-K cosmic background is included in the measured emission. It would not have been possible to make this image with purely interferometric telescopes.

Figure 5 shows a horizontal cut in RA through the data of Figure 4, close to Dec= 0. The strong peak is the Galactic plane, several degrees north of the Galactic Center. Note the small source near RA= 40°, which for illustration is shown in more detail in Figure 7.

Figure 6 shows the range of spatial frequencies present in the all-sky image of Figure 4. This was obtained simply from a Fourier transform of Figure 5. The horizontal axis is calibrated in reciprocal degrees; at the observing wavelength of 74 cm one reciprocal degree corresponds to a baseline of 42 meters. Note the logarithmic vertical scale, which measures the relative power in the

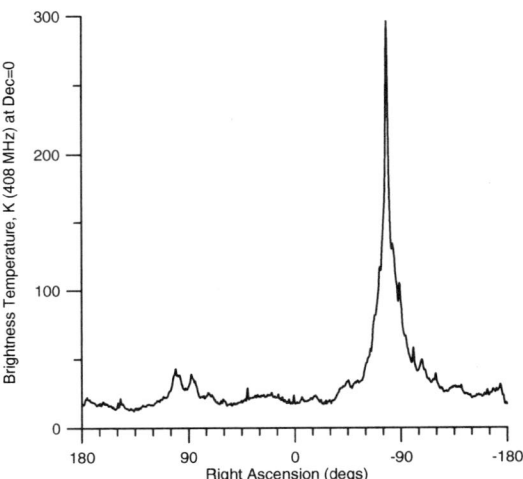

Figure 5. A cut through the data of Figure 4, close to Dec= 0. The strong peak is the Galactic plane.

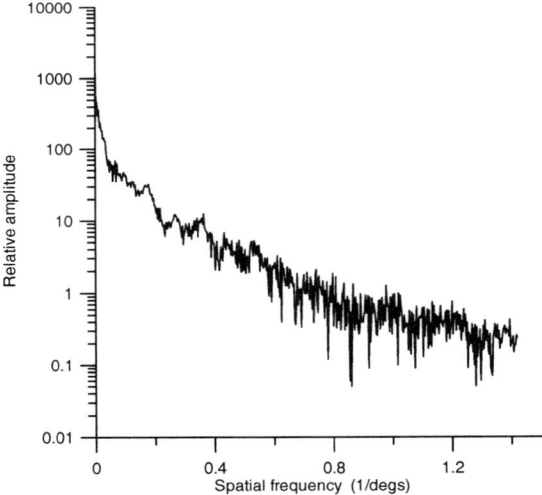

Figure 6. A Fourier transform of Figure 5, showing the relative amplitude of signals at different scale sizes. Note the logarithmic vertical scale of relative brightness (arbitrary units). The horizontal scale is in reciprocal degrees. The relative amplitude rises rapidly at larger scale sizes, at the left extreme of the plot. Although this is a unique image, including emission from the entire universe, the rapidly increasing signal at larger and larger scale sizes is the normal characteristic of maps of extended, complex sources.

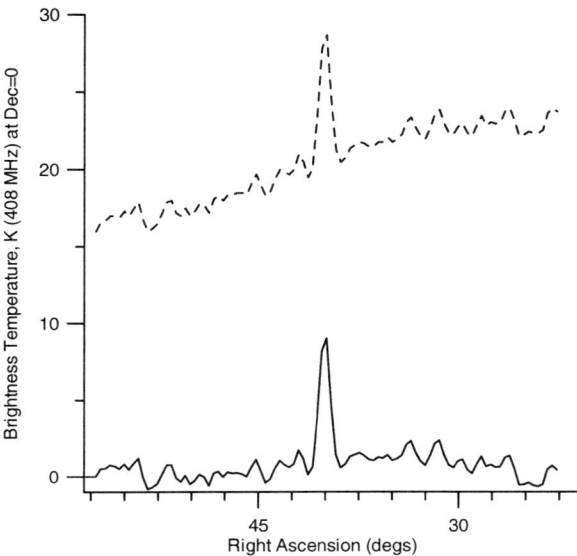

Figure 7. This is an expanded plot from Figure 5, centered on RA $\sim 40°$. The upper, dashed line is taken directly from Figure 5. The strong central peak shows emission from M77, or NGC 1068. The offset, sloping baseline of this dashed line is a result of extended emission from the Galactic Loop-II feature. The lower plot, drawn with a solid line, is the same data, but with a linear baseline removed. Although a less accurate representation of the complete emission in this part of the sky, for study of the source NGC 1068 this is a more appropriate starting point. For such a study, interferometric data, which are insensitive to the large scale size of emission, although strictly less accurate may nevertheless be more desirable.

sky distribution at different spatial frequencies. The amplitude rises rapidly at lower frequencies, or larger scale sizes, towards the left of this plot. Although this all-sky image is clearly a unique source, containing emission from the entire Universe, including the 3-K cosmic background, the rapidly increasing signal seen at larger and larger scale sizes is a common characteristic of maps of extended, complex sources. Most of the flux is at low spatial frequencies, measured by the shortest antenna spacings.

Figure 7 expands the part of the plot shown in Figure 5, centered on RA$\sim$ 40°. The upper, dashed line is taken directly from Figure 5. The strong central peak corresponds to M77, or NGC 1068. The offset, sloping baseline is a result of extended emission from the Galactic Loop-II feature in this region of sky.

The lower plot, drawn with a solid line, is the same data but after removing a linear baseline. Although now a less accurate representation of the complete emission in this part of the sky, for a detailed study of the source NGC 1068 this would be a more appropriate starting point. The same result might have been

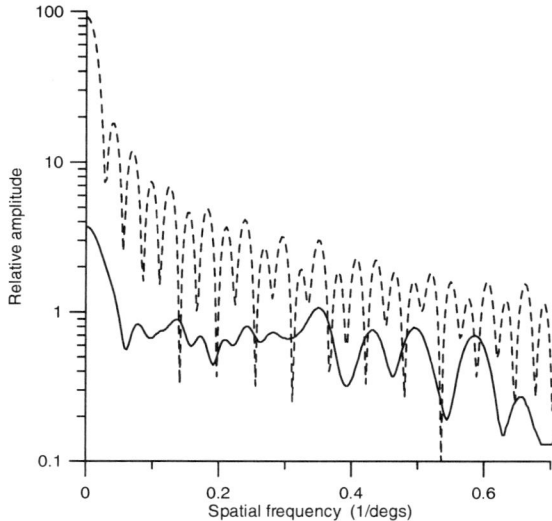

Figure 8. Plots of the Fourier transform of the data shown in Figure 7. The upper, dashed line is from the original data, shown dashed in Figure 7. The lower, solid line is from the data of Figure 7, which has had a linear baseline subtracted. Again, the vertical scale is logarithmic: the intensity of the largest scale size of emission has been reduced by a factor of about 25 by this baseline removal. Nevertheless, even after removing the extended emission by removing the linear baseline, note how the intensity is still relatively larger at spatial frequencies below ~ 0.1 deg^{-1}. This large-scale would likely be filtered out if observations using only interferometers were used - depending on the science being studied, that filtering out may be a good or a bad thing.

achieved by filtering out the largest scale structure from the original data shown above. The filtering might, perhaps fortuitously, have been carried out by making an aperture synthesis observation, omitting any complementary single-dish data. The interferometer's lack of sensitivity to large scale structure, although giving a less accurate representation of the sky, might be used to advantage and allows a clearer interpretation for some science.

The plots shown in Figure 8 were derived from a Fourier transform of the data shown in Figure 7. The upper, dashed line is from the original data. The lower, solid line is from the data with linear baseline removed (the solid line in Figure 7). Note the logarithmic vertical scale; one effect of the baseline subtraction is to reduce the intensity of the largest scale structure within this limited map by a factor of about 25. However, even after removing the extended emission via baseline subtraction, the intensity is still relatively large at spatial frequencies below ~ 0.1 deg^{-1} (i.e. scale sizes $> 10°$). This remaining large-scale emission might also have been filtered out if observations using only interferometers had been used. Depending on the science being performed, such filtering might be a good or a bad thing.

To summarize the results illustrated above; most of the flux of an extended, complex source is seen at lower spatial frequencies, equivalent to the larger structure scale sizes, and observable with the shorter baselines. If the science of interest does not require knowledge of the lower spatial frequencies, the natural filtering inherent in interferometric observations may be desirable, even though a less accurate representation of the region of sky is obtained. It is in any case usual to apply some form of baseline removal to single-dish data, which itself has the effect of filtering out large scale structure. If however, the science requires precise knowledge of the extended emission, then there is no choice but to use a single-dish instrument, perhaps combined with data from an interferometric array.

5. Practical Details

So far, I have concentrated on the fundamental differences between single-dish and interferometric observations. However, there are many technical and practical "details" which may well become the overriding considerations.

It has to be admitted that the instrumentation of a single-dish telescope is usually simpler than that of an interferometric array. There is one focal point, with in principle only one set of receivers to cover the desired range of wavelengths. With an aperture synthesis array of N telescopes, there have to be, of course, N sets of receivers, and a cross-correlator able to correlate $N(N-1)/2$ individual baselines, providing perhaps a thousand-point spectrum for each of these baselines. Both the frontend and the backend are much more technically manageable with a single-dish telescope, and easier to upgrade as technology advances.

In principle, both aperture synthesis and single-dish instruments could make use of multibeaming, and so make themselves more efficient collectors of photons. But again, complexity becomes an overriding consideration; while there are now several single-dish instruments outfitted with focal plane arrays, I know of no aperture synthesis instrument similarly equipped. In principle there should be no difference, but there is a limit as to how much complexity synthesis array designers are prepared to accept.

Single-dish telescopes at short mm and sub-mm wavelengths often use bolometers - these are incoherent detectors, with very high continuum sensitivity due primarily to the large bandwidth they are able to accept. In principle bolometer arrays could be adapted to interferometry - this is sometimes done with interferometry at IR and optical wavelengths, in the form of CCD detector arrays. However, although in principle possible at mm-wavelengths, many of the technological advantages of interferometer instrumentation would be given up. The ability to use large focal plane bolometer arrays has to count as a definite "plus" for single-dish instruments at the moment.

For some experiments, angular resolution or ranges of scale size just are not an issue; point-source sensitivity may be all that counts. Pulsar, and transient phenomena are examples of that. In such cases, the only thing that counts is the effective collecting area of the instrument. Assuming that receivers are all equivalent, it makes no difference whether this collecting area is made up of many individual dishes, or is in the form of one big instrument. Currently, the Arecibo telescope has the largest collecting area in the world, and so reigns supreme in this area. Following Arecibo, the VLA in principle has the next largest collecting area at short cm wavelengths, although it is closely followed by the GBT and the Effelsberg 100-meter telescopes. For the VLA to compete with Arecibo, it would need to upgrade from its current 27 × 25-meter antennas to about 150 antennas (just the ratio of antenna diameters, squared). It could be done, but probably is not likely in the near future!

General flexibility is a very big "plus" for single-dish instruments. The ability to upgrade instrumentation in response to technological developments, and even the ability to reconfigure instrumentation in some unforeseen way during an observing session, is there for single-dish instruments, but because of the sheer complexity an array just cannot compete.

It is interesting to note that large single-dish instruments do sometimes become elements of interferometric arrays - with VLBI, this happens frequently enough. This is perhaps another example of the flexibility of a single-dish instrument. I would hazard a guess that, while most large single-dish instruments have at some time been used for interferometry, very few arrays have been used for successful single-dish observing.

Another interesting example is interplanetary radar. The VLA has in fact been used as a "large single-dish" as a receiving element for some interplanetary radar experiments, but I do not know of any array that has been used as a transmitter in this context. Again, it is a matter of complexity. It seems to be easier to put a 1 MW transmitter at the focus of a single large antenna such as Arecibo, than, say, to mount a 37 kW transmitter at the focal point of each VLA antenna. [An exercise for the student: what power of transmitter at the focus of each of 27 25-meter antennas would really be required to equal the EIRP of a 1 MW transmitter at the focus of a 1000-ft telescope?] However, things could change in the future. Note that the trend at short wavelengths with high gain transmitting satellite antennas is indeed to use a phased array, with perhaps 1000 small transmitter amplifiers at each sub-element of a 1000-sub-element array. One motivation for the satellite engineer is the ease of rapid electronic steering of the transmit beam; this technology creates a bigger RFI problem for the radio astronomer, but that is another story. It may be that some future interplanetary radar will indeed use a phased array of hundreds of smaller dish antennas, in preference to the single huge transmitter at the focus of an enormous reflector.

6. Future Telescopes

Interferometer instruments and single-dish instruments have different demands on technology. For single-dish instruments the internal noise and amplifier gain stabilities requirements are extreme - for example, a gain stability of better than 1 part in 10^4 in a second may be required. This is very difficult to achieve. Conversely, the demands on gain stability of the interferometer instrumentation are not so extreme, but of course phase stability is critical. So far, major instru-

Figure 9. *Does this array make single-dish observations or is it a multi-element interferometer?* The image is an artist's impression of the planned ALMA radio telescope. This telescope is scheduled to be completed in the second decade of this century. It is to operate at mm and sub-mm wavelengths, from a site at 5000 m altitude at Chajnantor in northern Chile. Current plans call for 64 × 12-meter antennas, capable of operating from ∼30 to ∼950 GHz. As well as operating as a multi-element interferometer, the antennas and their instrumentation are designed to operate in single-dish mode - for the science goals of this instrument, both single-dish and interferometric observations are essential, and both types of data will routinely be combined together. Hopefully the combination will be seamless to the observer. Image courtesy of ESO.

ments have been designed either for single-dish operation or for interferometry, but not optimized for both. In some cases the antenna elements and instrumentation of synthesis arrays have later been adapted for single-dish operation, but this has usually ended up as a severe compromise. With a new generation of telescopes this situation may change.

At short wavelengths, measured in mm or less, the field of view of "reasonable" antennas is extremely small. For example, at a wavelength of 1 mm, or 300 GHz, a 12-meter antenna may have a field of view of only $\sim 20''$. For an aperture synthesis array using such elements, "large scale structure" means anything approaching $20''$ or more. Such an array is the planned ALMA instrument, which will consist of about 64 antennas, each 12 meters in diameter, to be built at an altitude of 5000 m in northern Chile; the specifications call for operation between about 30 and 950 GHz. Figure 9 shows an artist's impression of the instrument, which is scheduled to be completed in about 2012.

It has been recognised from the beginning that, with such a small field of view, many - probably most - observations with the telescope will require a combination of single-dish and interferometric data. Figure 3 showed how, with 2 interferometer dishes nearly touching each other, the sensitivity to spatial frequencies actually overlaps with the spatial frequency response of an identical single antenna, used in single-dish mode. So, it has been planned that the 64 antennas of the ALMA array will sometimes be used as 64 individual single-dishes, to obtain the "large" scale structure to complement the higher resolution of the interferometric observations. This has severe implications on the electronic design, but the requirement has been put there in the specifications of the instrument.

So, with future generations of radio telescopes, especially operating at higher frequencies, the distinction between single-dish observing and interferometric observing may simply disappear. In submitting your observing proposal, you will specify the range of scales sizes you need - from the largest to the smallest - and the telescope software and scheduling will take care of the rest. You may not even know which parts - i.e. which scale sizes - of your data were observed with an interferometer, and which with single-dish operation. The glue that combines the different data is "just" software. [One may argue about whether this is a good trend or not, but the trend is there.]

7. Conclusions

The only fundamental difference between single-dish telescopes and aperture synthesis arrays is the range of spatial frequencies to which the telescope is sensitive. In very many - but not all - observations, in principle either an array or a single-dish could do the job. For the highest angular resolution, arrays are the only choice. For measurements of the intensity of the 3-K Cosmic Background (but not necessarily of its angular structure), single-dish instruments or their derivatives are the only choice.

In practice, the implicit relative simplicity of a single-dish instrument inevitably gives a great deal more flexibility than is possible with an array telescope. It is easier to keep the single-dish instrumentation at the state of the art, and to adapt the telescope to novel experiments. This often makes the single-dish telescope the instrument of choice.

In future, the appearance of large mm-wave and sub-mm-wave array telescopes, with their inevitable small field of view, has lead to greater effort being applied to the appropriate and seamless combination of single-dish and interferometric data. To cover the needed dynamic range of spatial scale sizes, this is essential. Today, we have to be sure to choose the right instrument for the job, to answer our scientific questions in the most effective way. In future, - with ALMA being the prime example - telescopes may be built that combine both techniques in a way more or less transparent to the user; the distinction between single-dish and interferometric telescopes will become much less obvious.

References

Haslam, C. G. T., Salter, C. J., Stoffel, H., & Wilson, W. E. 1982, A&AS, 47, 1
 (Numerical data obtained from http://skyview.gsfc.nasa.gov)

Paul Goldsmith illustrating the size of the Arecibo reflector.

Radio Telescopes and Measurements at Radio Wavelengths

Paul F. Goldsmith

National Astronomy and Ionosphere Center, Department of Astronomy, Cornell University, Ithaca, New York 14853-6801, USA

Abstract. We present an overview of how single–dish radio telescopes collect energy in an astronomical context. We begin with a discussion of key radio astronomical terminology, including specific intensity, flux density, antenna temperature, and point–like and extended sources. In order to analyze radio telescope system performance, we discuss single mode transmission lines, the feed system which couples the receiver to the antenna, and the concept of the antenna as a phase transformer. Using the distribution of the electric field in the aperture plane, we calculate the sensitivity of the radio telescope to a point source. This discussion is based on the antenna aperture efficiency and includes its dependence on the characteristics of the feed system, on the blockage, on systematic errors due to feed or secondary reflector defocus, and on random errors. The aperture plane field distribution is also the starting point for the Fourier transform relationship used to compute the far–field response, or power pattern, of the antenna. We analyze the effects of the telescope characteristics on the beam width and sidelobe level of the antenna's power pattern, and on the coupling efficiency to an extended source. Understanding this basic behavior allows us to determine what measurements are needed to calibrate the antenna, and to determine whether the system is performing properly.

1. Introduction

Several different approaches can be used to define *radio* astronomy, beyond saying that it involves collecting and analyzing electromagnetic radiation. The traditional definition has been in terms of wavelength, with the *radio window* defining the frequency or wavelength range in which radio astronomy is done. The low–frequency end of the radio window (for terrestrial radio astronomy) is defined by the cutoff imposed by the Earth's ionospheric plasma. The cutoff frequency is typically between 4 MHz (1 MHz = 10^6 Hz) and 12 MHz, depending on solar activity and time of day (since the ionization is produced by solar flux, both increased solar activity and solar illumination raise the charge density and thus the cutoff frequency. Karl Jansky's observations of radio waves from the Milky Way, which inaugurated radio astronomy, were carried out at a frequency of 20.5 MHz, modestly above this cutoff. The high-frequency limit of radio astronomy is not such a sharp one, but is a combination of the increasing attenuation of the absorption of the atmosphere at wavelengths less than 1 mm (frequencies above 300 GHz; 1 GHz = 10^9 Hz) and the increasing technological difficulties that have hindered development of these higher frequencies. Thus, for decades, there was a cap on the radio window (from sea level), again at a frequency of about 300 GHz. The ionospheric cutoff can be avoided by going into space, and there has indeed been a modest amount of very low frequency radio astronomy carried out using satellites. Similarly, now that technical barriers have been lifted, it is possible to observe from high, dry mountain sites, from airplanes, and from satellites, and the radio window blends in with the *submillimeter* (or *far-infrared*) region of the electromagnetic spectrum. A reasonable definition is that radio astronomy encompasses the range of frequencies between 10 MHz and 1000 GHz, or a range of five decades.

The second significant definition of radio astronomy is in terms of its distinguishing technological aspects. The radio astronomy frequency range is distinguished by:

- **Diffraction–Limited telescope performance**, and angular resolution in particular;

- **Single Mode transmission media** and consequent coupling to a single mode of the radiation field;

- **Coherent signal processing** preserving amplitude and phase of the incident signal.

Each of these aspects plays a definite role in defining "how radio astronomy is done", but as we shall see, the distinctions between radio astronomical techniques and those employed in other branches of astronomy are becoming increasingly blurred.

Diffraction–limited operation of a telescope is necessary to achieve the highest possible angular resolution, but it limits the "sharpness" obtainable with a telescope of diameter D operating at wavelength λ to $\simeq \lambda/D$ (radians). As will be discussed in Section 4.2., this requires that the surface accuracy of the antenna (measured as a root mean square variation from desired shape), should be no more than approximately 1/20 wavelength. Wavelengths in radio astronomy are large, in particular relative to other branches of astronomy, so it has always been a challenge to achieve the desired angular resolution. In partial compensation, it is relatively easier to achieve the surface accuracy required for diffraction–limited operation. Until recently, this issue was not relevant for optical astronomy, where atmospheric phase variations ("seeing") impose a limit on achievable angular resolution which is much worse than can be obtained with a telescope of very modest diameter. With the advent of optical telescopes in space and the use of adaptive optics on the ground, the seeing limit is becoming less important, and diffraction–limited telescope systems are becoming relevant at all wavelengths used for astronomy.

A single mode system transmits electromagnetic energy with a unique configuration of the electric and magnetic fields. For example, within a coaxial cable or a waveguide, at a given frequency, **E** and **B** must have a specific geometrical form, or mode, if the boundary conditions are to be satisfied. There is only a single variable – the field amplitude – which scales the strength of **E** and **B** everywhere. This is very different from situation commonly encountered at shorter, *e.g.* infrared or optical wavelengths, where radiation can propagate in a variety of directions, with varying degrees of collimation, and different polarization states. The requirement for single mode propagation is that the transverse dimensions of the transmission medium be on the order of the wavelength. This is relatively easy to achieve at longer radio wavelengths, but does become a challenge and ultimately a limitation for the highest radio frequencies where single mode structures must be only hundreds of microns in size. The use of single mode transmission has definite implications for radio astronomical systems in that only a single polarization state can be handled, and the specific field configuration leads to restrictions on how we can utilize the telescope's collecting area, as discussed further in Section 3.

Radio astronomy systems employ amplifiers to increase the strength of the very weak signals that are collected, and use mixers to change the frequency of the signal being processed, if this is desired. Both of these components operate on the voltage in a single-mode transmission medium caused by the incident electric field, and produce an output whose amplitude is proportional to that of the input, and with an output phase which is related to that of the input signal. In contrast to a photomultiplier, or CCD, a coherent system preserves the oscillatory nature of an electromagnetic signal. It is only at the output of the radio astronomy receiver system that we "detect" the signal, meaning that we measure its power. In a single-dish system, this is generally done by an operation which amounts to squaring the input signal, but it is critical that up to this final operation, the coherent system has preserved the phase of the input signal. As a result, the coherent system is subject to restrictions imposed by quantum mechanics; the uncertainty principle limits how well we can know the signal intensity if we know something about the phase. The coherent nature of radio astronomy systems imposes various limits on their sensitivity, even if we do not ultimately do anything with the phase information.

At shorter radio wavelengths, incoherent (or direct) detectors are seeing increasing use for broadband measurements. This again is blurring the distinction between radio and other branches of astronomy, but inevitably, the urge to exploit the entire electromagnetic spectrum results in blending together and cross-fertilization of techniques from one wavelength range to another. So while the three distinguishing technical characteristics of radio astronomy enumerated above remain largely correct, there is no doubt that the borders are increasingly fuzzy. Optical telescopes will be closer to diffraction limited, and radio receivers employ arrays of incoherent detectors. However the majority of considerations explained in some detail in what follows will likely remain of relevance in defining how radio astronomers obtain, calibrate, and interpret their data.

2. Definitions and Measurement of Radiation

2.1. Fundamentals

We are not going to include in this section several important aspects of electromagnetic radiation. The first is *polarization*, which we omit only because it is covered in a different lecture. For our purposes, we will consider electromagnetic radiation to be effectively a scalar quantity, and ignore how sources and propagation, as well as behavior of radio telescopes, depend on polarization. A second aspect is the *coherence* of the astronomical signal received. We will consider incident radiation to be *incoherent*, as is characteristic of thermal sources. This is typically not a significant issue for radio astronomy, and greatly simplifies analysis of collection and measurement of radiation.

With these restrictions, we can characterize the electromagnetic energy at frequency ν coming from a certain range of solid angle $\delta\Omega$ around a direction defined by usual polar angles θ and ϕ, in terms of the *specific intensity* (also sometimes called the brightness). The unit of specific intensity, denoted I_ν is energy per unit time per unit area per unit solid angle per unit frequency. The energy that flows through an infinitesimal aperture of area δA and bandwidth $\delta\nu$ in time δt from specific intensity I_ν coming at angle θ from normal incidence is

$$\delta E = I_\nu \ \delta t \ \cos\theta \ \delta A \ \delta\Omega \ \delta\nu \ . \qquad (1)$$

The power, P, flowing through the aperture is just the energy flowing per unit time. Radio astronomers generally, but not always, employ the mks system, so energy is measured in joules (denoted J), power is measured in watts (W) and area in square meters (m^2), while solid angle is measured in steradians (sr). Thus, we can write the power flowing as

$$P(\text{W}) = I_\nu(\text{W m}^{-2} \text{ sr}^{-1} \text{ Hz}^{-1}) \ \cos\theta \ \delta A(\text{m}^2) \ \delta\Omega(\text{sr}) \ \delta\nu(\text{Hz}) \ . \qquad (2)$$

The preceding definition evidently allows for radiation to be incident from a range of angles, which is often the case in radio astronomy. In some cases, we may have a perfectly collimated beam of radiation, which effectively has a single direction of propagation. If we integrate I_ν over a delta function in solid angle, we eliminate the "per solid angle" and have a new unit with units energy per unit time per unit area per unit frequency, or power per unit area per unit frequency. This quantity, called the flux density and denoted S_ν, refers to energy propagating in a specific direction. Thus, the units of flux density are Wm^{-2}Hz^{-1}, and the power flowing through an aperture is

$$P = S_\nu \ \cos\theta \ \delta A \ \delta\nu \ . \qquad (3)$$

Radio astronomers have defined a special unit of flux density, the Jansky (denoted Jy). The definition is 1 Jy = 10^{-26} W m^{-2} Hz^{-1}. Its small size reflects the weakness of the signals characteristic of radio astronomy.

The distinction between having to consider specific intensity, and the simpler situation of dealing only with flux density, depends on the angular size of the source being observed compared to the range of angles over which the radio telescope in question accepts signals. So Sections 4.2. and 5.5. below are really distinguished by having the range of incidence angles from the source being **small** compared to the range of angles accepted by the antenna in the former case, which we refer to as observing a *point–like source*, and **large** in the latter case in which we are observing an *extended* or a *distributed source*.

2.2. Thermal sources

Radio astronomers observe an enormous variety of objects, ranging from asteroids to galaxies. While differing enormously, many of these share the quality of emitting like a *thermal source*. What this means is that in analogy to Equation 1, we can consider the power that passes through a unit surface area dA heading towards an element of solid angle $\delta\Omega$ in direction defined by angles θ and ϕ, in the range of frequencies $\delta\nu$ about frequency ν. This quantity has the same units as the specific intensity, and the power traveling through the element of surface is given by Equation 2. However, if the object in question is a true thermal source, the specific intensity which it emits is uniquely determined by the so-called "blackbody brightness". This means that the source is defined to absorb 100 percent of radiation incident on it, and thus, through Kirchoff's law, is defined to have an emissivity equal to unity. It thus emits the maximum specific intensity possible for a thermal source at its temperature, T, which is

$$I_\nu(\text{thermal}) = \frac{2h\nu^3}{c^2} \frac{1}{\exp(h\nu/kT) - 1}, \qquad (4)$$

where h is Planck's constant (6.626×10^{-34} Js) and k is Boltzmann's constant (1.38×10^{-23} J K^{-1}). The emission from a blackbody is unpolarized, and the above formula gives the specific intensity summed over both polarizations.

Radio astronomy, having started at relatively low frequencies, naturally tended to be characterized by $h\nu \ll kT$, the Rayleigh-Jeans (R–J) limit. Another way of looking at this is that if we define an equivalent temperature by $T_{\text{eq}} = h\nu/k$, we find that $T_{\text{eq}} = 0.048$ K $\times$ ν(GHz). Thus, for frequencies up to several GHz, the equivalent temperature is certainly below any relevant physical temperature. Assuming, then, that the Rayleigh–Jeans limit is applicable, we find

$$I_\nu(\text{thermal}, \text{R}-\text{J}) = \frac{2kT\nu^2}{c^2} = \frac{2kT}{\lambda^2}. \qquad (5)$$

In the R–J limit, the specific intensity emitted by a source is proportional to its temperature, and inversely proportional to the square of the wavelength.

2.3. Radio telescopes, transmission lines, and the antenna theorem

It is appropriate to present here some of the basic definitions associated with radio telescopes. In analogy with the situation associated with Equations 1 and 2, we can consider a radio telescope collecting radiation defined by its specific intensity. Radio astronomers typically consider the source to be unpolarized (for purposes of definition, even if not for research). As will be discussed in more detail in Section 3., single mode transmission lines used for radio astronomy receivers accept only a single polarization. Consequently, if we have an unpolarized plane wave incident on a radio telescope, we can define its *effective area* A_e in terms of the power received, P_{rec} through

$$P_{\text{rec}} = \frac{1}{2} A_e S_\nu \delta\nu. \qquad (6)$$

The factor of 1/2 reflects the coupling of the unpolarized wave to the single polarization transmission line and receiver. The effective area is related to the telescope's physical area, A_p (projected geometrical area) by the *aperture efficiency*, ϵ_a, through the Equation

$$A_e = \epsilon_a A_p \ . \tag{7}$$

A radio telescope is not sensitive only to radiation propagating in a single direction. Any system collecting electromagnetic radiation is subject to diffraction; this is the basis for the behavior discussed in Sections 4.2. and 5. The effect is that the effective area applies to radiation progating in the *boresight* direction, which is by definition the direction in which the radio telescope is most sensitive. For radiation incident from other directions, the relative sensitivity of the telescope is given by the *normalized power pattern*, $P_n(\theta, \phi)$, where we define θ to be the angle from boresight, and ϕ the azimuthal angle. By definition, $P_n(0,0) = 1$, and P_n is independent of ϕ for an axially symmetric telescope. The power pattern for a radio telescope is analogous to the relative sensitivity factor $\cos \theta$ for a simple aperture seen in Equations 1 through 3. The radiation received from a plane wave incident from direction (θ, ϕ) is thus

$$P_{\rm rec} = \frac{1}{2} A_e \ P_n(\theta, \phi) \ S_\nu(\theta, \phi) \ \delta\nu \ . \tag{8}$$

We now consider the telescope observing an extended source of radiation of specific intensity I_ν defined in a coordinate system relative to the telescope boresight direction, each infinitesimal element of solid angle produces a flux density

$$S_\nu(\theta, \phi) = I_\nu(\theta, \phi)\delta\Omega \ , \tag{9}$$

so that Equation 8 integrated over solid angle becomes

$$P_{\rm rec} = \frac{1}{2} A_e \delta\nu \int \int P_n(\theta, \phi) I_\nu(\theta, \phi) d\Omega \ . \tag{10}$$

Equation 10 is very important in its own right for relating the reponse of the telescope to the extended radiation source, but it also leads to a very valuable insight into radio astronomy calibration. We suppose that a radio telescope is observing a thermal source in the Rayleigh–Jeans limit which fills the entire power pattern of the telescope at a uniform temperature T_s. Then using Equations 5 and 10 we see that

$$P_{\rm rec} = \frac{1}{2} A_e \delta\nu \left[\frac{2kT_s}{\lambda^2} \int \int P_n(\theta, \phi) d\Omega \right] \ . \tag{11}$$

The integral here defines the *antenna solid angle*, Ω_a:

$$\Omega_a = \int \int P_n(\theta, \phi) d\Omega \ . \tag{12}$$

There is a very general theorem, which applies to any lossless diffraction-limited system fed by a single mode transmission system, which relates the effective area, the antenna solid angle, and the wavelength. The *antenna theorem* states that

$$A_e \Omega_a = \lambda^2 \ . \tag{13}$$

It means that to whatever type of antenna you couple your single mode system, you cannot arbitrarily change the integral of the power pattern (Ω_a) and the effective area (A_e); their product is fixed. So if you do something to broaden the power pattern and increase the antenna solid angle, you will inevitably reduce the effective area. Similarly, if you make an antenna which has a very large effective area, it will inevitably have a narrow beam. You cannot have an antenna that simultaneously has a large effective area **and** a broad beam. Despite the fact that such an antenna would be exactly what is needed for large–area high–sensitivity surveys, the antenna theorem says this is not realizable. With the antenna theorem, our expression for the power received by the radio telescope and coupled to the transmission line becomes much simpler:

$$P_{\rm rec} = kT_s \delta\nu \frac{A_e \Omega_a}{\lambda^2} = kT_s \delta\nu \ . \tag{14}$$

One important characteristic of single mode transmission lines concerns the power they transmit from a thermal source. To study this, we first have to consider that a transmission line can be *terminated* by a *matched load*. This means that by putting a component having the right impedance (relationship between electric and magnetic field, or voltage and current) at the end of the transmission line, a wave traveling along the transmission line will be completely absorbed. Conversely, any signal emitted by the matched load will travel without reflection down the transmission line. A matched load, or *termination*, is thus simply a transmission line blackbody. Matched loads are resistors at very low frequencies, but become more elaborate for high–frequency coaxial lines and waveguides. If we consider a matched load at temperature T, the power that is emitted flows down the transmission line in bandwidth $\delta\nu$ in the R–J limit is given by

$$P_{\rm trans} = kT\delta\nu \ . \tag{15}$$

This is closely related to that given in Equation 5, and the difference is that here we have only a **single** mode and a **single** polarization. You can define the equivalent blackbody temperature in terms of the power flowing down the transmission line, $P_{\rm trans}$ through

$$T_{bb} = P_{\rm trans}/k\delta\nu \ . \tag{16}$$

Now applying this relationship to the power received by the telescope (Equation 14), we find that

$$T_{bb} = T_s \ . \tag{17}$$

The equivalent blackbody temperature, one that would produce the same power as was collected by the telescope and coupled to the transmission line, is identical to the temperature of the (presumably distant) blackbody filling the antenna's power pattern. This leads to defining this equivalent temperature as the *antenna temperature*, T_a. This temperature simply is another way to define the power flowing in the transmission line. It can be, and is used whether or not the antenna is observing a thermal source, and whether or not the source fills the antenna power pattern. In a completely general sense, then, we define

$$T_a = P_{\rm rec}/k\delta\nu \ . \tag{18}$$

The above equations can be rewritten in terms of the antenna temperature as

$$T_a = \frac{A_e}{2k} P_n(\theta,\phi) S_\nu(\theta,\phi) \ , \tag{19}$$

[plane wave from arbitrary direction]

$$T_a = \frac{A_e}{2k} S_\nu(0,0) \ , \tag{20}$$

[plane wave from boresight (0,0) direction]

$$T_a = \frac{A_e}{2k} \int\int P_n(\theta,\phi) I_\nu(\theta,\phi) d\Omega \ , \tag{21}$$

[extended source expressed in terms of specific intensity], and

$$T_a = \frac{1}{\Omega_a} \int\int P_n(\theta,\phi) T_s(\theta,\phi) d\Omega \ , \tag{22}$$

[extended source expressed in terms of temperature][1].

A source having angular size larger than that of the antenna's power pattern is said to be *resolved*, while a source which is comparable to or larger than the power pattern is referred to as an *extended* source. A source of angular size much smaller than that of the power pattern is called an *unresolved* or *point-like* source. The solid angle subtended by the source, Ω_s, for a circular source of angular radius θ_s is

$$\Omega_s = \int_0^{\theta_s} \int_0^{2\pi} d\phi \sin\theta d\theta = 2\pi(1 - \cos\theta_s) \ . \tag{24}$$

In the limit $\theta_s \ll 1$, we obtain

$$\Omega_s = \pi\theta_s^2 \ , \tag{25}$$

and if the unresolved source (lying in the direction of maximum response of the antenna, and in the R–J limit) has uniform temperature T_s, the integral in Equation 22 becomes

---

[1] If we are not in the Rayleigh–Jeans limit, we must use the full expression for the specific intensity of a thermal source given in Equation 4, which yields

$$T_a = \frac{1}{\Omega_a} \int\int \frac{h\nu/k}{e^{h\nu/kT_s(\theta,\phi)} - 1} P_n(\theta,\phi) d\Omega \ . \tag{23}$$

This can be important starting at millimeter wavelengths ($h\nu/k = 5.5$ K at the frequency of the $J = 1 \to 0$ transition of carbon monoxide, which can be produced in molecular clouds with temperatures as low as $\simeq 10$ K).

$$T_a = \frac{\Omega_s}{\Omega_a} T_s \ . \qquad (26)$$

It is important to appreciate the fundamental relation between the antenna temperature and the single mode transmission line, for with the antenna theorem, we have a method to calibrate measurement of any radio source in terms of thermal sources. Additional discussion of basic formulae relating to radiation and radio telescopes can be found in Kraus (1966). Ulaby, Moore, and Fung (1981) present a discussion of calibration in terms of thermal emission.

2.4. The reciprocity theorem

The *Reciprocity Theorem* (see Kraus 1988 for a good discussion and Terman 1943 for discussion and additional references) is a valuable aid in analysis of radio telescope system performance. The theorem assumes that the medium between the antennas is itself reciprocal (propagation properties independent of direction of propagation). It basically states that if you have two radio telescopes (or *antennas*, in engineering parlance) in any relative orientation, the power received from the first by exciting the second with a source of known strength is equal to the power received from the second with the same source coupled to the transmission line of the first antenna. Since this applies for any arbitrary orientation of the antennas (which can be identical), and you can make one of them appear as a point–like source by moving it sufficiently far away, this is equivalent to saying that the angular response of any radio telescope when used as a transmitting antenna, is the same as when it is used as a receiving antenna.

The utility of the Reciprocity Theorem is that while we use radio telescopes almost exclusively as receiving antennas (Arecibo being one of the very few exceptions), it is often easier conceptually, to analyze the performance of an antenna as a transmitter. In this way, we often envision replacing the receiver with a transmitter, and follow the signal through the transmission line, to the feed system, and finally to the antenna itself. The effective area (A_e) and the power pattern (P_n) that we so derive are, according to the Reciprocity Theorem, the same as is relevant when the radio telescope is used as a receiving antenna. In the more detailed discussion of the performance of radio telescopes which follows, we will often make use of the Reciprocity Theorem and imagine an antenna with a RF power source rather than a receiver, but any results obtained can also be found without resort to this approach as it is only really a convenience in analyzing antenna systems.

3. Components of Radio Telescope Systems

3.1. Overview

The present lecture is supposed to focus on "single–dish" radio telescopes, which can be interpreted to mean antennas which collect radiation by using a reflecting surface. The only real alternative is at low frequencies, where conductors of appropriate dimensions can present a relatively large collecting area to an incident wave. We thus restrict ourselves to so–called *aperture antennas*, meaning ones involving an aperture which is at least moderately large when measured in wavelengths (as contrasted with *wire antennas* which are characteristically no more than one wavelength in size). We take an astronomical single–dish radio telescope to include (considering things in the context of the reciprocity theorem):

1. A **single mode transmission medium** (guided-wave propagation of a single polarization, connecting the receiver to the feed system);

2. A **feed system** (mode transformer from single mode to divergent free-space propagation and optionally a polarization transformer);

3. A **reflector antenna** (a phase transformer operating on the beam from the feed system).

3.2. The transmission line

It is important to emphasize again that the single mode transmission media used to construct radio frequency amplifiers and mixers support only a single polarization state [2].

If you wish to observe a specific polarization (*e.g.* *l.h.* circular), you can make a polarizer to transform the polarization state of the transmission medium to *l.h.* circular. Ideally, this can be done with perfect efficiency, but the real–world situation, as will be explained in a subsequent lecture, is less ideal. If you want to observe two polarization states simultaneously, this can be done, but only by separately processing the two signals in individual transmission media. Thus, you can have two transmission lines carrying power deriving from incident *l.h.* and *r.h.* circularly polarized signals, or vertically and horizontally linearly polarized signals, to two independent receivers. In any case, the field configuration in each transmission line corresponds to **its own** mode configuration.

3.3. The feed system

The feed system

1. carries out the desired polarization processing, and

2. transforms power propagating in a single mode transmission line to illuminate the antenna.

The feed system design determines the sequence and method of these two functions. For example, at relatively low frequencies, the single-mode transmission lines may be waveguides, typically with a rectangular cross section and

---

[2] One has to be a little careful, due to the fact that while **most** single-mode transmission media by choice of dimensions support only a single polarization state, there are some, such as circular waveguide, which over a certain frequency range support only a single mode (field configuration) but in which due to symmetry, this configuration can have any orientation. This means that, for example, circular waveguide propagating the TE_{11} mode can have any polarization with the electric field perpendicular to the direction of propagation, which is the axis of the waveguide. So in this type of transmission medium you can have a pair of linearly polarized, or a pair of circularly polarized signals, or any combination thereof. For purposes of simplicity, we will consider our single mode transmission systems to have only a single polarization state, and consider the multi-polarization single mode transmission media to be part of the feed system.

the electric field perpendicular to the direction of propagation and to the larger waveguide wall. This field is linearly polarized, and you obviously cannot combine two single mode waveguides together at a given frequency efficiently; beyond being impossible from an electromagnetics viewpoint, this would violate thermodynamics. The two waveguides **can** be combined in a more complex, multimode transmission structure; there are many designs and names for such devices, and these will be discussed in some detail in the lecture on receiver systems. From our point of view, we need simply to know that it is possible to combine two individual single–mode transmission line systems into a single guided wave transmission medium[3].

In the preceding example (starting from the singe–mode interface), the feed system first combined the polarizations, and then transformed them to a beam producing the desired illumination of the antenna. At higher frequencies, the order of these two steps is often reversed, and we find systems with a pair of independent feeds – one for each orthogonal polarization state. The feed system transforms the single–mode transmission line signal into the appropriate beam for illuminating the antenna. The two polarizations are combined and, if desired transformed, while propagating in free space.

The second function of the feed system is to produce the desired illumination of the antenna. In most radio astronomy systems, the antenna is in the *far field* of the feed system ($L \gg 2D_{\text{feed}}^2/\lambda$, where L is the distance between the feed and the antenna, and D_{feed} is the diameter of the feed). This means that we can consider the radiation produced by the feed to be a diverging spherical wave (in terms of phase), with a specific angular distribution of power. The radiation from the feed can be considered to *illuminate* the antenna (continuing to use the reciprocity theorem to allow us to consider an antenna operating as a transmitter).

3.4. The antenna (reflector)

The essential function of the antenna itself is that of a *phase transformer*. That is, to change the spherical wavefront produced by the feed system into a plane wave. It is evident that while the output of an antenna can be a plane wave in terms of phase distribution, it cannot be a true plane wave, which by definition is infinite in extent transverse to its direction of propagation. The "plane wavefront" produced by an antenna is thus at best a truncated plane wave (having an extent equal to the diameter of the antenna). More generally, as will be discussed below, the illumination of the antenna will be non–uniform, but typically centrally concentrated.

The antenna itself can be of many types, but most radio telescopes fall into categories of prime focus antennas (parabolic primary reflector only), Cassegrain antennas (parabolic primary reflector plus hyperbolic secondary reflector) and Gregorian antennas (parabolic primary reflector plus ellipsoidal secondary). For

---

[3]Since the two signals corresponding to the two polarization states are handled by the coherent radio astronomical receiver system independently, and preserving the phase of each, all of the information about the polarization state of the incident signal can be recovered from the amplitude and phase information. This is quite different than if only the intensity of variously polarized components of the signal can be measured, as is the case in the optical range.

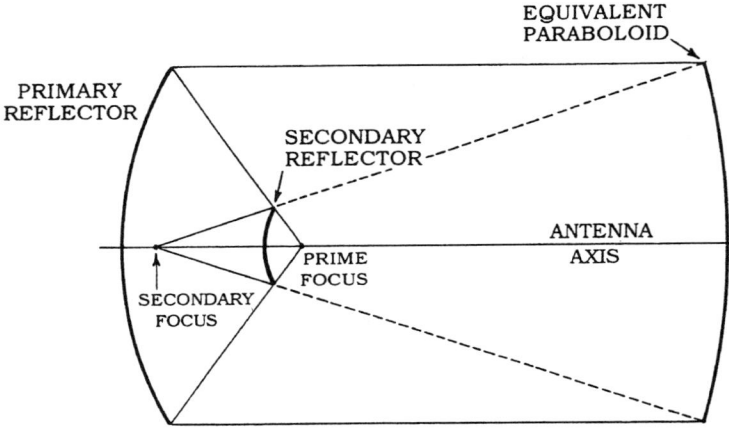

Figure 1. Equivalent paraboloid of Cassegrain radio telescope.

example, the Green Bank Telescope (GBT) operates as a prime focus instrument at low frequencies, and as a Gregorian at higher frequencies. The FCRAO 14-m telescope is a Cassegrain instrument, as will be the 50-m diameter Large Millimeter Telescope (LMT). Arecibo is unusual in employing a spherical primary reflector, and the "Gregorian" corrector system includes two quasi–ellipsoidal reflectors. These systems are all designed to transform the spherical wave produced by the feed system into a plane wave. In analyzing a specific antenna system, the concept of the *equivalent paraboloid* (Hannan 1961) is very useful. As illustrated in Figure 1, it allows us to treat a multi–reflector antenna as equivalent to a single antenna, having diameter equal to that of the primary (or main) reflector, D, and focal length given by the effective focal length, F_e, of the system. In this picture, you do not have to worry about the action of the various reflectors independently, but can consider only their overall effect. Consider a geometry in which a ray from the focal point of the system which hits the edge of the secondary reflector also hits the edge of the primary reflector. The *edge angle* defined by this ray thus determines the edge of the antenna that can be usefully illuminated. It is also the angle to the edge of the equivalent paraboloid, which we consider to act as a pure phase transformer, and thus the bundle of rays diverging from the focal point are *collimated* by the phase–transforming properties of the reflector. We can then consider these rays to travel in a bundle, parallel to each other, to the *aperture plane*[4] The phase distribution in the aperture plane is ideally uniform but, as we will discuss at greater length in

---

[4]The location of the aperture plane is not critical due to the collimation, but it is conceptually placed just beyond the physical aperture of the antenna.

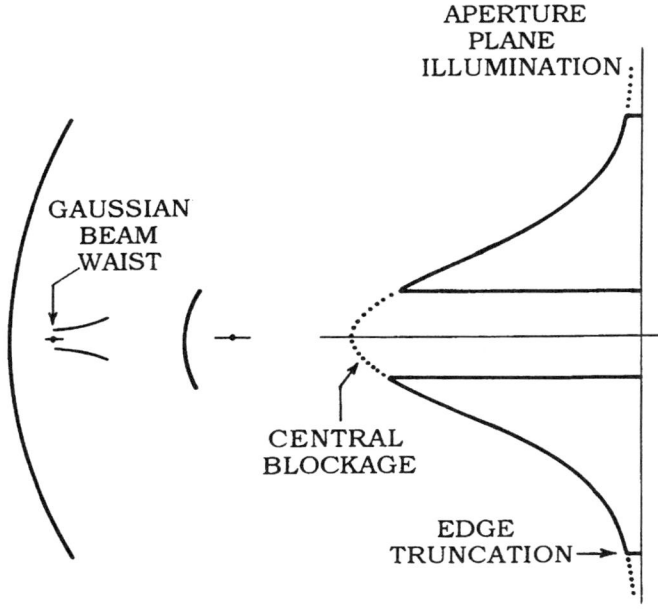

Figure 2. Schematic of the aperture plane with a Cassegrain reflector system.

what follows, a variety of effects including defocusing and surface errors modify the ideal situation.

4. Coupling of the Antenna to Incident Radiation

4.1. The aperture plane

For the large telescopes commonly encountered in radio astronomy, the aperture plane field distribution is 100's to 1000's of wavelengths in size. The aperture plane is a plane in free space over which the electric field is specified, and the magnetic field is everywhere perpendicular to the electric field and to the direction of propagation. Given the large size (in terms of λ) of the illuminated region in the aperture plane, we make only a very small error by ignoring "edge effects" that are produced at the perimeter of the reflector surface, and assuming that the illumination is the amplitude pattern of the feed modified in amplitude by the geometrical constraints of the antenna, and in phase by its acting as a phase transformer. The aperture plane is shown schematically in Figure 2.

The aperture plane electric field distribution is defined as $E_a(r_a)$, where r_a defines a point in the aperture plane; this can be conveniently put in polar coordinate system for an axially symmetric antenna, and a rectangular coordinate system for more general application. The electric field is treated as a scalar quantity, as we assume that the feed system produces a specified polarization over the entire aperture. The aperture field distribution is real only for a perfectly uniform phase distribution, but in general is a complex quantity due to phase variations.

4.2. Coupling to a plane wave

Although the power pattern of a radio telescope and coupling to a distributed source are determined by diffraction and thus involve significant additional calculations, the coupling to a plane wave can be calculated directly from the aperture field distribution. This is because the aperture plane represents a field distribution which couples perfectly to the feed system; what we want to know is what fraction of a plane wave incident on the antenna is coupled to that distribution[5]. In the aperture plane, we have three components to the electric field. These are: (1) the feed distribution modified in phase by the antenna; (2) the spillover of the feed past the edge of the reflector which has not been affected by phase transformation properties of the antenna (if we have a Gregorian or Cassegrain system this does end up in the aperture plane; for a prime focus system it does not); and (3) radiation *blocked* by feed support legs and/or *scattered* by these structural elements, if the antenna is not of offset geometry. All of these contribute to the radiation pattern of the antenna, but in general only component (1) makes a significant contribution to the coupling to a plane wave, due to the very different phase distribution of the other components.

To calculate the coupling between the aperture field distribution, E_a and the plane wave, it is convenient to define a *truncated plane wave*, which is just one that extends only over the projected area of the antenna. Clearly, this is what is appropriate for normalizing the power in a plane wave that **could** be coupled to the antenna. Denoting this as E_{pw}^t, the amplitude coupling between the plane wave and the radio telescope can be expressed as

$$c = \frac{\int\int E_a^* E_{\text{pw}}^t dA}{[\int\int E_a^* E_a dA \int\int E_{\text{pw}}^{t*} E_{\text{pw}}^t dA]^{0.5}} . \qquad (27)$$

It is convenient to denote the *overlap integrals* in this expression with $\langle a|b\rangle$ representing $\int\int a^* b \, dA$, which gives us

$$c = \frac{\langle E_a | E_{\text{pw}}^t \rangle}{[\langle E_a | E_a \rangle \langle E_{\text{pw}}^t | E_{\text{pw}}^t \rangle]^{0.5}} , \qquad (28)$$

and

---

[5] We have here used conceptually and will use in detail, the reciprocity theorem to calculate the aperture plane distribution based on operation as a transmitter, while we will consider operation as a receiver to analyze the coupling, although we could equally well deal with coupling to a **radiated** plane wave when considering operation as a transmitter.

$$\epsilon_a = |c|^2 = \frac{|\langle E_a | E_{\text{pw}}^t \rangle|^2}{\langle E_a | E_a \rangle \langle E_{\text{pw}}^t | E_{\text{pw}}^t \rangle} \qquad (29)$$

since, as seen in Equations 6 and 7, the aperture efficiency refers to power coupling rather than field coupling.

The limits of the integrals involving the truncated plane wave are just those to include the projected area of the antenna, so for an axially symmetric antenna of diameter D we have $0 \leq r \leq D/2$, and $0 \leq \phi \leq 2\pi$. The integral $\langle E_a | E_a \rangle$ in the denominator reflects the radiation pattern of the feed; by extending the outer limit to infinity we include all of the power radiated by the feed, which normalizes the coupling integral to include the effect of any spillover. In the numerator, however, we include **only** that portion of the feed radiation pattern which is intercepted by the antenna and subsequently has its phase distribution appropriately modifed to be (one hopes) close to uniform. The integral in the numerator should also exclude any regions blocked by the secondary reflector or feed support legs , in which the electric field (in the geometrical optics *shadow region*) is zero.

We can divide the aperture efficiency into the product of factors which reflect different aspects of the feed and antenna performance:

$$\epsilon_a = \Pi_i \epsilon_i = \epsilon_t \epsilon_s \epsilon_b \epsilon_p , \qquad (30)$$

where ϵ_t, called the *taper efficiency* indicates how effectively the aperture is being used, ϵ_s, the *spillover efficiency*, is the fraction of power from the feed which is usefully intercepted by the antenna, ϵ_b is the fractional efficiency due to *blockage* produced by the secondary reflector and the feed support structure, and ϵ_p is the fractional efficiency resulting from phase errors. To appreciate these contributions in detail, we need to use a specific model of a feed system and antenna. Many excellent feed horns used at centimeter through millimeter wavelengths have radiation patterns which can be well–represented by Gaussian distributions (*e.g.* Aubrey & Bitter 1975; Goldsmith 1997) and antenna analysis can profitably be carried out in the framework of Gaussian illumination (Lamb 1986; Goldsmith 1987).

The Gaussian beam characterizing the distribution of the electric field pattern radiated by a feedhorn can be written as

$$E(\theta)/E(0) = \exp[-(\theta/\theta_0)^2] , \qquad (31)$$

where θ_0 is the *beam divergence angle* of the Gaussian beam. We assume that the phase distribution of the feedhorn beam is a spherical wave. Ideally, the antenna coverts the spherical wavefront from the feed into a uniform phase front, leaving the amplitude distribution unaffected. The aperture plane electric field is thus given by

$$E_a = \exp[-(r/w)^2] , \qquad (32)$$

where r is the radius from the axis of symmetry, and w is the *beam radius* of the Gaussian beam amplitude distribution[6], respectively. The on–axis power density relative to the power density at the edge of the aperture is called the *edge taper*, T_e. For a circular aperture of radius $D/2$ illuminated by a Gaussian beam having beam radius w, T_e is equal to $e^{2\alpha}$, where

$$\alpha = (D/2w)^2 . \qquad (33)$$

The edge taper expressed in decibels (dB) is given by

$$T_e(\text{dB}) = 10\log_{10}(e^{2\alpha}) = 8.686\alpha , \qquad (34)$$

so that

$$w = \frac{D}{2}\sqrt{\frac{8.686}{T_e(\text{dB})}} . \qquad (35)$$

The larger the edge taper, the more centrally concentrated is the illumination, and the smaller is the beam radius w relative the the aperture radius $D/2$.

Consider a circular, unblocked aperture with perfect phase performance ($\epsilon_b = \epsilon_p = 1.0$). We obtain the taper efficiency ϵ_t by deliberately ignoring the spillover of the feed illumination pattern, and do this by setting the upper limit to the first integral in the denominator of Equation 29 to $D/2$, giving us

$$\epsilon_t = \frac{|\int_0^{D/2} E_a dA|^2}{\int_0^{D/2} |E_a|^2 dA \ \int_0^{D/2} dA} . \qquad (36)$$

For Gaussian illumination, this yields

$$\epsilon_t = \frac{2}{\alpha}\frac{(1-e^{-\alpha})^2}{1-e^{-2\alpha}} . \qquad (37)$$

This function has a maximum of $\epsilon_t = 1$ for $\alpha = 0$, corresponding to uniform illumination of the antenna. This is an understandable result, in that it is reasonable that we get the best coupling to a plane wave if the antenna illumination is also a plane wave over its finite extent. However, if we consider a realistic feed system, we see that this cannot be achieved without penalty; the cost is that the spillover will be relatively large.

The spillover efficiency is the ratio of the aperture efficiency including the spillover (letting the integral normalizing the feed power extend to infinity) to that ignoring the spillover (upper limit equal to $D/2$). Thus,

$$\epsilon_s = \frac{\int_0^{D/2} |E_a|^2 dA}{\int_0^{\infty} |E_a|^2 dA} , \qquad (38)$$

---

[6]For dual reflector systems having moderately large focal ratios, the mapping between angle and radius is linear and given just by the effective focal length. In this case, r and w are equal to the effective focal length times θ and θ_0.

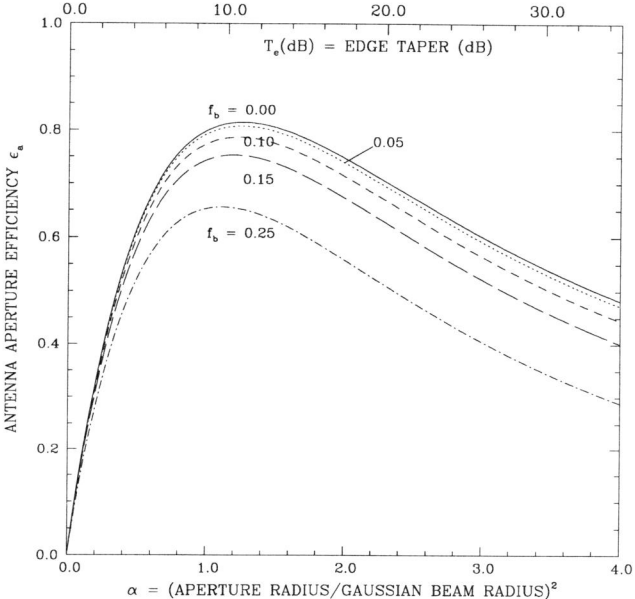

Figure 3. Aperture efficiency for Gaussian–illuminated circular aperture as a function of the edge taper in decibels, $T_e(\text{dB})$ or equivalently, α. The different curves are for different values of the fractional radius for the central blockage, f_b, with the unblocked aperture corresponding to $f_b = 0.0$.

which for Gaussian illumination yields

$$\epsilon_s = \frac{\int_0^{D/2} e^{-2(r/w)^2} 2\pi r dr}{\int_0^{\infty} e^{-2(r/w)^2} 2\pi r dr} = 1 - e^{-2\alpha} \ . \tag{39}$$

The spillover efficiency thus approaches 0 for $\alpha \to 0$, corresponding to uniform illumination, and approaches unity for large α, corresponding to centrally concentrated illumination[7].

It thus is appropriate to consider the product of ϵ_t and ϵ_s, and for a Gaussian–illuminated zero phase error unblocked circular antenna, the aperture efficiency is given by

$$\epsilon_a = \frac{2}{\alpha}(1 - e^{-\alpha})^2 \ . \tag{40}$$

This is also shown in Figure 3 as the curve for $f_b = 0.0$.

---

[7] A Gaussian that uniformly illuminates an antenna of finite size has $w \to \infty$, so that the fraction of the power in the Gaussian intercepted by the antenna $\to 0$.

Under these conditions, the aperture efficiency has a maximum value equal to 0.815, which occurs for $\alpha = 1.25$, or an edge taper equal to 10.9 dB. This illustrates a general result: the maximum aperture efficiency for a reasonable feed pattern is a tradeoff between spillover efficiency and taper efficiency, and typically requires an edge taper on the order of 10 dB. The GBT should be able to approach this value. The Gregorian system at Arecibo employs two aspheric reflectors to achieve a relatively uniform illumination over most of the illuminated area, with the power dropping rapidly at the edges, which can in principle achieve even higher aperture efficiencies, but of course the complete system does suffer from significant blockage produced by the suspended structure.

The issue of blockage has been introduced, but it is a difficult subject to treat generally as it can take so many forms. The simplest is to consider an on-axis circular antenna employing a secondary reflector which subtends a fraction f_b of the radius of the primary reflector. The effect of the *central blockage* that results can be treated to first order by realizing that in the aperture plane, a fraction f_b of the radius will not be illuminated, and in consequence we have a lower limit $f_b D/2$ for the overlap integral in the numerator of Equation 29. Thus

$$\epsilon_b = \frac{|\int_{f_b D/2}^{D/2} E_a r dr|^2}{|\int_0^{D/2} E_a r dr|^2} \ . \tag{41}$$

For Gaussian illumination, the resulting blockage efficiency is given by

$$\epsilon_b = \frac{(e^{-f_b^2 \alpha} - e^{-\alpha})^2}{(1 - e^{-\alpha})^2} \ . \tag{42}$$

The effect on the aperture efficiency can be seen in Figure 3. For $\alpha \to 0$ corresponding to uniform illumination

$$\epsilon_b = (1 - f_b^2)^2 \ , \tag{43}$$

which for small values of f_b becomes

$$\epsilon_b = 1 - 2f_b^2 \ . \tag{44}$$

In this limit, the reduction in aperture efficiency is **twice** the fraction of the area blocked. This is a result of the fact that the blockage affects the overlap integral of the electric field, and thus enters into the aperture efficiency raised to the second power. This aspect of the effect of blockage is independent of the form of the blockage, as long as the aperture illumination is uniform. Thus, it applies to feed support legs as well as the secondary reflector. In most systems, the blockage of the support legs is actually greater than that of the secondary reflector, and consequently their effect is greater [8]. General calculations of the

---

[8] It is important to calculate the effect of blockage from the feed support legs by including the shadow produced in the highly divergent beam propagating from the secondary reflector towards the primary reflector. If the feed support legs intercept the primary reflector at radii significantly less than $D/2$, the shadowed region they produce will be much larger than their

effect of arbitrary blockage with a given non–uniform illumination pattern must be carried out numerically.

The term *phase errors* refers to any deviation of the phase in the aperture plane from its ideal constant value. Phase errors can result from errors in the antenna surface; if these are not too small (measured in wavelength) we can imagine the rays representing the field propagating from the secondary to the primary as being perturbed in phase by an amount proportional to the displacement of that section of the surface from its desired location[9]. These surface errors can be *large scale* if their characteristic size is comparable to that of the primary reflector, or *small scale* if they are much smaller in scale than the reflector. The former can be due to gravitational deformation, while the latter can result from mis–adjustment of surface panels comprising the reflector, of which Arecibo has, for example, more than 38,000. Another source of phase errors can be a focus error; since the radius of curvature of the spherical wave heading towards the primary is proportional to the distance from its focal point, any change in separation between primary and secondary from its nominal value, or a change in the focal length of the system, produces a phase error which varies quadratically as the radius in the aperture plane.

The effect of any of these types of phase errors is to reduce the aperture efficiency. If we write the complex aperture plane field in terms of an amplitude E_a and a phase error $\delta\phi$, we obtain

$$E_a = E_a^0 e^{i\delta\phi(r_a)} , \qquad (45)$$

where r_a denotes a location in the aperture plane. The *phase efficiency* is given by

$$\epsilon_p = \frac{|\int E_a^0 e^{i\delta\phi(r_a)} dA|^2}{|\int E_a^0 dA|^2} . \qquad (46)$$

To evaluate this expression exactly requires detailed knowledge of the phase errors in the aperture plane. Sometimes this information is available, and then the degradation in aperture efficiency due to the phase errors can be determined. One case in which this is possible is the aforementioned one of a defocused antenna, in which the aperture plane phase has the form

$$\delta\phi(r) = \beta(2r/D)^2 , \qquad (47)$$

where β is the *edge phase error*, which is directly proportional to the feed displacement[10]. In this situation, with Gaussian illumination, Equation 46 can be evaluated, giving

---

geometrical area. Also, the effective cross section of a structural beam of width s at long wavelengths is never much smaller than the wavelength, so for $\lambda \gg s$, the blockage will again be much larger than expected from purely geometrical considerations.

[9] The same applies to the secondary reflector, but this is generally so much smaller than the primary reflector that it can be made to have errors which are negligible compared to those of the primary.

[10] For a system with effective focal length f_e and offset δ, the edge phase error is given by $\beta = \pi\delta/[4\lambda (f_e/D)^2]$ (Goldsmith 1997), where δ is the change in spacing between the feed and the reflector relative to its nominal value.

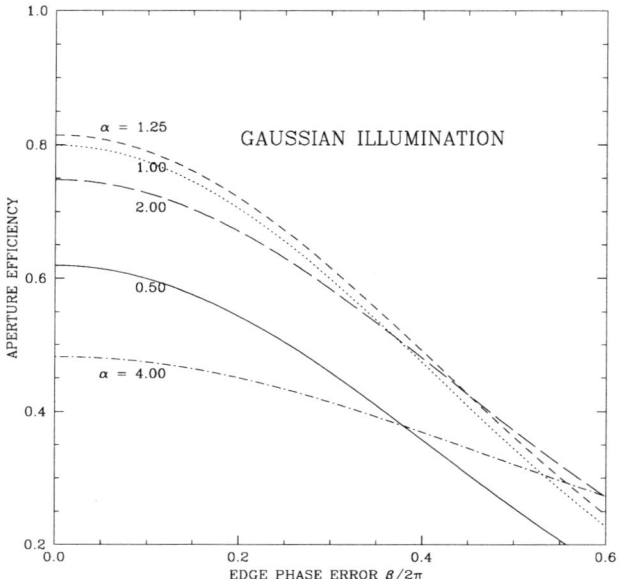

Figure 4. Effect of defocus on the aperture efficiency of a Gaussian–illuminated unblocked circular antenna.

$$\epsilon_p = \frac{\alpha^2}{\alpha^2 + \beta^2} \frac{1 + e^{-2\alpha} - 2e^{-\alpha}\cos\beta}{(1 - e^{-\alpha})^2} . \tag{48}$$

As can be seen in Figure 4, the effect of the defocusing is always to reduce ϵ_p and thus the aperture efficiency. The situation is simplified if we consider uniform illumination, ($\alpha = 0$) for which we find $\epsilon_p = 2(1 - \cos\beta)/\beta^2$, which in the limit of small β becomes

$$\epsilon_p = 1 - \frac{\beta^2}{12} . \tag{49}$$

This allows us to see how feed motion or change in antenna focal length will affect the aperture efficiency, or conversely, to set tolerance on allowable motion of the feed and/or the secondary reflector.

In the general case of *small phase errors*, the exponential in Equations 45 and 46 can be expanded in a power series, $(e^{i\delta\phi} \simeq 1 + i\delta\phi - \delta\phi^2)$, and again assuming uniform illumination, the integral gives us

$$\epsilon_p = 1 + \overline{\delta\phi}^2 - \overline{\delta\phi^2} , \tag{50}$$

where the bar indicates the average over the aperture plane. This is conveniently expressed in terms of the *rms phase error*, $\delta\phi_{\rm rms}$ as

$$\epsilon_p = 1 - \overline{\delta\phi^2_{\text{rms}}} \ . \tag{51}$$

For the defocused antenna with uniform illumination, it is straightforward to determine that $\overline{\delta\phi} = \beta/2$ and $\overline{\delta\phi^2} = \beta^2/3$, yielding $\delta\phi^2_{\text{rms}} = \beta^2/12$, exactly recovering Equation 49. This illustrates a general result, that for small phase errors, their distribution is not important: what matters is the *rms* of the aperture plane phase distribution.

There are cases where we do not wish to restrict the analysis to small phase errors. One example is that of surface errors arising from such causes as fabrication errors and panel adjustment inaccuracies. This situation is generally modeled by assuming that there is a random component to the aperture phase distribution, and that we know only the rms surface error. It is a general tendency of radio astronomers to try and use radio telescopes at shorter and shorter wavelengths, so that the aperture plane phase error will **not** necessarily be small, and it is of interest to understand how the performance of the telescope will vary as a function of wavelength. Analysis of the effect of random errors on radio telescope performance has been carried out by Ruze (1966). The first parameter is the *rms* phase variations, and the second is the correlation length, c, of the variations. In the limit of very small correlation length $c \ll D$, the result is

$$\epsilon_p = e^{-\delta\phi^2_{\text{rms}}} \ . \tag{52}$$

The relationship between the phase error and the surface error depends on the telescope geometry, and the *effective rms* depends on the illumination pattern as well. For not too curved reflecting surfaces, the phase shift is related to the surface error δ by $\delta\phi = 4\pi\delta/\lambda$, so that defining the *rms surface error* to be ϵ, we obtain the *Ruze formula*

$$\epsilon_p = e^{-(4\pi\epsilon/\lambda)^2} \ . \tag{53}$$

This gives the often–cited result that the aperture efficiency is reduced by 3 dB (a factor of 2) for a *rms* surface error ϵ equal to $\lambda/16$. This is certainly a useful indication of general effect of random surface errors, but the details of the degradation of the performance depend on other parameters, and it should also be borne in mind that the surface errors may not actually be random with small correlation length. In the case of finite correlation length, the situation is considerably more complex. Several specific situations were modeled by Ruze (1966) who showed that the result is a slightly smaller reduction in ϵ_a, but an additional component (often called the error pattern) is added to the normalized power pattern corresponding to the radiation from the finite–sized constituent segments of the reflector.

5. The Radiation Pattern and Coupling of a Radio Telescope to a Distributed Source

5.1. The electric field and power patterns

In order to analyze the coupling of a radio telescope to an extended source, we need to know the normalized power pattern as well as the aperture efficiency or antenna solid angle (as seen in Equations 21 and 22). The aperture plane electric field distribution is again the basis for these calculations. For an antenna with $D \gg \lambda$, the *far–field* or *Fraunhofer* radiation pattern is given by a relatively simple diffraction integral (Silver 1949). We take the distance from the antenna to be R, and as we are in the far–field, the overall electric field amplitude varies as R^{-1} and has spherical phase fronts with phase variation e^{-ikR}, where $k = 2\pi/\lambda$. Beyond this, the details are contained in what is essentially a sum of contributions of each element of the aperture with its relative phase. The most convenient way to write the latter depends on the geometry of the antenna. If we define a point in the aperture plane by coordinates (x_a, y_a), and define angles θ_x and θ_y aligned with the appropriate axes, the far–field electric field amplitude is

$$E_{ff}(\theta_x, \theta_y) = \frac{i}{\lambda} \frac{e^{-ikR}}{R} \int\int E_a(x_a, y_a) e^{ik(x_a\theta_x + y_a\theta_y)} dx_a dy_a \ . \qquad (54)$$

The integral is to be carried out over the aperture plane, or at least over the portion of the aperture plane with nonzero field, and we have restricted ourselves to small angles, and so approximated $\cos\theta \simeq 1$ and $\sin\theta \simeq \theta$. A key result seen here is that there is essentially a Fourier transform relationship between the aperture plane distribution (as a function of x_a and y_a) and angular field distribution (as a function of θ_x and θ_y) in the far–field. This is an important guide to thinking about various effects, particularly antenna size (measured in wavelengths), illumination, and blockage, and is extensively exploited in analyses of telescope performance. For example, if we add a linear phase gradient of s radians per wavelength to the aperture illumination, we have in the aperture plane $\delta\phi_a = e^{2\pi s x_a/\lambda}$ and it is evident from Equation 54 that E_{ff} will be unchanged in form, but will be *shifted* by an angle s. This "shift theorem" and other valuable insights gained from consideration of the Fourier relationship between E_a and E_{ff} are discussed at length by Christiansen and Högbom (1985).

Using angles θ for the off–axis angle and ϕ for the azimuthal angle in the far–field, we find

$$E_{ff}(\theta,\phi) = \frac{i}{\lambda} \frac{e^{-ikR}}{R} \int\int E_a(x_a, y_a) e^{ik\sin\theta(x_a\cos\phi + y_a\sin\phi)} dx_a dy_a \ . \qquad (55)$$

If we have an axially symmetric antenna system, it is more convenient to use a polar coordinate system (r_a, ϕ_a) to describe a point in the aperture plane, and we obtain

$$E_{ff}(\theta,\phi) \simeq \int\int E_a(r_a, \phi_a) e^{ik\sin\theta r_a(\cos\phi_a\cos\phi + \sin\phi_a\sin\phi)} r_a dr_a d\phi_a \ , \qquad (56)$$

but if the aperture illumination is also independent of ϕ_a, expressing the trigonometric terms as $\cos(\phi - \phi_a)$ and integrating over ϕ_a, we obtain

$$E_{ff}(U) \simeq \int E_a(\rho_a) J_0(\rho_a U) \rho_a d\rho_a \ , \qquad (57)$$

where U is the *normalized angular coordinate* given by

$$U = \pi D \sin\theta/\lambda \,, \tag{58}$$

J_0 is the Bessel function of order zero, and ρ_a is the *normalized radial aperture coordinate* defined by $\rho_a = r_a/(D/2)$. This is a *Hankel* rather than a Fourier transform relationship, but the general relationships between aperture plane and far–field distributions are not very different. As can be seen from Equation 57, the pattern depends only on U and for small angles $U \simeq \theta D/\lambda$. Thus, all antenna characteristics scale in angle as $\theta \simeq \lambda/D$. This behavior is, of course, a general feature of diffraction–limited operation[11]. For all of the above situations, the terms outside the integrals are not critical, since in order to obtain the normalized field pattern, we divide $E_{ff}(\theta)$ by its value at $\theta = \phi = 0$. For the completely azimuthally symmetric situation, as $J_0(0) = 1$, we obtain the *normalized field pattern*

$$E_n(U) = \frac{\int E_a(\rho_a) J_0(\rho_a U) \rho_a d\rho_a}{\int E_a(\rho_a) \rho_a d\rho_a} \,. \tag{60}$$

To obtain the normalized power pattern, we simply take the squared magnitude of E_n, giving us

$$P_n(U) = \frac{|\int E_a(\rho_a) J_0(\rho_a U) \rho_a d\rho_a|^2}{|\int E_a(\rho_a) \rho_a d\rho_a|^2} \,. \tag{61}$$

In addition to being immediately highly relevant for an unblocked, circular aperture such as the GBT, this serves to illustrate the general properties of antenna radiation patterns. Field and power patterns can be presented in many forms, but are obviously simplified if you have only a single coordinate to deal with. A popular method is to display power patterns in polar form, with the radius from the origin proportional to the logarithm of the normalized power. As indicated in Figure 5, the most prominent feature is the *main lobe* (or *main beam*, with maximum value of $P_n = 1$. The term generally used to define the angular width of the main lobe of the power pattern is the *FWHM beam width*, which is the angular width between the half power points of P_n. It is most often written as $\Delta\theta_{\text{fwhm}}$. Of course, giving a single value for this quantity is implicitly assuming an azimuthally symmetric power pattern, which may be close to reality for the GBT, but is certainly not the case for Arecibo, where by design,

---

[11] It may be of interest that the divergence of the antenna beam obtained from diffraction theory as expressed by *e.g.* Equation 54 can also be obtained from quantum mechanics. To do this, we consider a photon, travelling in the z direction, which encounters an aperture in the xy plane. We will deal only with one transverse coordinate, and assume the aperture to have size D in the y direction. By passing the photon through the aperture, we are constraining its y position, and from the Heisenberg uncertainty principle, we know that this implies an uncertainty in the y–component of its momentum $\delta p_y \simeq h/D$. If we assume this to be small compared to the original z–component of the photon's momentum, we have an uncertainty in the direction of the photon given by $\Delta\theta = \delta p_y/p_z$. Since the z–component of the momentum is equal to h/λ, we obtain

$$\Delta\theta = (h/D)/(h/\lambda) = \lambda/D \,, \tag{59}$$

a result which does not involve Planck's constant, and which thus can also be derived from classical arguments.

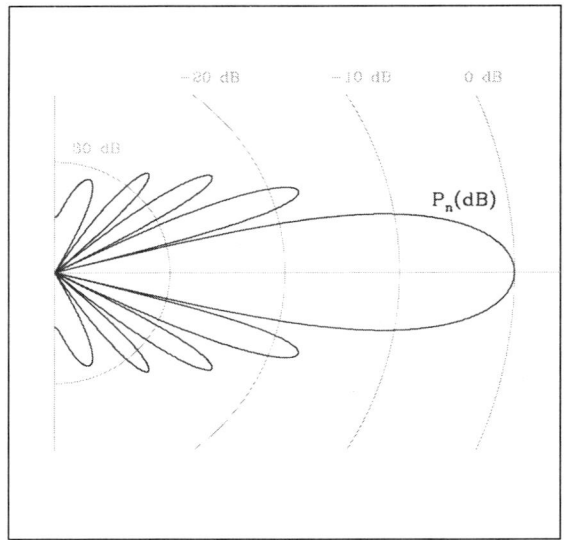

Figure 5. Polar diagram of normalized power pattern for a uniformly illuminated aperture having diameter $D = 5\lambda$.

the illuminated area in the aperture plane is elliptical and in consequence so is the far–field radiation pattern. The normalized field pattern for most aperture plane distributions has a sequence of zeros, and oscillates with diminishing amplitude as a function of off–axis angle. Results for Gaussian-illuminated circular apertures are discussed in Goldsmith (1997), and a sample are shown in Figure 6.

5.2. Uniformly illuminated aperture

Let us consider in more detail the radiation pattern from a circular aperture illuminated by a feed providing uniform amplitude and constant phase (while recognizing that this is somewhat unrealistic, it gives results that are not atypical of what can be achieved by a high–quality system). In this case the integrals in Equation 60 can be evaluated analytically, and we find

$$E_n(U) = \frac{2J_1(U)}{U}, \tag{62}$$

and

$$P_n(U) = \frac{4J_1^2(U)}{U^2}. \tag{63}$$

This is familiar as the radiation pattern from an optical telescope or from a simple circular aperture illuminated by a plane wave[12]. Each zero–crossing

---

[12]It has the same form as the radiation distribution in the focal plane of a lens illuminated by a plane wave, as the lens has the same Fourier–transformation characteristics as the propagation

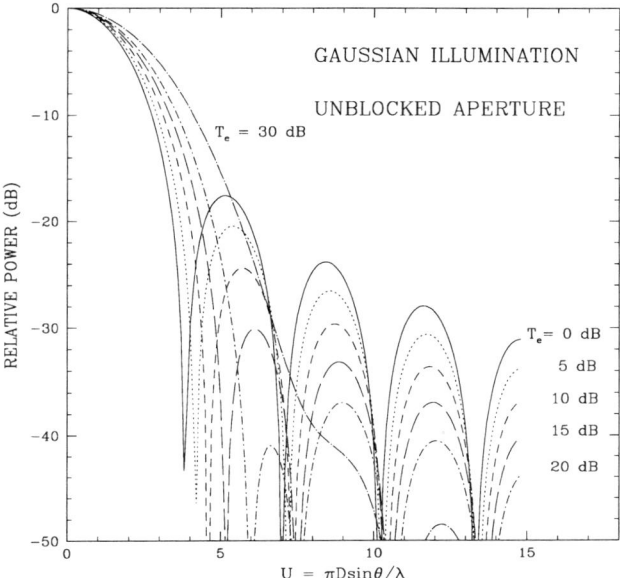

Figure 6. Normalized radiation patterns of unblocked Gaussian-illuminated circular apertures as a function of edge taper.

defines a *null* in the telescope's radiation pattern, and each successive maximum defines a *sidelobe* or *secondary maximum*. The fact that for the electric field pattern, successive regions between zero–crossings have different signs reflects the fact that there is a phase change of π radians between successive sidelobes. This is understandable intuitively from the fact that the diminution of the magnitude of the normalized field pattern as you go off axis is due to the change in the relative phase of different parts of the telescope aperture: while on axis, the entire aperture radiates in phase (assuming E_a has uniform phase), but at a null, the contribution of some parts of the aperture is out of phase, and the sum (or integral) is zero. Beyond the null, the contributions add up to a non–zero value, but the sum is out of phase with that in the main lobe.

Having the radiation pattern of the uniformly illuminated circular aperture let us demonstrate an example of the antenna theorem introduced in Section 2.3. The antenna solid angle is the integral of the normalized power pattern over all solid angle, which here is

$$\Omega_a = \int \int \frac{4J_1^2(\pi D \sin\theta/\lambda)}{(\pi D \sin\theta/\lambda)^2} \sin\theta d\theta d\phi \ , \tag{64}$$

which if we restrict ourselves to small angles (appropriate for $D \gg \lambda$) yields

---

from aperture plane to infinity. Thus, if you start with a plane wave incident on the lens, the transverse distribution in the focal plane is the transform of the truncated uniform illumination.

$$\Omega_a = \frac{4}{\pi}(\frac{\lambda}{D})^2 . \tag{65}$$

The uniformly–iluminated antenna (assuming no spillover loss, nor any blockage or phase errors) has aperture efficiency equal to unity, and thus effective area

$$A_e = \frac{\pi D^2}{4} . \tag{66}$$

Combining the two preceding equations indeed verifies that the antenna theorem holds for this case, *i.e.* that $A_e \Omega_a = \lambda^2$.

5.3. Choice of aperture illumination pattern

Two important aspects involved in choosing the illumination pattern are the sidelobe level and beam width that result. While the details depend on the exact form of the aperture field, for Gaussian illumination the first sidelobe level is –17.6 dB for $T_e = 0$ dB (uniform illumination), –20.5 dB for $T_e = 5$ dB, –24.5 dB for $T_e = 10$ dB, and –30 dB for $T_e = 15$ dB. In terms of the beam width, a convenient formula is (Goldsmith 1987)

$$\Delta\theta_{\text{fwhm}} = [1.02 + 0.0135 T_e(\text{dB})]\frac{\lambda}{D} . \tag{67}$$

Both of these characteristics – decreasing sidelobe level and increasing beam width as a function of increasing edge taper – are expressions of the Fourier relationship between the aperture plane and far–field distributions. Increasing the edge taper lowers the edge illumination and thus reduces the effect of truncating the illumination at the edge of the antenna. In a Fourier transformation, a sharp change produces "ringing" which here means sidelobes. Increasing edge taper also means that the illumination pattern is more centrally concentrated. As can be seen in all of the expressions for the far–field patterns, the far–field characteristic angle is proportional to λ/D, so that compressing the illumination pattern is equivalent to reducing the telescope diameter, and consequently increasing the beam width.

5.4. Defocus and spillover effects

A focus error, as discussed before, produces a quadratic phase error in the aperture plane field distribution. When put into the integrals for the far–field pattern, this results in; (1) an increase in the width of the main lobe of the radiation pattern, accompanied by (2) a blending together of the sidelobes (loss of sharp nulls in radiation pattern), with a general increase in their level. These effects, accompanying the reduction in the aperture efficiency given by Equation 48 for Gaussian illumination, indicate that maintaining correct focus is important for observations of extended as well as point–like sources. The changes in the beam shape produced by focus variations also complicate use of any reconstruction algorithms, and so are particularly harmful if details of the source structure are being sought.

When considering the radiation pattern and coupling to extended sources in detail, it is important to be aware of the **total** aperture plane field distribution.

Thus, in the preceding examples, we have considered only the illumination of the feed system as modified in phase by the antenna. But there is also a contribution in dual–reflector systems from the spillover, for this radiation reaches the aperture plane, but is unmodified by the antenna. The aperture plane illumination that results from the spillover looks like the feed pattern itself, but with the central part cut out. Its phase distribution is still that of the spherical wave produced by the feed, and the result is that it diverges into a hollow cone subtending angles between the edge angle of the subreflector as seen from the feed, and several times this value. This energy, which is typically a few percent of that intercepted by the antenna, thus diverges into a solid angle which can be 10^6 times larger than Ω_a, and in consequence is very difficult to measure directly. Thus, if you add the *spillover radiation pattern* to that predicted from the antenna illumination, you change Ω_a by only a few percent, but the value of P_n at large angles from boresight can be dominated by the spillover radiation pattern. This can make a real difference when you consider coupling to extremely extended sources.

Another consideration for the aperture plane field is that of blockage produced by secondary reflector and feed support legs. As for the aperture efficiency, we can first treat the blockage as simply producing regions in which the aperture plane field is zero. This can be handled relatively simply for axially symmetric systems (*e.g.* secondary reflector only) simply by changing the lower limit of the integrals in Equations 60 and 61. A very useful approach to this type of problem is to realize that the aperture plane illumination with blockage is equal to the unblocked illumination **minus** the illumination of the blocked region. Since the Fourier relationship between field distributions in aperture plane and far–field is linear, the far–field pattern of the blocked aperture is equal to that of the unblocked aperture **minus** that of the blocked region, as shown in Figure 7. This indicates that since the blockage represents a small fraction of the aperture diameter, its radiation pattern will be broader than that of the original aperture by the reciprocal of this fraction, and that its contribution will be out of phase. When the two are combined, the on–axis response is reduced, and successive sidelobes close to the main lobe are alternately increased and reduced, due to the relative phasing of the unblocked and blockage electric field patterns (note that it is the electric field patterns which are linear, so they are subtracted and the result squared to obtain the power pattern of the blocked aperture).

The case of the uniformly illuminated circular aperture is again useful due to its simplicity. As illustrated in Figure 8, the blockage pattern has exactly the same form as that of the unblocked aperture, but on an angular scale expanded by a factor f_b^{-1} and has relative normalization factor equal to f_b^2, giving for the normalized power pattern

$$P_n(U) = \frac{1}{(1-f_b^2)^2} \left[\frac{2J_1(U)}{U} - f_b^2 \frac{2J_1(f_b U)}{f_b U} \right]^2 . \tag{68}$$

While by definition, P_n is always normalized to have maximum value equal to unity, it is accompanied by a multiplication of the aperture efficiency by a factor $(1 - f_b^2)^2$ as given in Equation 43.

Thus, the energy removed from the main lobe as a result of the blockage ends up in the increased sidelobes. For uniform illumination, the contribution to

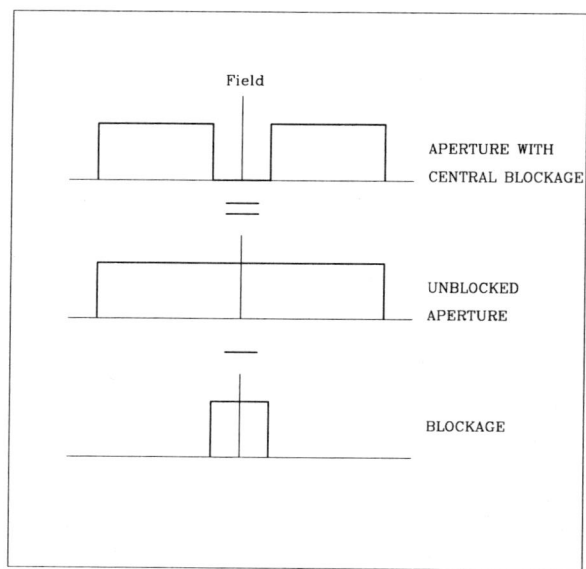

Figure 7. Schematic of cut through aperture plane, showing how the blocked, uniformly–illuminated distribution can be represented as that of the unblocked aperture minus the illumination of the blocked region.

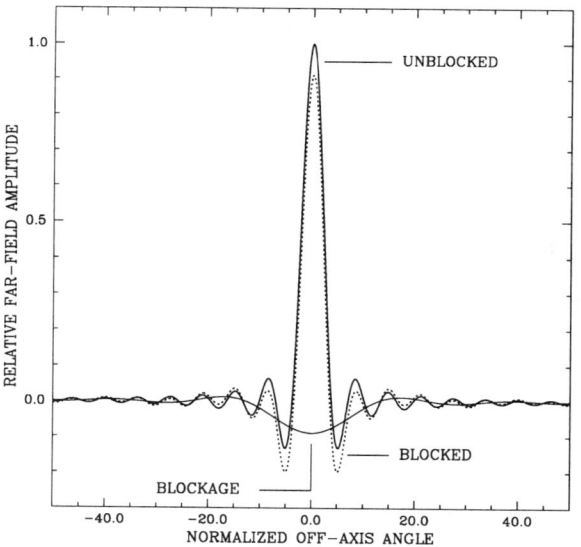

Figure 8. Relative amplitude (field) pattern for uniformly illuminated circular aperture with radial blockage fraction $f_b = 0.3$. No normalization factor is included.

a sidelobe at a specific off–axis angle from the central blockage remains comparable to that of the sidelobe produced by the unblocked aperture, for all angles. However, for a telescope with strongly tapered illumination, the effect of the blockage can be more dramatic, since it will result in a sharp discontinuity in the aperture plane illumination, with consequent increase in the sidelobe level at large angles. Including the blockage of feed support legs makes the problem significantly more complicated and difficult to treat analytically. Since the blocked area can be quite substantial (often exceeding that of the secondary reflector by a factor of a few), and the radiation is relatively concentrated, the feed leg blockage can have a relatively dramatic impact on the radiation pattern of the radio telescope.

Finally, we should mention that we are not including at all the effect of scattering from feed legs, nor from the edge of the secondary reflector. These both require more advanced electromagnetic theories for analysis, but they can both result in particular directions having noticeably enhanced radiation compared to that from the unblocked aperture. These effects (if present), combined with the more easily analyzed ones mentioned above, result in making detailed calculation of the radiation pattern to high accuracy a challenge. Their combination to produce the *stray radiation pattern* makes proper interpretation of observations of very extended emission noticeably difficult.

5.5. Coupling efficiency to extended source

Calculation of the coupling efficiency is in principle straightforward if we know the power pattern of the radio telescope and the distribution of specific intensity or temperature of the source, and can evaluate *e.g.* Equation 21 or 22. Recalling that we need to include all of the components of the electric field in the aperture plane to get the correct form for P_n and the correct value for ϵ_a, we see that the situation is, in fact, quite complex. Some progress can be made if we restrict ourselves to axially symmetric situations. We can also consider a uniform source which in fact, to a reasonable degree, characterizes the planets and represents conceptually, at least, some extended radio sources. In this case, the antenna temperature we measure is given by Equation 22 with a uniform source of temperature T_s and angular radius θ_s:

$$T_a(\theta_s) = \frac{2\pi T_s}{\Omega_a} \int_0^{\theta_s} P_n(\theta) \sin\theta d\theta , \qquad (69)$$

since we are assuming that P_n is independent of ϕ. The *coupling efficiency* is defined as the antenna temperature actually measured relative to what would be measured if the source filled the entire antenna solid angle. Thus, the coupling efficiency for a source of angular radius θ_s is given by

$$\epsilon_c(\theta_s) = \frac{\int_0^{\theta_s} P_n(\theta) \sin\theta d\theta}{\int_0^{\pi} P_n(\theta) \sin\theta d\theta} . \qquad (70)$$

With knowledge of P_n, the integrals can be evaluated. Some examples are presented in Figure 9. The coupling efficiency is a monotonically increasing function of the source size, increasing as each successive sidelobe is included within θ_s, while leveling out as each null of the power pattern is crossed. A

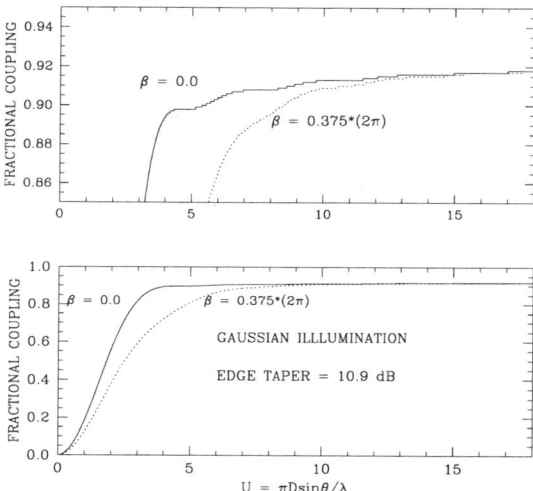

Figure 9. Coupling efficiency for circular antenna with Gaussian illumination and 10.9 dB edge taper. Each panel has a solid curve for a perfectly–focused antenna (edge phase error $\beta = 0.0$) and a dashed curve for moderately defocused antenna ($\beta = 0.375$). The upper panel is an expanded view of the lower panel.

particularly important value of the coupling efficiency is that for a source which exactly fills the main lobe of the radiation pattern of the radio telescope; this is called the *main beam efficiency*. Note that the source here does **not** fill the main lobe to its half power points (which would mean $\theta_s = \Delta\theta_{\rm fwhm}/2$). Rather, assuming the antenna pattern has a well–defined null or at least minimum, defining θ_m as the angle to the first minimum of the pattern:

$$\epsilon_{mb} = \epsilon_c(\theta_m) = \frac{\int_0^{\theta_m} P_n(\theta) \sin\theta d\theta}{\int_0^{\pi} P_n(\theta) \sin\theta d\theta} \ . \tag{71}$$

We can also define the solid angle associated with the main beam, denoted $\Omega_{\rm mb}$, as

$$\Omega_{\rm mb} = \int\!\!\!\int_{\rm main\ beam} P_n(\theta,\phi) d\Omega \ , \tag{72}$$

and we see that a general definition is

$$\epsilon_{\rm mb} = \frac{\Omega_{\rm mb}}{\Omega_a} \ . \tag{73}$$

The main beam solid angle gives an idea of the region over which a radio telescope has significant sensitivity.

From a practical viewpoint, we have to consider the fact that the contribution to P_n from the antenna's diffraction pattern stops contributing significantly to ϵ_c beyond a certain angle, but that the value of ϵ will be less than 1 there. At this point, we have effectively included the contributions from the aperture plane illumination by the main reflector and the effects of subreflector blockage. The error pattern is often distributed over a quite wide angular range, depending on the correlation length of the errors. If the error is due to panel misalignments, and the panel size is 1/100 of the primary reflector diameter, the error pattern will extend to $\simeq$ 100 times the angular diameter of the diffraction–limited main lobe of the antenna pattern. The spillover pattern may start to make a contribution to ϵ_c at angles of several degrees off boresight in dual–reflector systems, while the effects of blockage and scattering from subreflector support legs can extend over a significant portion of the forward hemisphere. Thus, the fractional coupling (ϵ_c) shown in Figure 9 rises only to $\simeq 0.92$, since the spillover radiation is not being coupled to the source even for the largest angles included.

6. Determination of Important Antenna–Related Parameters

The type of astronomical observation for which the radio telescope is being used determines, to a significant extent, which antenna parameters are important. The simplest categorization divides observations into (1) those of point–like sources and (2) those of extended sources.

6.1. Observations of point–like sources

In this situation, the quantity you are likely to want is the flux density of the source. Different telescopes perform their calibrations differently. At millimeter wavelengths, the receiver and feed are generally directly calibrated using thermal loads. This means that a measurement is fairly directly obtained in terms of the antenna temperature. To obtain the flux density, assuming you have verified that you are properly pointed at the source (it is on boresight), the quantity needed is the effective area. This may be a "tabulated" quantity, or else can be determined by measuring a source of known flux density, and simply inverting the process. For telescopes with obvious geometry and physical area, the aperture efficiency is relatively easily determined. For the Arecibo Gregorian system, the illuminated area is by design much less than the physical area, but it is always meaningful to think in terms of the effective area of the radio telescope.

Using the relationship between antenna temperature and flux density (Equation 20), you can define an "effective area", A_e^*, in terms of the antenna temperature per unit flux density. The actual conversion factor is just equal to $A_e/2k$, but you may hear a telescope referred to as having A_e^* equal to some number of *Kelvins per Jansky*. This is equivalent to saying that A_e (m^2) $= 2.76 \times 10^3 A_e^*$ (K/Jy). For Arecibo, we see values of A_e^* as high as 10 K/Jy with the Gregorian, indicating an effective area of 2.8×10^4 m^2. The illuminated area, which is equivalent of A_p, is close to 4×10^4 m^2, indicating an aperture efficiency of 0.69. This includes all effects, such as blockage (ϵ_b) and phase errors (ϵ_p). It is likely that this can be improved by further adjustment of the surface panels.

It is often the case, particularly at longer wavelengths, that the calibration signals are "defined" in units of flux density. This is really a consequence of how the calibration procedures are carried out, but this gives you the ability to convert an observed signal to a flux density.

For observations of point–like sources, the beam size and beam solid angle may still be relevant, particularly because they give you a feeling of the chance of there being another source in your beam, in addition to the one you are observing. This is especially an issue at higher sensitivity levels, and is part of the broader *confusion* problem. Having a reasonable idea of $\Delta\theta_{\rm fwhm}$ and $\Omega_{\rm mb}$ is the starting point for assessing the severity of this problem. These quantities are often determined as part of the observing procedure employed to verify the pointing of the telescope. By doing a "cross scan" or at least a five–point map, the pointing offsets are determined, but along with them generally come the beam size in two orthogonal directions.

6.2. Observations of extended sources

Obtaining from a single observation of an extended source accurate information about the source properties is impossible unless you have *a priori* knowledge, such as that it is symmetric, uniform, for example. If you make a *map* of the source, you can *reconstruct* more information about the source, but this is the topic of another lecture. What we can indicate here is how, in a very simple way, you can attack the problem.

In many cases, you have some idea about the size of the source. If you are willing to assume the source is uniform and symmetric, and you have an idea of θ_s, you can at least use the appropriate value of $\epsilon_c(\theta_s)$ (Equation 70) to convert the measured antenna temperature to a source temperature. Of course, if you have an idea that the source has a Gaussian (or even some other) form, you can work out the coupling for that source to your antenna beam, and use the value of the coupling efficiency so derived.

It is certainly a good idea to verify the size of the beam for observations of extended sources, whether because you want to understand how well the telescope you are using is resolving the source, or because you need to chose the spacing of points in a map. The beam size is useful in a different way, which is related to the telescope efficiency. Although the various issues, such as antenna illumination, blockage, and defocus, all affect the coupling and radiation pattern in various ways, it is a fair statement that they tend to leave the *shape* of the main lobe of the beam pattern unchanged, while affecting its amplitude (through the aperture efficiency) as well as its amplitude relative to sidelobes, and blockage, scatter, and error patterns. It also is a fairly empirical rule, that the main lobe generally is quite Gaussian in form; this is obviously not exactly what is predicted by diffraction integrals, but it is in many cases a good representation of the normalized power pattern, down to the -10 dB and sometimes even -15 dB level. If we are willing to make the assumption that the antenna pattern can be divided into an axially symmetric Gaussian main lobe and a much larger sidelobe/error pattern, we can simplify determination of the main beam efficiency. This is not of negligible value, since obtaining this quantity from first principles requires measuring the normalized power pattern over the whole sky, or at least over a large solid angle, which can take a great deal of time.

With the assumption that the main lobe has the Gaussian form $P_{\rm mb} = \exp[-(\theta/\theta_a)^2]$, then if θ_a is small (as will be the case for a large antenna),

$$\Omega_{\rm mb} = 2\pi \int_0^\infty e^{-(\theta/\theta_a)^2} \theta d\theta = \pi \theta_a^2 \ . \tag{74}$$

The angle θ_a is related to the fwhm beam width as $\theta_a = \frac{\theta_{\rm fwhm}}{2\sqrt{ln2}}$. This yields

$$\Omega_{\rm mb} = \frac{\pi}{4\ln 2}\theta_{\rm fwhm}^2 = 1.13\theta_{\rm fwhm}^2 \ . \tag{75}$$

Observing **any** point–like source, by accurately measuring the fwhm beam width, we can obtain the main beam solid angle[13]. Using a point source of known flux density, we can obtain the effective area, and using the antenna theorem, we get the antenna solid angle, Ω_a. Finally, using Equation 73 we obtain the main beam efficiency. Thus, by observations exclusively of point sources, we can get a good idea of the coupling to a moderately extended source.

If we have a source with considerable structure, or one that is extended in such a manner that it couples significantly to specific sidelobes or other components of the beam pattern, the situation is obviously more complex. However, determining the main beam efficiency, and the coupling efficiency to a relatively large source such as the moon, help bracket the coupling efficiency to sources of intermediate size, and are at least a start at determining the coupling efficiency for an arbitrary source.

Appendix: The Antenna Theorem

This derivation of the antenna theorem is divided into two parts. The first shows that any antenna has the same value of $A_e\Omega_a$, and the second utilizes a uniformly illuminated aperture to determine this value.

Consider an antenna with a transmitter attached to its feed system, with power radiated per unit solid angle given by $U(\theta, \phi)$. The integral of this quantity over all angles is W, the total power radiated. We thus can write

$$W = U(0,0)\int\int \frac{U(\theta,\phi)}{U(0,0)} d\Omega \ , \tag{76}$$

and defining the integrand as the normalized power radiated per unit solid angle, $U_n(\theta, \phi)$, we see that

$$W = U(0,0)\int\int U_n(\theta,\phi) d\Omega \ . \tag{77}$$

From the reciprocity theorem, we may infer that the normalized pattern of an antenna acting as a transmitter is the same as that when it is acting as a receiver. This means that

$$U_n(\theta,\phi) = P_n(\theta,\phi) \ , \tag{78}$$

---

[13] If we have an elliptical Gaussian beam, Equations 74 and 75 can be used if we replace θ_a^2 by $\theta_{a_x}\theta_{a_y}$ and $\theta_{\rm fwhm}^2$ by $\theta_{{\rm fwhm}_x}\theta_{{\rm fwhm}_y}$, where the subscripts x and y denote beam properties in a Cartesian angular system as used in Equation 54.

where the second quantity is the usual ("receiving") normalized antenna pattern. The integral in Equation 77 is thus the integral of the normalized power pattern, which is just the antenna solid angle, Ω_a. We can then write the on–boresight power transmitted as

$$U(0,0) = \frac{W}{\Omega_a} . \tag{79}$$

We now consider the situation in which antenna 1 has transmitter of power W at wavelength λ attached to its feed system, and antenna 2 is receiving. We assume that the two antennas are separated by distance R placing them in each others' far field, and that their boresight directions are aligned. The power received by antenna 2 from antenna 1 is

$$P_{12} = U_1(0,0)\frac{A_{e2}}{R^2} = \frac{W A_{e2}}{\Omega_{a1} R^2} . \tag{80}$$

If we now consider the same transmitter to be attached to antenna 2, and that antenna 1 is receiving, the power received by antenna 1 from antenna 2 is

$$P_{21} = U_2(0,0)\frac{A_{e1}}{R^2} = \frac{W A_{e1}}{\Omega_{a2} R^2} . \tag{81}$$

The reciprocity theorem guarantees, as long as the intervening medium is reciprocal, that $P_{12} = P_{21}$. This implies that $A_{e2}/\Omega_{a1} = A_{e1}/\Omega_{a2}$, and hence that

$$A_e \Omega_a = constant, \tag{82}$$

for any antenna operating at wavelength λ. The question that remains is the value of this constant.

To complete the analysis we choose an extremely simple antenna, namely a uniformly illuminated square aperture. We assume that it is sufficiently large that we can make appropriate small angle approximations for the radiation pattern. Using Equation 54, we find for an square aperture of size a with unit field amplitude that

$$E_{ff}(\theta_x, \theta_y) = \frac{ia^2}{\lambda} \frac{e^{-ikR}}{R} \frac{\sin(\pi a \theta_x/\lambda)}{(\pi a \theta_x/\lambda)} \frac{\sin(\pi a \theta_y/\lambda)}{(\pi a \theta_y/\lambda)} . \tag{83}$$

This yields the normalized power pattern

$$P_n(\theta_x, \theta_y) = \frac{\sin^2(\pi a \theta_x/\lambda)}{(\pi a \theta_x/\lambda)^2} \frac{\sin^2(\pi a \theta_y/\lambda)}{(\pi a \theta_y/\lambda)^2} . \tag{84}$$

Integration over solid angle is facilitated by the fact that the power pattern is separable in the two coordinates, and we obtain the result that

$$\Omega_a = \int\int P_n d\Omega = (\lambda/a)^2 . \tag{85}$$

For a uniformly illuminated lossless antenna, the aperture efficiency is unity, and the effective area here is $A_e = a^2$. Thus, the product of the antenna solid angle and the effective area for any lossless antenna is

$$A_e \Omega_a = \lambda^2 , \qquad (86)$$

and combined with the preceding discussion, we see that this result applies to any lossless antenna operating at this wavelength.

The derivation leading to Equation 86 can be generalized to include ohmic loss. As long as the antenna system loss (absorption) does not alter the *form* of the aperture plane illumination, it is evident that the product $A_e \Omega_a$ is multiplied by a factor ϵ_r, where $1 - \epsilon_r$ is the fraction of the incident power absorbed. This is certainly the case for ohmic loss in the transmission line, or loss uniformly distributed over the surface of the antenna.

Turning back to our example, the effective area of a uniformly illuminated aperture antenna is equal to its projected physical area only if the antenna is lossless, and it will be less for finite loss. The antenna solid angle Ω_a is unchanged, as long as we avoid a pathological distribution of loss across the aperture which significantly changes the form of the aperture plane field distribution. Thus, for reasonable cases, the antenna theorem including antenna loss can be written

$$A_e \Omega_a = \epsilon_r \lambda^2 . \qquad (87)$$

Acknowledgments. I would like to thank Dr. Edith Falgarone and her colleagues at the Ecole Normale Supérieure, Paris for their warm hospitality during a stay there during which much of this lecture was written. I also gratefully acknowledge numerous discussions with Drs. Germán Cortés Medellín and L. Baker about various aspects of antenna performance. I am indebted to Dr. J. Hagen for sharing valuable information about the reciprocity theorem and outlining this derivation of the antenna theorem, and to Dr. C. Salter for a careful reading of the manuscript which resulted in many improvements.

References

Aubrey, C. & Bitter, D. 1975, Electron. Lett., 11, 154

Christiansen, W. N. & Högbom, J. A. 1985, Radio Telescopes (Cambridge: Cambridge University Press)

Goldsmith, P. F. 1987, Int. J. Infrared and Millimeter Waves, 8, 771

Goldsmith, P. F. 1997, Quasioptical Systems (New York: IEEE), Ch. 6

Hannan, P. W. 1961, IEEE Trans. Ant. Propag., AP-9, 140

Kraus, J. D. 1966, Radio Astronomy (New York: McGraw–Hill), 212

Kraus, J. D. 1988, Antennas, (2nd edition; New York: McGraw–Hill), 410

Lamb, J. W. 1986, Int. J. Infrared and Millimeter Waves, 7, 1511

Ruze, J. 1966, Proc. IEEE, 54, 633

Silver, S. 1949, Microwave Antenna Theory and Design (New York: McGraw–Hill), Ch. 6

Terman, F. E. 1943, Radio Engineers' Handbook (New York: McGraw–Hill), 787

Ulaby, F. T., Moore, R. K., & Fung, A. K. 1981, Microwave Remote Sensing I, (Norwood: Artech House), 186

Working hard deep into the night. Don Campbell et al.

Measurement in Radio Astronomy

Donald B. Campbell

National Astronomy and Ionosphere Center, Department of Astronomy, Cornell University, Ithaca, New York 14853, USA

Abstract. This lecture covers the fundamentals of observations with a filled aperture antenna, basic receiver systems and sources of noise. A simple derivation is provided for the radiometer equation, which determines a telescope's ability to detect radio sources. The lecture ends with a discussion of antennas as filters on the spatial Fourier components of the sky brightness distribution.

1. Antenna - Sky Coupling

In the lecture by Goldsmith, it was shown that for a radio antenna with a normalized power response pattern, $P_n(\theta, \varphi)$ observing a sky brightness distribution $I_\nu(\theta, \varphi)$, the received power is given by:

$$P_{rec}(\nu) = 1/2 \; A_e \int_\Omega I_\nu(\theta, \varphi) \; P_n(\theta, \varphi) \; d\Omega \quad \text{watts Hz}^{-1}, \tag{1}$$

where ν is the frequency, A_e is the antenna's effective area and the factor of $1/2$ is to account for polarization. The incoming radiation is assumed to be unpolarized, but each receiving element on an antenna is sensitive to only one linear or one circular polarization. The total received power in a bandwidth $\Delta \nu$ is

$$P_{rec} = \int_{-\Delta\nu/2}^{\Delta\nu/2} P_{rec}(\nu) \; d\nu \quad \text{watts}, \tag{2}$$

where $\Delta\nu$ is assumed to be $\ll \nu_0$; ν_0 is the central frequency in the band.

From Planck's law, the brightness of a black body at a temperature T is given by

$$I_\nu(\theta, \varphi) = \frac{2h\nu^3}{c^2} \frac{1}{e^{h\nu/kT(\theta,\varphi)} - 1} \quad \text{watts m}^{-2} \text{ ster}^{-1} \text{ Hz}^{-1}, \tag{3}$$

where h is Planck's constant,
 k is Boltzmann's constant,
 c is the velocity of light.

Invoking the Rayleigh-Jeans approximation when $h\nu \ll kT$

$$I_\nu(\theta, \varphi) = 2kT(\theta, \varphi)/\lambda^2, \tag{4}$$

where λ is the wavelength. Note that; 1) $T(\theta,\varphi)$ is not necessarily the physical temperature – it is the equivalent black body temperature and depends on parameters such as the emissivity and optical depth, which may depend on ν; 2) for mm wavelengths and low T, $kT \approx h\nu$ making the Rayleigh-Jeans approximation inapplicable. Substituting, we have

$$P_\nu = \frac{kA_e}{\lambda^2} \int_\Omega T(\theta,\varphi)\, P_n(\theta,\varphi)\, d\Omega \quad \text{watts Hz}^{-1}. \tag{5}$$

The power per unit frequency available at the terminals of a resistor is related to the resistor's temperature T by (Nyquist, 1928)

$$W = kT \quad \text{watts Hz}^{-1}. \tag{6}$$

If we replace the antenna with a resistor at a temperature such that the power available at the resistor's terminals is P_ν, then we can define an equivalent antenna temperature T_A such that

$$\begin{aligned} P_\nu &= kT_A \quad \text{watts Hz}^{-1} \\ \text{and} \quad T_A &= \frac{A_e}{\lambda^2} \int_\Omega T(\theta,\varphi)\, P_n(\theta,\varphi)\, d\Omega \\ &= \frac{1}{\Omega_A} \int_\Omega T(\theta,\varphi)\, P_n(\theta,\varphi)\, d\Omega, \end{aligned} \tag{7}$$

since for a lossless antenna the beam solid angle $\Omega_A = \lambda^2/A_e$ (see lecture by Goldsmith).

If $T(\theta,\varphi)$ is constant over the solid angle subtended by $P_n(\theta,\varphi)$, then

$$T_A = T \quad \text{since} \quad \int_\Omega P_n\, d\Omega = \Omega_A. \tag{8}$$

The implication of this result is that for an antenna looking at an extended source with a uniform brightness temperature T, the measured antenna temperature in the Rayleigh-Jeans limit is equal to T (assuming no ohmic losses in the antenna) independent of the size of the antenna. This, of course, implies that P_ν is also constant.

2. Flux Density

For a discrete source of radio emission (i.e. one that has a clear boundary), the total flux density at the antenna location is

$$S = \int_{source} I_\nu(\theta,\varphi)\, d\Omega \quad \text{watts m}^2\text{ Hz}^{-1}. \tag{9}$$

When the source is observed with an antenna having power pattern $P_n(\theta,\varphi)$, the observed flux density is

$$S_o = \int_{source} I_\nu(\theta,\varphi) \, P_n(\theta,\varphi) \, d\Omega \quad \text{watts m}^{-2} \text{ Hz}^{-1} \tag{10}$$

$\leq S$ depending on the angular extent of the source relative to $P_n(\theta,\varphi)$.

If the source is much larger than $P_n(\theta,\varphi)$ and $I_\nu(\theta,\varphi)$ does not vary significantly over $P_n(\theta,\varphi)$, then

$$S_o = I_\nu(\theta,\varphi) \int P_n(\theta,\varphi) \, d\Omega \approx I_\nu(\theta,\varphi) \, \Omega_A \quad \text{watts m}^{-2} \text{ Hz}^{-1}. \tag{11}$$

When the source is much smaller than the antenna beam, then

$$S = S_o = P_n(0,0) \int_{source} I_\nu(\theta,\varphi) \, d\Omega \tag{12}$$

$$= \int_{source} I_\nu(\theta,\varphi) \, d\Omega \quad \text{watts m}^{-2} \text{ Hz}^{-1}.$$

The standard unit of flux density used in radio astronomy is the Jansky; 1 Jy = 10^{-26} watts m^{-2} Hz^{-1}.

3. Antenna Sensitivity

From Equations 1, 7 and 10 we can write

$$P_\nu = kT_A = 1/2 A_e S_o \quad \text{watts Hz}^{-1} \tag{13}$$

and $T_A/S_{oJ} = A_e \, 10^{-26}/2k$ Kelvins/Jy,

where S_{oJ} is the flux density in Janskys.

Each telescope's "Kelvins per Jansky (K/Jy)" is a measure of its sensitivity. An A_e of 2,760 m^2 is required to give a sensitivity of 1.0 K/Jy. Note that A_e is normally a function of frequency due to frequency dependent illumination functions, surface r.m.s. errors, etc.

4. One-Dimensional Generalization

Consider a one-dimensional antenna beam, $P_n(\theta)$, scanning a one-dimensional brightness distribution, $I_\nu(\theta)$. The expressions for $P_\nu(\theta)$ and $T_A(\theta)$ can then be written as:

$$P_\nu(\theta) = 1/2 \, A_e \int I_\nu(\theta') \, P_n(\theta' - \theta) \, d\theta'. \tag{14}$$

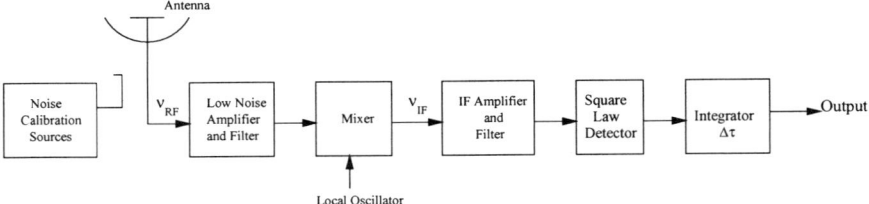

Figure 1. A schematic diagram of a super-heterodyne receiving system.

This expression is very close to that of a convolution and if we make a change of variable $\tilde{P}_n(\theta) = P_n(-\theta)$, the new expression

$$P_\nu(\theta) = 1/2 \; A_e \int I_\nu(\theta') \, \tilde{P}_n(\theta - \theta') \, d\theta' \quad \text{watts Hz}^{-1}, \qquad (15)$$

has the form of a convolution of $I_\nu(\theta)$ with $\tilde{P}_n(\theta)$. $\tilde{P}_n(\theta)$ is the mirror image of $P_n(\theta)$ and is identical to $P_n(\theta)$ for a symmetric beam pattern. The equivalent expression for the antenna temperature is

$$T_A(\theta) = \frac{1}{\Omega_A} \int T_\nu(\theta') \, \tilde{P}_n(\theta - \theta') \, d\theta'. \qquad (16)$$

There are two issues that need to be addressed:

1. Given that a radio source produces an antenna temperature of T_A K, what determines our ability to actually detect this (perhaps very small) change in the power that appears at the output terminals of the telescope?

2. How well does our measurement of the sky brightness distribution (represented by $T_A(\theta)$) represent the actual sky brightness distribution?

5. Receiving Systems

Figure 1 shows a super-heterodyne receiving system very similar to what is used in most radio telescopes today. Its main components are:

- A low noise amplifier (LNA) and filter. For frequencies greater than ~300 MHz, the LNA is normally cryogenically cooled to about 15K.

- A mixer to convert ν_{RF}, the center of the band of frequencies $\Delta\nu_{RF}$ of interest, to a more convenient intermediate frequency ν_{IF}, that may be used by all the receiving systems on the telescope. The local oscillator signal is almost always phase locked to the Observatory's master oscillator, normally a hydrogen maser frequency standard.

- An intermediate frequency (IF) amplifier and filter which limits the bandwidth to $\Delta\nu_{IF}$.

- A square law detector (i.e. its output voltage is proportional to the input power) followed by an integrator which averages the detector output for $\Delta\tau$ secs. This stage may be used for continuum observations (i.e. measuring the received power at ν_{RF} over the bandwidth ν_{IF}), but for other types of observations it may be replaced by such devices as a spectral processor, a digital sampling system for, say, observations of pulsars or a recording system for very long baseline interferometric (VLBI) observations.

For this receiving system, the answer to the first question, "what determines our ability to detect small changes in the power that appear at the output terminals of our telescope due to a radio source which produces an antenna temperature of T_A K", is dependent on the level of additive noise power, which is usually represented by a temperature, the system noise temperature T_{sys}, the pre-detection bandwidth, $\Delta\nu_{IF}$, and the post-detection integration time, $\Delta\tau$.

The sources of noise which contribute to T_{sys} are:

1. The general background "sky" temperature, T_{sky}. For frequencies above about 1 GHz, this is dominated by the 2.73 K cosmic microwave background temperature. At frequencies below 1 GHz non-thermal emission from electrons spiraling in the magnetic field of our Galaxy and its halo is increasingly important, the lower the frequency. It has a frequency dependence of approximately $\nu^{-0.7}$.

2. Emission from the Earth's atmosphere, T_{atm}, which typically has a value of a few K. Above 10 GHz, emission due to water vapor is increasingly important and is the dominant contribution to T_{sys} at mm wavelengths.

3. Radiation from the surrounding 300 K terrain can contribute an amount T_{scat} to T_{sys} via spillover of the horn or secondary reflector radiation pattern over the edge of the primary reflector or via scattering from parts of the antenna such as tripod legs that are in the horn or secondary reflector radiation pattern. The amplitude of T_{scat} is very dependent on the design of the telescope. For the Arecibo telescope's Gregorian system it is estimated to be about 4 K.

4. Additive noise from the low noise amplifier, T_{rec}. For LNA's that have been cooled to about 15 K, T_{rec} can range from 2 to 20 K at cm-wavelengths depending on the quality of the amplifier and losses in the other associated components such as the horn, cabling, etc. It is considerably higher at mm wavelengths.

The combined noise power can be represented by the temperature T_{sys} of a resistor at the output terminals of the antenna.

$$T_{sys} = T_{sky} + T_{atm} + T_{scat} + T_{rec} . \qquad (17)$$

6. The Radiometer Equation

How do we detect the presence of the power due to a radio source as represented by T_A, in the presence of the additive noise power represented by T_{sys}?

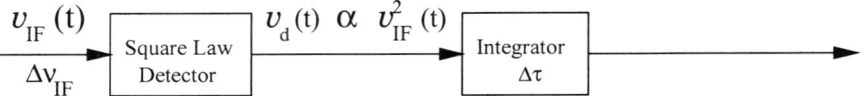

Figure 2. A schematic diagram showing the last two stages of the receiving system.

At the output of the integrator (see Figure 2), the voltage is proportional to the sum of the input powers due to the source and the additive noise, which are proportional to T_A and T_{sys}, respectively. This requires that the input voltages due to T_A and T_{sys} are uncorrelated, which is always the case. Hence, the expected value of the integrator's output voltage due to the source is

$$\langle v_A(t) \rangle = G T_A \text{ where } G \text{ is the total system gain between the} \quad (18)$$
input of the LNA and the output of the integrator;
$\langle \, \rangle$ denotes the time averaged value of the output;

and the expected value due to the mean noise power is

$$\langle v_n(t) \rangle = G T_{sys}. \quad (19)$$

The signal-to-noise ratio at the output of the integrator is given by

$$S/N = \langle v_A(t) \rangle / \langle v_n(t) \rangle = T_A/T_{sys}. \quad (20)$$

However, what dictates our ability to detect the mean source power, $\langle v_A(t) \rangle$, in the presence of $v_n(t)$ is not the mean value of $v_n(t)$ but rather its r.m.s. fluctuation level or standard derivation. The desired quantity is

$$\sigma_n = [\langle v_n^2(t) \rangle - \langle v_n(t) \rangle^2]^{1/2}. \quad (21)$$

We will use an intuitive method to derive σ_n. A rigorous derivation can be found in Rohlfs & Wilson (1999). The integrator is an averaging device that can be thought of as averaging n independent estimates of the output voltage of the square law detector and forming a mean. The standard deviation of this mean is $\sigma_n = \sigma/\sqrt{n}$ (see e.g. Papoulis 1965), where σ is the standard deviation of the output voltage of the square law detector. To find σ_n, we need to know n and σ.

The input to the square law detector is a noise-like voltage obeying Gaussian statistics. For this case, the standard deviation of the output voltage of the square law detector, $v_d(t)$, is equal to its mean value (see e.g. Ulaby et al. 1981).

$$\begin{aligned} i.e. \quad \sigma &= [\langle v_d^2(t) \rangle - \langle v_d(t) \rangle^2]^{1/2} \\ &= \langle v_d(t) \rangle \\ &= G T_{sys} \quad \text{"off" source} \\ &= G(T_{sys} + T_A) \quad \text{"on" source}, \end{aligned} \quad (22)$$

since there is a contribution to σ from the (noise-like) source. What is n? A signal with a bandwidth $\Delta\nu$ can fluctuate on a time scale, Δt, given by

$$\Delta\nu\,\Delta t = 1. \tag{23}$$

Voltages samples separated by time Δt can be regarded as independent. Hence, the number of independent samples, n, averaged by the integrator is $\Delta\tau/\Delta t = \Delta\tau\,\Delta\nu_{IF}$. The signal-to-r.m.s. noise ratio (the Radiometer Equation) can now be written as:

$$S/N_{rms} = \langle v_A(t)\rangle/\sigma_n = \frac{T_A}{T_{sys}}(\Delta\tau\,\Delta\nu_{IF})^{1/2}. \tag{24}$$

For unity signal-to-r.m.s. noise ratio, the equivalent r.m.s. fluctuation in the antenna temperature is given by

$$\Delta T_A = T_{sys}(\Delta\tau\Delta\nu_{IF})^{-1/2}. \tag{25}$$

The normal requirement for detectability is

$$\frac{T_A}{T_{sys}}(\Delta\tau\,\Delta\nu_{IF})^{1/2} \geq 5. \tag{26}$$

These expressions assume that $T_A \ll T_{sys}$ since the contribution to σ_n from the (noise-like) source is being neglected. However, typical values of $\Delta\nu_{IF}$ and $\Delta\tau$ are 10 MHz and 1 sec so that $(\Delta\tau\,\Delta\nu_{IF})^{1/2} \approx 3.10^3$. For this case, the minimum detectable T_A is given by:

$$T_A \approx T_{sys}/0.6.10^3 \ll T_{sys}. \tag{27}$$

7. Performance Measures

From Equation 13, the power at the output terminals of an antenna due to a point radio source with flux density S (watts m^{-2} Hz^{-1}) is given by

$$P_\nu = k\,T_A = 1/2\,A e\,S \quad \text{watts Hz}^{-1} \tag{28}$$

$$\text{or} \quad T_A = A e\,S/2k.$$

We have just seen that the r.m.s. noise level is

$$T_{sys}/(\Delta\nu_{IF}\,\Delta\tau)^{1/2}, \tag{29}$$

so that the signal-to-noise ratio (S/N) when observing the point source is

$$S/N_{rms} = 1/2\,\frac{A e}{T_{sys}}\,\frac{S}{k}\,(\Delta\nu_{IF}\,\Delta\tau)^{1/2}. \tag{30}$$

For a telescope observing a point source at a specific frequency, Ae and T_{sys} are fixed parameters and Ae/T_{sys} is a measure of the telescope performance at that frequency.

An alternative measure of telescope performance is the System Equivalent Flux Density (SEFD). Consider a 1.0 Jy point source; the antenna temperature due to this source is just the K/Jy value defined in Section 3. The SEFD is the point source flux density (in Janskys) which produces $T_A = T_{sys}$, i.e.

$$\text{SEFD} = T_{sys}/(K/Jy). \tag{31}$$

For an antenna with a sensitivity equal to 2 K/Jy and a system noise temperature equal to 20 K, the SEFD is equal to 10 Jy.

8. Beam Patterns and Spatial Fourier Components

The second issue "how well does our measurement of the sky brightness distribution represent the actual sky brightness distribution" can be addressed by examining the relationship between the antenna temperature as a function of sky position, $T_A(\theta)$ in one dimension, and the true sky brightness temperature distribution $T_\nu(\theta)$. From Equation 16, we saw that:

$$T_A(\theta) = \frac{1}{\Omega_A} \int T_\nu(\theta') \tilde{P}_n(\theta - \theta') \, d\theta'. \tag{32}$$

Since $P_n(\theta)$ normally has a finite value for only small values of θ, we can extend the range of integration so that

$$T_A(\theta) = \int_{-\infty}^{\infty} T_\nu(\theta') P'(\theta - \theta') \, d\theta', \tag{33}$$

where $\tilde{P}_n(\theta)/\Omega_A$ has been replaced by $P'(\theta)$ and

$$\int P'(\theta) \, d\theta = 1. \tag{34}$$

Equation 33 is now a convolution integral and can be written in the form

$$T_A(\theta) = T_\nu(\theta) * P'(\theta), \tag{35}$$

where * indicates convolution.

Invoking the convolution theorem (see e.g. Bracewell 1965),

$$\bar{T}_A(u) = \bar{T}_\nu(u) \times \bar{P}'(u), \tag{36}$$

$$\text{where} \quad \bar{T}_A(u) \longleftrightarrow T_A(\theta), \text{etc.} \tag{37}$$

and $\longleftrightarrow$ indicates a Fourier transform.

$\bar{T}_\nu(u)$ for different values of u are the spatial Fourier components of the (one-dimensional) sky brightness distribution. The above expressions can easily be expanded to the two-dimensional case with $\bar{T}_\nu(u,v)$ being the two-dimensional distribution of Fourier components of the sky brightness.

Equations 35 and 36 are the standard expressions for a filter, with $P'(\theta)$ being its impulse response and $\bar{P}'(u)$ the frequency response, operating on the sky

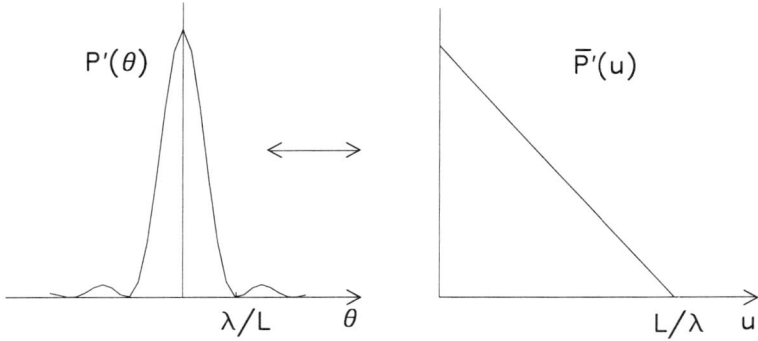

Figure 3. The beam pattern and spatial frequency response of a uniformly illuminated one-dimensional aperture of length L.

brightness distribution and its spatial Fourier components. That is, the antenna is a filter on the spatial Fourier components of the sky brightness distribution.

For an (one-dimensional) antenna with a uniform aperture distribution, we have (see Section 7 in the lecture by Goldsmith)

$$P'(\theta) = \frac{\lambda}{L}\left[\frac{sin\,\pi\theta L/\lambda}{\pi\theta}\right]^2, \qquad (38)$$

where L is the aperture size and we have made the approximation $sin\,\theta \approx \theta$ over the range of θ of interest for large L/λ. As can be seen from Figure 3, $\bar{P}'(u)$, the Fourier transform of $P'(\theta)$, is zero for values of u greater than L/λ. Hence the maximum spatial frequency that this antenna will pass is $L/\lambda\,\mathrm{rad}^{-1}$.

More generally, the far-field electric field amplitude distribution, E_{ff}, is proportional to the Fourier transform of the aperture plane field distribution (see Section 5.1 in the lecture by Goldsmith). Thus, the far-field power pattern is proportional to the square of the Fourier transform of the aperture plane field distribution. Assuming a symmetric beam shape then

$$P'(\theta) \propto E_{ff}(\theta) \times E_{ff}(\theta) \text{ and} \qquad (39)$$

$$E_{ff}(\theta) \longleftrightarrow E_a(x_a)\,, \qquad (40)$$

where x_a denotes the coordinate in the aperture plane. Taking the Fourier transform of both sides and again invoking the convolution theorem, we obtain

$$\bar{P}'(u) \propto E_a(x_a) * E_a(x_a)\,. \qquad (41)$$

That is, the spatial frequency response of the antenna's power pattern is proportional to the autocorrelation function of the aperture plane field distribution. Consequently, the maximum spatial frequency passed by a (one-dimensional) filled aperture antenna is determined by the maximum extent of the antenna in wavelengths independent of the aperture illumination. In the two-dimensional case, the spatial frequency response is determined by the size of the antenna in wavelengths in the direction of interest.

The concept of an antenna as a filter on the spatial frequencies of the sky brightness distribution applies to any antenna whether a filled aperture, an interferometer, or any of the other types of radio telescopes that have been built over the past 50 years. A simple two-element interferometer consisting of two parabolic antenna with diameters L spaced D apart has a low frequency response with a maximum frequency of L/λ rad^{-1} determined by the size of the individual elements and a band pass response centered on D/λ rad^{-1} determined by the distance between the two elements. This sensitivity of a two-element interferometer to a small band of spatial frequencies determined by the spacing between the elements underlies the use of multi-element interferometer systems, such as the 27-element Very Large Array in New Mexico, to synthesize images of the sky brightness distribution with angular resolution determined by the maximum extent of the array in wavelengths.

References

Bracewell, R. 1965, The Fourier Transform and Its Applications (New York: McGraw Hill)

Nyquist, H. 1928, Phys.Rev.A, 32, 110

Papoulis, A. 1965, Probability, Random Variables and Stochastic Processes (New York: McGraw Hill)

Rohlfs, K. & T. L. Wilson 1999, Tools of Radio Astronomy (3rd edition; New York: Springer)

Ulaby, F. J., R. K. Moore, & A. K. Fung 1981, Microwave Remote Sensing: Active and Passive, Volume I: Fundamentals and Radiometry (Norwood: Artech House)

The Receiver System – cm Regime

Roger D. Norrod

National Radio Astronomy Observatory, P. O. Box 2, Green Bank, West Virginia 24944, USA

Abstract. The receiver front-end of a radio telescope is generally considered to encompass components that amplify, filter, and frequency convert signals, provided by the antenna, to a level and frequency range appropriate for detection. This presentation will discuss critical parts of the centimeter wave radio astronomy front-end, and factors impacting the design and performance.

The feed efficiently converts propagating electromagnetic fields near a reflector antenna's focal point to a guided wave in coax or waveguide. Some types of feeds inherently detect and separate polarizations; other types require an orthomode transducer to deliver orthogonal polarizations to separate channels. Low-noise amplifiers, usually cryogenically cooled, amplify the signal and set the receiver noise level, and are followed by filters, mixers, and additional amplification. All the passive and active components add electrical noise to the signal, and models used during receiver design will be presented, explaining why loss and noise introduced in the early stages of the receiver are critical. The linear operating range of active components is limited by their power handling capacity, and how these limitations are considered will be discussed. We will also discuss stability of the receiver, and practical means of achieving the required performance.

1. Introduction

Figure 1 shows a very simplified block diagram of a typical super-heterodyne receiver front-end. ("Front-end" generally excludes the part of a receiver that involves data acquisition, the "back-end"). A super-heterodyne receiver involves at least one frequency conversion, or mix, and is by far the most common type of front-end for radio astronomy in the centimeter regime. This paper will provide an introduction to the major components in such a front-end and the system aspects typically faced by a designer.

2. Feeds

The feed is placed at or near the focal point of a reflector. The angle subtended by the reflector as seen by the feed strongly influences the types of feeds that may be used, and the details of their design. Generally, smaller angles require larger feeds in units of wavelengths.

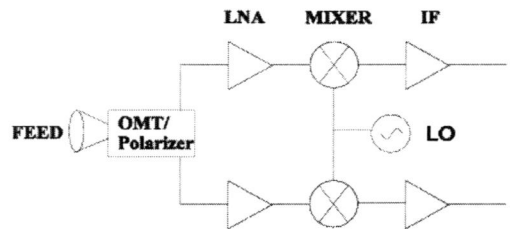

Figure 1. A typical centimeter-wave radio astronomy receiver.

The feed interfaces in an efficient manner between EM fields at the focal point and a guided transmission line. Compromises involve bandwidth, efficiency, size, polarization purity, beam shape, and other performance considerations. The gain required and practical considerations dictate the type of designs that may be considered.

Dipoles in a cavity or with a combination of reflectors are common at long wavelengths and at prime focus where their low gain is not a problem, and their small size in wavelengths is an advantage. Corrugated horns typically provide high gain and superior performance in most applications. Good circular symmetry of the beam, good polarization performance, and high efficiency are achieved (Olver et al. 1994; Love, Rudge, & Olver 1982).

3. Orthomode Transducer and Polarizer

The orthomode transducer (OMT) provides dual polarization reception, necessary in order to gather all the photons. OMTs come in many forms, depending on the wavelength and bandwidth needed (Rosenberg, Uher, & Bornemann 1993). In general, they have a square or circular waveguide port that connects to the feed horn and supports orthogonal linear polarizations, and two output ports separating the polarizations into two channels. By adding to the OMT input or output a phase shifter that delays one polarization by 90 degrees with respect to the other, circular polarization may be received or transmitted, Figure 2. It is difficult to design a broadband phase shifter, and this component currently limits the bandwidth of many receivers.

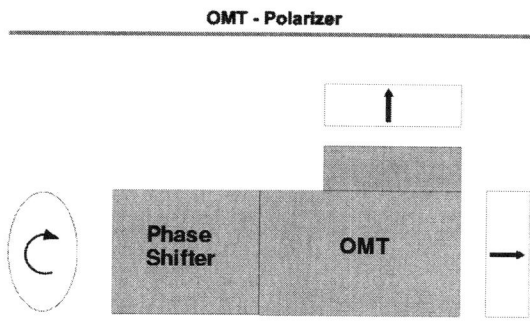

Figure 2. Orthomode transducer and phase shifter (shown here at the OMT port) form a polarizer.

4. Thermal Noise in Receivers

Thermal noise, also known as Johnson noise, arises from random motion of free electrons in a conductor due to thermal agitation (Paczkowski & Whelehan 1987; Johnson 1928; Nyquist 1928). Figure 3 shows symbolically an electrical resistance at physical temperature T, connected to a band-limiting filter. The open-circuit rms voltage that will be measured at the output terminals is given in the figure. The quantities h and k are Planck's and Boltzmann's constants respectively, and f is frequency. If $hf/kT \ll 1$, the integral reduces to B, the form commonly used in microwave system analysis. However, the inequality needs to be checked if $f/T > 1$ GHz/Kelvin. Common electrical circuit theory says that for any voltage source, maximum power is available when load resistance is equal to the source resistance. Under this condition, the available power from the thermal noise source is kBT. The concept of equivalent source temperature is used for convenience when the statistics of any signal resembles gaussian noise. The true source is replaced with an equivalent thermal noise source that produces a noise power at the same level. The equivalent noise temperature, Ts, may be a function of frequency, but should be flat over the filter bandwidth B.

If we connect a thermal noise source to a microwave amplifier input, and measure the amplifier output power at various source temperatures, we will find that the output power is the noise source power times the amplifier gain plus a constant term, Figure 4. This constant is noise added by the amplifier, and we model that noise by thermal noise at the amplifier input.

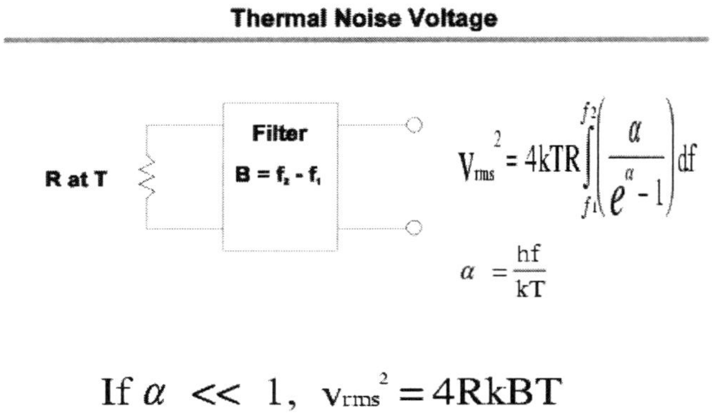

Thermal Noise Voltage

R at T — Filter B = $f_2 - f_1$

$$V_{rms}^2 = 4kTR \int_{f_1}^{f_2} \left(\frac{\alpha}{e^\alpha - 1}\right) df$$

$$\alpha = \frac{hf}{kT}$$

If $\alpha \ll 1$, $V_{rms}^2 = 4RkBT$

Figure 3. Model of band-limited thermal noise in a conductor.

Amplifier noise is not entirely thermal. That is, it is not directly proportional to the amplifier temperature, and is generally a function of frequency. GaAs or InP HFET (HEMT) devices are now used almost exclusively to build low-noise microwave amplifiers. In addition to being inherently low-noise, high gain, and stable, they can be designed to cool well - that is, much of their noise is thermal and decreases when cooled.

In a cascade of amplifiers, the noise contribution of a particular stage is divided by the gain of preceding stages. Therefore, one wants the first stages of a receiver to have high gain and low noise. Most radio astronomy receivers have cryogenically cooled amplifiers for the first stages of gain. When losses exist at the input of a receiver, two negative effects plague the system. An ohmic loss adds noise at a level proportional to its temperature, plus it attenuates the input signal. The net effect is to both add noise, and to multiply the effective noise temperature of following stages.

5. Frequency Conversion

Figure 5 shows a receiver using a mixer for frequency conversion. The mixer has at its output multiple instances of the input radio frequency (RF) and local oscillator (LO) signals shifted up and down in frequency by integer multiples. Use of filters allows selection of the m, n coefficients for the output signal. The top plot in Figure 5 illustrates a simple upper sideband (RF band above the LO) down-conversion where the RF band is shifted to a lower intermediate frequency (IF) frequency. The lower plot illustrates lower sideband (RF band below the

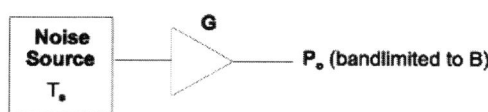

Amplifer Noise Temperature

$P_o = GkBT_s + K$

Define $K = GkBT_e$

Then, $P_o = GkB(T_s + T_e)$

T_e is the amplifier *Equivalent Input Noise Temperature*

Figure 4. Amplifier equivalent input noise temperature.

LO) down-conversion. Note that in this case, the RF spectrum is inverted in frequency at the IF output. Use of mixing provides several advantages:

Tunability – By using a narrow IF filter and a variable LO frequency, the detected RF frequency can be varied over a wide range.

Cost – Amplifiers, transmission lines, and other components are generally less expensive at IF frequencies.

Performance – Having massive amounts of gain at one frequency can cause instability due to leakage from the output circuits coupling back to the receiver input stages. Filtering out unwanted signals is simplified – filter responses scale by the ratio of frequencies, and the down-conversion process greatly increases the ratio of frequencies in the passband. Hence fewer filter resonators are required for a given level of rejection. Also, amplifiers tend to be less sensitive to temperature and other environmental effects.

6. Receiver Stability

Figure 6 shows a greatly simplified diagram of a total power receiver. The receiver chain is represented by G, and typically consists of many amplifiers, mixers, and other microwave and IF components with net gain well over 100 dB. The IF signal presented to the detector is band-limited gaussian noise with effective noise bandwidth of B. The detected output is then integrated with time constant τ. The output voltage V_0, proportional to the detected power,

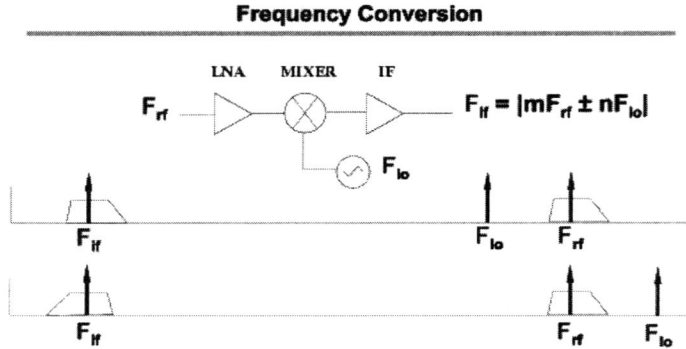

Figure 5. Frequency down conversion in a receiver, both upper and lower sideband.

has both DC and AC components. The ratio of these two components is inversely proportional to the square root of the product of B and τ (Kraus 1966), assuming the receiver gain and noise temperature is constant. A radiometer's sensitivity is often defined as the change in antenna equivalent temperature which results in a change of V_0 equal to the rms value of V_{ac}. However, since a change in G also results in a change in V_0, the receiver gain must be much more stable than V_{ac}/V_{dc} for periods much greater than τ.

Gain stability is achieved by temperature control, both active and passive, design of components that are not sensitive to vibration and shock, and avoidance of stress on cables and connectors by careful alignment of components. Even so, sophisticated switching techniques are often required to achieve adequate sensitivity.

7. Receiver Linearity

Any practical amplifier, mixer, or other active device has limited power-handling capability. At some level of output power, the gain begins to become non-linear, and eventually the output power saturates. For sinusoidal inputs, the output voltage becomes clipped and approaches a square wave, producing harmonics in the output frequency spectrum. Sum and difference terms arise in the output spectrum if multiple frequencies are present at the input (Maas 1988).

The most troublesome intermodulation products are usually odd-order difference terms. The strongest of these are the third-order products, Figure 7.

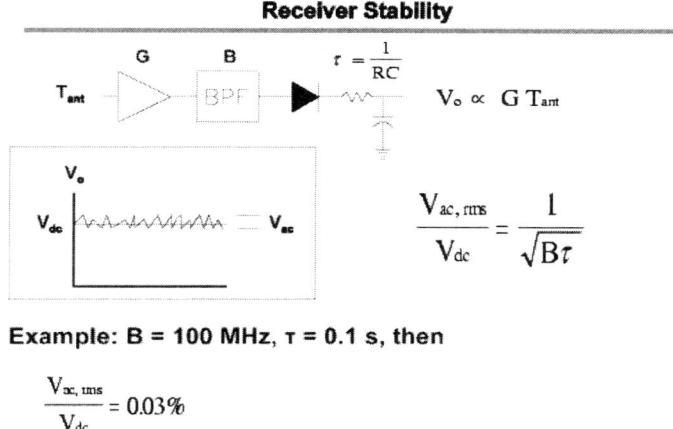

Figure 6. The output signal of a simple total-power receiver.

For two closely spaced input tones F1 and F2, the 2F1-F2 and 2F2-F1 products fall near F1 and F2 and are often impossible to filter out. A log-log plot of the third-order product level has a slope of 3:1. For example, if the level of F1 and F2 is dropped by 10 dB, the third-order products drop by 30 dB, a relative improvement of 20 dB.

The third-order performance of a component is usually specified by the "third-order intercept", a fictitious power level where the fundamental and third-order curves intersect. The device is usually not capable of actually producing that level of power, but knowing this point and the slopes, the relative third-order levels can be calculated for any reasonable fundamental output power. For example, if the fundamental tones are 20 dB below the intercept point for a particular amplifier, the third-order products will be 60 dB below the intercept, or 40 dB below the fundamental tones.

8. Summary

The most common topology for centimeter-wave radio-astronomy receivers is dual-polarization super-heterodyne type with cooled HEMT low-noise amplifiers. The feed, OMT, polarizer, and first amplifiers are often the most critical components. Gain stability has a large effect on data quality and observing efficiency and must be considered and measured during receiver design and construction. The linearity and power handling of the receiver determines its per-

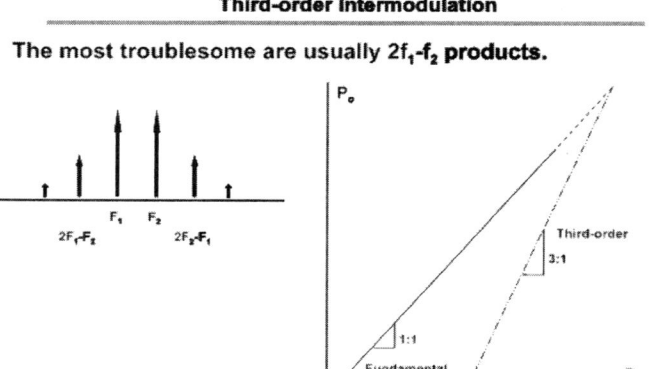

Figure 7.   Intermodulation products result from receiver non-linearities.

formance in the presence of strong unwanted signals such as interference from man-made terrestrial or space-borne sources.

References

Johnson, J. B. 1928, Phys.Rev.A, 32, 97

Kraus, J. D. 1966, Radio Astronomy (New York: McGraw-Hill), Ch. 7

Love, A. W., Rudge, A. W., & Olver, A. D. 1982, in Handbook of Antenna Design, 1, ed. A. W. Rudge et al. (London: P. Peregrinus), 338

Maas, S. A. 1988, Non-Linear Microwave Circuits (Norwood: Artech House), 1

Nyquist, H. 1928, Phys.Rev.A, 32, 110

Olver, A. D., Clarricoats, P. J. B., Kishk, A. A., Shafai, L. 1994, in IEE Electromagnetic Waves Series 39, Microwave Horns and Feeds, ed. P. J. B. Clarricoats, Y. Rahmat-Samii, & J. R. Wait (New York: IEEE Press)

Paczkowski, H. C., Whelehan, J. 1987, IEEE MTT Newsletter, No. 118

Rosenberg, U., Uher, J., & Bornemann, J. 1993, Waveguide Components for Antenna Feed Systems: Theory and CAD (Norwood: Artech House)

The Receiver System–mm Regime

John M. Payne

National Radio Astronomy Observatory, 949 N. Cherry Avenue, Campus Bldg. 65, Tucson, Arizona 85721-0655, USA

Abstract. The millimeter-wavelength radio astronomy band is now generally taken to include frequencies from approximately 60 GHz to around 300 GHz, while the so-called sub-mm band extends this up to frequencies of around 1000 GHz (1 THz). Unlike centimeter-wave ground-based radio astronomy, the mm/sub-mm frequency range is limited by the properties of the atmosphere, which is briefly described. The difficulties of constructing highly sensitive receivers for these high frequencies are described along with the commonly adopted solutions. Particular emphasis is given to modern receivers such as those for the ALMA interferometric array now being planned for installation on a 5,000-meter-altitude site in the Atacama desert in Chile. These receivers use superconducting devices which require temperatures of around 4 K to operate satisfactorily. Brief descriptions of the various sub-systems needed to construct such a receiver will be given.

1. Introduction

There are perhaps two major differences between radio astronomy receivers for the cm regime and those designed for mm/sub-mm use. Those differences will be mentioned here and described more fully below.

1.1. The atmosphere

The atmosphere imposes limitations on the frequencies available for use in the mm/sub-mm part of the radio spectrum. This has several major consequences both for the receiver design and the location of telescopes. High dry sites are preferable to minimize transmission loss through the atmosphere and, due to the transmission being marred by several broad absorption lines of oxygen and water, receivers for these wavelengths invariably are designed around "atmospheric windows".

1.2. Technical difficulties

These are many and varied. The very short wavelengths involved result in the obvious difficulty of constructing components in waveguide as is done routinely at longer wavelengths. The neglect of the mm/sub-mm parts of the spectrum by strong commercial interests has not provided the healthy boost to the development of these systems as experienced for cm-wave devices. Consequently, the

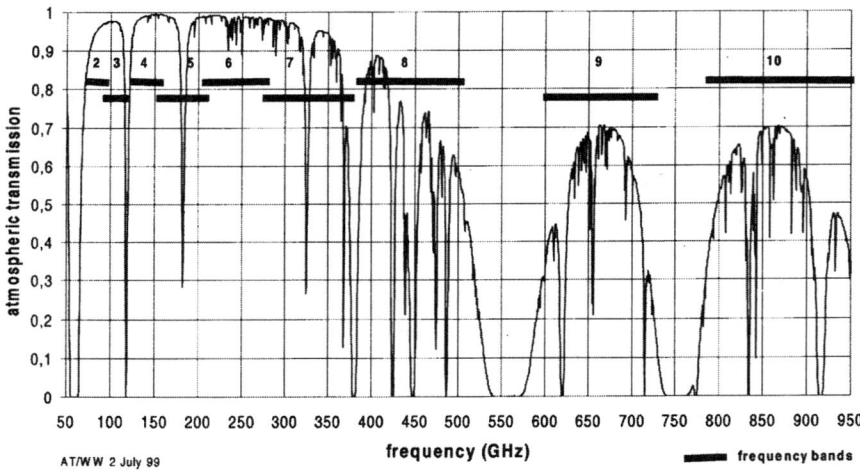

Figure 1.   Atmospheric transmission at Chajnantor, PWV=0.5 mm.

development has been a little slower, although the stage has now been reached where the quantum limit in noise performance is being approached. In a short lecture such as this, it is not possible to delve too deeply into these questions but it is hoped that this summary will prompt further study and reading. References will be given to help this process.

2. General Considerations

As mentioned previously, receivers for use in this frequency range generally have maximum sensitivity in bands that coincide with the various atmospheric windows. A plot of the atmospheric transmission is shown in Figure 1 for one of the best sites known; the ALMA site at an altitude of 5000 m in the Atacama desert (see ALMA Web Site, http://www.alma.nrao.edu). The transmission is shown for one value of precipitable water vapor, 0.5 mm, and the proposed 10 receiver bands for ALMA are marked on this figure. The highest frequency band is Band 10 at around 900 GHz, although there is evidence to suggest that this site may be useable at frequencies over 1 THz.

We should now consider the desirable characteristics of a receiver for these high frequencies. First it should be mentioned that there are two general classes of observation that are commonly used: spectral line and continuum. In spectral line observations, the exact frequency of observation is critical: emission features of just a few kHz in bandwidth at a well defined frequency may be of interest. Continuum measurements are concerned with the measurement of the total amount of power emitted over a given bandwidth, usually made as large as possible.

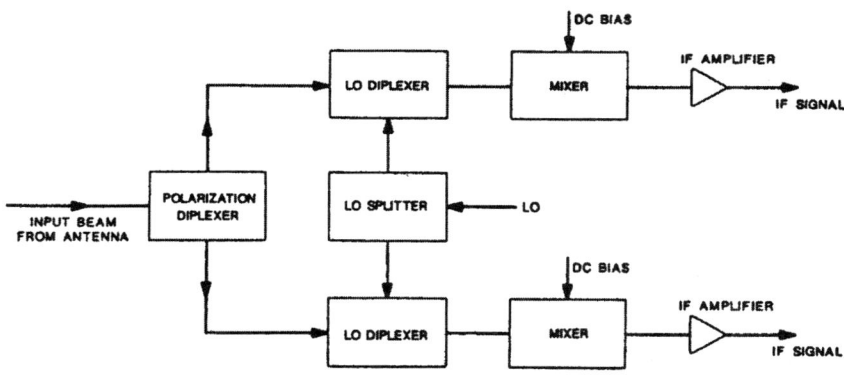

Figure 2. Elemental block diagram of a mm/submm wave receiver.

Incoherent receivers find application as highly sensitive continuum receivers and generally consist of an element known as a bolometer which changes its temperature and so its electrical resistance when subjected to an incident photon. Such receivers generally operate at very low physical temperatures and can be either single elements or incorporated into focal plane arrays (see Lecture by Holland, Duncan & Griffin in this volume). In this short summary, it is not possible to go into further detail of these systems.

Generally, coherent receivers have received more attention and the field of mm-wave spectral-line astronomy has been one of the most active in all astronomy over the past 25 years.

Before considering the desirable characteristics of a receiver designed for spectral line observations, it is as well to mention that the fundamental configuration of spectral line receivers has remained unchanged over the past several decades. However, there have been improvements in components and the overall improvement in sensitivity has been remarkable. In 1970, a receiver noise temperature of 1500 K at the important frequency of 115 GHz was considered excellent. Today, at that same frequency, a noise temp of 50 K is usual. As the time taken to reach a given sensitivity in an observation is inversely proportional to the square of the system temperature, the improvement in system performance is immediately obvious.

The configuration of receivers has changed little over the decades. Due to the lack of RF amplifiers at these high frequencies, the strategy in the past has been to reduce the frequency to be analyzed to a manageable frequency as soon as possible. In general, the signals received will not be polarized and a factor of 2 in observation time may be realized by combining the uncorrelated signals after the detection process. So, it is common to build a receiver for a given frequency band that is sensitive to both orthogonal polarizations and to combine these after detection. An elemental block diagram of a receiver is shown in Figure 2.

The input from the antenna is first split into the two orthogonal linear polarizations. At the lower mm-wave frequencies, this may be achieved by a

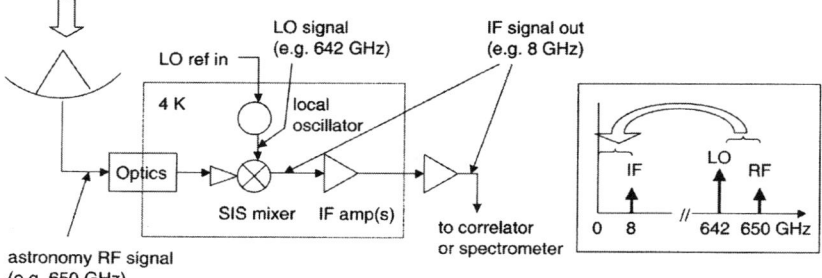

Figure 3. One channel of a mm/submm radio astronomy receiver.

waveguide device. At the higher frequencies, a "quasi optical" device such as a grid of wires wound so as to form a polarizer may be used. Each of the two channels may now be treated separately as shown in Figure 3, which is a simple block diagram for a receiver for the 650-GHz band for ALMA.

This uses the "heterodyne principle" in which two frequencies are "mixed" to produce a difference frequency. In this case, the input frequency is mixed with the "local oscillator" (LO) frequency to produce the "intermediate frequency" (IF). The IF is chosen to be some convenient lower frequency where high sensitivity amplifiers are easily obtainable.

In a modern receiver, the components will be cooled to reduce the noise contributions. The mixing element is a superconductor-insulator-superconductor (SIS) junction that uses the sharp non-linearity in the quasi-particle tunneling between two superconductors separated by an oxide insulating layer. These SIS mixers exhibit very low intrinsic noise and require only low levels of LO power for good operation. The development of SIS mixers for use at mm and sub-mm wavelengths is fully covered in the literature (Tucker & Feldman 1985; Kerr & Pan 1996).

3. A Practical Receiver

Perhaps the easiest way of illustrating the general principles of mm/sub-mm receivers is to use an existing receiver and describe each part of the system. First, let us consider the characteristics of an ideal system.

- Receiver noise contribution far lower than Atmospheric Emission noise.

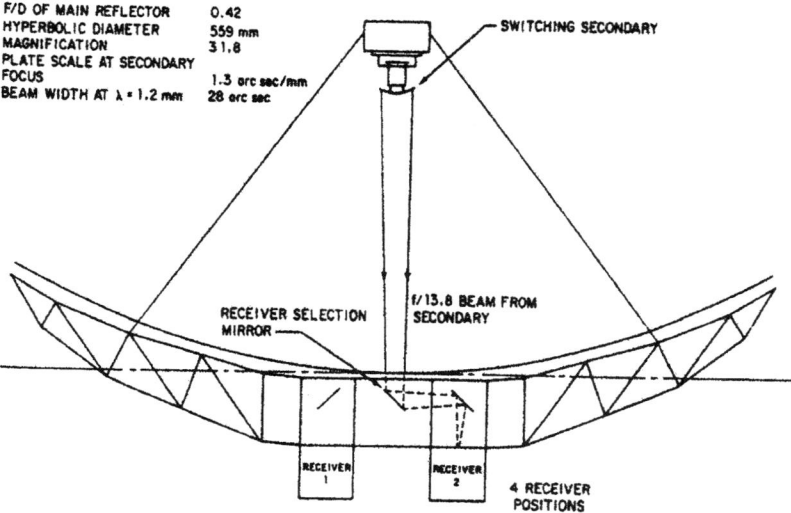

Figure 4.   Optical arrangement of the 12-m telescope.

- Single sideband response to eliminate both emission noise in the unused sideband and confusion from image response.
- Wide I.F. bandwidth.
- Dual linear polarization. Two channels per signal frequency.
- Cover all atmospheric windows at a particular site.
- Rapidly tuneable remotely.
- Continuous operation. Closed cycle refrigeration.
- Good stability.
- High reliability.

The receivers developed several years ago for the National Radio Astronomy Observatory's 12-m radio telescope on Kitt Peak serve as an example. The sole deviation from the above list is covering all available windows with one receiver and, in fact, the telescope installation permits the installation of up to four separate receivers. For various reasons, the Cassegrain focus is preferred; perhaps the most compelling being the ease of mounting the inevitably large receivers behind the main reflector. The mounting arrangement used on the 12-m telescope is shown in Figure 4.

In this arrangement, different receivers may be selected by the rotating flat mirror, as shown in the diagram.

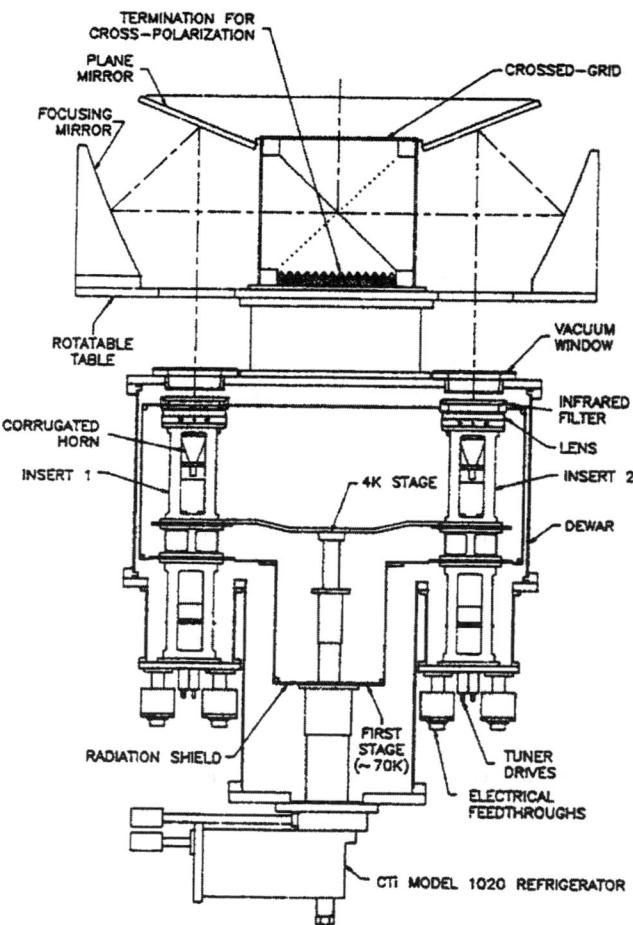

Figure 5. Diagram of receiver used to cover the 3-mm (67-115 GHz) and 2-mm (130-170GHz) bands.

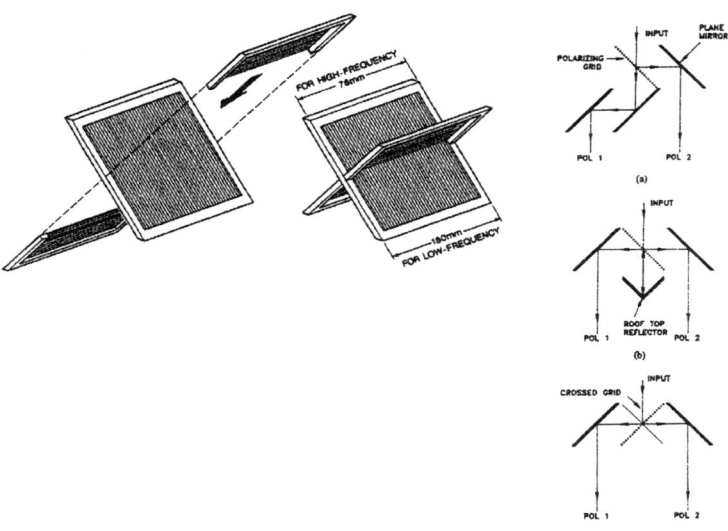

Figure 6. Polarization diplexers.

A diagram of the receiver used to cover the 3-mm (67-115 GHz) and 2-mm (130-170 GHz) bands is shown in Figure 5. As mentioned previously, it is possible to obtain an increase in overall sensitivity by constructing two receivers sensitive to orthogonal linear polarizations and combining the outputs after detection. So, the first component in the receiver is a device to separate the two linear polarizations. A suitable component for this is a grid of wires arranged to reflect the component parallel to the wires and pass the component orthogonal to the wires. Various configurations are possible as shown in Figure 6. For this receiver, the "crossed grid" configuration was chosen.

The beam matching between the telescope beam and the feed horns of the two receivers is achieved in two stages: first, an offset parabolic mirror outside the cryogenic enclosure; second, a teflon lens at around 4 K.

The receivers are constructed as separate "stand-alone" assemblies that may be tested in a separate dewar before installation in the main dewar. This is shown in Figure 7. A separate pair of these elemental receivers may be selected by rotating the assembly consisting of the crossed-grid polarizer and offset parabolas on top of the receiver. A full description of this receiver may be found in Payne et al. (1994). A second receiver, the high frequency receiver, (also fully referenced in Payne et al. (1994) is shown in Figure 8. Of particular interest here is the image-terminating scheme shown in detail in Figure 9 (also referenced in Payne et al. 1994).

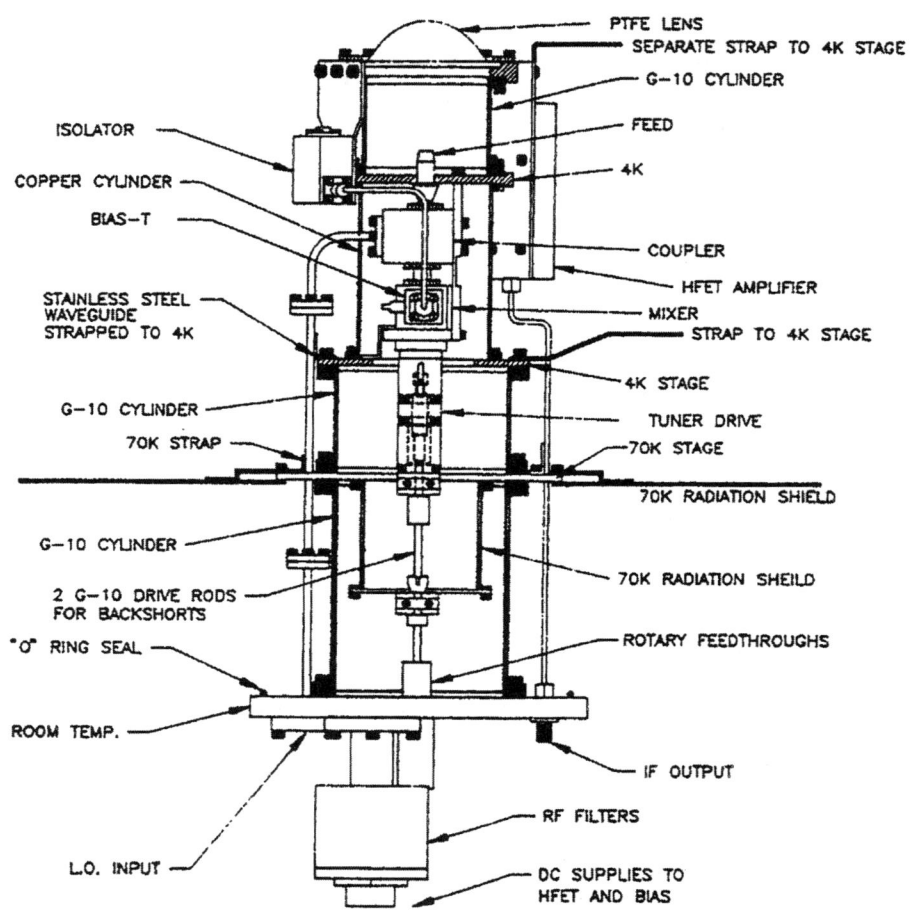

Figure 7. A receiver insert.

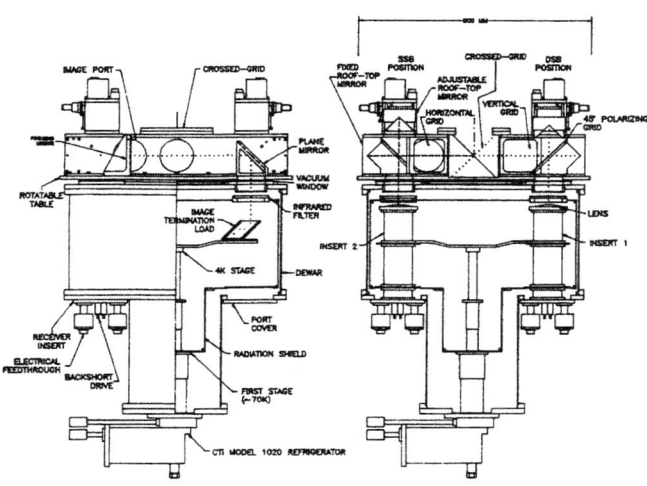

Figure 8.   Layout of the high-frequency receiver.

4. Major Topics

There are, of course, many areas of the design of mm/sub-mm receivers that cannot be covered in detail in this short lecture. These will be listed here, together with the appropriate references, where available.

4.1. Quasi-optical analysis

One of the characteristics of this wavelength regime is that the dimensions of components tend not to be many wavelengths in diameter. It turns out that the so-called "Gaussian beam analysis" is extremely useful for analyzing mm/sub-mm systems (Goldsmith 1982).

4.2. Double sideband / Single sideband

This is a complex subject that has been the subject of much debate within the ALMA project. Generally speaking, double sideband mixer systems are simpler than single sideband systems but (for spectral line work) incur a penalty of increasing the system noise due to atmospheric noise entering the system via the unwanted sideband. This is, obviously, site dependent and the issues become quite complex. A good reference is the ALMA Memo series (Lamb 2000). An illustration of a practical system to terminate the unwanted sideband is shown in Figure 9.

Figure 8 is a diagram of the so-called "high frequency receiver" in use on the 12-m telescope in Arizona and follows the modular form of the "low frequency receiver". Figure 9 shows the details of the scheme used to terminate the unwanted sideband in a cold load situated within the main dewar. The optical arrangement is known as a Martin Puplett interferometer, a device that has found many applications in mm/submm receivers (Martin & Puplett 1969).

4.3. Local oscillator sources

A vital component in any receiver is the local oscillator. This should be spectrally pure and locked in frequency to a frequency standard. It is usual to also lock in phase, a vital requirement for an interferometer.

In the early days of mm astronomy, reflex klystrons were the only components available for generating the required frequencies of 100 GHz or so. With the high voltages involved, phase locking the device to a low frequency standard proved to be something of an engineering adventure. In recent years, solid state devices (Gunn oscillators) have become available and obtaining a source of several tens of milliwatts at frequencies around 100 GHz is now routine. To obtain higher frequencies, it is usual to use a non-linear device fed with this fundamental source. The modern SIS mixers require only a microwatt or so of LO power, and this method has been used up to around 1 THz.

An interesting alternative that is being worked on by many groups today involves the mixing together of two infrared lasers. It has been shown that the beat frequency between two such lasers may be locked quite simply to a low frequency standard, and several groups report the generation of power up to 1 THz (Payne et al. 1998).

4.4. Local oscillator injection

Due to the difficulties of machining a fundamental mode wave-guide at these short wavelengths, the tendency in the past has been to use quasi-optical methods. These can be broadly divided into two categories:

1. *Broad Band High Loss*: An example of this is a dielectric sheet placed in front of the receiver feed that couples a small amount of the available LO power into the receiver. The disadvantages are fairly obvious: poor coupling efficiency and a loss of signal power.

2. *Quasi-Optical Diplexers*: These are tuneable devices that have several advantages; the most significant being low loss to the LO and the rejection of noise on the LO signal at the signal frequency. The disadvantages are the need for mechanical tuning and the restricted intermediate frequency bandwidth. Payne (1989) deals with this.

4.5. Dielectric materials

In this frequency regime, it is usual to use lenses and mirrors in a rather similar manner to how these components are employed at optical wavelengths. Complications ensue when the materials are used at cryogenic temperatures and this is well dealt with in Lamb (1996). Also to be considered is the anti-reflection coating. In this case, the usual method is to reduce the refractive index at the surface of the dielectric by removing some fraction of the material, usually by grooving or partially drilling with small holes. Polarization effects are important here.

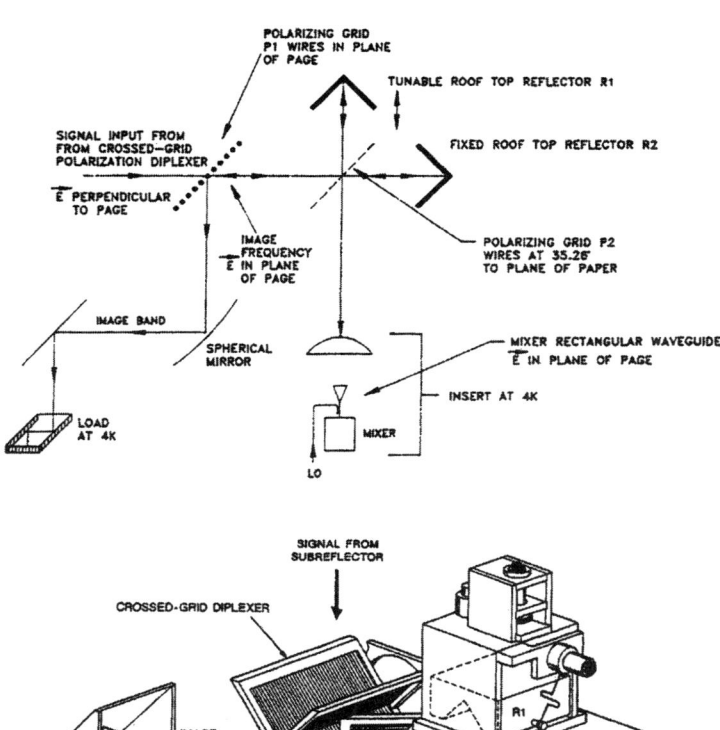

Figure 9.   Image termination scheme.

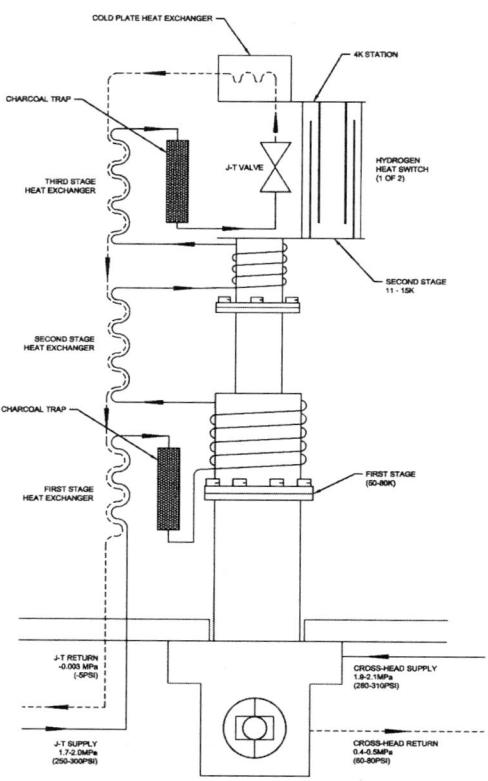

Figure 10. Block diagram of a 4 K refrigerator using a Gifford McMahon commercial refrigerator plus a Joule-Thomson expansion valve.

4.6. Refrigeration methods

Modern receivers for this frequency range inevitably use temperatures around 4 K to ensure that the super-conducting devices operate properly. There are many ways of achieving these temperatures, although, for large receivers operated continuously, a closed-cycle system is obviously preferred. In the past, these closed-cycle systems have generally been limited to two-stage Gifford-McMahon (G-M) machines with a separate Joule-Thomson (J-T) system installed as a third stage. Recently, however, three-stage G-M machines have become available with the temperature stability and capacity well suited to the requirements for a complex receiver such as those required for ALMA. A typical G-M + J-T system is shown in Figure 10.

References

Goldsmith, P. F. 1982, in Infrared and Millimeter Waves, ed. K. J. Button (New York: Academic Press), 277

Kerr, A. R. & Pan, S.-K. 1996, ALMA Memo 151

Lamb, J. W. 1996, Int. J. Inf. Mm. W., 17, 1977

Lamb, J. W. 2000, ALMA Memo 301

Martin, D. H. & Puplett, E. 1969, Infrared Physics, 10, 105

Payne, J. M. 1989, Proc. IEEE, 77, 993

Payne, J. M., D'Addario, L., Emerson, D. T., Kerr, A. R., & Shillue, B. 1998, in Proc. SPIE, 3357, Advanced Technology MMW, Radio, and Terahertz Telescopes, ed. T. G. Phillips (Bellingham, Wash.: SPIE), 143

Payne, J. M., Lamb, J. L., Cochran, J. G., & Bailey, N. 1994, Proc. IEEE, 82, 811

Tucker, J. R. & Feldman, M. J. 1985, Rev. Mod. Phys., 57, 1055

Wendy Lane and Karen O'Neil in a serious discussion while Jo Ann Eder, Kristine Spekkens and Maria Soledad del Rio work hard on hands-on data reduction.

Single-Dish Radio Astronomy: Techniques and Applications
ASP Conference Series, Vol. 278, 2002
S. Stanimirović, D. R. Altschuler, P. F. Goldsmith, and C. J. Salter

Back-ends

J. R. Fisher

National Radio Astronomy Observatory, P. O. Box 2, Green Bank, West Virginia 24944, USA

Abstract. The final stages of receiver electronics are designed to extract information about the intensity of cosmic signals as a function of time, frequency, and polarization. The required signal processing can be as simple as a total power detector or as complex as a pulsar search machine that looks for periodic, dispersed pulse signatures in the time and frequency domain. This mini-lecture gives a brief overview of square-law detectors, FFT spectrometers, polarimeters, and pulsar processors. It touches on calibration issues and the synchronous control of front-end calibration signals and beam and load switches.

1. Introduction

Radio telescopes have traditionally been divided into three subsystems: the antenna, the radio frequency (RF) and intermediate frequency (IF) electronics (front-end), and the "back-end". The first two are common to all types of observing, but a back-end is usually specific to continuum, spectroscopic, or pulsar measurements. Signal processing and astronomical information extraction that are unique to a particular observing type typically reside in the back-end.

Nearly all astronomical observations can be summarized as the measurement of radiation intensity as a function of five independent variables: two position coordinates (e.g. right ascension and declination), time, frequency, and polarization. All other parameters (distance, pulsar period, radial velocity, etc.) are derived quantities. The back-end is directly involved in the measurement of intensity as a function of time, frequency and polarization, and it is also responsible for the coordination of calibration and other switching signals with data integrators.

2. System Overview

Figure 1 shows the relationship of the back-end to the rest of the telescope system. The association of intensity with sky coordinates is usually done between the telescope control system and the data recording computers and, hence, does not directly involve the back-end.

A very simplified back-end is drawn in Figure 1 to show only how the signal intensity accumulators are synchronized with the calibration and front-end switches. A reference load switch is shown, but it could also be a beam switch, polarization switch, or something else. In this illustration the four accumulators

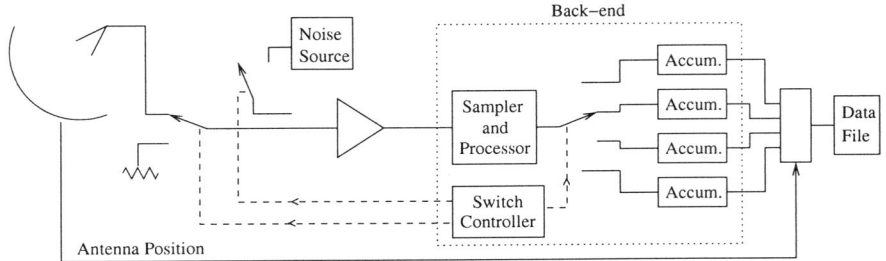

Figure 1. The back-end's place in the telescope system.

could be sequenced as shown in Table 1. The details of the sampler and processor block will be discussed in later sections. We assume here only that the output of this block is an intensity value proportional to the system noise power plus power from the radio source in the antenna beam. The accumulators are set to integrate the signal for a chosen amount of time, often through many switch cycles, and then their values are recorded with the antenna position at the center of the integration in the output data file. The data file usually contains many successive accumulator integrations. Interpretation of the accumulator values is then the task of the data analysis software.

Table 1. Example accumulator, calibrator, and front-end switching sequence in Figure 1.

Active Accumulator	Calibration Noise Source	Load Switch
First	On	Antenna
Second	Off	Antenna
Third	On	Load
Fourth	Off	Load

Some front-ends and some types of observing do not use a reference switch so the switching cycle would use only two accumulators for calibration noise source on and off. If the noise source were left on or off for the duration of a scan, only one accumulator would be used. Any precise synchronization that is required between the back-end and other parts of the system is usually driven by the timing circuitry in the back-end.

Since many electronic switches take a finite time to change states or may generate transient noise during the state change, it is best to avoid accumulating data during a switch transition. Most back-ends provide for a "blanking" interval at the beginning of each switch transition when all accumulators are inactive. The length of the blanking interval depends on the type of switch being used, but it is typically on the order of tens of microseconds to a few milliseconds. If the switch involves a physical movement with significant inertia, such as a mirror or reference load motion, the switching command to the front-end might be sent slightly in advance of the accumulator switch change to maximize the useful accumulation time on each switch state and minimize the required blanking time.

3. Samplers and Detectors

The output of the RF/IF section of the receiver is a band-limited voltage. By that we mean that the receiver noise and cosmic signals have been restricted in frequency by bandpass or low-pass filters and the output of the last amplifier is a voltage proportional to the amplitude (not intensity) of the combined signals. The average value of this voltage is zero. To measure the power of the receiver output this voltage must be squared. The average value of this squared signal is not zero, and it may be accumulated for long periods of time to obtain a better estimate of its average value.

Because all cosmic signals are noise-like with no *a priori* knowledge of their emitted phases, the only measure of these signals available to astronomers is the temporal and spatial nature of their intensity or power. However, there are many signal processing operations that must be performed on the voltage output from the front-end, such as further band limiting, spectral decomposition, and correlation with other copies of the same signal, before the final squaring is done to measure the power. We will discuss a few of these signal processing operations in later sections. The important point here is to keep in mind the difference between the voltage and power signal domains. Voltages can add coherently. Powers add only statistically.

3.1. Square-law detector

The simplest telescope output is the direct squaring of the voltage from the front-end. This constitutes what is often called a continuum receiver. A natural power detector is a calorimetric device where the signal is used to heat a resistor, and the resistor's temperature rise is proportional to the power. This is the principle used by a bolometer. Unfortunately, the response time of such a device is too slow for many radio telescope measurements. The most frequently used alternative is a semiconductor diode whose exponential voltage-current transfer characteristic is used to approximate a square-law detector over a limited signal power range. Its output voltage is proportional to the square of its input voltage, and the output voltage may be further processed as a quantity proportional to the input power.

Figure 2 shows three methods for averaging or accumulating the output of a square-law detector. The first method uses a low-pass filter on the output of the detector. The effective integration time is approximately equal to the inverse of the bandwidth of this filter ($\tau \approx (1/B)$) and the analog-to-digital converter is expected to sample the filter output slightly faster than this interval. Each integration-time selection requires a different filter and sampling time.

The second method in Figure 2 uses just one low-pass filter and A/D sampling rate consistent with the shortest integration time required. Longer integrations are performed by accumulating the samples in an arithmetic logic unit. This requires considerably less hardware for a wide selection of integration times.

Power Detection and Accumulation

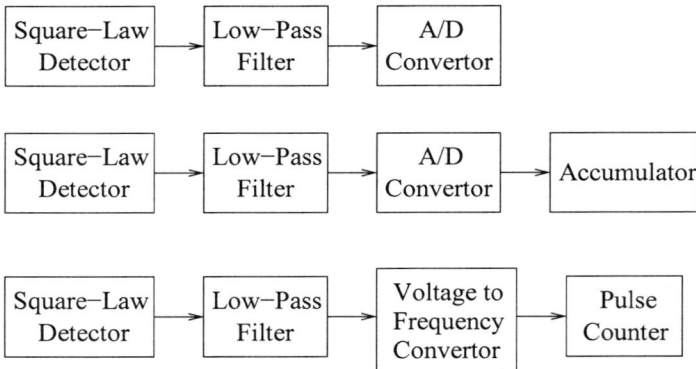

Figure 2. Three methods for power averaging following a square-law detector.

One disadvantage of both this and especially the first method is that for wide pre-detection filter bandwidths the A/D converter must have sufficient amplitude resolution to resolve the low-pass filter output noise on top of a large DC offset plus adequate dynamic range for signals stronger than the system noise. A/D output words at least as wide as 16-bits may be required, which can be rather expensive.

A third method of square-law detector output accumulation is shown at the bottom of Figure 2. The low-pass filter output drives a voltage-to-frequency converter, running at a nominal frequency of about one megahertz, which in turn drives a pulse counter. The pulse counting interval is equal to the chosen integration time. This scheme has the property that the amplitude resolution is proportional to the integration time. Long integration times and wide pre-detection bandwidths are easily accommodated. For a 1 MHz pulse rate and an integration time of 0.1 seconds the amplitude resolution is one part in 10^5, which is equivalent to a 20-bit A/D, assuming a factor of ten headroom for strong signals.

3.2. Voltage sampling

To retain the spectral and possibly the phase information in the signal from the front-end for further processing, the voltage signal must be sampled. This is done with a fast A/D converter connected directly to the front-end amplifier output. To prevent aliasing of the spectral information the A/D sampling rate must be twice the bandwidth of the RF signal, i.e. a 100 MHz bandwidth signal must be sampled at 200 megasamples per second. The first stages of digital processing following the A/D must handle the same data rate.

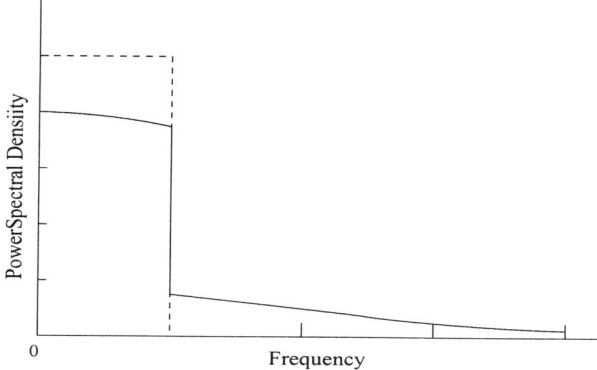

Figure 3. Output spectrum of a one-bit (two-level) sampler with a band-limited, white noise input whose spectrum is shown by the dashed line. This figure is adapted from Figure 8.3 of Thompson, Moran, & Swenson (2001).

To construct the widest bandwidth digital spectrometers and signal correlators at an affordable cost most machines have taken advantage of the fact that the signal from a radio astronomy receiver is almost always dominated by system noise with the astronomical signal adding an only small fraction to the total. This permits the use of an extremely simple sampling device and subsequent processing arithmetic with only one or two bits of input voltage resolution. Such a sampler would severely distort a non-noise-like signal, producing many harmonics and cross-products of the frequencies present in the spectrum. However, when random noise is the dominant signal component, all of the harmonics and cross-products are converted into broadband noise that can usually be tolerated in an astronomical measurement.

Figure 3 shows the spectral content of the output of a one-bit (two-level) voltage sampler. The dashed line shows the band-limited input noise spectrum. The power lost from the input noise frequency range is due to harmonics and cross-products being generated by the sampler quantization, as can be seen by the long spectral tail extending to higher frequencies. This spectral tail will be aliased back into the input frequency range, if the sampler is clocked at twice the bandwidth of the input spectrum. The harmonic power loss and aliasing results in the sensitivity of a one-bit sampler being only 64% ($2/\pi$) of a perfect sampler. A faster sampling clock frequency will reduce the aliased noise power and improve the sensitivity somewhat. Using more sampling levels also helps. Most spectrometers and correlators in use today use two-bit (three-level) samplers as a compromise between sensitivity loss and hardware complexity.

As you might surmise, a one-bit sampler loses all amplitude information. This must be restored by measuring the total spectral noise power with a separate square-law detector to renormalize the sampled spectrum. A two-bit sampler retains the amplitude information in the statistics of the three sampler level counts over a limited amplitude range, so a separate power detector is not necessarily required. However, for best sensitivity and accurate spectral power recovery, the input level to a two-bit sampler must be well controlled and carefully monitored.

At the time that this lecture is being written we are entering a transition to multi-bit samplers that will no doubt result in many receiver functions currently implemented with analog electronics (mixers, local oscillators, filters, etc.) being taken over by digital signal processors. The enormous increase in digital computational speed, financed by the personal computer revolution, makes this all possible at frequencies and bandwidths of interest to radio astronomy. Multi-bit samplers increase sensitivity and amplitude dynamic range, but they also allow higher performance filters and much more stable signal processing in the digital domain.

4. Spectrometers and the Fourier Transform

Nearly all spectrometers in radio astronomy use the Fourier transform principle of spectral decomposition. They can have one of two forms, either a direct Fourier transform of the temporal series of voltage samples, $V(t)$:

$$A(f) = \int V(t) e^{i2\pi ft} dt, \tag{1}$$

$$P(f) = A^2(f), \tag{2}$$

or a transform of the autocorrelation function, $ACF(\Delta t)$,

$$P(f) = \int ACF(\Delta t) e^{i2\pi f \Delta t} d\Delta t, \tag{3}$$

where $A(f)$ is the voltage frequency spectrum of $V(t)$, $P(f)$ is the power spectrum, and Δt is the autocorrelation function delay. The relationship between the voltage samples and the autocorrelation function is

$$ACF(\Delta t) = \int V(t) V(t + \Delta t) dt. \tag{4}$$

The advantage of an autocorrelation spectrometer is that the relatively complex computation of the Fourier transform needs to be performed only after that data have been averaged for a considerable length of time as is implied by the integral in Equation 4. The direct Fourier transform of the voltage samples must be done at the rate dictated by the sampling speed. For example, if we wanted to compute a 10-MHz, 1024-point spectrum averaged over one second, the direct transform method requires 10,000 transforms while the autocorrelation method requires only one. The one transform per second can be done in a general purpose computer while the much faster transform must be implemented in more specialized hardware. For one- and two-bit samplers the hardware implementation of an autocorrelator is relatively inexpensive.

The hardware block diagram of an autocorrelator is shown in Figure 4. With a one-bit sampler the Δt delay chain is a simple binary shift register, the multipliers are binary 'exclusive or' gates, and the accumulators are simple

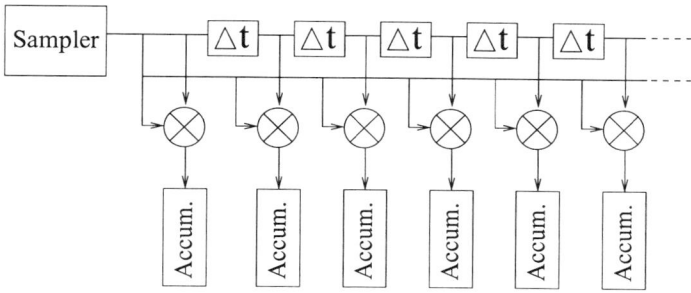

Figure 4. Hardware block diagram of an autocorrelator. The Δt boxes represent time delays and the crosses represent multiplications.

counters. The linear architecture of the autocorrelator also fits well into a hardware design. A two-bit sampler requires somewhat more hardware complexity, but only by a factor of three or four. A multi-bit sampler requires a multi-bit shift register, an arithmetic logic unit for each multiplier and accumulator and multi-bit memory associated with each accumulator.

The direct Fourier transform spectrometer is inherently a multi-bit arithmetic device because the complex exponential term in Equation 1 must be represented with sufficient precision, typically 12 to 16 bits, to prevent the generation of spurious products due to phase and amplitude errors in the computational process. Small input word lengths offer no advantage. However, the Fourier transform, in its fast Fourier transform (FFT) implementation (Cooley & Tukey 1965), does have an inherent computational advantage over the autocorrelation calculation. The number of arithmetic operations per computed spectrum in an FFT is proportional to $N \log_2(N)$ while an autocorrelator requires N^2 operations, where N is the number of points in the spectrum. For large N this is a considerable advantage.

For the same economic reasons that mixers, filters, and other traditionally analog functions are now possible in digital hardware, the direct FFT spectrometer is now feasible with bandwidths approaching 1 GHz. This will very likely increase the role of FFT devices in future back-end designs. The details of a hardware version of an FFT are somewhat more than can be covered in this lecture. For more information see one of a number of digital signal processing texts, such as Rabiner & Gold (1975).

Notice in Equation 1 that the output of a direct Fourier transform spectrometer, $A(f)$, is a complex voltage spectrum as opposed to a real-valued power spectrum, $P(f)$, from the autocorrelation function in Equation 3. Each frequency channel of the voltage spectrum is exactly like a voltage sample of a much narrower passband, and it may be used in exactly the same way for further signal processing. All phase information is lost in the power spectrum so, in many cases, the autocorrelation operation is necessarily the final step in any real-time signal processing chain.

Alert readers will notice that I have casually glossed over the difference between real and complex voltage samples when dealing with the Fourier transform spectrometer. This is a key issue that must be addressed in the FFT design, but it is not a fundamental limitation. One can build complex (sine and cosine) voltage samplers or add a stage to the FFT computation to account for the fact that real samples have been fed to the inherently complex FFT operation.

5. Signal Correlation

Any radio signal, even random noise, is always correlated with itself. This fact is exploited in the autocorrelation spectrometer. Two copies of a signal may travel through different antenna and receiver electronics, such as two orthogonal polarizations or two adjacent focal plane antenna elements. These two copies are also correlated.

When we say that signals are correlated we mean that their time-averaged vector product is not zero. Mathematically, this is the average of one signal times the complex conjugate of the other

$$C = \langle V_1(t) V_2^*(t) \rangle . \tag{5}$$

Since $V_1(t)$ and $V_2(t)$ are copies of the same signal,

$$V_1(t) = aVe^{i2\pi ft}, V_2(t) = bVe^{i2\pi ft+\phi}, \tag{6}$$

and

$$C = abV^2 e^{i\phi}. \tag{7}$$

The correlation product, C, is a vector with units of power and a phase equal to the phase difference between the two copies of the signal.

A good example of the utility of this correlation value in single-dish work is for the measurement of polarization. The self and cross (correlation) products of the signals from two orthogonal feed polarization outputs contain all there is to know about a polarized radio source. A few specific cases of fully polarized radiation using orthogonal linear feed outputs are

Linear pol'n aligned with 'b' output........ $a = 0.000, b = 1.000, \phi = n/a$
Linear pol'n at 45 degrees to 'a' and 'b'.. $a = 0.707, b = 0.707, \phi = 0$
Circular pol'n.. $a = 0.707, b = 0.707, \phi = 90$

These values are normalized to the total signal power and phase referenced to the feed output. The receiver electronics will modify the relative phase and amplitude of the two signal paths, and these receiver effects need to be calibrated away. A partially polarized signal will produce only partial correlation between the two feed polarization outputs. In general one needs to solve for the correlated and uncorrelated components of the receiver outputs, but the basic idea is the same. A much more complete treatment of this subject will be covered elsewhere in this compendium.

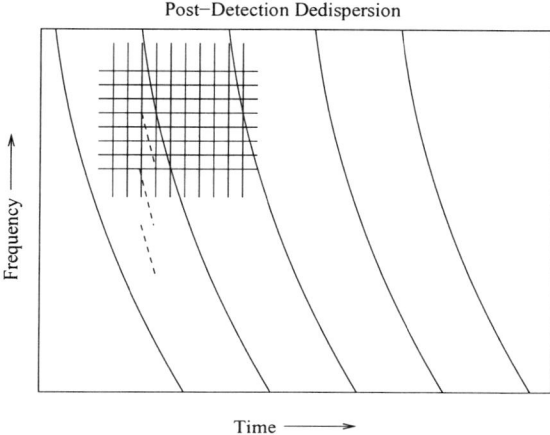

Figure 5. Schematic illustration of pulsar dispersion. The curved lines represent the loci of peak pulse radiation. The rectangular grid indicates how a pulsar spectrometer bins the time and frequency samples. The short dashed lines show how a pulsar may be dedispersed with post-detection sample shifts.

6. Pulsar Processing

Pulsar observations are some of the most demanding in terms of back-end signal processing because high resolution is required in both the time and frequency domains. Pulsar pulses are dispersed in the sense that a pulse arrives at high frequencies earlier than at low frequencies as is illustrated schematically in Figure 5.

The most common method for removing pulse dispersion is to bin the power samples in both time and frequency, as shown by the grid in Figure 5, and advance the lower frequency channel outputs to match the arrival times of the higher frequency channels as shown by the dashed line segments. This works well for low and moderate dispersions where the time and frequency bin sizes may be adjusted to provide as much resolution as needed to resolve the time structure within one pulse. However, for short period and high dispersion pulsars one runs into a fundamental limit that the transit time across one frequency bin can be shorter than the response time of the frequency channel that is set by its inverse bandwidth ($\tau \approx 1/B$).

This time resolution limitation may be overcome with pre-detection, sometimes called coherent, dedispersion as illustrated in Figure 6. In this case the broadband voltage time samples are transformed into the frequency domain with an FFT the same way as one would do for a direct Fourier transform spectrometer. Instead of squaring the output channel voltages to generate power samples in each time and frequency bin, the complex voltages are given a phase offset with progressively larger offset values across the frequency band. A linear phase

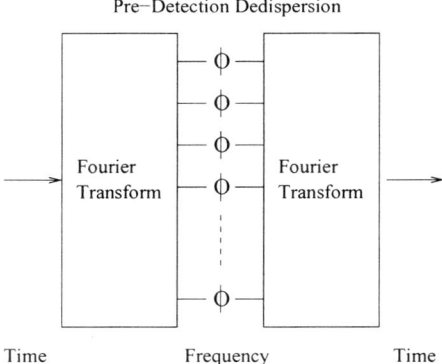

Figure 6. A pre-detection pulse dedispersion scheme for achieving higher time resolution on short period and high dispersion pulsars. A phase offset is introduced to each frequency path.

gradient as a function of frequency is exactly equivalent to a simple delay in the time domain. A non-linear phase gradient is equivalent to dispersion or dedispersion according to its sign. The phase-modified spectrum is then transformed back into the time domain to produce a dedispersed time series with high pulse time resolution. An equivalent pre-detecton dedisperser can be constructed with a delay-domain digital filter instead of the back-to-back FFT's.

Maybe the most demanding type of observing in terms of signal and data processing load is a search for pulsars. Since none of the parameters of the pulsars (sky position, period, and dispersion measure) are known ahead of time, spectra must be recorded continuously at about 200 microsecond intervals for the duration of the survey. These data are then searched in pulse period and dispersion measure at each position on the sky with very capable off-line computers. If the spectra are 1000 channels of 4-bit data, one hour of observing requires 9 gigabytes of storage, and the data rate is 2.5 megabytes per second.

References

Cooley, J. W. & Tukey, J. W. 1965, Mathematics of Computation, 19, 297

Rabiner, L. W. & Gold, B. 1975, Theory and Application of Digital Signal Processing (Englewood Cliffs: Prentice-Hall)

Thompson, A. R., Moran, J. M., & Swenson, G. W., Jr. 2001, Interferometry and Synthesis in Radio Astronomy (New York: John Wiley & Sons, Inc.)

Spectrometry and Autocorrelation

Jon Hagen

National Astronomy and Ionosphere Center, Arecibo Observatory, HC 3 Box 53995, Arecibo, Puerto Rico 00612, USA

Abstract. The spectral power density, at a frequency f, of an electrical signal is simply the average power per unit bandwidth when the signal is passed through a lossless narrow bandpass filter centered at f. In the limit that the bandwidth goes to zero, this can be taken as the definition of the spectral power density. In radio astronomy, however, it is common to use a digital processing method wherein uniformly-spaced samples of the signal are used to estimate an average of the signal's autocorrelation function. This function is then Fourier transformed to produce an estimate of the spectral power density. In the limit that the averaging time goes to infinity, this is an equivalent definition of the spectral power density. The equivalence is reviewed below.

1. Operational Definition of Spectral Density

Just as the light entering an optical telescope contains a spectrum of optical frequencies (colors), the electromagnetic waves intercepted by a radio telescope also span a band of radio frequencies. Analyzing, i.e. displaying, the spectrum allows us to see spectral lines, to study other spectral structure, and to measure Doppler shifts. Light is commonly analyzed using narrow transmission filters. In radio work we do not have to put filters over the aperture of the radio telescope. The antenna converts the incident E field (volts/meter vs. time) into a voltage at the antenna terminals (volts vs. time), and we can install an electrical bandpass filter anywhere along the line connecting the antenna to the receiver, or at any point in the receiver before the detector. If we use a tunable narrow bandpass filter and a square-law detector or a bank of filters, each with its own square-law detector, we can measure the incoming power vs. frequency by averaging the output of each detector. (Remember that power is proportional to the *square* of voltage). If a lossless bandpass filter centered at f has a bandwidth of B Hertz and its power output is P watts, the input spectral density at f must be $S(f) = P/B$ watts Hz^{-1}. This is an operational "filter" definition of power spectral density. The basic bandpass filter/squarer/averager is shown in Figure 1.

2. Spectral Density Defined in Terms of the Signal's Time History

Since the signal from the antenna is simply a time-varying voltage, the spectral density function must also be definable directly in terms of observed sequences

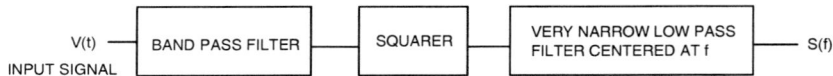

Figure 1. Basic filter channel used to estimate spectral power density at one frequency.

of voltage vs. time. Such a definition is needed in order to implement digital processors for spectral analysis. Let us jump ahead and first present the definition indirectly, in the form of a recipe for doing spectral analysis via the signal's *autocorrelation function*. This function, $R(\tau)$, is defined as the time average of the "lagged products" $V(t)V(t-\tau)$, i.e.

$$R(\tau) = \langle V(t)V(t-\tau) \rangle , \qquad (1)$$

where the brackets $\langle\ \rangle$ denote averaging over t. If we have sequences of receiver output voltage values, we can calculate and estimate this function, $R(\tau)$. We can also build hardware autocorrelators that operate in real time to calculate simultaneous estimates of $R(\tau)$ for many successive values of τ. The spectral density is the Fourier transform of the autocorrelation function:

$$S(f) = \int_{-\infty}^{+\infty} R(\tau) e^{-2\pi i f \tau} d\tau . \qquad (2)$$

In autocorrelation spectrometry, a hardware autocorrelator does the heavy computing work. After a suitable integration time, it dumps the lagged product averages into a general purpose computer which, with a single Fourier transform, converts this estimated $R(\tau)$ into an estimate of the spectral density function. The frequency resolution is inversely proportional to $\tau_{\max}$, the maximum delay for which R is estimated. Note that the inverse of the transform in Equation 2 is given by:

$$R(\tau) = \int_{-\infty}^{+\infty} S(f) e^{2\pi i f \tau} df . \qquad (3)$$

To see that Equation 2 (which is often used as the *definition* of spectral density) agrees with our operational understanding of spectral density, the following relation can be derived (see, for example, Thomas 1969) by substituting Equation 1 into Equation 2:

$$S(f) = \lim_{T \to \infty} \langle |\, X_T(f)\,|^2 \rangle , \qquad (4)$$

where

$$X_T(f) = \frac{1}{2T} \int_{-T}^{+T} V(t) e^{-2\pi i f t} dt . \qquad (5)$$

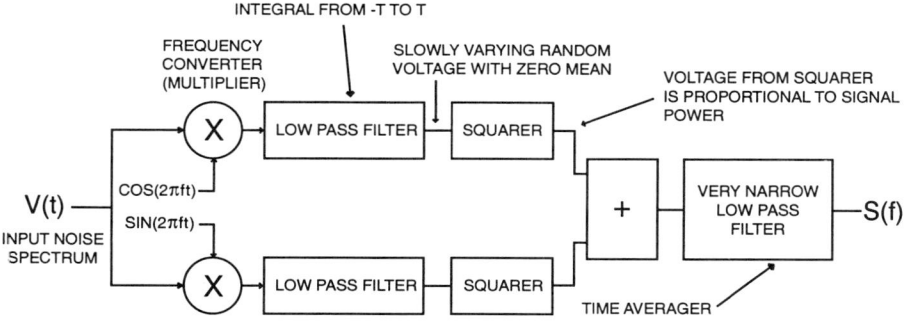

Figure 2. Circuit equivalent to the operations specified by Equations 4 and 5.

In engineering terms, Equations 4 and 5 can be seen as a frequency conversion (the multiplication by $e^{-2\pi i f t}$ in Equation 5) which shifts the spectrum so that the frequency in question becomes dc, followed by a "box-car" lowpass filter (the integration from $-T$ to T), which passes only a narrow frequency range near dc. Successive evaluations of the integral in Equation 5 produce a sequence of random numbers (random and with zero mean, no matter how large T becomes), equivalent to a sequence of output voltages from a narrow bandpass filter. These are squared and averaged in Equation 4 to estimate the power output of the filter. Finally, the $T \to \infty$ limit makes the analyzing filter infinitesimally narrow around the frequency f. These operations are shown in Figure 2. Note that Equations 4 and 5 provide a recipe for FFT spectral analysis, except that a finite T would be used; its value determines the spectral resolution. Likewise, when the circuit of Figure 2 is used for the channels of an analog filter bank, the lowpass filters have finite width.

3. Equivalent Circuit Using a Bandpass Rather Than Two Lowpass Filters

While analog filters and filter banks are sometimes implemented at base band, as shown in Figure 2, it is more common to use bandpass filters. Consider the circuit of Figure 3, which is equivalent to the circuit of Figure 1.

To see the equivalence, note that the baseband outputs (signals that extend down to dc) of the two low pass filters are converted back up to the original center frequency, f, to produce the signals $I\cos(2\pi f t)$ and $Q\sin(2\pi f t)$, where $I(t)$ and $Q(t)$ denote the slowly varying outputs of the lowpass filters. These two signals are added to produce $I\cos(2\pi f t) + Q\sin(2\pi f t)$. The square of this voltage contains the term $I^2\cos^2(2\pi f t) + Q^2\sin^2(2\pi f t)$, whose average, $1/2(\langle I^2 \rangle + \langle Q^2 \rangle)$, is the same as the output of the circuit of Figure 2. The other terms in the square contain double frequency components that average to zero. The components inside the dashed box in Figure 3 are equivalent to a simple bandpass filter. This can been seen by considering the overall response at any single frequency.

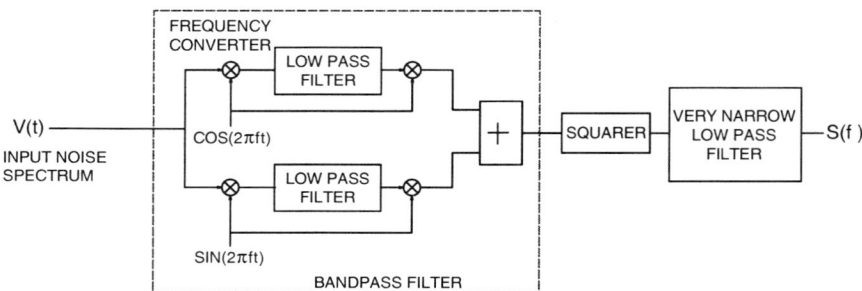

Figure 3. An alternate circuit, equivalent to the circuit of Figure 1. The circuitry in the dashed box is equivalent to a bandpass filter.

4. Properties of the Autocorrelation Function

Given that the Fourier transform of the autocorrelation function is, indeed, the power spectral density, consider the following properties: The "zero-lag" value of the autocorrelation function, $R(0)$, is just the average square of the voltage, i.e. the power. No lag will have an absolute value greater than the value of the zero lag. The autocorrelation function, $R(\tau)$, is real and even (as is the spectral density function) and contains all the information available in the signal. Higher moments of the received signal (which is a Gaussian process, as explained below), contain no additional information, because they can be expressed in terms of the second moment function, $R(\tau)$. For example,

$$\langle V(t)V(t-\tau)V(t-\tau')V(t-\tau'')\rangle = \langle V(t)V(t-\tau)\rangle\langle V(t-\tau')V(t-\tau'')\rangle + \langle V(t)V(t-\tau')\rangle\langle V(t-\tau)V(t-\tau'')\rangle + \langle V(t)V(t-\tau'')\rangle\langle V(t-\tau)V(t-\tau')\rangle . \quad (6)$$

5. Gaussian Nature of the Signal

In radio astronomy, the incoming signal will always be a random process whose amplitude has a zero mean (i.e. no dc) and whose amplitude probability distribution is Gaussian. That is, if we take a series of independent voltage samples (samples taken with adequate spacing between them) and plot a histogram, normalized to have unit area, the result will be:

$$f(V) = \frac{1}{\sqrt{2\pi\sigma^2}}e^{-\frac{V^2}{2\sigma^2}} , \quad (7)$$

where σ^2 is the average square of the voltage. This always-Gaussian amplitude distribution results from the fact that the signal voltage is always the superposition of voltages from a great number of individual physical radiators. The central limit theorem states that if a random process is a sum of random processes with any given amplitude distribution or distributions, then the sum process will have a Gaussian amplitude distribution. You probably have a computer program that can generate random numbers, x_i, uniformly distributed between -1 and 1. You can easily show that the average value of x^2 is $1/3$. Sum 100 of these numbers to create a new random variable, y. Do this sum a few thousand times and plot a histogram of the values of y. You will see that it has a Gaussian shape. The point is that you could have predicted this Gaussian by calculating the expected value of y^2 (the variance of y). To calculate $\langle y^2 \rangle$, write $\langle y^2 \rangle$ as a double sum over i and j. The only terms that have non-zero expectations are the terms where $i = j$. There are 100 such terms, so $\langle y^2 \rangle = 100 \langle x^2 \rangle = 100 \times 1/3 = 33.3$. Use this value, $\sigma^2 = 33.3$, to plot the Gaussian function of Equation 7 to see how closely it agrees with the histogram.

Since our signals always have a Gaussian amplitude distribution, it is not absolutely necessary that we average the square of the voltage in order to estimate the power. We might average the absolute value of the voltage, since its expectation (using Equation 6) is given by $(2/\pi)^{1/2}\sigma$. Or, we might simply measure the fraction of the time that the absolute value of the voltage exceeds some threshold, α, since the expectation of this fraction is easily calculated (again using Equation 6) as a function of σ. In this case, we are able to make an accurate estimate of the power, even though the samples of the signal have been quantized very coarsely to only three levels. As you might expect, it turns out that the autocorrelation function can also be accurately estimated when the signal has been coarsely digitized. The extreme case is two-level (one-bit) sampling, where only the polarity of the voltage is measured. In this case

$$R(\tau)/R(0) = \frac{\pi}{2} \sin \langle \text{sign} V(t) \, \text{sign} V(t - \tau) \rangle. \quad (8)$$

This equation, containing both the trigonometric sine function and the algebraic sign function was first derived by J.H. VanVleck (Van Vleck & Middleton 1966), who investigated the spectrum of clipped noise during WWII. Jamming transmitters could be much more efficient if they were allowed to clip (limit) the peaks of the transmitted noise. VanVleck showed that communication and radar receivers could be jammed as effectively with clipped Gaussian noise as with undistorted Gaussian noise.

The advantage of using coarse quantization is hardware simplicity. The disadvantage is the addition of quantization noise, requiring longer integrations to produce estimates of R with statistical fluctuations no greater than if precise many-bit quantization had been used. Another minor disadvantage, in the case of one-bit sampling, is that all amplitude information is lost, so the power, $R(0)$, must be estimated separately.

6. Hardware Autocorrelators

Evaluating a lagged product requires a multiplier and averager (adder and accumulator). If the signal has one-bit quantization, the one-bit x one-bit multiplier can be a simple exclusive-or logic gate and the adder/accumulator can be just a digital counter. A multichannel autocorrelator, as shown in Figure 4, consists of many multiplier/averager "lag units" plus a digital delay line so that each lag unit receives the present ("prompt") signal and a delayed signal. The delay line provides successively delayed versions of the prompt signal.

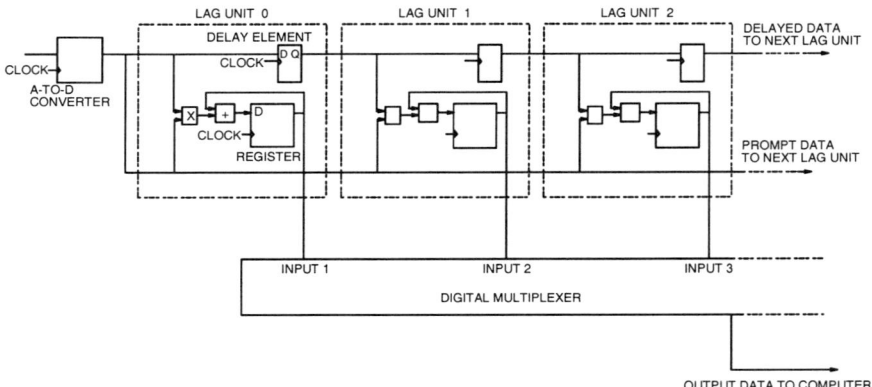

Figure 4.   Autocorrelator Block Diagram.

Three-level quantization (the digitizer output values can be assigned the values -1, 0 and $+1$) requires only a little more complexity and introduces less quantization noise than two-level quantization. Autocorrelator architecture is very simple, especially when coarse quantization is used. It can be expanded in several ways so that chips containing N lag units can be arranged to implement correlators with any number of lags and with overall effective clock speeds that are multiples of the chip's maximum clock speed. The FFT lacks this flexibility. Nevertheless, at some high number of spectral channels, the intrinsic higher efficiency of the FFT algorithm ($N \log N$ operations vs. N^2 operations, where N is the number of spectral bins) should make FFT spectrometers more economical than autocorrelation spectrometers.

Appendix: Autocorrelation function and spectrum of a pure sine wave

The case of an infinitely narrow spectral line, a sine wave, requires some care. Let the input voltage be a sine wave at frequency f_0: $V(t) = \sin(2\pi f_0 t)$. From Equation 1, it is readily shown that $R(\tau) = {}^1\!/_2 \cos(2\pi f_0 \tau)$. The power, $R(0)$, is therefore $^1\!/_2$, which was to be expected, since $\langle \sin^2(2\pi f_0 t) \rangle = {}^1\!/_2$. If we calculate the spectral density by substituting this autocorrelation function into Equation 2, we get infinity; the line is infinitesimally narrow so the spectral density at that point goes to infinity. To approach this infinity in an orderly way, we would give our sine wave some spectral width, via modulation controlled by an adjustable parameter. Then we would calculate the autocorrelation function and its transform, the spectral density. Finally we would adjust the parameter to narrow the line and find that the spectrum approaches a sum of two delta functions. This result is easy to guess: since the total power is $^1\!/_2$, the spectrum must be given by $S(f) = \delta(-f_0)/4 + \delta(f_0)/4$. If we substitute this spectrum in Equation 2, we do indeed get $R = {}^1\!/_2 \cos(2\pi f_0 \tau)$.

References

Thomas, J. B. 1969, An Introduction to Statistical Communication Theory, (New York: John Wiley & Sons Inc.)

Van Vleck, J. H., & Middleton, D. 1966, Proc. IEEE, 54, 2

Larry Solanch assists Carl Heiles to demonstrate the polarization properties of a telephone cord!

A Heuristic Introduction to Radioastronomical Polarization

Carl Heiles

Astronomy Department, University of California, Berkeley, California 94720-3411, USA

Abstract. Radio sources are often polarized. Accurate measurement of simply the flux density of a radio source requires a basic understanding of polarization and its measurement techniques. We provide an introductory, heuristic discussion of these matters with an emphasis on practical application and avoiding pitfalls.

1. Introduction

Many astronomers wish nothing more than to measure the total flux density of a source. Radio sources are often polarized, so even this basic measurement requires a basic understanding of polarization. Astronomers whose vision is so limited should read Sections 3.5. and 5.4., and then turn to some other activity.

Astronomers who are interested in astronomical magnetic fields and esoteric scattering geometries need to go further and measure polarization. Magnetic fields are a very important force in astrophysics, rivalling gravity and gas pressure in some regions such as interstellar space. Some extragalactic edge-on disks obscure light from the central black hole, but the highly polarized scattered light reveals not only the radiation from the black hole but also the properties of the scattering medium.

Synchrotron radiation is linearly polarized perpendicular to the magnetic field with fractional polarization typically $\sim 70\%$; pulsars are mainly linearly and partly circularly polarized; Faraday rotation, caused by the intervening magnetoionic gas, rotates the position angle of linear polarization; weak Zeeman splitting of spectral lines produces two circularly polarized components, and strong splitting also produces linear polarization; scattered spectral lines and continuum radiation are linearly, and sometimes weakly circularly, polarized.

The basic reference for our discussion of the fundamentals is the excellent book on astronomical polarization by Tinbergen (1996) and references therein. A more mathematical and fundamental reference is Hamaker, Bregman, & Sault (1996), which the theoretically-inclined reader will find of interest. Our discussion below will be heuristic in nature, avoiding proofs and mathematical detail. We will make several unproven statements and assertions; the explanations and justifications can be found in the abovementioned references. Practical details of calibration and application to real telescopes are in the Arecibo Technical and Operations Memo Series by Heiles and his collaborators (Heiles et al. 2000a and b) and, also, in a forthcoming set of articles in the PASP (Heiles et al. 2001a and b; Heiles 2001); all of these are listed in the references.

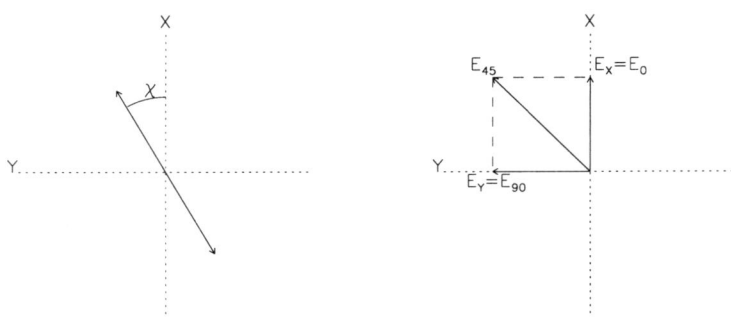

Figure 1. *(a, left)*, linear polarization at position angle χ; *(b, right)*, E_{45} is equivalent to the (non-vector) sum $(E_X + E_Y)/\sqrt{2}$.

2. The Jones Vector For the Electric Field

2.1. Polarization of an oscillating telephone cord

It is fun to take a long coiled telephone cord, tie one end to a fixed point, and wiggle the other end to excite standing waves. These waves are characterized by polarization, just as electromagnetic waves are.

If you wiggle back-and-forth vertically, it is vertically *linearly* polarized, and ditto with horizontally. Let us call these directions X and Y, as in Figure 1a. (Contrary to usual, we denote the horizontal direction with Y!). More generally, define the *position angle of polarization* χ to be measured from the vertical in the counterclockwise direction; then when you wiggle at angle χ, the position angle of linear polarization is also χ. The peak amplitude in the X direction is $A_{pX} = A_p \cos \chi$ and in the Y direction $A_{pY} = A_p \sin \chi$. So generally, your wiggling produces amplitudes in both directions. Of course, the stronger you wiggle, the larger the amplitude A_p. So your wiggling in linear polarization mode can be specified by two quantities, the amplitude and the position angle.

χ is periodic in π, not 2π like an ordinary angle. This means that linear polarization does not have a *direction*; rather, it has an *orientation*. Often one represents its position angle on the sky by short lines, like iron filings. These lines are sometimes called vectors—incorrectly, because vectors do have a direction.

You can be fancier and wiggle with a circular motion, and this can be either clockwise or counterclockwise. This is *circular polarization*, and the two directions are called left-hand and right-hand circular. Here, too, you have amplitudes in both directions.

In what basic essence does the circular mode differ from the linear mode at $\chi = 45°$? It is not that the X and Y directions have different *amplitudes*. Rather, they have different *phases*. Specifically, suppose you describe your vertical wiggle with $A_{pX} = A_p \cos(\omega t)$; then you would describe the horizontal wiggle with $A_{pY} = A_p \sin(\omega t)$. One is a cosine, the other a sine. Now, there is only one difference between a cosine and a sine: it is the phase angle. Remember the trig identity

$$\sin(\omega t) = \cos\left(\omega t - \frac{\pi}{2}\right). \tag{1}$$

In other words, you can think of the Y and X amplitudes as being identical in all respects except that they differ in phase by $90°$. The phase angle can be positive or negative: positive produces clockwise rotation, negative counterclockwise. More generally, the X and Y amplitudes can differ in phase by an arbitrary angle ϕ; when $\phi \neq 0°$ or $90°$, we have *elliptical* polarization. Elliptical polarization can also be described by different X and Y amplitudes.

2.2. Vectorial representation with trig notation for time

When you wiggle with linear polarization, you can describe X and Y amplitudes with identical trig functions, like $A_Y(t) = A_{pY}\cos(\omega t)$ and $A_X(t) = A_{pX}\cos(\omega t)$. It is convenient to ignore the time dependence and write the peak amplitudes using vector notation:

$$\mathbf{A_p} = \begin{bmatrix} A_{pX} \\ A_{pY} \end{bmatrix} \tag{2}$$

and in this case, with $A_{pX} = A_p \cos\chi$ and $A_{pY} = A_p \sin\chi$, we write

$$\mathbf{A_p} = A_p \begin{bmatrix} \cos\chi \\ \sin\chi \end{bmatrix}. \tag{3}$$

The meaning of this is perfectly clear without explicitly writing the time dependence. However, suppose we are dealing with circular polarization. Then, with ordinary trig notation, we need to explicitly include the time dependence and write

$$\mathbf{A(t)} = \begin{bmatrix} A_{pX}\cos(\omega t) \\ A_{pY}\cos(\omega t + \phi) \end{bmatrix}, \tag{4}$$

which is awkward: our description of the wiggling does not really need the time dependence, which only specifies in a cumbersome way that the motion is periodic with a certain frequency. What is really important is the phase angle. To reflect this, often an equation like the above is simplified by writing

$$\mathbf{A_p} = \begin{bmatrix} A_{pX} \\ A_{pY}\angle\phi) \end{bmatrix}. \tag{5}$$

The meaning of this is clear, but mathematically one cannot manipulate the angle sign so it is not useful in a mathematical sense.

2.3. Complex exponential notation

Complex exponential notation is handy because it eliminates the need to explicitly write the time dependence in terms of (ωt), it allows one to specify phase angles, and the notation can be mathematically manipulated. We recall that the complex plane has a real and imaginary axis. The cosine and sine functions are the projections of a complex exponential on the real and imaginary axis, respectively. We have

$$e^{i\theta} = \cos\theta + i\sin\theta, \tag{6}$$

which leads to

$$\cos\theta = Re\ [e^{i\theta}], \tag{7}$$

$$\sin\theta = Im\ [e^{i\theta}] = Re\ [-ie^{i\theta}] \tag{8}$$

and, most importantly it allows the *addition* of a phase angle to be replaced by the *multiplication* of its complex exponential,

$$\cos(\theta + \phi) = Re\ [e^{i(\theta+\phi)}] = Re\ [e^{i\theta}e^{i\phi}], \tag{9}$$

$$\sin(\theta + \phi) = Re\ [-ie^{i(\theta+\phi)}] = Re\ [-ie^{i\theta}e^{i\phi}]. \tag{10}$$

This last is important for our purposes because we can use exponential notation to replace Equation 4 with

$$\mathbf{A} = \begin{bmatrix} A_{pX} \\ A_{pY}e^{i\phi} \end{bmatrix}, \tag{11}$$

which tells the essence, namely that A_Y lags A_X by phase ϕ.

All aspects of the above discussion carry over to the electric field in electromagnetic waves; we simply replace the amplitude A with the electric field E. Its vector representation is called the *Jones vector*.

3. Polarized Power and the Stokes Vector

In astronomy we are almost never interested in the electric field because we measure the *power*. Power is the time average of the square of the electric field. Or, rather, the time average of the product of the electric field with its complex conjugate; this takes care of any difficulties with phases.

3.1. How many parameters are required?

We made a heuristic argument above that the polarization is specified by three parameters, namely the three in Equation 11. However, there is one additional wrinkle. We were describing a 100% polarized wave in which the amplitude has a single frequency and, correspondingly, a single polarization mode, which is generally elliptical. Any monochromatic wave exists forever, so its polarization never changes.

However, there can also exist a superposition of sine waves of different frequencies within some bandwidth. In fact, this is *always* the case of natural radiation like sunlight, the 21-cm line, or even astronomical masers. In nature, they are packed tightly together in frequency with infinitely small separations. They produce an electric field that varies randomly with time. These fields can all be polarized in the same sense, just like a monochromatic wave.

But the polarizations can also be randomly distributed with frequency. In this case we have unpolarized radiation. The Jones vector, which treats a single sine wave and has only three parameters, is not adequate to describe this case.

This extra possibility, that the electric field can have a time-random unpolarized component, turns the three parameters into four: the fourth tells the fraction of power that is nonpolarized. These four parameters are combinations of polarized power called *Stokes parameters*.

Stokes parameters are linear combinations of power measured in *orthogonal polarizations*. We measure power in a particular polarization by constructing an antenna that responds to that polarization, meaning that the incoming electric field generates a corresponding voltage in a cable. With a radio astronomy "dish", the antenna is called a *feed*. Here we will think of a feed's antenna as a probe in a waveguide that extracts linear polarization. More generally, feeds can be made to sample linear, circular, or even arbitrary elliptical polarization.

We describe radiation in terms of electric fields having particular polarizations; what we mean is that we have constructed an antenna sensitive to that polarization and when we write the electric field E we really mean voltage in the cable that was induced by the field.

3.2. Linear polarization: Stokes Q and U

It is intuitively obvious that orthogonal *linear* polarizations have χ differing by $90°$: for example, vertical (X) and horizontal (Y) polarizations are orthogonal and have $\chi = (0°, 90°)$, respectively.

The power in the (X, Y) directions is just (E_X^2, E_Y^2), respectively[1]. We can linearly combine these two powers by adding and subtracting them, and in the process we generate the first two Stokes parameters I and Q:

$$I = E_X^2 + E_Y^2 = E_{0°}^2 + E_{90°}^2 , \qquad (12)$$

$$Q = E_X^2 - E_Y^2 = E_{0°}^2 - E_{90°}^2 . \qquad (13)$$

Let us reflect on these quantities for a moment.

The sum: Stokes I The *sum* represents the *total power* in the incoming radiation. It makes intuitive sense that, by sampling two orthogonal polarizations, you pick up all of the incoming power. It may not make intuitive sense, but is

---

[1] To be more precise, one must realize that the E's have phases and are therefore complex, so the proper expressions for power are $(\langle E_X \overline{E}_X \rangle, \langle E_Y \overline{E}_Y \rangle)$, where the $\langle \rangle$ denotes time averages and the bar denotes complex conjugate. We ignore these complications in this introductory portion.

nevertheless true, that it does not matter *which* two orthogonal polarizations you measure and add together. The orthogonal circulars, or two orthogonal linears at any pair of angles $(\chi, \chi + 90°)$, or even orthogonal ellipticals always sample all of the power and their summed powers always gives the total intensity I.

This is easy to see for the particular case of linear polarization at $\chi = 45°$. You can express $E_{0°}$ in terms of $(E_{45°}, E_{-45°})$ (see Figure 1b):

$$E_{0°} = \frac{E_{45°} + E_{-45°}}{\sqrt{2}}, \tag{14}$$

$$E_{90°} = \frac{E_{45°} - E_{-45°}}{\sqrt{2}} \tag{15}$$

and when you take the sum of the squares, you find—naturally enough—that $E_{0°}^2 + E_{90°}^2 = E_{45°}^2 + E_{-45°}^2$.

The difference: Stokes Q The difference tells about the polarization. Suppose the incoming electric field is vertically polarized (the X direction); then $Q = I$. If it is horizontally polarized, then equation 13 says $Q = -I$. If it is coming in at $\chi = 45°$, then $Q = 0$. In fact, more generally,

$$\frac{Q}{I} = p_{QU} \cos(2\chi), \tag{16}$$

where p_{QU} is the total fractional linear polarization (which we discuss below). So Q is a very valuable diagnostic of the linear polarization! But also it is *not complete*: for example, for $\chi = 45°$ we have $Q = 0$ and, with this parameter alone, we would not suspect that the signal is polarized.

Another difference: Stokes U We need one more parameter to completely define the linear polarization. That parameter is Stokes U, and is equal to

$$U = E_{45°}^2 - E_{-45°}^2. \tag{17}$$

With a little reflection it becomes clear that

$$\frac{U}{I} = p_{QU} \sin(2\chi). \tag{18}$$

The combination (Q, U) completely specifies the linear polarization of the signal. The combination $(Q^2 + U^2)^{1/2}$ is the total linearly polarized power and is independent of χ. Generally, even for a partially polarized signal, the fraction of linear polarization and its position angle are given by

$$p_{QU} = \left[\left(\frac{Q}{I}\right)^2 + \left(\frac{U}{I}\right)^2 \right]^{1/2}, \tag{19}$$

$$\chi = 0.5 \tan^{-1} \frac{U}{Q}. \tag{20}$$

3.3. Circular polarization: Stokes V

There are only two circular polarizations, which are called right- and left-handed, or RCP and LCP, and they are orthogonal. So one can derive two Stokes parameters from them: one is I, which is the same as discussed above; you can include the 90° phase difference in Equations such as 14 and 15 to prove this for yourself.

The difference is Stokes V. If you ever work with circular polarization, you have to be careful about sign. Physicists use one definition, radio astronomers use another (the IEEE definition, reflecting our EE heritage), and optical astronomers use both, sometimes without bothering to specify exactly which they are using. The IEEE definition is

$$V = E_{LCP}^2 - E_{RCP}^2 . \tag{21}$$

LCP is generated by transmitting with a left-handed helix, so from the transmitter the E vector appears to rotate counterclockwise. From the receiver, LCP appears to be rotating clockwise.

We define the fractional circular polarization just as we do for linear polarization:

$$p_V = \frac{V}{I} . \tag{22}$$

V can be positive or negative, and one can retain the sign in the definition of p_V if one wishes, as we have done here.

3.4. The Stokes vector and total polarized power

We have four Stokes parameters, and it will be convenient to write them in vector format, the 4-element *Stokes vector*

$$\mathbf{S} = \begin{bmatrix} I \\ Q \\ U \\ V \end{bmatrix} . \tag{23}$$

The total fractional polarization is just

$$p = \left[\left(\frac{Q}{I}\right)^2 + \left(\frac{U}{I}\right)^2 + \left(\frac{V}{I}\right)^2 \right]^{1/2} . \tag{24}$$

If both p_{QU} and p_V are nonzero, then the polarization is *elliptical*, which is the general situation. Every elliptical polarization has its orthogonal counterpart, and one can even build an elliptically polarized feed. One normally prefers pure linear or circular and tries to avoid the intermediate cases. However, Arecibo uses turnstile junctions for some receivers, which have the advantage that the polarization can be adjusted to pure circular with exquisite accuracy—but the polarization becomes increasingly elliptical, changing to linear and back again, with increasing departure from the design frequency (see Heiles et al. 2000b, 2001b)!

3.5. If you do not remember anything else, remember THIS!

Often you find yourself needing to combine polarizations. For example, if you measure the polarization of some object several times, you need to average the results.

There is only one *proper* way to combine polarizations, and that is to use the Stokes parameters. The reason is simple: because of conservation of energy, powers add and subtract. But it is definitely wrong to average fractional polarizations or angles. Consider that fractional polarizations are always positive, so they can never average to zero. And angles are even worse! Consider averaging two measurements having angles of $0°$ and $179°$—angles that differ by only $1°$ because of the π periodicity of position angle. The straight average of the angles gives about $90°$—the orthogonal polarization!

What you must do is convert the fractional polarization and position angle to Stokes parameters, average the Stokes parameters, and convert back to fractional polarization and position angle.

4. Measuring Stokes Parameters in Radio Astronomy

In contrast to optical astronomers, radio astronomers can measure all Stokes parameters *simultaneously*. It may not be obvious how they do this: we have described Stokes parameters as differences between powers in various pairs of orthogonal polarizations, each pair "belonging" to a particular Stokes parameter, and we cannot simultaneously place six feed probes at the same physical location to simultaneously measure $(E_{0°}, E_{90°}, E_{45°}, E_{-45°}, E_{LCP}, E_{RCP})$ because all these antennas would interact with each other and make a total mess. Fortunately, we can generate a Stokes parameter not only by subtraction of its own orthogonal polarizations, but also by *multiplying* electric fields of two orthogonal polarizations belonging to a *different* Stokes parameter[2].

4.1. Example: generating Stokes U from $(E_{0°}, E_{90°})$

This is easy to see for the case of deriving Stokes U (which is $E_{45°}^2 - E_{-45°}^2$) from its non-belonging brethren $(E_{0°}, E_{90°})$. Referring again to Figure 1b, it is clear, graphically speaking, that the product $(E_{0°} E_{90°})$ is related to Stokes U. As the E field at $\chi = 45°$, which belongs to Stokes U, oscillates, it induces in-phase fields in the $0°, 90°$ directions, each smaller by a factor of $\sqrt{2}$. The E-fields in these two directions are therefore *correlated*. If you measure the time average product $\langle E_{0°} E_{90°} \rangle$, it is identical to $\frac{\langle E_{45°}^2 \rangle}{2}$.

In particular, if you begin with Equations 14 and 15 you can easily show that

$$U \equiv E_{45°}^2 - E_{-45°}^2 = 2 E_{0°} E_{90°} . \tag{25}$$

---

[2] We do not have to multiply; we can add and subtract, as in Figure 1b. But then we lose the advantage of crosscorrelation discussed in Section 5.2.

If you throw in a phase factor of 90° in the above equation, you will recover Stokes V—this makes sense because the only difference between linear and circular polarization is, in fact, the phase factor.

4.2. A general expression for Stokes parameters

One can write Stokes parameters in terms of electric fields of any two orthogonal polarizations. Here we provide the version in which one measures (E_X, E_Y) with linearly polarized antennas at $\chi = 0°$. For this case,

$$I = E_X \overline{E_X} + E_Y \overline{E_Y} , \qquad (26)$$

$$Q = E_X \overline{E_X} - E_Y \overline{E_Y} , \qquad (27)$$

$$U = E_X \overline{E_Y} + \overline{E_X} E_Y , \qquad (28)$$

$$iV = E_X \overline{E_Y} - \overline{E_X} E_Y . \qquad (29)$$

The overbar indicates the complex conjugate. These products are time averages; we have omitted the indicative $\langle \rangle$ brackets to avoid clutter. And remember, these equations apply to the voltages induced into the antenna as well as to the original electric fields, because they are proportional; below, we are always referring to voltages even though we will write E.

These equations make it clear how to measure all four Stokes parameters simultaneously. Namely, begin with orthogonal polarizations; any pair will do, but our equations are written for orthogonal linears. Then digitize the resulting voltages and perform the above products in a computer. The aspects of digitizing and computing are a story all in themselves, but we leave that for another time.

4.3. The need for calibration

The above Equations 26-29 are simple in theory but not so simple in practice because the radioastronomical receiving system produces unwanted modifications in the astronomical polarization. Feeds are almost never perfect, so their polarizations are only approximately linear or circular; and generally speaking, no feed has two outputs that are perfectly orthogonal.

Most important in practice is the electronics system, which introduces its own relative gain and phase differences between the two linearly polarized channels. Figure 2a shows the important elements of the system for this discussion. Two orthogonal feed probes in a waveguide convert the incoming electric field to voltages. These travel through cables having lengths (L_X, L_Y) to a directional coupler, where the correlated noise source is injected (through cables of different length—a fact we ignore for the sake of simplicity). (L_X, L_Y) cannot be exactly identical, and the difference produces a phase shift between the two polarization channels. The two polarization channels are amplified with gains that are inherently complex, meaning that each introduces its own phase shift. The signals are multiplied in a mixer by a local oscillator, again injected through cables whose lengths cannot be precisely equal, leading to an additional phase offset between the channels. The resulting i.f. signals are sent down to the digital correlator through cables from the feed; these also have different lengths and losses.

The total gains in the (left, right) channels are $(G_X G_A, G_Y G_B)$. If the *magnitudes* of these gains differ, then the difference between the two channels

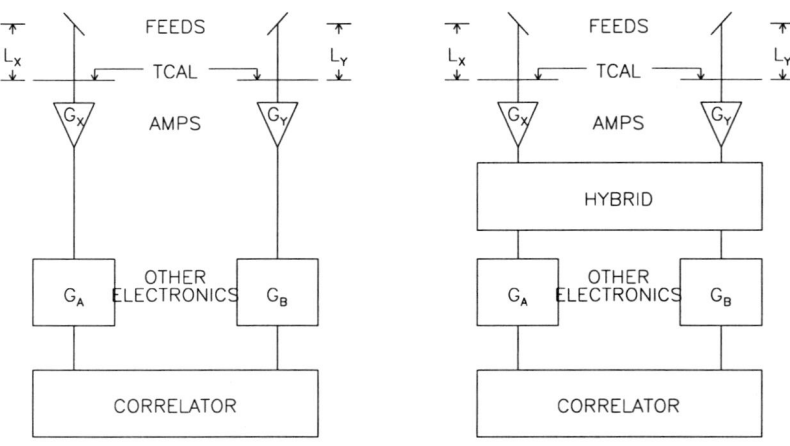

Figure 2. Two versions of a radio astronomical receiver: *(a,left)*, **good**; *(b, right)*, **bad** (see Sections 5.3. and 5.4.).

is nonzero for an unpolarized source, making the source appear to be *linearly* polarized. If the *phases* differ, then a linearly polarized source appears to be partially *circularly* polarized. These electronics gains and phases must be calibrated relatively frequently because they can change with time. This calibration is most effectively done by injection of a correlated noise source into the two feed outputs.

5. Calibration: Jones and Mueller Matrices

The *modification* of the derived polarization by the system components is most generally described by matrices. The *Jones matrix* describes the modification of the Jones vector, and the *Mueller matrix* describes the modification of the Stokes vector.

Return to our wiggling telephone cord. Suppose we wiggle one end with linear polarization in the vertical direction, making $Q = I$ and $(U, V) = 0$. Now a Martian spy starts to move the other end around in a small circle at the same frequency. This would change the original pure vertical linear polarization into elliptical polarization, taking some power from Q and putting it into U and V.

Feeds and electronic devices are like the Martian, modifying the electric field's polarization. Generally, a device can couple an arbitrary fraction of the Y voltage into X with an arbitrary phase; and also *vice-versa*. The easy way to write these mutual perturbations is with a matrix transfer function that relates the output voltages to the input voltages. This matrix operates on the

Jones vector, so it is naturally called the *Jones matrix*. With two orthogonal polarizations, there are two voltages; thus the Jones matrix is 2×2.

If the Jones matrix is unitary, then it produces no modification in the original polarization. Unitary matrices are not very interesting in astronomy, because we need lots of amplification for the tiny voltages induced at the feed! But we can imagine a system in which the Jones matrix is diagonal; that would mean there is no coupling between the X and Y channels. Large diagonal elements would increase the voltages so that we can measure them, and if the two diagonal elements were equal, it would also keep the polarization—and thus the Stokes parameters—unchanged.

More generally, the Jones-matrix-modified voltages change the polarization state, and thus change the Stokes parameters. So every Jones matrix has its Stokes-parameter counterpart, which is called the *Mueller matrix*. The Mueller matrix relates the output Stokes vector of Equation 23 to the input Stokes vector:

$$\mathbf{S_{out}} = \mathbf{M} \cdot \mathbf{S_{in}} \, . \tag{30}$$

There are four Stokes parameters, so the Mueller matrix is 4×4; in general, all elements may be nonzero, but they are not all independent. In the usual way, we write

$$\mathbf{M} = \begin{bmatrix} m_{II} & m_{IQ} & m_{IU} & m_{IV} \\ m_{QI} & m_{QQ} & m_{QU} & m_{QV} \\ m_{UI} & m_{UQ} & m_{UU} & m_{UV} \\ m_{VI} & m_{VQ} & m_{VU} & m_{VV} \end{bmatrix} . \tag{31}$$

Each matrix parameter is the coupling of the two Stokes parameters indicated by its subscripts.

Each system component has its own Jones and Mueller matrices, and the total system matrices are the products of the individual component's matrices. The Mueller matrix for the feed has complicated off-diagonal elements. Fortunately, feeds are usually well-designed and these off-diagonal elements are small, meaning the instrumental polarization is small.

Here, we will restrict our detailed discussion to the two most important specific components, the electronics chain and the coupling of the telescope to the sky; fortunately they are also the simplest. For a complete treatment that includes the feed, see Heiles et al. (2000a, 2000b).

5.1. The Jones and Mueller matrices for the electronics chain

The two polarization channels go through different amplifier chains as in Figure 2a, and we can safely assume that there is no coupling between the channels in the electronics system; this, in turn, means that the nondiagonal elements of the Jones matrix are zero. Suppose the two channels have *voltage* gain (g_A, g_B), *power* gain $(G_A, G_B) = (g_A^2, g_B^2)$, and phase delays (ψ_A, ψ_B). Clearly, the Jones matrix is

$$\begin{bmatrix} E_{A,out} \\ E_{B,out} \end{bmatrix} = \begin{bmatrix} g_A e^{i\psi_A} & 0 \\ 0 & g_B e^{i\psi_B} \end{bmatrix} \begin{bmatrix} E_{A,in} \\ E_{B,in} \end{bmatrix} . \tag{32}$$

You calculate the Mueller matrix $\mathbf{M_A}$ from the Jones matrix by the laborious procedure of applying Equations 26-29. The result is surprisingly complicated. However, for a well-designed and calibrated system, we can assume that the two gains (g_A, g_B) are nearly identical, and for simplicity assume that their average is unity. That is, we can assume that

$$\Delta G \equiv G_A - G_B \ll 1. \tag{33}$$

We then carry the algebra only to first order in $g_A g_B$, meaning we take $g_A g_B = 1$. With this first-order approximation, the electronics Mueller matrix becomes

$$\mathbf{M_A} = \begin{bmatrix} 1 & \frac{\Delta G}{2} & 0 & 0 \\ \frac{\Delta G}{2} & 1 & 0 & 0 \\ 0 & 0 & \cos\psi & -\sin\psi \\ 0 & 0 & \sin\psi & \cos\psi \end{bmatrix}. \tag{34}$$

Here we have set $\psi \equiv \psi_A - \psi_B$: the difference is all that matters because it is always the *relative* phase between the two channels that counts. The matrix consists of two submatrices, the upper left and the lower right. Let us reflect on these submatrices.

The upper left submatrix represents coupling between Stokes I and Q. The relative gain error directly affects these two parameters because they are the sum and difference of the X and Y powers. In contrast, these powers are completely unaffected by the phase difference, so ψ does not appear in this submatrix. The two diagonal elements are unity because we have assumed $\Delta G \ll 1$. Consider the specific example of an unpolarized source: a gain difference makes it look polarized, with Q nonzero, and it also affects Stokes I.

The lower right submatrix represents coupling between Stokes U and V. The relative phase directly affects these two parameters because they are the correlated products $E_X E_Y$, and the departure of the correlation's phase angle from zero directly reflects the degree of circular polarization, Stokes V.

5.2. A very important property of correlated voltages

Embodied in Equation 34 is a highly important principle: *gain errors do not affect correlation products when there is no polarization*. This is important because amplifier gains fluctuate with time and their calibration is subject to measurement error; in contrast, amplifier phase delays and cable lengths tend to change only slowly with time.

Consider the case of a nonpolarized source. Equation 34 shows that the error in the measured Stokes Q is directly proportional to ΔG. However, Stokes U and V are completely independent of ΔG, so a nonpolarized source *cannot* produce fake nonzero (U, V). This is because if $E_X E_Y = 0$, as it is for an unpolarized source, then the product is zero even with a gain error.

In practice, this makes the measurement of small polarizations more accurate when using correlated products. This fundamental fact appears again and again in precision radioastronomical measurements of small quantities, and is also the basis for interferometry. The corollary is the somewhat nonintuitive fact: *If you want to measure small circular polarizations accurately, then use* linearly polarized feeds; if you want to measure small linear polarizations accurately, then use circularly polarized feeds!

5.3. Carrying correlation too far: using a hybrid

Suppose you want to measure weak linear polarization. As we discussed above, the best technique is to crosscorrelate orthogonal circulars. But suppose your telescope only has a linearly polarized feed!

You may be tempted to modify the system block diagram to include a 90° hybrid, as shown in Figure 2b. The hybrid inserts a 90° phase shift into one channel and then adds them. This turns the dual linear system [polarizations (X, Y)] into a dual circular one [polarizations (A, B)]. You can then use the crosscorrelation technique to generate the dual linears.

However, this system does *not* provide the abovementioned advantage of crosscorrelation in measuring small signals. The reason is simple: the (X, Y) signals are combined *after* the first amplifier and have been multiplied by the gains (G_X, G_Y) with their corresponding uncertainty and time variability. Thus, the combined (A, B) signals are pure circular polarization only to the extent that $G_X = G_Y$—and, of course, this includes their complex portions, the phases, as well. And we are ignoring the inevitable imperfections in the hybrid. After the hybrid, the complex channel gains (G_A, G_B) operate on the signal, as usual.

We now have *four* combinations of gain to worry about: $(G_X G_A, G_X G_B, G_Y G_A, G_Y G_B)$. In contrast, the straightforward system without the hybrid has only two combinations: $(G_X G_A, G_Y G_B)$. This makes the calibration process for the hybrid system more complicated, requiring turning on the correlated cal not only when it is connected to both channels simultaneously but also when it is connected to each one individually, one at a time. The details are discussed by Heiles & Fisher (1999).

The hybrid *completely* removes the cross correlation advantage: with the hybrid, there is *no* Stokes parameter that is unaffected by ΔG.

5.4. Stokes I and the hybrid

The hybrid even creates problems measuring Stokes I because of the more complicated calibration procedure described above. Unfortunately, however, astronomers rather traditionally prefer to measure Stokes I using dual circular polarization instead of dual linear. The reason is partly scientific, partly historical. Historically, the first receiving systems had only a single polarization channel: it was hard and expensive enough to make a single low-noise receiver, let alone two. Classical radio sources, such as quasars, often exhibit significant linear polarization, but very little circular. Thus, to obtain reliable, repeatable flux density measurements—particularly at low frequency, where ionospheric Faraday rotation is important—the single polarization of choice was circular. Similarly, in historic single-polarization VLBI with its differing ionospheric Faraday rotations at the different stations, circular polarization was preferred. And finally, pulsars are more highly linearly than circularly polarized. When faced with a single polarization system and sources that are linearly polarized, the polarization of choice is circular because one needs only to multiply the measured flux by two to get Stokes I.

This traditional emphasis on circular polarization persists in dual-polarized receivers. Many astronomers who want to measure nothing more than Stokes I, when faced with a dual-linear system, insert a hybrid to convert the system to dual-circular. But they do not carry through with the extra steps of calibration required.

To bring the point home that using a hybrid is inappropriate, consider the extreme case when the Y-polarization amplifier is turned off. The astronomer who uses a hybrid points the telescope to the source of interest and sees both channels (A, B) respond. Then he[3] turns on the cal and sees both channels respond. He has *no idea* that one channel is dead. He might wonder why the levels are 3 db lower than usual, but astronomers usually do not pay attention to power levels, ascribing them to the domain of the receiver engineer.

In historical times, radio astronomers often did use a hybrid to generate the circular polarization(s) from a dual linear feed, but placed the hybrid *before* the first amplifier. This is far better, because then one is reduced to the simpler situation of having only two sets of gains to determine, $(G_X G_A, G_Y G_B)$. The problem with this approach is that hybrids have some loss, and therefore introduce noise. In those historical times receiver temperatures were high enough that this extra noise could be tolerated. Today's receiver temperatures are too low for this approach unless the hybrid is cooled.

The moral: *do not ever use a post-amp hybrid unless you* **really need** *to change the polarization for a special, specific purpose!* To measure Stokes I, use the native feed polarization; calibrate and measure the two polarization channels independently, and add the results.

5.5. The Mueller matrix for the sky

A linearly polarized astronomical source has Stokes (Q_{src}, U_{src}) parameters defined with respect to the north celestial pole (NCP). It is these quantities that we want to measure.

The first device encountered by the incoming radiation is the telescope ($\mathbf{M_{SKY}}$). These days, all major telescopes are alt-az mounted. This means that the feed mechanically rotates with respect to the sky as the dish tracks the source. The angle of rotation is called the *parallactic angle* PA_{az}. It is defined to be zero at azimuth 0° and increase towards the east; for a source near zenith, which is always the case at Arecibo, $PA_{az} \sim az$, where az is the azimuth angle of the source. The Stokes parameters seen by the feed are (Q_{sky}, U_{sky}); the conversion between $\mathbf{S_{src}}$ and $\mathbf{S_{sky}}$ is given by the Mueller matrix[4]

$$\mathbf{M_{SKY}} = \begin{bmatrix} 1 & 0 & 0 & 0 \\ 0 & \cos 2PA_{az} & \sin 2PA_{az} & 0 \\ 0 & -\sin 2PA_{az} & \cos 2PA_{az} & 0 \\ 0 & 0 & 0 & 1 \end{bmatrix}. \qquad (35)$$

---

[3]Sexism here is intentional: female astronomers are presumably smart enough to avoid such foolishness.

[4]One can also write the corresponding Jones matrix, should one so desire.

The central 2×2 submatrix is, of course, nothing but a rotation matrix. $\mathbf{M_{SKY}}$ does not change I or V, which makes sense.

For an equatorially-mounted telescope, the feed *does not* rotate on the sky as the source is tracked. This fact might still be of interest to optical astronomers, but with the demise of the last of the great equatorial telescopes—the NRAO 140-footer—this fact recedes into the fog of history for us radio astronomers.

5.6. The total system Mueller matrix

Heiles et al. (2000a, 2001a) define seven parameters that specify the complete system Mueller matrix. Two of these refer to the cal gain and phase with respect to the sky (Equation 34 above), four refer to the ellipticity and nonorthogonality of the feed, and the seventh is a rotation angle. In contrast, the 4×4 Mueller matrix contains sixteen elements; not all of the elements are independent. For illustrative purposes, we write the full system matrix, without the sky correction, in terms of the first five parameters:

$$\begin{bmatrix} 1 & (-2\epsilon \sin\phi \sin 2\alpha + \frac{\Delta G}{2} \cos 2\alpha) & 2\epsilon \cos\phi & (2\epsilon \sin\phi \cos 2\alpha + \frac{\Delta G}{2} \sin 2\alpha) \\ \frac{\Delta G}{2} & \cos 2\alpha & 0 & \sin 2\alpha \\ 2\epsilon \cos(\phi + \psi) & \sin 2\alpha \sin\psi & \cos\psi & -\cos 2\alpha \sin\psi \\ 2\epsilon \sin(\phi + \psi) & -\sin 2\alpha \cos\psi & \sin\psi & \cos 2\alpha \cos\psi \end{bmatrix}.$$

$\mathbf{\Delta G}$ is the error in relative intensity calibration of the two polarization channels. It results from an error in the relative cal values (T_{calA}, T_{calB}).

ψ is the phase difference between the cal and the incoming radiation from the sky (equivalent in spirit to $L_X - L_Y$ on our block diagram.

α is a measure of the voltage ratio of the polarization ellipse produced when the feed observes pure linear polarization.

ϵ is a measure of imperfection of the feed in producing nonorthogonal polarizations (false correlations) in the two correlated outputs.

ϕ is the phase angle at which the voltage coupling ϵ occurs. It works with ϵ to couple I with (Q, U, V).

6. Calibrating and Using the Matrix Parameters

6.1. The role of the correlated cal

In practice, the amplifier gains and phases are calibrated with a correlated noise source (the "cal"). Thus, our amplifier gains (G_A, G_B) in Equation 34 have nothing to do with the actual amplifier gains. Rather, they represent the gains as calibrated by specified cal intensities, one for each channel. If the *sum* of the specified cal intensities is perfectly correct, then the absolute intensity calibration of the instrument is correct for an unpolarized source (i.e. Stokes I is correctly measured in absolute units). Above, we have assumed $G_A + G_B = 2$, which means that we are dealing with fractional polarizations and neglecting the absolute calibration of intensity.

The difference between the amplifier *phases* is also referred to the cal. Thus the phase difference $\psi = \psi_A - \psi_B$ represents the phase difference that exists between a linearly polarized astronomical source and the cal and has nothing to do with the amplifier chains.

We assume the cal to be constant. Thus, the measured Stokes vector is referred to the cal. This means that all artifacts of the electronics chain, which change with time, are removed by referring the measured Stokes vector to the cal, which is constant in time.

It remains to relate the cal to the sky. This must be done by astronomical observations that determine the Mueller matrix of the calibrated system. In other words, the calibrated system's Mueller matrix multiplies the incoming Stokes vector from the sky and produces the measured Stokes vector.

6.2. Determining the Mueller matrix astronomically

Astronomical radio sources exhibit signficant linear polarization but usually negligible circular polarization. We determine the matrix astronomically by tracking a linearly polarized radio source over a wide range of parallactic angle PA. As PA changes, Stokes Q and U from the source are modified by $\mathbf{M_{SKY}}$ in Equation 35. In contrast, any PA dependence of the measured Stokes V must reveal nonzero matrix elements coupling Stokes (Q, U) into V, namely (m_{QV}, m_{UV}) and their two counterparts.

Figure 3 shows a calibration observation for Arecibo's dual-linear LBW feed. The crosses show the PA dependence of $X - Y$—the difference between the two linears, which is the measured Stokes Q; the diamonds show XY, the measured Stokes U. They follow sine and cosine waves with comparable amplitudes, as they should (Equations 16, 18). The smallness of the departure from these conditions is expressed by the tiny values for (m_{QU}, m_{UQ}). The amplitude of the sine/cosine curves gives the linear polarization of the source, $\sim 7\%$.

However, the sine wave for the crosses is displaced above zero by about 0.06. This reflects coupling of Stokes I into Q, m_{QI}; this is the effect of nonzero ΔG, an error in the relative cal values. In contrast, m_{UI} is very small, consistent with its derivation from crosscorrelation; the fact that $m_{UI} \neq 0$ reflects cross coupling in the feed, which is described by the parameter ϵ above.

Finally, Stokes V exhibits a small PA dependence, which indicates nonzero values for (m_{UV}, m_{QV}); this results from an error in the relative phase of the cal with respect to the sky, $\Delta \psi$. The small departure of Stokes V from zero could result either from nonzero m_{VI} or from nonzero circular polarization of the source; one needs to observe several sources to be sure.

6.3. Using the matrix to correct data

Once the system's matrix parameters are determined it is a simple matter to correct the data: one simply multiplies the measured Stokes vector by the inverse of the system Mueller matrix. This must, of course, include the sky rotation portion. Matrices are noncommutative and you have to be careful about constructing the system Mueller matrix; details are in Heiles et al. (2000a, 2001a).

6.4. Jones and Mueller matrices for antenna arrays

When using an array of antennas, such as the VLA, the output of each interferometer pair is equivalent to the output of the single dish described above. The pair's measured Stokes parameters can be corrected by a Mueller matrix,

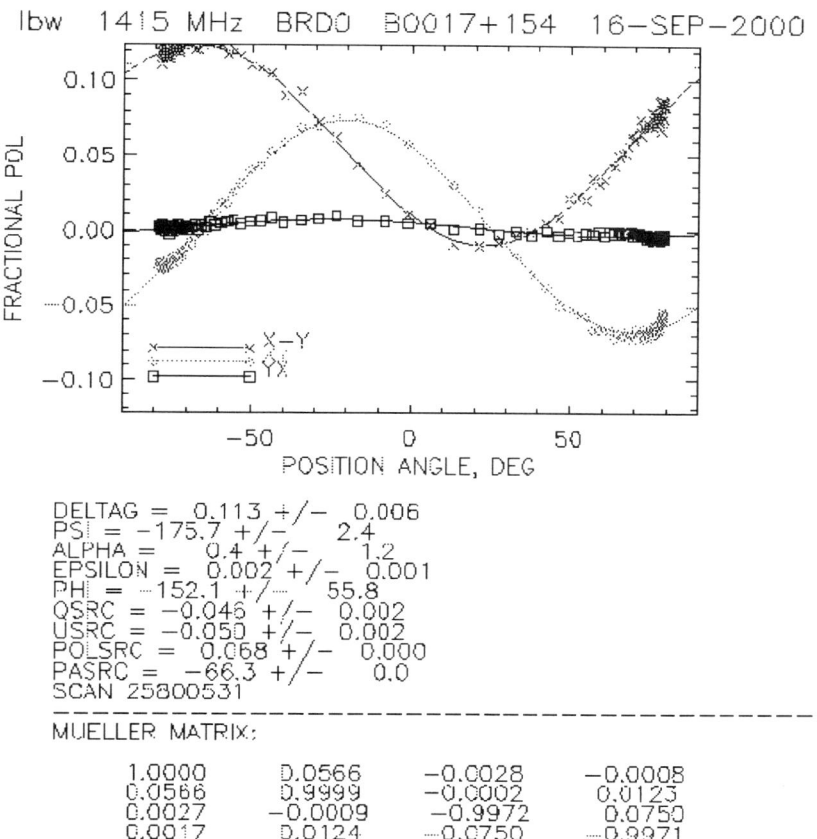

Figure 3. A calibration observation of the source 0017 + 154 for Arecibo's dual-linear LBW feed. At the top, the measured Stokes parameters, uncorrected for the system's and sky Mueller matrices, versus PA. In the middle and bottom we have the derived values for the instrumental parameters for the total Mueller matrix, and also the Mueller matrix itself (Section 5.6.).

whose elements can be derived by tracking a polarized source, in a similar manner to that described above. Each antenna pair is characterized by a different Mueller matrix. Correcting the output of each pair with its Mueller matrix is a *baseline-based* scheme.

However, for large arrays it is much more efficient to use an *antenna-based* calibration scheme. With this, you characterize each antenna by its Jones matrix; you obtain the Mueller matrix for any antenna pair from the two Jones matrices. The Jones matrices for the individual antennas can be derived from the PA dependence of the Stokes vectors for all the baselines using a least square technique.

7. Polarized Beam Structure

We have left unspecified the implicit fact that we have been describing the Mueller matrix corrections on the axis of the main beam, as we would measure for a pulsar or a small radio source. You may be surprised to learn that the telescope beam contains unavoidable, intrinsic polarization structure. The type of structure depends on the Stokes parameter.

The fundamental cause of the polarized structure is the curved reflector surface, which slightly changes the direction of an incident linearly polarized electric vector upon its reflection. On the main beam axis these distortions cancel, but off-axis they do not. The distortions add in fundamentally different ways for linear and circular polarization because, when a source is off-axis, the path lengths to the source from different portions of the reflecting surface are not all equal. The distortions increase with curvature, and hence become more serious with decreasing focal ratio; beam squint varies $\left(\frac{f}{D}\right)^{-1}$ (Troland & Heiles 1982). Radio telescopes have small $\frac{f}{D}$, so the effects become very significant.

We do not have the space to delve into the somewhat esoteric details of these distortions; see Tinbergen (1996, Section 5.5.5) and quoted references. These descriptions are usually given for prime-focus fed paraboloids; with their different geometries, Arecibo and the GBT differ in detail but not in fundamental principle.

7.1. Main beam linear polarization

With linear polarization, there are two sources of distortion. One relates to the feed: for a feed probe sampling the X direction, the waveguide nature of the feed tends to make the feed's illumination pattern broader in X than in Y. After reflection from the dish surface, the telescope HPBW is broader in Y than X. The second is the abovementioned dish curvature, which also produces a similar distortion. We call these differing beamwidths *beam squash*.

Both effects produce the same result, namely different beamwidths in orthogonal polarizations. When these two polarizations are subtracted to produce the Stokes (Q, U) parameters, one obtains a four-lobed "cloverleaf" structure in the (Q, U) beam responses. Figure 4 shows an example, taken at Arecibo for a source near the telescope's maximum zenith angle 20° where some additional distortions are introduced and the sidelobe is somewhat accentuated. The main

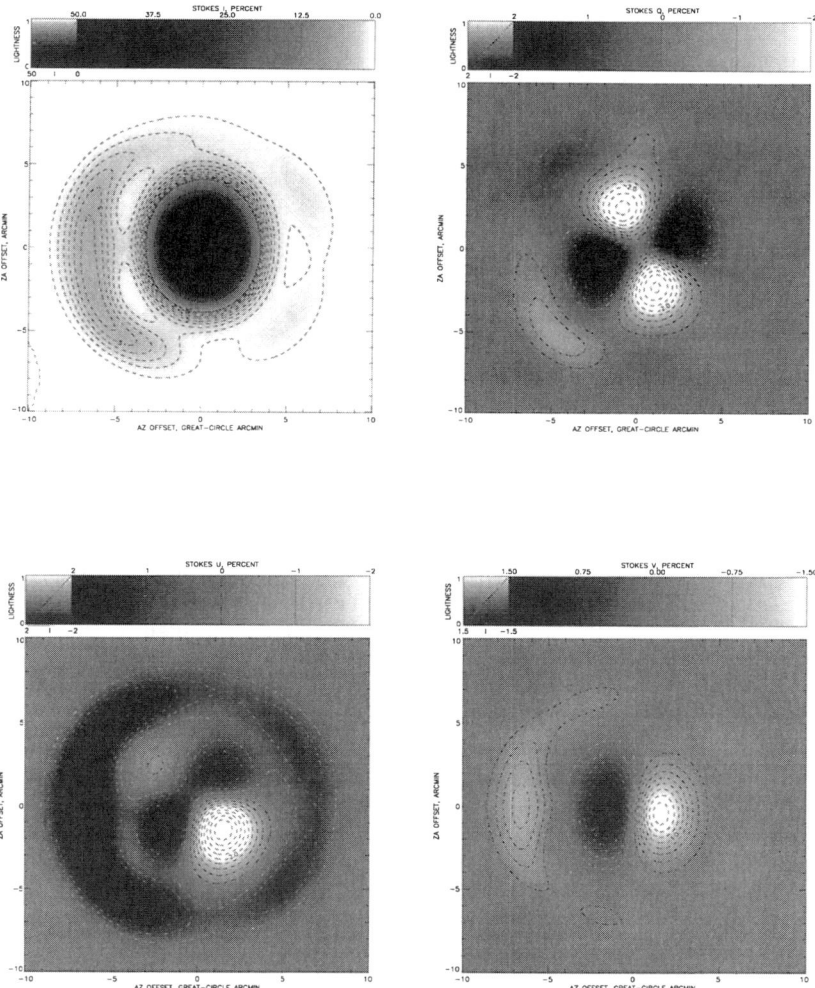

Figure 4. Stokes (I, Q, U, V) grey-scale/contour images of Arecibo's main beam and first sidelobe near zenith angle $20°$. For I, solid (white) contours are $(0.1, 0.2, \ldots)$ of the peak; dashed (black) contours are $(0.01, 0.02, \ldots)$. For the others, black contours are for areas with negative (Q, U, V) with the grey scale tending towards white; white contours are positive (Q, U, V) with the greyscale tending towards black. Contours are in percent of Stokes I at beam center and spaced by 0.4% for (Q, U) and 0.2% for V; the 0% contours are omitted. U is aligned with the azimuth arm.

beam exhibits not only the expected squash, but also squint and higher-order distortions (Heiles 1999; Heiles et al. 2000b, 2001b).

7.2. Main beam circular polarization

With circular polarization and a paraboloid having the feed aligned perfectly at the focal point, everything is circularly symmetric so there can be no beam structure in Stokes V. In practice, however, such perfection can never be achieved. If the feed points slightly away from the vertex of the paraboloid, say in the X direction, then the two circularly polarized beams point in slightly different directions along the Y direction. When the two circulars are subtracted to produce Stokes V, there is a two-lobed structure in the Stokes V beam response. We call this *beam squint*; see Figure 4.

Arecibo and the GBT both employ an asymmetrically fed design in which the feed is both located off-axis and does not point towards the center of symmetry. Thus, both have intrinsic beam squint built into their designs. One can minimize the polarization structure by appropriate design of the subreflector geometry. This optimization was done for both telescopes and, at least at Arecibo, the predicted main beam Stokes V performance is close to what is measured; the difference in direction ~ 2 arcsec at L-band.

7.3. Sidelobe polarization

Figure 4 shows the effects of beam squash and squint. They also show that the first sidelobe is highly polarized. As one moves away from the beam center one encounters higher order sidelobes, the effects of distortion increase and the sidelobe polarizations increase. These effects are not well studied or understood.

Arecibo has a large central blockage produced by the suspended "triangle" structure from which the moving feed hangs. This produces large sidelobes (Heiles et al. 2000b, 2001b), and these have high polarization. The GBT, with its unblocked aperture, has exceedingly low sidelobes, so effects arising from sidelobe polarization are minimized.

7.4. The effect on astronomical polarization measurements

Suppose one is observing a large-scale feature where the brightness temperature T_B varies with position. One can express this variation by a two-dimensional Taylor expansion. Beam squint, by its nature, responds to the first derivative and only slightly to the second; beam squash responds primarily to the second.

The polarized beam structure produces fake results in the *polarized* Stokes parameters (Q, U, V) that arise from spatial gradients in the *total intensity* Stokes parameter I. The effects are exacerbated by the polarized sidelobes, which are further from the beam center.

Heiles et al. (2000b, 2001b) calculated these effects for the Arecibo beam at 1.4 GHz, including both the main beam and first sidelobe but no additional sidelobes. Consider a total intensity gradient of 1 K arcmin^{-1} and second derivative 1 K arcmin^{-2}, values which are not necessarily realistic but are convenient for practical use. For these particular values the fake contributions from the first and second derivatives are comparable. They yield fake results for Stokes $Q, U \sim 0.3$ K. For Stokes V the contributions are about ten times smaller, ~ 0.03 K.

The fractional polarization of extended emission is often small (but not always), so spatial gradients in I can be very serious. Consider, for example, measuring Zeeman splitting of the 21-cm line in emission, which involves measuring Stokes V of the 21-cm line. If the central velocity of the 21-cm line has a spatial gradient $\frac{dv}{d\theta} = 1$ km s^{-1} deg^{-1} — a not-uncommon value as measured with a 36 arcmin beam (Heiles 1996) — we get $B_{fake} \sim 1.1$ μG. Gradients might be larger when measured with smaller $HPBW$. Typical values of B are in the μG range, so this effect can be — but is perhaps not always — serious!

In principle, these effects can be corrected for. Correcting for them at Arecibo is a complicated business because of the PA variation with azimuth and zenith angle. It is also an uncertain business, especially for (Q, U) and somewhat less so for V, because these variations are unpredictable and must be determined empirically. Presumably, corrections at the GBT will be much more straightforward.

8. Summary

After a brief introduction to the astrophysical significance of polarization measurement, Section 2 began by introducing the Jones vector, which describes the polarization state of a sine wave. Section 3 went on to define the Stokes parameters and the Stokes vector, which are required to completely describe the polarization state of natural radiation, which is always at least partly randomly polarized. Section 3 also used the Stokes parameters to define the conventionally used quantities, fractional polarization and position angle, and cautioned against their use in arithmetic operations.

Section 4 related the Jones vector to the Stokes vector; this relationship tells exactly how radio astronomers measure all four Stokes parameters simultaneously. However, the receiving system modifies the incoming polarization with instrumental effects, which must be measured and corrected for; Section 5 described the quantitative aspects of this correction using Jones and Mueller matrices. It detailed the specific cases of the amplifier chain and the mechanical rotation of the telescope on the sky as instructive and most important examples. The Mueller matrix for the amplifier chain leads naturally to a discussion of the advantages of cross correlation for measuring small effects. It also shows how one must not use this falsely, as is often done using a post-amplifier hybrid.

Section 6 discussed practical details, including noise-diode calibration and measuring the matrix elements astronomically, and presented results for Arecibo's "L-band wide" feed as an example.

Finally, Section 7 described the major polarization effects in the main beam, namely beam squint and beam squash, and illustrated these effects again using the L-band wide feed as an example. Arecibo has a prominent first sidelobe, and Section 7 also discussed its rather severe polarization properties. It concluded by discussing the effects of this beam structure, interacting with spatial derivatives in Stokes I, on contaminating measurements of polarized Stokes parameters.

Acknowledgments. This work was supported in part by NSF grants 95-30590 and AST-0097417.

References

Hamaker, J. P., Bregman, J. D., & Sault, R. J. 1996, A&AS, 117, 137
Heiles, C. 1996, ApJ, 466, 224
Heiles, C. 1999, Arecibo Technical and Operations Memo 99-02
Heiles, C. & Fisher, J. R. 1999, NRAO Electronics Division Internal Report 309
Heiles, C., et al. 2000a, Arecibo Technical and Operations Memo 2000-05
Heiles, C., et al. 2000b, Arecibo Technical and Operations Memo 2000-04
Heiles, C. 2001, PASP, 113, 1243
Heiles, C., et al. 2001a, PASP, 113, 1247
Heiles, C., et al. 2001b, PASP, 113, 1274
Tinbergen, J. 1996, Astronomical Polarimetry (Cambridge, New York: Cambridge University Press)
Troland, T. H. & Heiles, C. 1982, ApJ, 252, 179

Part 2
Single-Dish Observing Disciplines and Associated Techniques

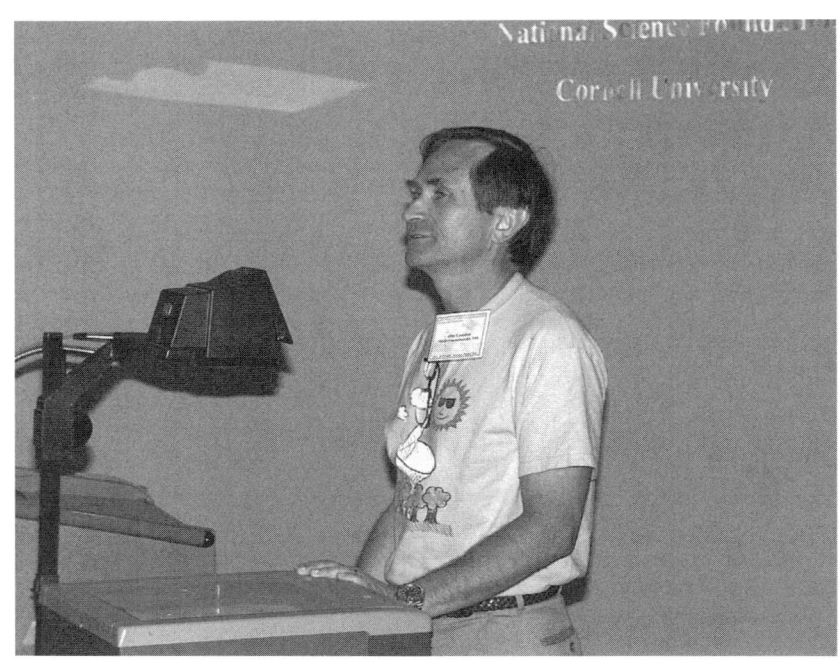

Jim Condon awaits the next question.

Continuum 1: General Aspects

J. J. Condon
National Radio Astronomy Observatory, 520 Edgemont Road, Charlottesville, Virginia 22903, USA

Abstract. Single dishes have been superseded by interferometers for much continuum observing, but they are still used to image very extended objects, especially at higher frequencies where confusion is negligible and a filled aperture is needed to detect low-brightness emission. Even spectral-line and pulsar astronomers normally observe continuum sources to check telescope performance, calibrate their data, and make real-time pointing corrections. This introduction to single-dish continuum describes simple observing techniques and the most important observational errors, supplemented by references to fundamental papers for more detailed explanations and to data sets useful for planning and making observations.

1. Introduction

Active galaxies, quasars, normal galaxies, planetary nebulae, H II regions, supernova remnants, planets, and even stars are continuum radio sources. Their emission mechanisms include synchrotron radiation by relativistic electrons in magnetic fields, free-free emission by hot electrons in H II regions, and thermal emission from interstellar dust. All three mechanisms contribute to the radio spectrum of the nearby starburst galaxy M82 as shown in Figure 1. Radio astronomers study continuum sources in their own right and use them as position, intensity, and polarization calibrators for pulsar or spectral-line observations. Lacking the rapid time variability of pulsars and the narrow spectral signatures of line emitters, astronomical continuum sources are distinguishable from receiver baseline drifts, atmospheric emission fluctuations, ground radiation, and each other only by their positions on the sky. These competing "signals" often exceed radiometer noise for single-dish observations. In addition, telescope gain uncertainties and pointing errors affect even strong sources. The continuum observer must understand both these noiselike and intensity-proportional errors to obtain the best possible data and to make reliable error estimates for measured source parameters. Section 2 outlines the use of strong point sources to calibrate and correct single-dish observations. Section 3 describes continuum observations of faint sources and Section 4, large-scale images and sky surveys. Future uses of single-dishes in a field now largely dominated by interferometers are considered in Section 5.

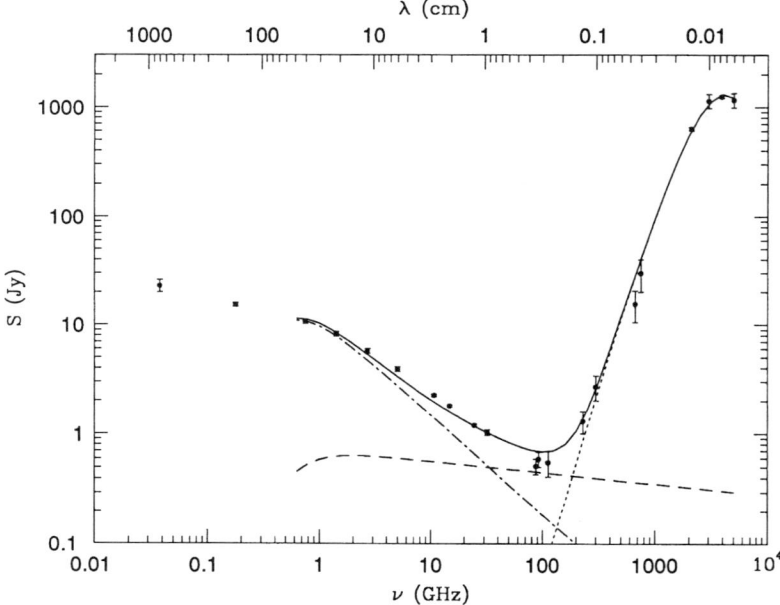

Figure 1. The continuum radio emission from the nearby starburst galaxy M82 is broadband noise resulting from the superposition (heavy curve) of three physically and spectrally distinct mechanisms. Steep-spectrum synchrotron radiation (dot/dash line) from supernova remnants and interstellar cosmic-ray electrons accelerated by them is strongest below $\nu \approx 30$ GHz. Flat-spectrum free-free emission (dashed line) is stronger in "normal" and starburst galaxies over the frequency range $30 < \nu < 200$ GHz, and free-free absorption is important at the lowest frequencies. Thermal emission from dust (dotted line) emerges at higher rest frequencies. The sharp rise of the dust spectrum, as steep as $S \propto \nu^4$, favors radio detection of luminous starburst galaxies at high redshifts.

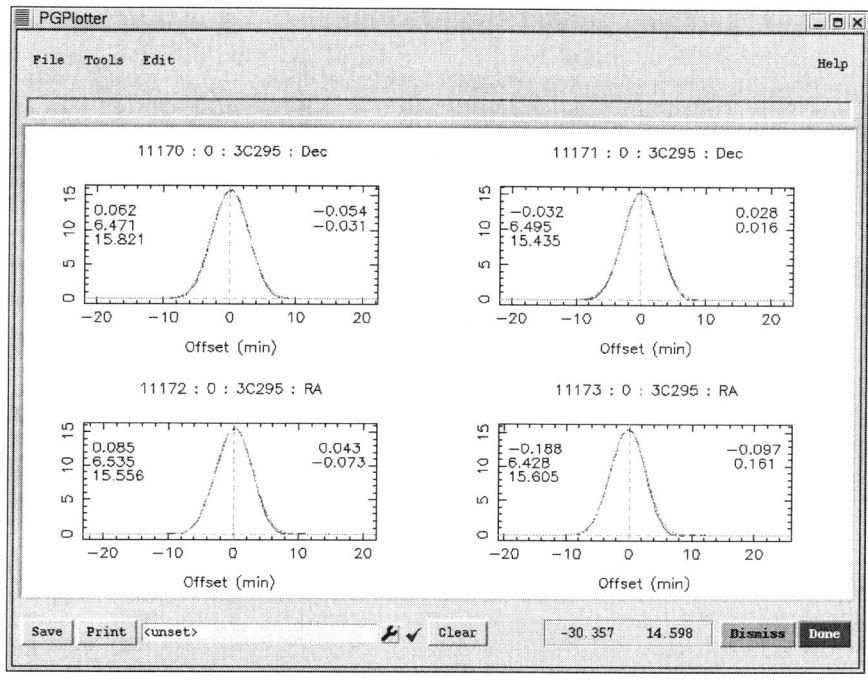

Figure 2. AIPS++ DISH display of the data and Gaussian fits from a cross scan on 3C 295 made with new 100-m Green Bank Telescope (GBT) at $\nu = 2$ GHz. They yield the declination offset, full width between half-maximum points (FWHM), and maximum antenna temperature (+0.062 arcmin, 6.471 arcmin, and 15.821 K for the first scan).

2. Using Continuum Sources as Observing Tools

Figure 2 illustrates a self-correcting cross scan used to measure the position, angular size, and peak antenna temperature of the unresolved calibration source 3C 295. This type of scan can be used to calibrate most single-dish observations and to make real-time pointing corrections. It is also useful for verifying telescope and receiver performance at the beginning of any observing run. First, the telescope beam was scanned up and down in declination and the apparent declination of the source was measured. Then the beam was scanned in the orthogonal coordinate (right ascension) at the corrected declination to determine the right-ascension offset and a more accurate antenna temperature at the beam peak. Scans in elevation and azimuth could have been made instead.

Single-dish telescopes have roughly Gaussian beams, so Gaussians are traditionally fitted to scan data. The maximum antenna temperature T_A of the fitted Gaussian is proportional to the source "peak flux density" S_p. This deceptively named quantity is not a flux density (typical units of flux density are 1 Jy $\equiv 10^{-26}$ W m^{-2} Hz^{-1} or 1 mJy $\equiv 10^{-3}$ Jy) but a brightness (usually expressed in units of Jy beam^{-1}, where "beam" is shorthand for the beam solid angle Ω_b). Only for a point source is the flux density in Jy even numerically equal to the peak flux density in Jy beam^{-1}. The beam solid angle Ω_b of an elliptical Gaussian beam having FWHM major and minor diameters θ_M and θ_m is

$$\Omega_b = \frac{\pi \theta_M \theta_m}{4 \ln 2} \,. \tag{1}$$

The volume under the elliptical Gaussian determined by the perpendicular cuts of a cross scan is proportional to the source's integrated flux density S (Jy). The ratio (T_A/S) produced by an unpolarized point source is a common measure of telescope effective aperture A_e:

$$\frac{T_A}{S} = \frac{A_e}{2k} \,, \tag{2}$$

where $k \approx 1.38 \times 10^{-23}$ J K^{-1} is the Boltzmann constant. A sensitivity of $(T_A/S) = 1$ K Jy^{-1} corresponds to an effective collecting area $A_e \approx 2.76 \times 10^3$ m^2. The sensitivities of the Arecibo Gregorian systems and the 100-m GBT are about 10 K Jy^{-1} and 2 K Jy^{-1}, respectively, at low frequencies and about half those values at the highest usable frequencies. You should begin every single-dish observing run with a cross scan on a continuum calibrator to verify that the receiver is operating correctly, the telescope pointing is accurate, and the system sensitivity is normal.

2.1. Intensity and polarization

The absolute flux density scale at cm wavelengths is based on the supernova remnant Cas A and a few other radio sources too strong and/or too extended to be practical for calibrating large telescopes. Most observers rely instead on the Baars et al. (1977) list of compact secondary flux density calibrators or its revisions (e.g. Ott et al. 1994). The absolute accuracy of this flux density scale is about 5% rms and the internal consistency is about 1%. Some of the calibrators (notably 3C 286) may be linearly polarized by up to 12%, so results from scans made with linearly polarized feeds must be corrected appropriately. The degree of circular polarization is much smaller ($\ll 1\%$) and can usually be ignored. Continual revisions of the calibrator flux densities are necessary because the secondary calibrators vary on time scales of years in both total intensity and polarization, with the possible exception of 3C 295. Updated calibrator lists are maintained on-line at the Effelsberg Observatory
http://www.mpifr-bonn.mpg.de/div/effelsberg/index_e.html
and the VLA
http://www.aoc.nrao.edu/~gtaylor/calib.html.
In the far south the principal flux calibrator is PKS B1934−638; see
http://wwwnar.atnf.csiro.au/calibrators/
for more information.

How often is it necessary to observe one of these flux density calibrators? A modern receiver should be stable enough that a single calibration per day is sufficient to monitor its gain. [The gain fluctuations that produce $1/f$ noise

(Section 3.2) on total-power receiver baselines are much smaller than other calibration errors.] Telescope gain may vary more rapidly near the short-wavelength limit, conventionally the wavelength at which the (Ruze 1966) reflector efficiency

$$\eta_{\rm r} \approx \exp\left[-\left(\frac{4\pi\sigma_{\rm r}}{\lambda}\right)^2\right] \tag{3}$$

falls to 0.5, or $\lambda \approx 16\sigma_{\rm r}$, where $\sigma_{\rm r}$ is the rms deviation of the reflector from its ideal shape. Thermal deformations can occur on time scales ~ 1 hour, especially when sunlight heats the reflector or feed-support structures. Gravitational deformations of telescopes with movable reflectors normally depend only on zenith angle, and their effects on gain are plotted as "gain-elevation curves" provided by the observatory staff. The gain of the Arecibo telescope depends on both zenith angle and azimuth in a more complex way since different portions of its fixed reflector are illuminated as the feeds move. At mm wavelengths, even the atmospheric opacity can change significantly.

Correcting these short-term gain changes requires frequent monitoring of any fairly strong, unresolved, steep-spectrum source that is near on the sky to the program source. Conventionally, sources with spectral indices $\alpha > 0.5$ are called "steep-spectrum", where

$$\alpha \equiv -\frac{d \ln S}{d \ln \nu} \tag{4}$$

or

$$\alpha(\nu_1, \nu_2) = -\frac{\ln(S_1/S_2)}{\ln(\nu_1/\nu_2)}, \tag{5}$$

for a power-law spectrum measured at any two frequencies ν_1 and ν_2. Flat-spectrum ($\alpha < 0.5$) sources are not reliable flux/gain monitors because some are intraday variables. Any steep-spectrum pointing calibrator (Table 1) can serve simultaneously for monitoring pointing, gain, and often polarization. Beware that the old flux densities given in Table 1 are only approximate, particularly at 5 GHz. Thus the current flux density of the local monitor must be determined once per observing run by comparison with one of the known flux density calibrators, a process called "bootstrapping." For simple polarization calibration techniques, see
http://www.aoc.nrao.edu/~gtaylor/calman/polcal.html
in the VLA calibration manual.

2.2. Pointing

Pointing errors are often more important than surface deformations in limiting the high-frequency performance of large radio telescopes. Suppose a point source of known position is tracked by a telescope having a Gaussian beam with FWHM θ and a Gaussian pointing-error distribution. Let the rms pointing error in each sky coordinate be $\sigma_\alpha \approx \sigma_\delta = \epsilon\theta$, so ϵ is the pointing error in one coordinate expressed as a fraction of the beamwidth. Then the normalized telescope gain G ($G \equiv 1$ at the beam center) on the source position will have the very skewed probability distribution

$$P(G) = [(8\ln 2)\epsilon^2 G]^{-1} \exp\left[\frac{\ln G}{(8\ln 2)\epsilon^2}\right] \tag{6}$$

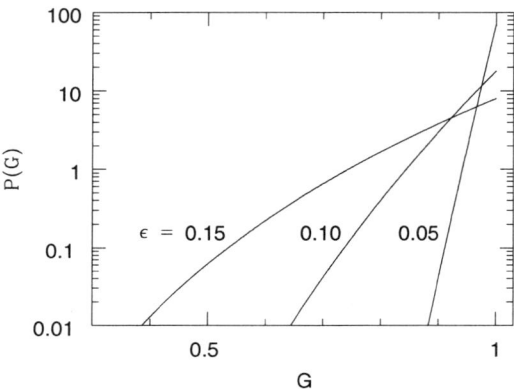

Figure 3. Random pointing errors reduce the normalized telescope gain G on the position of a point source. This causes a variable reduction of sensitivity and intensity-proportional errors in the observed flux density. The probability distribution $P(G)$ is shown for three values of ϵ, the rms pointing error normalized by the FWHM beamwidth. Pointing errors larger than $\epsilon \sim 0.1$ seriously degrade many observations.

shown in Figure 3. The gain variance is $\sigma_G^2 = \langle G^2 \rangle - \langle G \rangle^2$, where $\langle G^2 \rangle = [1 + (16 \ln 2)\epsilon^2]^{-1}$. Consequently, the mean observed source intensity is reduced by the factor

$$\langle G \rangle = [1 + (8 \ln 2)\epsilon^2]^{-1} \qquad (7)$$

and the rms intensity-proportional flux error caused by pointing errors is

$$\frac{\sigma_G}{\langle G \rangle} = (8 \ln 2)\epsilon^2 + [(8 \ln 2)\epsilon^2]^2 \; . \qquad (8)$$

For example, the rms nonrepeatable pointing error at Arecibo is currently $\sigma_\alpha \approx \sigma_\delta \approx 5''$ and the X-band beamwidth is $\theta \approx 40''$, so $\epsilon \approx 0.125$. Thus you should expect an 8.7% loss of signal and a 9.5% rms intensity-proportional flux uncertainty during "blind" tracking of a point source.

Fortunately, most pointing errors vary slowly enough with both time and position on the sky that offset pointing corrections based on frequent observations of a nearby pointing calibrator can significantly reduce tracking errors on the program source. The largest nonrepeatable pointing errors are often caused by thermal deformations having time scales ~ 1 hour (Figure 4), so updating the pointing offsets twice per hour is probably adequate. A new catalog (Condon & Yin 2001) of 3399 strong, compact, and unconfused sources suitable for pointing large radio telescopes has been extracted from the 1.4 GHz NRAO VLA Sky Survey (NVSS, Condon et al. 1998). The NVSS positional accuracy is $\sigma_\alpha \approx \sigma_\delta \approx 0\rlap{.}''5$. These sources are uniformly distributed on the sky, so the average angular distance $\langle \phi \rangle$ from any point with declination $\delta > -40°$ to the nearest

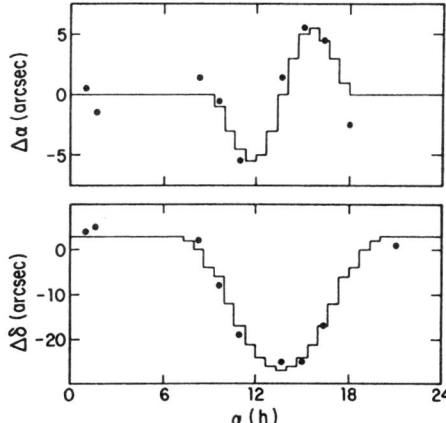

Figure 4. Differential expansion resulting from solar heating made the former 300-foot telescope in Green Bank bend away from the Sun. These characteristic pointing errors were measured when the Sun was located near $\alpha \approx 13^{\rm h}20^{\rm m}$ (Condon, Broderick, & Seielstad 1989).

calibrator is only $\langle\phi\rangle \approx 0.03$ rad (Figure 5). Table 1 is a partial list of the calibrators and the maximum FWHM beamwidths $\theta_{\rm max}$ with which their positions can be measured with rms errors less than 1% of the beamwidth ($\epsilon < 0.01$).

The effects of noise errors on Gaussian fitted parameters such as source position, size, and flux density have been evaluated for both one-dimensional (Kaper et al. 1966) and two-dimensional (Condon 1997) fits. The rms uncertainty in a fitted coordinate caused by rms noise σ is $\epsilon \approx \sigma/(2 S_{\rm p})$, where $S_{\rm p}$ is the source peak flux density. Thus a signal-to-noise ratio of at least 50 on an offset pointing calibrator is needed to reach $\epsilon \approx 0.01$.

One exception to the "slowly varying" rule for pointing errors is "anomalous refraction," a rare phenomenon caused by large tropospheric water-vapor fluctuations drifting through the telescope beam. It can cause apparent source positions to vary by several arcsec on time scales of about a minute (Altenhoff et al. 1987), too short for correction by offset pointing. For long integrations, anomalous refraction effectively broadens the telescope point-source response and reduces aperture efficiency.

Remember that errors in the assumed position of your program sources also contribute to the gain errors described in this section and are not corrected by offset pointing. For best results, determine their positions with uncertainties $\epsilon < 0.1$ before observing.

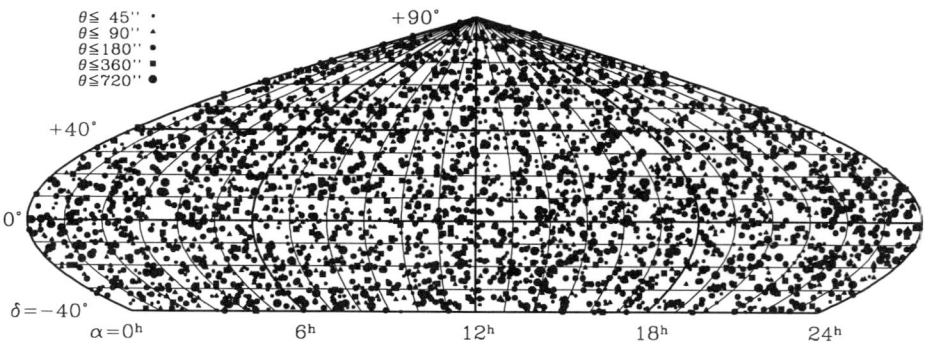

Figure 5. This dense grid of pointing calibrators can be used to reduce tracking errors north of $\delta = -40°$. Symbols indicate the maximum beamwidth for which each calibrator is useable.

Table 1. Pointing Calibration Sources

Right Ascension (J2000)	Declination (J2000)	θ_{max} (arcsec)	$S(1.4\ \mathrm{GHz})$ (Jy)	$S(5\ \mathrm{GHz})$ (Jy)
00 00 20.33	−32 20 59.0	180	0.52	0.54
00 00 20.45	+55 39 08.6	360	1.52	0.47
00 00 53.06	+40 54 01.9	360	1.30	0.60
00 01 07.23	+60 51 23.4	90	0.55	0.25
00 01 24.49	−04 38 01.4	180	0.63	0.25
00 01 30.95	+11 01 40.6	180	0.69	0.20
00 01 43.47	−30 57 31.4	45	0.52	0.18
00 03 22.00	−17 27 11.4	45	2.42	0.89
00 03 27.24	−15 47 06.3	90	0.53	0.21
00 04 09.83	+17 17 51.3	180	0.50	0.13
00 05 30.54	+69 23 58.2	180	1.10	0.48
00 05 55.71	+31 39 48.8	180	0.78	0.25
00 05 57.14	+38 20 15.3	90	0.57	0.57
00 06 13.87	−06 23 35.2	180	2.05	2.46
00 06 22.60	−00 04 25.1	720	3.90	1.48
...	...	...	...	...

The full table is available electronically from Condon & Yin (2001) or as PCALS2.2 from ftp://ftp.cv.nrao.edu/NRAO-staff/jcondon.

3. Continuum Observations of Faint Sources

Most continuum sources of astronomical interest are faint enough that noise-like errors are larger than calibration errors. Noiselike errors produce additive fluctuations in the data that are independent of source strength. Their causes include confusion from background radio sources, receiver noise, baseline drift, and fluctuating atmospheric emission. The rms amplitude of pure receiver noise is given by the classic radiometer equation

$$\sigma = \frac{T_s}{(B\tau)^{1/2}}, \qquad (9)$$

where T_s is the system noise temperature, B is the predetection bandwidth, and τ is the integration time. The total noise in many pulsar, spectral-line, and interferometric observations actually does decline as $\tau^{-1/2}$ for integration times as long as hours. This is *not* true for single-dish continuum observations—other noiselike sources often dominate, even after only 1 s of integration! Consequently, continuum observers must be aware of them and design their observations to minimize their effects. At low frequencies ($\nu \lesssim 5$ GHz) confusion (Section 3.1) is the biggest limitation. At higher frequencies, $1/f$ noise (Section 3.2) becomes important and forces the astronomer to make many short observations instead of a single long integration.

3.1. Confusion

"Confusion" is the term for sky brightness fluctuations caused by numerous faint radio sources blended in the telescope beam (Figure 6). It is present even if receiver noise is negligible. Confusion from extragalactic sources is ubiquitous and severely limits the sensitivity of all continuum observations with single-dish telescopes at frequencies $\nu \lesssim 5$ GHz. Every bump in Figure 6 is "real" in the sense that repeated observations would always give the same result, but a subimage made with higher angular resolution (Figure 7) reveals that some of the smaller bumps are actually superpositions of two or more physically distinct radio sources that just happen to be close in projection onto the sky.

The confusion amplitude distribution, historically called the "$P(D)$ distribution", specifies the probability P that any point on the sky will have an observed brightness D, usually expressed as a peak flux density. The observed $P(D)$ distribution depends on the sky density of sources and on the telescope beam. Since the differential number of extragalactic sources per steradian is roughly a power law in flux density [$n(S) \approx kS^{-\gamma}$, where k depends only on the frequency ν and γ is a constant], the $P(D)$ distribution is almost scale-free. The calculated $P(D)$ distribution (Condon 1974) for $\gamma = 2.1$ and $k\Omega_b = 1$ is shown in Figure 8. The counts of faint ($S \sim 1$ mJy at $\nu = 1.4$ GHz) extragalactic sources (Mitchell & Condon 1985) have an integral slope $\gamma - 1 \approx 1$ in the intensity range relevant to large single-dishes. Consequently:
(1) At any frequency, the rms confusion expressed in units of peak flux density (mJy beam^{-1}) is directly proportional to the telescope beam solid angle Ω_b. It is independent of beam area if expressed as a sky brightness temperature (K).
(2) For any beam area, the rms confusion scales with frequency following the median spectral index, $\langle \alpha \rangle \approx -0.7$ (flux density) for extragalactic sources at

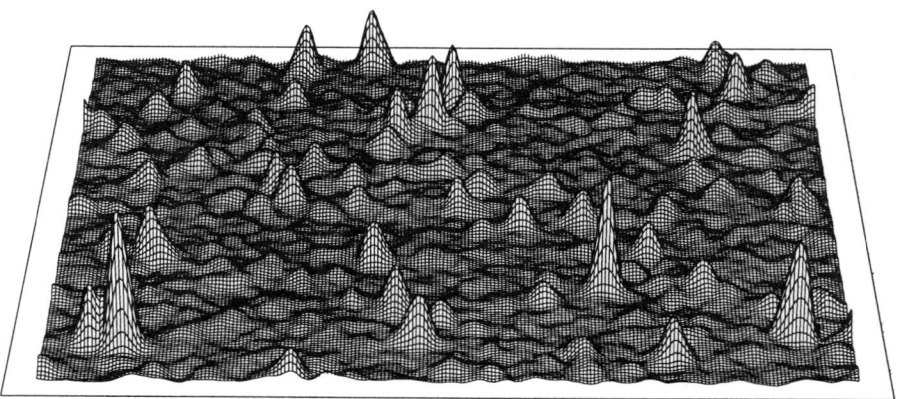

Figure 6. This profile plot shows 45 deg² near the North Galactic Pole imaged with $\theta \approx 12'$ FWHM resolution at $\nu = 1.4$ GHz by the former 300-foot telescope. The strongest source shown has $S \approx 1.5$ Jy. I deleted the intensity and angular scales to emphasize the scale-free character of confusion from sources with power-law flux-density distributions—all confusion-limited images are qualitatively similar.

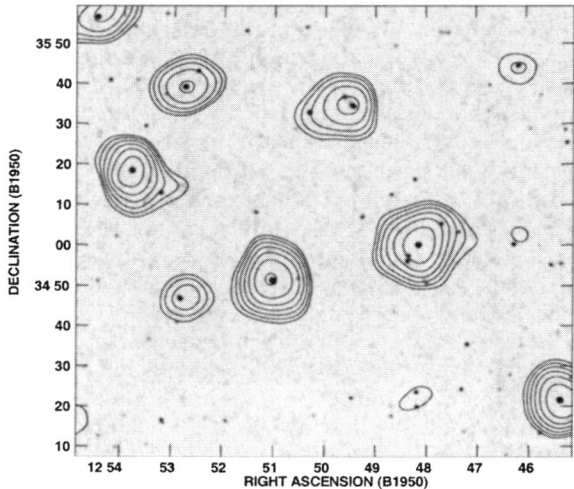

Figure 7. Zooming in on 4 deg² of Figure 6, with the NVSS ($\theta = 45''$) gray-scale image under the single-dish data contours, reveals source blending. The lowest contour is $2\sigma_c = 45$ mJy beam^{-1} and successive contours are spaced by factors of $\sqrt{2}$ in brightness.

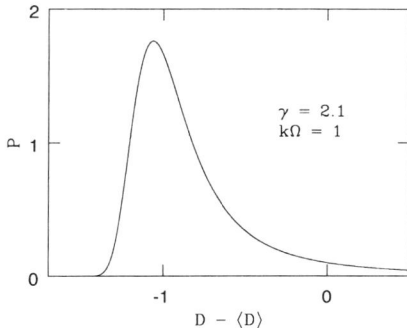

Figure 8. A typical confusion $P(D)$ distribution. Abscissa: "Deflection" D above the mean $\langle D \rangle$. Ordinate: Normalized probability P.

cm wavelengths. At frequencies approaching $\nu \sim 30$ GHz free-free emission from starburst galaxies is expected to flatten the confusion spectrum, and the confusion spectrum rises sharply at mm wavelengths owing to thermal emission from dusty starburst galaxies (Eales et al. 2000; Takeuchi et al. 2001).

Although the rms confusion σ_c is not the best statistic for describing the width of the $P(D)$ distribution, it is widely used. At cm wavelengths, the rms confusion in a Gaussian beam can be approximated by

$$\left(\frac{\sigma_c}{\text{mJy beam}^{-1}}\right) \approx 0.2 \left(\frac{\nu}{\text{GHz}}\right)^{-0.7} \left(\frac{\theta_M \theta_m}{\text{arcmin}^2}\right) \quad (10)$$

or

$$\left(\frac{\sigma_c}{\text{K}}\right) \approx 0.07 \left(\frac{\nu}{\text{GHz}}\right)^{-2.7}. \quad (11)$$

Only sources stronger than about $5\sigma_c$ can be detected reliably because the $P(D)$ distribution has a long tail. There are about 25 beam areas per source stronger than $5\sigma_c$, so extracting more than one source per 25 beam areas from a confusion-limited image is dangerous. Over the years, numerous astronomers have succumbed to temptation and published detections of individual sources fainter than $5\sigma_c$. Most have later regretted it.

Figure 8 also shows why subtracting the mean baseline (attempting to determine $\langle D \rangle$) is a poor choice for confusion-limited scans. Instead, one typically subtracts either the running median (cf. Condon et al. 1989) or the mode [the value of D that maximizes $P(D)$] to minimize the bias and scatter introduced by rare strong sources.

Approximating the beamwidth by $\theta \approx 1.2\lambda/d$ for a telescope having illuminated diameter d at wavelength λ results in the Arecibo ($d \approx 220$ m) and GBT ($d = 100$ m) confusion limits plotted in Figure 9. Confusion is much larger than receiver noise even for short ($\tau < 1$ s) integrations below $\nu \approx 5$ GHz. Thus single-dishes cannot compete with interferometers for observations of weak point sources at lower frequencies. For example, at 1.4 GHz the NVSS covered

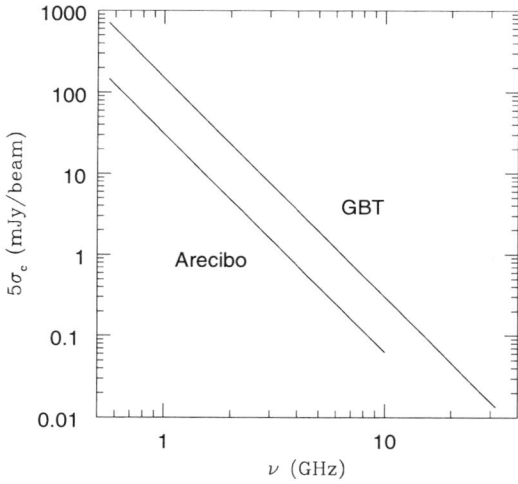

Figure 9. The $5\sigma_c$ extragalactic confusion limits for Arecibo (illuminated diameter $d \approx 220$ m) and the GBT ($d = 100$ m). Fainter continuum "sources" may be blends of unrelated objects. Confusion in certain areas of sky (e.g., near the center of our Galaxy) can be significantly higher.

the sky north of $\delta = -40°$ with $\theta = 45''$ resolution and detected most sources with $S \geq 2.5$ mJy, well below Arecibo's $5\sigma_c \approx 13$ mJy confusion limit. Single-dishes remain competitive above 5 GHz, particularly for observations of diffuse, low-brightness sources and for large sky surveys.

Confusion can also be a problem for spectral-line observers who want to select source-free "off" positions to minimize standing waves during position-switching observations. The NVSS images available on-line at http://www.cv.nrao.edu/nvss/ and http://nedwww.ipac.caltech.edu/ are very convenient for finding good "off" positions before or during your observations.

3.2. $1/f$ Noise

Pure thermal noise (Equation 9) from the receiver, atmospheric emission, ground radiation, etc. has a flat postdetection power spectrum and can be reduced by long integrations. Very small changes ΔG_r, ΔT_s in receiver gain G_r or system noise temperature T_s independently add noiselike fluctuations to the output of a total-power receiver (Wollack 1995):

$$\sigma = T_s \left[\frac{1}{B\tau} + \left(\frac{\Delta G_r(f)}{G_r} \right)^2 + \left(\frac{\Delta T_s(f)}{T_s} \right)^2 \right]^{1/2}, \qquad (12)$$

where f is the postdetection frequency. The fluctuations resulting from the second and third terms in Equation 12 are sometimes called $1/f$ noise because their postdetection spectrum follows a declining power-law (see Figure 10). In

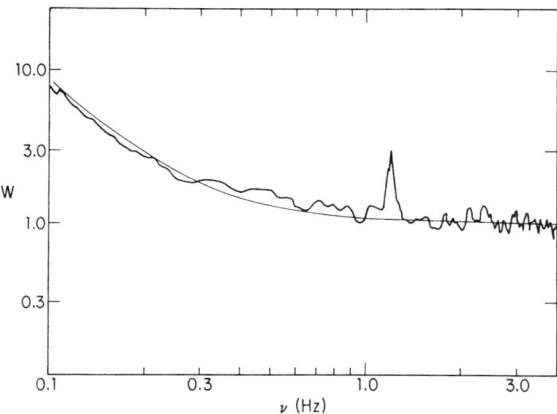

Figure 10. The postdetection power spectrum of the total-power receiver used to make the Green Bank 4.85 GHz sky survey. Pure radiometer noise would have produced a (noisy) horizontal line at $W = 1$. The spike at ≈ 1.2 Hz reveals a microphonic response of the receiver to mechanical vibrations from the cryogenic pump, a common phenomenon called "refrigerator modulation." Below ≈ 0.3 Hz, $1/f$ noise is stronger than thermal noise, so the receiver output baseline drifts noticeably on time scales longer than ≈ 3 s.

the time domain, the signature of $1/f$ noise is an erratic baseline drift that increases with long integration times. Even a good cm-wavelength receiver has fractional fluctuations $\Delta G_{\rm r}(f)/G_{\rm r} \approx 10^{-4}$ on time scales of seconds, so total-power integrations with $B\tau > 10^8$ are actually counterproductive. Individual scans should be made as *quickly* as possible, then repeated and averaged to reduce noise. For example, the Green Bank 4.85 GHz sky surveys were made by scanning the old 300-foot telescope at its slew rate, $10°$ min$^{-1} \approx 3$ beamwidths per second, with $B \approx 600$ MHz and $\tau = 0.1$ s. Running-median baselines 4 s long were subtracted to suppress slow drifts.

Fluctuating atmospheric emission in the telescope beam also causes baseline fluctuations which have a $1/f$ spectrum and favor short, repeated scans to build up sensitivity. The primary culprit is water vapor, which is poorly mixed in the troposphere (see Thompson, Moran, & Swenson 1986 for a detailed discussion). Emission from its pressure-broadened 22 GHz $6_{16} - 5_{23}$ transition is negligible for $\nu < 2$ GHz even during rainstorms, degrades total-power observations at 5 GHz under cloudy or humid conditions, and makes total-power observations impractical above $\nu \approx 10$ GHz at most sites.

Differential measurements (e.g., beam switching) can greatly reduce baseline fluctuations having $1/f$ spectra. The costs are (1) a loss of $2^{1/2}$ to 2 in sensitivity and (2) some suppression of extended source structures (Emerson, Klein, & Haslam 1979).

4. Surveys and Large Images

Single-dish telescopes can rapidly image very large areas of sky, and sky surveys illustrate many aspects of continuum observing. For example, the Green Bank 4.85 GHz survey covering $0° < \delta < +75°$ (Condon et al. 1989) was made with a seven-beam feed array and a 2 polarization × 7 beam = 14 channel total-power receiver. Not only does multibeaming multiply the sky coverage rate, it continuously yields comparative receiver diagnostics and greatly simplifies the detection/removal of impulsive radio-frequency interference (RFI) since only RFI can produce an impulse in all beams simultaneously. Using a short integration time and oversampling the data also aids recognition and excision of impulsive RFI. Making continuum observations with a spectrometer, deleting corrupted spectral channels, and averaging the clear ones can sometimes avoid narrowband RFI.

The old 300-foot telescope was scanned as rapidly as possible ($\pm 10°$ min^{-1}) in elevation so that point sources passed through the beam in about 0.3 s and running-median baselines 4 s long could remove drifts caused by $1/f$ noise, which exceeded thermal noise at frequencies $f < 0.3$ Hz. Large but repeatable baselines (see Figure 11) caused by ground radiation were also subtracted. Figure 12 shows the output from all 14 channels for a single scan after baseline subtraction. The outputs are quite flat, indicating that most $1/f$ noise has been eliminated. These techniques are effective for finding relatively compact sources but suppress smooth emission extended by more than half the baseline length, about 20 arcmin in this case. Longer baselines for imaging very extended sources can be obtained by "basketweaving", raster scanning in nearly orthogonal directions (Sieber, Haslam, & Salter 1979; Emerson & Gräve 1988). Scans in one direction are "tied together" by points from the perpendicular scans.

Flux density calibration sources were observed every few days. No real-time pointing calibrations were made, but the positions of strong pointing calibrators in the survey area were used to correct for both systematic pointing errors (e.g., hysteresis) and the primarily thermal time-dependent pointing errors (Figure 4).

5. Whither Single-dish Continuum?

Most continuum radio astronomy is now done with interferometers, which have several advantages over single-dishes:
(1) Higher angular resolution for mapping compact source structures and for reducing confusion by unrelated background sources. Single-dishes can resolve the internal structure in only a few extragalactic sources and their continuum sensitivity is strongly limited by confusion below $\nu \sim 5$ GHz.
(2) Better positional accuracy since interferometric positions are nearly independent of mechanical antenna pointing errors and the plane-parallel component of atmospheric refraction.
(3) Suppression of atmospheric emission fluctuations, receiver baseline drifts, and some RFI by correlation interferometry.
(4) Higher dynamic range with the use of self-calibration and CLEAN, although CLEANing single-dish data may also yield some improvement in dynamic range

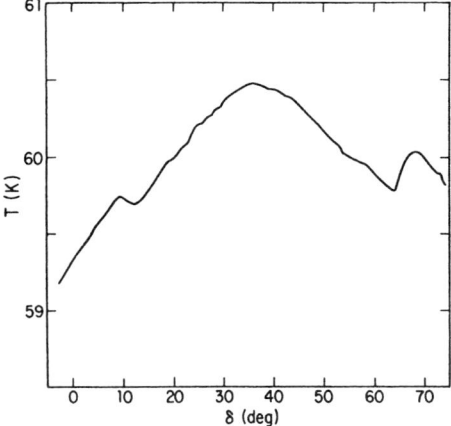

Figure 11. The total-power output from one receiver channel during a 4.85 GHz survey scan covering $0° < \delta < +75°$. The average system noise temperature was $T_s \approx 60$ K. On this are superposed repeatable changes with amplitude $\Delta T_s \sim 1$ K caused by atmospheric emission, spillover, ground radiation leaking through the reflector mesh, etc.

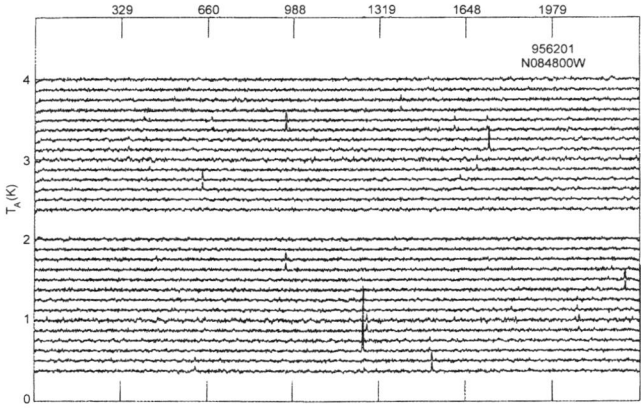

Figure 12. The output from all 14 channels = 2 polarizations × 7 beams of the NRAO 4.85 GHz multibeam receiver after running-median baselines were subtracted. Radiometer noise with amplitude $\sigma \approx 60 \text{ K}/(6 \times 10^8 \text{ Hz} \times 0.1 \text{ s})^{1/2} \approx 8$ mK and discrete sources (pairs of spikes on the outputs of the left- and right-circular receivers on a single feed) are now visible. The remaining baselines are quite flat, a sign that $1/f$ noise has largely been removed.

(Klein & Mack 1995).

Single-dish continuum observing continues for several reasons:
(1) Filled apertures are more sensitive to sources with low surface brightness such as nearby spiral galaxies at cm wavelengths (Klein & Mack 1995), halo and relic sources in clusters of galaxies, and mm-wavelength dust emission from distant ultraluminous galaxies (Eales et al. 2000).
(2) Large single dishes can determine accurate continuum spectra of sources detected by interferometers at low frequencies because they have matching beam sizes at higher frequencies. For example, the NVSS made with the VLA D-configuration has $\theta = 45''$ resolution at $\nu = 1.4$ GHz. The VLA cannot match this low resolution at higher frequencies (there is no E configuration), but Arecibo at $\nu \approx 7$ GHz or the GBT at $\nu \approx 17$ GHz have $\theta \approx 45''$. These single dishes would be optimum for measuring the spectral indices of slightly extended NVSS sources (e.g., planetary nebulae, nearby normal galaxies).
(3) Single dishes can be used alone or in combination with interferometers to provide the missing short-spacing data needed to image very extended galactic sources (Sieber et al. 1979).
(4) A multibeamed single dish can image very large areas of sky even at high frequencies, permitting all-sky surveys (e.g., Condon et al. 1989) or synoptic surveys covering limited areas (e.g., repeated GBT surveys to discover transient radio sources in the galactic plane).
(5) Historically, single dishes have been more flexible, allow wider continuum bandwidths, and have more complete frequency coverage than interferometers, especially at the highest frequencies.

Many of these factors tilt the balance in favor of using single dishes for continuum observations of extended sources or regions at higher frequencies. I anticipate increased single-dish continuum observing as the GBT and upgraded Arecibo begin operating at frequencies above $\nu \approx 5$ GHz and with multibeam receivers.

References

Altenhoff, W. J., Baars, J. W. M., Downes, D., & Wink, J. E. 1987, A&A, 184, 381
Baars, J. W. M., Genzel, R., Pauliny-Toth, I. I. K., & Witzel, A. 1977, A&A, 61, 99
Condon, J. J. 1974, ApJ, 188, 279
Condon, J. J. 1997, PASP, 109, 166
Condon, J. J., Broderick, J. J., & Seielstad, G. A. 1989, AJ, 97, 1064
Condon, J. J., Cotton, W. D., Greisen, E. W., Yin, Q. F., Perley, R. A., Taylor, G. B., & Broderick, J. J. 1998, AJ, 115, 1693 (NVSS)
http://www.cv.nrao.edu/nvss/
Condon, J. J. & Yin, Q. F. 2001, PASP, 113, 362
Eales, S., Lilly, S., Webb, T., Dunne, L., Gear, W., Clements, D., & Yun, M. 2000, AJ, 120, 2244

Emerson, D. T. & Gräve, R. 1988, A&A, 190, 353

Emerson, D. T., Klein, U., & Haslam, C. G. T. 1979, A&A, 76, 92

Kaper, H. G., Smits, D. W., Schwarz, U., Takakubo, K., & van Woerden, H. 1966, BAN, 18, 465

Klein, U. & Mack, K.-H. 1995, in ASP Conf. Ser. 75, Multi-feed Systems for Radio Telescopes, ed. D. T. Emerson & J. M. Payne (San Francisco: ASP), 318

Mitchell, K. J. & Condon, J. J. 1985, AJ, 90, 1957

Ott, M., Witzel, A., Quirrenbach, A., Krichbaum, T. P., Standke, K. J., Schalinski, C. J., & Hummel, C. A. 1994, A&A, 284, 331

Ruze, J. 1966, Proc. IEEE, 54, 633

Sieber, W., Haslam, C. G. T., & Salter, C. J. 1979, A&A, 74, 361

Takeuchi, T. T., Kawabe, R., Kohno, K., Nakanishi, K., Ishii, T. T., Hirashita, H., & Yoshikawa, K. 2001, PASP, 113, 586

Thompson, A. R., Moran, J. M., & Swenson, G. W. 1986, Interferometry and Synthesis in Radio Astronomy (New York: Wiley)

Wollack, E. J. 1995, Rev. Sci. Instrum., 66, 4305

Chris Salter talked about continuum observations after the distressing news of yet another Watford defeat.

Single–Dish Radio Astronomy: Techniques and Applications
ASP Conference Series, Vol. 278, 2002
S. Stanimirović, D. R. Altschuler, P. F. Goldsmith, and C. J. Salter

Continuum 2: Specific Applications

C. J. Salter

National Astronomy and Ionosphere Center, Arecibo Observatory, HC 3 Box 53995, Arecibo, Puerto Rico 00612, USA

Abstract. Following the previous lecture on the "General Aspects of Continuum Observing", this presentation attempts to address some of the more specific aspects of the discipline. In particular, a few thoughts on living with radio frequency interference as a continuum observer are presented. Then, following up the conjecture of the previous lecture that continuum mapping at centimeter and shorter wavelengths faces a likely period of growth, much of the rest of the lecture is dedicated to the problem of overcoming the negative effects of fluctuating atmospheric emission on high-frequency continuum imaging. In conclusion, some remarks are made concerning the inter-comparison of single-dish continuum images. These remarks also apply to comparing continuum distributions with other forms of astronomical information, such as spectral-line data cubes, infra-red, optical and X-ray images.

1. Single-Dish Continuum Observing and RFI

One of the biggest problems with using a single-dish telescope for continuum observing is the effect that radio frequency interference (RFI) has on the data collected. While it has been a major headache for many years at decimetric and longer wavelengths, encroachment by RFI has been moving inexorably to higher and higher frequencies. At opposite extremes in the characterization of these interferences are: a) those that are broad-band in nature, and often impulsive; b) others that are confined to a narrow band of frequencies. The former could be caused by such things as transient arcing, motor-vehicle ignition systems, power-line discharge in damp weather, etc., and an example is shown in Figure 1. If the interference is truly broad-band, then the only thing to do is to excise data acquired when it is present. This is, of course, a very good reason to record your data at a rate that greatly oversamples it. Impulsive RFI is spiky, and for raster mapping can usually be easily located by application of a suitable computer algorithm after filtering out the spatial frequencies that are passed by the telescope as it scans the sky (i.e. $< D/\lambda \, \text{rad}^{-1}$). If, as in Figure 1, the duration of the burst of RFI is less than the Nyquist sampling time, it can be excised without significant loss of information. If it lasts longer than the Nyquist time, then the whole interfered section has to be excised. This is, of course, a good reason to scan fast, and make a number of coverages of your region in order to fill in any "holes" caused by RFI in individual coverages. (Should your RFI

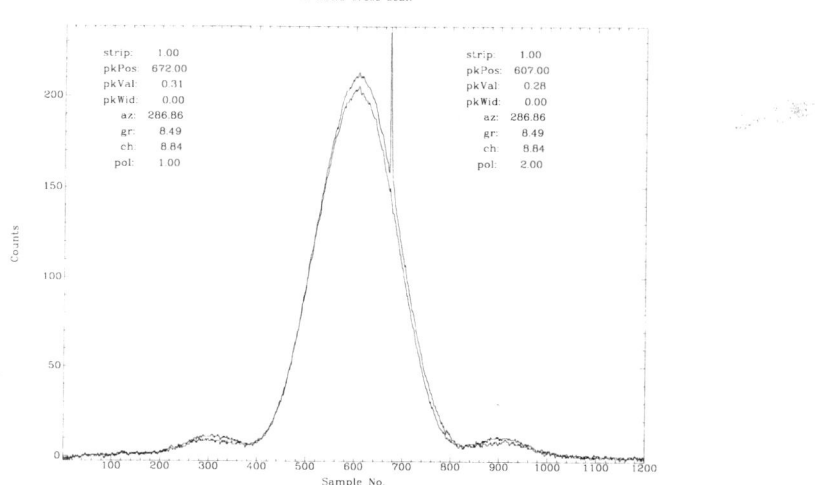

Figure 1. An impulsive RFI spike spoiling an otherwise clean continuum record. Both the vertical and horizontal polarizations are displayed, and the RFI is seen to be strongly linearly polarized.

always occur at the same celestial position, you are either incredibly unlucky, or have just rediscovered pulsars!)

On the other hand, if the RFI is narrow-band (e.g. caused by a radar, TV or radio transmissions, a satellite down-link, radio communication channel, etc.), then continuum observers have everything to gain by recording their data through a spectrometer of some variety, however coarse the resolution. Figure 2 illustrates this situation via a 430-MHz continuum scan of 12.5-MHz bandwidth taken recently at Arecibo with the telescope beam driven across a region of sky. A 64-channel spectrometer was used. The figure shows the receiver passband averaged along the entire scan, and it is seen that two narrow-band RFIs are present (in channels 18 and 48-49). In Figure 3, the data for the scan are shown as time series. From bottom to top: a) the signal is averaged over interference-free channels 23-43; b) the signal is shown separately for the individual channels 17, 18 and 19. Clearly, if at least channel 18 (plus channels 48 and 49) are excised from the data before averaging across frequency, then the data quality will be greatly enhanced over that for an all-inclusive frequency average.

2. Minimizing 1/f-Noise: The Dicke-Switched Receiver

In the previous lecture, it was pointed out that baseline fluctuations caused by small changes in the system gain and noise temperature add noise to the output of a total-power receiver (Figure 4a), and that the significance of this noise increases with the time scale considered (1/f-noise). It was further noted that above an observing frequency of about 2 GHz, fluctuating atmospheric emission

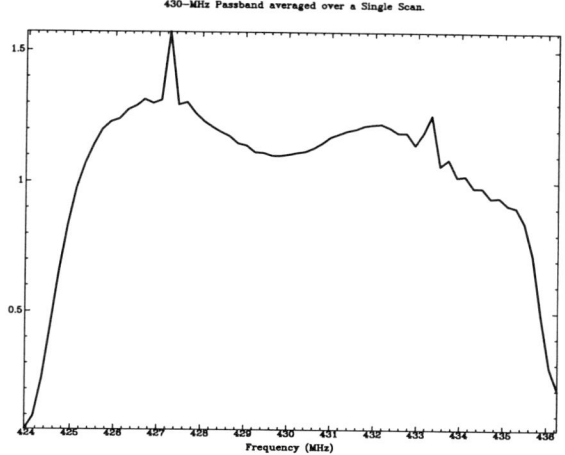

Figure 2. The 12.5-MHz receiver passband for a single 430-MHz continuum scan. There are 64 frequency channels over the spectrum, and a pair of narrow-band RFIs are seen in channels 18 and 48-49.

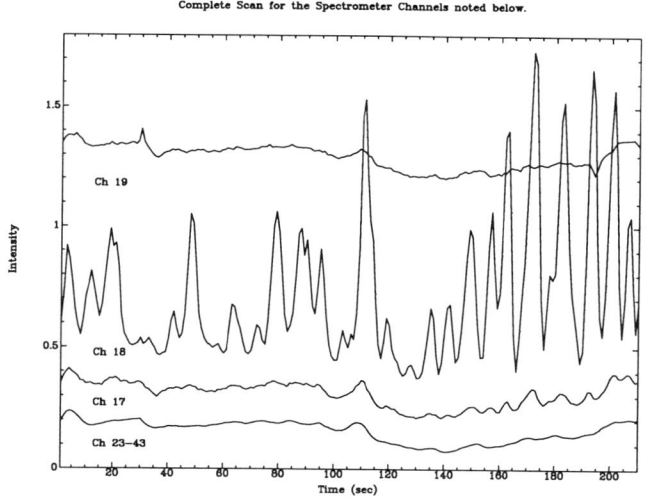

Figure 3. The data along the single continuum scan whose integrated passband is shown in Figure 2. The spectrometer channel numbers for each time series are marked below the plots. It is seen that while the data from channels 17, 18 (especially) and 19 individually show the effects of an RFI, the record representing the integrated signal across channels 23 – 43 is RFI-free, deviations along the scan representing true astronomical signals. Note that the Arecibo records are completely "confusion limited" (see the previous lecture) at 430 MHz.

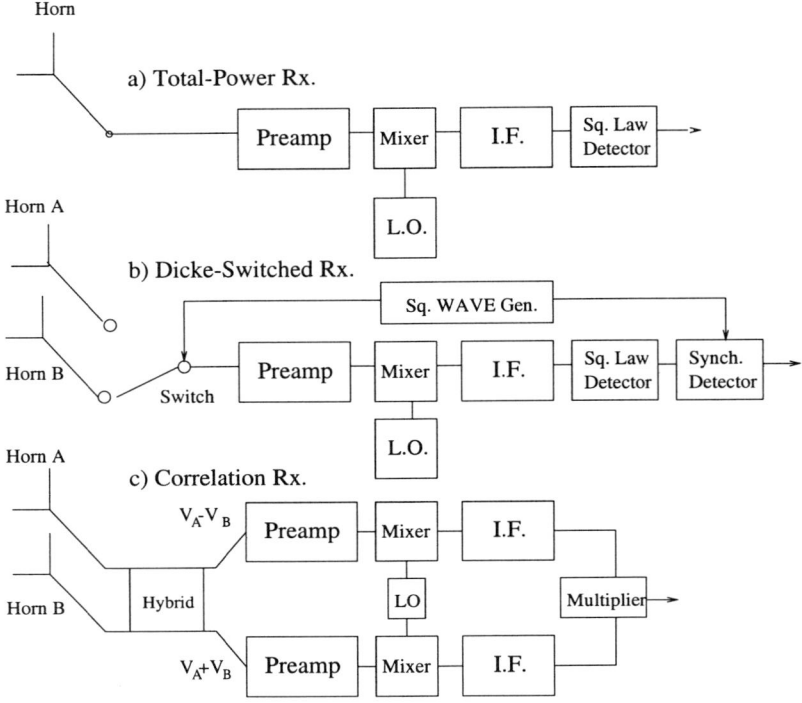

Figure 4. Schematic diagrams to illustrate, a) the total-power receiver, b) the Dicke-switched receiver, and c) the correlation receiver.

due to water vapor can cause extremely serious baseline fluctuations having the character of 1/f-noise. Interestingly, while this fluctuating atmospheric emission component is so troublesome to continuum observers that it can (and does) make observing with a total-power receiver impossible for them, it does not usually affect spectral-line observations seriously. This is because the atmospheric emission fluctuations have a very broad frequency spectrum and thus do not significantly corrupt spectral baselines for the relatively narrow range of frequencies over which spectral-line observers invariably make their measurements. For this reason, it is far from unusual that a single-dish continuum observer using short-centimeter or millimeter wavelengths will hand over the telescope to a spectral line back-up experiment when conditions become unduly damp.

However, continuum observers do have a line of defense against 1/f-noise. They can switch cyclically between the main telescope feed and another horn also situated in the focal plane that is pointing at a nearby "reference" region. If this switching is done sufficiently rapidly that its period is shorter than the time scales on which the 1/f-noise dominates, then subtraction of the signals from adjacent phases will minimize the effects of the 1/f-noise as signal from the two horns contain different "astronomy", but very similar atmospheric contributions (see below).

The simplest way of achieving such "beam switching" is via the so called Dicke-switched (DS) receiver, named after its inventor, Robert Dicke. In this system, the input to the receiver chain is typically switched at a rapid rate between a pair of horns separated in the focal plane of the antenna (Figure 4b). The switching is controlled by means of a square-wave generator, which also acts as the reference for a synchronous detector situated immediately after the square-law detector. In this way, the output of the receiver is the difference of the powers supplied by the two horns, i.e. Output Power, $P_O(DS) = Power_{Horn\ A} - Power_{Horn\ B}$. Now, if powers from the two horns represent antenna temperatures of T_A and T_B respectively, with the receiver having a noise temperature of T_R and a gain of g, then;

$$P_O(DS) = k(T_R + T_A)g\Delta\nu - k(T_R + T_B)g\Delta\nu , \qquad (1)$$

where, k is Boltzmann's constant, and $\Delta\nu$ is the bandwidth. Thus,

$$P_O(DS) = k(T_A - T_B)g\Delta\nu . \qquad (2)$$

Thus, the output power is proportional to the difference in the signal powers received from the two feeds. We can now compare this with the output of the total-power (TP) receiver, for which;

$$P_O(TP) = k(T_A + T_R)g\Delta\nu . \qquad (3)$$

For a gain change of magnitude Δg, the ratio of the change in the output power is;

$$\frac{dP_O(DS)}{dP_O(TP)} = \frac{(T_A - T_B)}{(T_A + T_R)} . \qquad (4)$$

As $T_A \sim T_B$ for measurements of weak sources, the ratio of the change in output powers is then $\ll 1$. This means that the Dicke-switched system is very much less susceptible to gain changes than a total-power receiver. In addition, provided that the receiver temperature, T_R, changes much less rapidly than the switching rate, then from Equation 2, we see that changes of receiver temperature do not affect the output of the Dicke-switched receiver significantly. Better still, while the two beams of the Dicke system are pointed in angularly nearby (but separate) directions in the far-field of the telescope, they hardly diverge from each other in its near field. In addition, the far-field distance, $(2D^2/\lambda)$, is well above the troposphere for any medium or large diameter telescope working at centimeter (or shorter) wavelengths. Thus, as long as the switching rate is sufficiently rapid (typically at least a few Hz) that significant fluctuations in the atmospheric emission vary more slowly than this, the Dicke system cancels the atmospheric emission to high order, and an essentially noise-limited performance is obtained.

In practice, we have not got something for nothing here. In an ideal world, the noise of the simple total-power receiver of Figure 4a is $T_{sys}/(\beta\tau)^{1/2}$, where β is the receiver bandwidth and τ is the integration time. Now, for the same total integration time, the Dicke-switched receiver only looks at the target source for $\tau/2$, but is recording noise for the whole time, τ. Thus, the noise on its output will be $2T_{sys}/(\beta\tau)^{1/2}$. Nevertheless, this factor of two is often well worth accepting, as without the use of switching the observations could well be impossible to make.

In practice, the situation can be improved still further, albeit at the cost of extra complexity, by having a second switch operating in antiphase with the first, and feeding its signal into a second, identical, receiver chain. Here, when one receiver chain is observing the target source, the other chain is looking at the reference region, and vice versa. For this "Double-Dicke System", after averaging the outputs of the two receiver chains, the noise level is $\sqrt{2}T_{sys}/(\beta\tau)^{1/2}$.

As drawn in Figure 4b, the Dicke system has the additional disadvantage of placing a lossy element, the electronic switch, at the very front end of the receiver, where it will certainly make an undesirable contribution to the system temperature. For some telescopes, this can be circumvented with beam switching being achieved by "nutating" or rocking some component of the telescope optics, often the subreflector. In addition, such an arrangement furnishes a simple way of beam switching a focal plane array of N receivers, giving the equivalent of N simple Dicke-switched receivers without recourse to the electronic complexity (and associated loss) of a separate front-end switch for each receiver pair. Alternatively, at millimeter wavelengths a quasi-optical rotating or oscillating chopper has sometimes been employed to achieve the beam switching.

In practice, when observing a small diameter source, many observers using a Dicke system choose to first record data for a given number of switching cycles with the target source located in the beam of one of the two horns. They then move the telescope such as to place the source in the beam of the second horn for the same number of cycles. This is repeated a number of times. From Equation 2, the output of the receiver is $kT_{source}g\Delta\nu$ when the source is targeted by Horn$_A$, and $-kT_{source}g\Delta\nu$ when Horn$_B$ is pointed at the source. Hence, the output signal is a "square-wave" of twice the deflection of a single pointing. In this way, the need to measure an "off-source zero level" is avoided.

3. The Correlation Receiver

An alternative method for achieving stable continuum observations in the presence of 1/f-noise is the Correlation Receiver (see Kraus 1986), shown in Figure 4c. Like the "Double Dicke System" this also uses two identical receiver chains, which in this case share a common "in-phase" local oscillator. By adding a 180°-hybrid ring at the receiver front end, the signals from the two horns are split and mutually combined such that they travel down one receiver chain in phase, and the other receiver chain in antiphase, i.e. the voltage in one arm is (V_A+V_B), but (V_A-V_B) in the other. When these two voltages are cross multiplied, the output is $\propto (V_A + V_B) \times (V_A - V_B) = (V_A^2 - V_B^2)$, which is proportional to the power difference between horns A and B, as for the Dicke-switched system, (Equation 2). As for the "Double Dicke System", the noise level is $\sqrt{2}T_{sys}/(\beta\tau)^{1/2}$.

In passing, the correlation receiver is easily transformed into a correlation continuum polarimeter. To achieve this, instead of connecting the signals from a pair of horns via the hybrid ring to the inputs of the identical receiver chains, the signals from the orthogonal polarizations of a single feed are connected to the receiver chains. These are then cross-multiplied in phase and in phase-quadrature in a pair of multipliers. If the input signals represent the opposite hands of circular polarization, then the outputs of the two multipliers are proportional to the "linear" Stokes parameters, U and Q (Kraus, 1986).

4. Software Beam Switching

Given the high level of stability of the modern total-power radio receiver, a post-recording approach to beam switching has been developed in recent years, especially at the MPIfR, Bonn (e.g. Morsi & Reich 1986). In this method, two or more feeds are placed in the focal plane, each having their own (dual-polarization) total-power receiver chain. Each receiver is equipped with a sophisticated calibration system with which to continuously monitor the system gain with high precision. The signals from each feed are calibrated off-line, and then mutually differenced to produce beam-switched total-intensity outputs. This technique has the great advantage that there are no extra lossy electronic components at the receiver front ends, such as switches or hybrids. Further, for an N element feed array, all $N \times (N-1)/2$ independent beam-switched combinations can be obtained simply. As the target and reference regions are both observed all the time, all else being equal, the noise level is again given by $\sqrt{2}T_{sys}/(\beta\tau)^{1/2}$.

5. Imaging with a Dual-Beam System

Single-dish telescopes are often called upon to survey regions of the sky via raster mapping. For this, the telescope beam is scanned along a series of parallel tracks in the sky, each spaced equidistantly from its nearest neighbors. Either all scans can be made driving the telescope in the same direction (with a "fly back" across the region before each new scan) or, to minimize wasted time at the scan turn around, by reversing the scanning direction between each row. In the latter case, reversal of scanning sense from row to row has to be taken into account during the subsequent map making.

If a dual-beam system (such as those described in Sections 3, 4 & 5) is used to map a region via raster scanning, each source in the region will appear in the data twice; once when crossed by the "main" beam, and again when traversed by the "reference" beam. As the output signal of a beam-switched receiver is the difference of the power between the two beams (Equation 2), a source will produce a positive deflection when in the main beam, and a negative deflection when in the reference beam. Of course, if the total size of the emission region is greater than the angular separation of the beams, the positive and negative responses produced by beam-switching will overlap and lead to a distorted image of the source structure.

The seminal study of this problem was made by Emerson, Klein & Haslam (1979; EKH), and the rest of this section will draw very heavily upon their results. In that work, the authors present 10.7-GHz images of Tycho's supernova remnant (SNR) made with the Effelsberg 100-m radio telescope using: (a) single-beam data (Figure 5), demonstrating the ravages caused by the 1/f-noise of atmospheric emission, and (b) dual-beam data taken in parallel, showing the ability of this techniques to cancel these "weather effects". However, the beam-

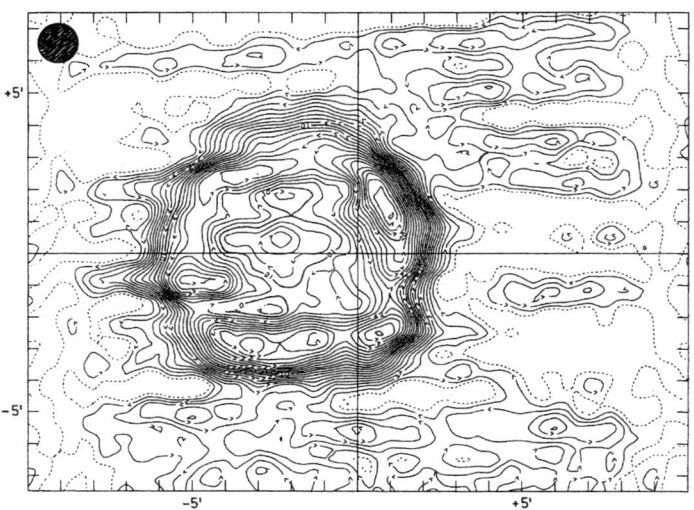

Figure 5. A 10.7-GHz single-beam map of Tycho's SNR made with the Effelsberg telescope. The effects of variable atmospheric emission are clearly seen in the image as (horizontal) stripes along the direction of scanning – in this case azimuth offset.

throw of the dual-beam data was little more than one half of the diameter of the SNR, and the distortion of the image caused by both beams being "on the source" simultaneously is clearly seen (Figure 6). The problem is how to process this dual-beam image such as to recover the brightness distribution that would have been observed by a single beam *without* the adverse influence of atmospheric emission. The direction of the raster scanning was along the line of separation of the two beams (azimuth in this case), making each individual scan an object for image restoration (or as it is often known, "deconvolution").

Looking at the problem in the abstract, the single-beam image is simply the convolution of the true sky distribution with the detailed (single-)beam pattern; see plots (i) and (ii) of the left hand pane of Figure 7a. The dual-beam response – plot (iv) – is, however, the convolution of the single-beam pattern (plot ii) with a "dual-beam function" consisting of a pair of delta functions of opposite sign – plot (iii). Thus, the actual dual-beam output, i.e. the convolution of plots (i) & (iv), is seen to be the same as the mutual convolution of the three quantities in plots (i) through (iii). Clearly, our problem is effectively how to "deconvolve" the dual-beam function of plot (iii) from this output. EKH pointed out that in the spatial frequency plane, the Fourier transform of the mutual convolution of plots (i) - (iii) is the mutual product of their Fourier transforms (plots v - vii of Figure 7a). In other words, the single-beam image could be recovered by multiplying (reweighting) the Fourier components of the dual-beam image by the reciprocal of plot (vii). The transform of the dual-

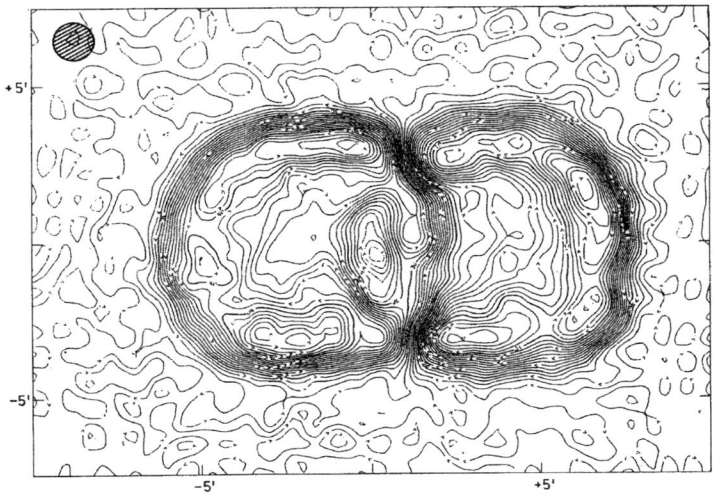

Figure 6. A dual-beam map of Tycho's SNR. This was made from data taken simultaneously with the single-beam picture in Figure 5. Although the atmospheric effects seen in Figure 5 have been essentially eliminated, the positive and negative responses to the SNR overlap on the image, distorting its true structure.

beam function, and the reciprocal of this are shown in the top two plots of Figure 7b. It is seen that this reciprocal is a cosecant function, and hence contains troublesome singularities. However, the equivalent of multiplying by this in the transform plane is to convolve the basic dual-beam scan by its Fourier transform, which has the simple form shown at the bottom of Figure 7b. This "restoring function" contains none of the singularities seen in its transform. As proof of its effectiveness, the restored map of the dual-beam image of Figure 6 is presented in Figure 8.

EKH also pointed out that a suitable combination of observations made of the same field with two or more *non-commensurate* beam spacings would permit greatly improved mapping of fields whose overall sizes are a relatively large number of dual-beam spacings by eliminating the nulls seen in the transform plane for observations made with a single spacing (see plot viii of Figure 7a), and which would have produced undesirably high noise for their spatial frequencies. A state-of-the-art simulation applying this approach to wide-field mapping is to be found in Emerson (1995).

In recent years, a number of other approaches to solving the image reconstruction problem for beam-switched data have been adopted, including the application of the Maximum Entropy Algorithm (Richer 1992) and matrix inversion approaches (Wright, Hinshaw, & Bennett 1996). A recent inter-comparison of the results of using these different techniques for the deconvolution of bolometer array data from the JCMT can be found in Johnstone et al. (2000).

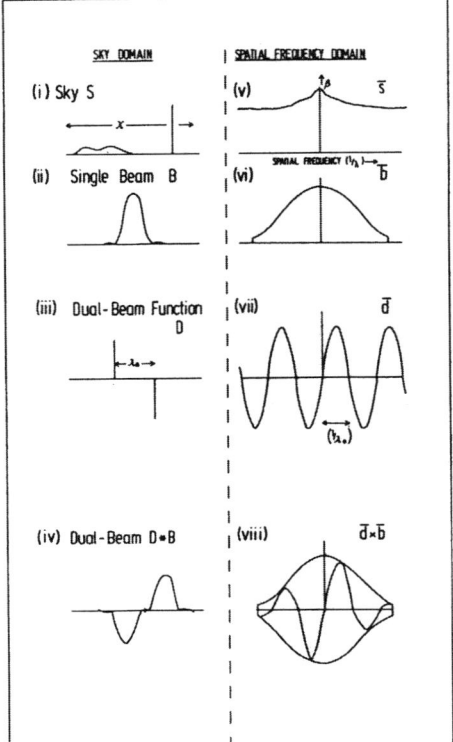 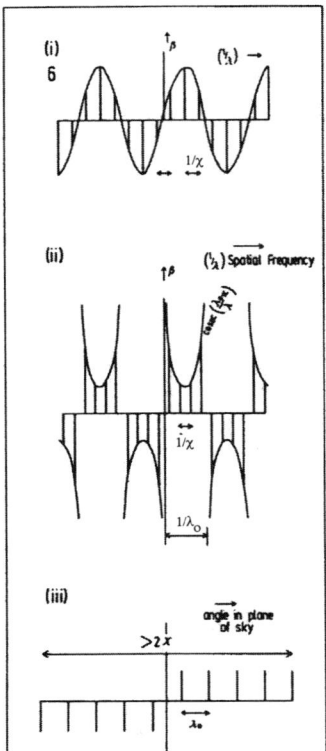

Figure 7. a) Left: a schematic representation of (i) the true brightness distribution, (ii) the single beam, (iii) the dual-beam function, and (iv) the actual dual beam. The respective Fourier transforms of these are shown in plots (v) - (viii); b) Right: shows the Fourier transform of the dual-beam function – as in plot (iii) of a) – in (i), its reciprocal in (ii), and the Fourier transform of this in (iii).

6. Comparison of Continuum Images

One reason for making a continuum image is to compare it with others made at different frequencies to achieve such objectives as determining the distribution of spectral index over a source, the separation of thermal and non-thermal emission in a region, or to derive the distribution of rotation measure and depolarization over a source from linear polarization data. It is generally good advice that one should only compare apples with apples. The radio-astronomical equivalent of this maxim is that if images made at two (or more) different frequencies are to be inter-compared, then before this can be meaningfully undertaken, the images should be brought to a common angular resolution. Inevitably, this will be that of the lowest resolution image used. Preferably, the images to be compared should not only have the same gross resolution (i.e the same main-beam

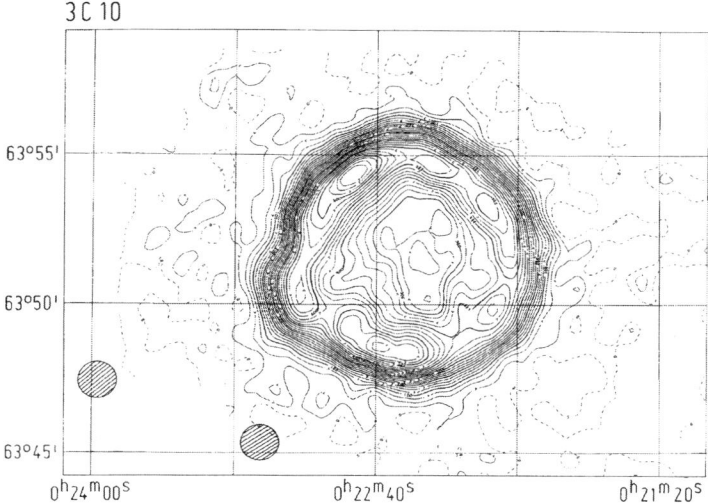

Figure 8. The deconvolved image of Tycho's SNR. This image has been derived from Figure 6 by application of the EKH algorithm, with the result transformed into equatorial coordinates.

FWHM), but appear as if observed with identical detailed beam patterns. This has sometimes been achieved via careful design of the observational equipment itself using the so-called "scaled array" approach. For this, the antennas at the different frequencies are constructed such as to be in every way identical, except that their dimensions are scaled by the ratio of the wavelengths (e.g. Webster 1974). It is to be noted that the VLA can achieve a good approximation to being a "scaled array" via its different size configurations which can be used to give similar resolutions at very different wavelengths.

Sadly, any particular single-dish does not have this ability to "stretch and contract", and so images made by a given telescope at different wavelengths have different resolutions (i.e. FWHM$\sim 1.2\lambda/D$). Fortunately, one great bonus of single-dishes is that they respond to all spatial frequencies between 0 and D/λ rad^{-1}. Hence, all the spatial frequency information that is needed to reduce the resolution of the higher frequency image to that at the lower frequency is present in the former.

The actual matching of resolutions between any two images can be achieved either by application of a suitable convolving function to the image made at the higher resolution, or via multiplying its Fourier transform by an appropriate transfer function. However, the ideal of achieving *exactly* the same effective beam pattern between two maps is not simple for images from a typical alt-azimuth telescope. During the time it takes to scan a given region, the detailed beam pattern rotates relative to the celestial plane by an amount equal to the change in parallactic angle. This can occasionally be circumvented by employing a very systematic observing procedure, such as the NODing technique used for the 408-MHz All Sky Survey (Haslam et al. 1982). For that survey, as for the

Green-Bank 300-ft surveys described in the previous lecture, all observing was performed with the telescope azimuth fixed at that of the prime meridian. The telescope was then driven only in elevation, so that the orientation of the telescope sidelobe pattern remained fixed relative to the celestial plane. Of course, the new Robert C. Byrd Green Bank Telescope (GBT) will offer unprecedentedly low single-dish sidelobe levels, with its unblocked aperture providing great symmetry to its (very low level) sidelobe structure. It should be a single-dish continuum observer's dream instrument. In practice, the issue of exactly matching the beam patterns of images before comparison is usually "fudged", with the smoothing of the higher resolution image being made using a Gaussian convolving function for simplicity. Providing that the frequencies are sufficiently separated, meaningful conclusions can often still be drawn. For example, if images separated in frequency by an order of magnitude are compared, say a future comparison between an Arecibo image at 2.3 GHz, and a 25-m telescope image at 23 GHz, then even a 10% error in the intensity estimated at one frequency would translate into an error in the derived spectral index of only $\Delta\alpha = 0.04$. However, let him beware who tries to derive spectral indices from images made at closely spaced frequencies; he needs to tread very carefully indeed.

A final cautionary word for newcomers to continuum polarization is to remind them that when they need to smooth linear polarization data for intercomparison, they will have to smooth the distributions of Stokes parameters Q and U separately and then combine these in order to recover the polarized intensity, I_p, and polarization angle, χ, via $I_p = (Q^2 + U^2)^{1/2}$ and $\chi = 1/2 \tan^{-1}(U/Q)$. To illustrate such a comparison, Figure 9 shows a study of the polarization properties of the SNR, G93.3+6.9 using data from the Effelsberg 100-m telescope at λ6- and 11-cm (Lalitha et al. 1984). The shorter wavelength data were smoothed to the resolution of the longer wavelength image before their combination to derive the displayed properties of the source.

References

Emerson, D.T. 1995, in ASP Conf. Ser. 75, Multi-Feed Systems for Radio Telescopes, eds. D.T. Emerson & J.M. Payne (San Francisco: ASP), 309

Emerson, D.T., Klein, U., & Haslam, C.G.T. 1979, A&A, 76, 92

Haslam, C.G.T., Salter, C.J., Stoffel, H., & Wilson, W.E. 1982, A&AS, 47, 1

Johnstone, D., Wilson, C.D., Moriarty-Schieven, G., Glannakopolou-Creighton, J., & Gregersen, E. 2000, ApJS, 131, 505

Kraus, J.D. 1986, Radio Astronomy (2nd edition; Powell, Ohio: Cygnus-Quasar Books)

Lalitha, P., Salter, C.J., Mantovani, F., & Tomasi, P. 1984, A&A, 131, 196

Morsi, H.W. & Reich, W. 1986, A&A, 163, 313

Richer, J.S. 1992, MNRAS, 254, 165

Webster, A.S. 1974, MNRAS, 166, 355

Wright, E.L., Hinshaw, G., & Bennett, C.L. 1996, ApJ, 458, L53

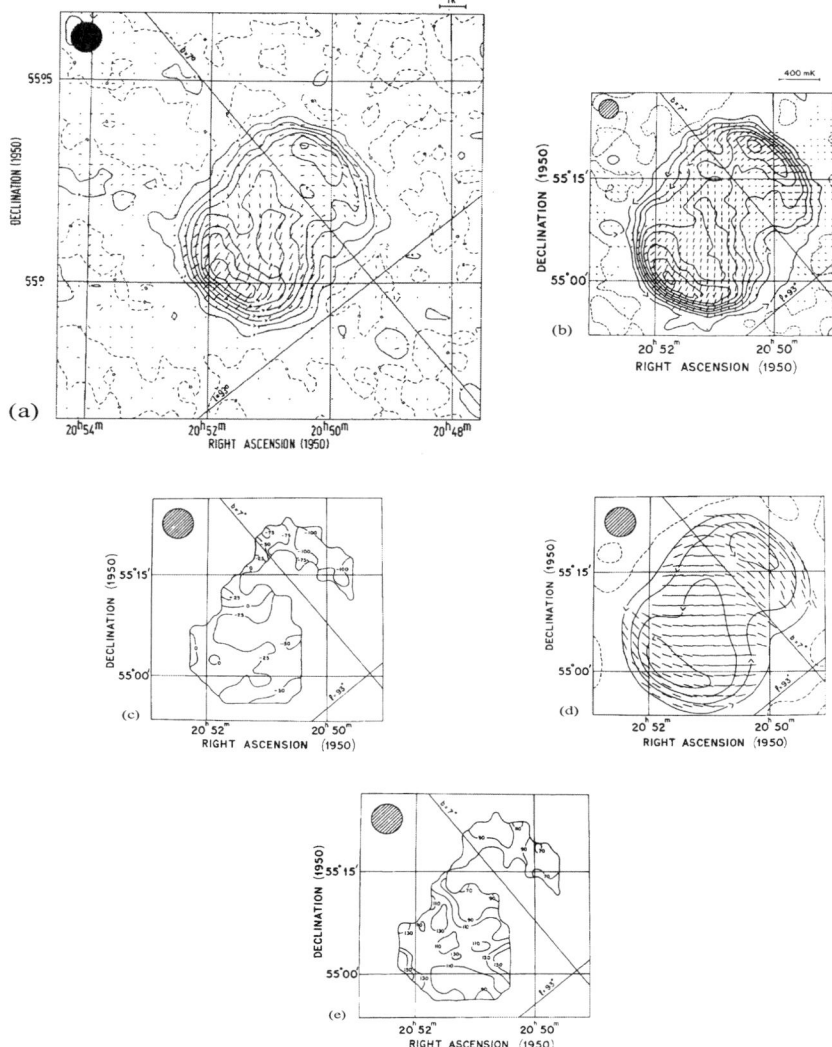

Figure 9. Continuum images of the SNR G93.3+6.9 and distributions of derived properties. Image a) is at $\lambda 11$cm with resolution 4.4 arcmin, and b) is at $\lambda 6$ cm with resolution 3 arcmin. Both are total-intensity images, with vectors representing the linearly polarized intensity and position angle overlaid. The derived quantities at a resolution of 4.4 arcmin are, c) the distribution of rotation measure in $\mathrm{rad\,m^{-2}}$, d) the projected magnetic field direction within the SNR overlaid on the smoothed $\lambda 6$-cm total-intensity contours, and e) the distribution of depolarization between the two wavelengths in percent.

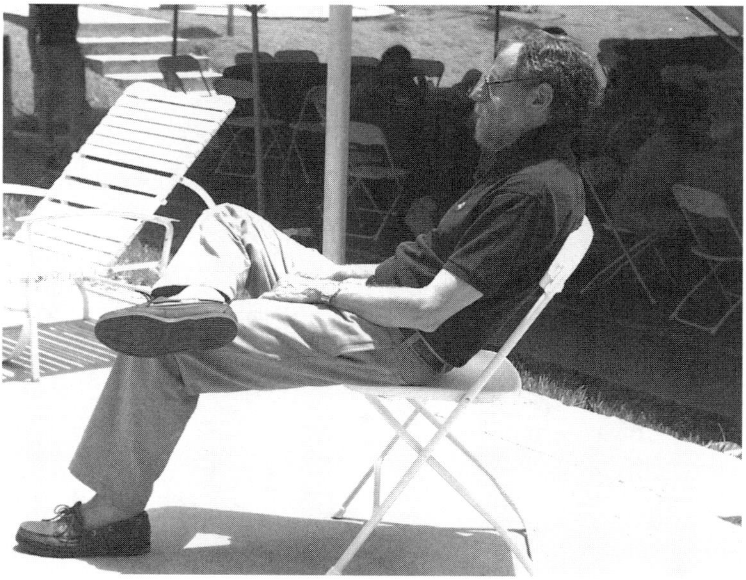

Thinking...

The Rudiments of Spectral Line Radio Astronomy

Harvey S. Liszt

National Radio Astronomy Observatory, 520 Edgemont Road, Charlottesville, Virginia 22903, USA

Abstract. The basics of spectral line work from A's to Z, dense but physically motivated.

1. Introduction

Spectral lines occur in nature because the internal structure of atoms and molecules is quantized and because the excited states of atomic systems spontaneously decay to less-excited ones. Spectral lines are studied at radio astronomical frequencies for a wide variety of reasons, partly having to do with atomic and molecular physics and partly having to do with the conditions in space. When conditions are conducive to the presence of atoms or molecules in the proper states, the radiation they produce may make the long journey to Earth and be of use to us. To understand the information which is contained in spectral line radiation, we must attempt to understand many different aspects of physics: the atomic structure which creates the energy levels; the thermodynamics and statistical physics which prepare the atoms and molecules to radiate; the transfer equations which describe how radiation interacts with intervening matter and propagates to us through space; the electrical engineering which describes how this radiation interacts with our telescopes; and the various computer manipulations which turn bits into data.

So there is an enormous amount of ground to cover. Here I have tried to put down as much as possible of what I have always wanted my summer students to know before coming to the National Radio Astronomy Observatory (NRAO) to work with me. I have tried to motivate the often dry subject matter with physical insight. Although the material may seem rather bloodless, it cannot be properly used except with feeling. This being the background for a lecture described in the course syllabus as 'Spectral Line I', the discussion is lamentably short of details on the astronomy which has been gleaned from our hard-won understanding of the subject. That I leave to the other lecturers and to the hands-on experience of the students during this course.

2. Radiative Processes and Thermodynamic Equilibrium

Einstein elaborated the basic radiative processes in atomic systems in 1905 well before the advent of new quantum mechanics or even the Bohr atom; they derive solely from the possibility of thermodynamic equilibrium and the form of Planck's law (whose meaning they clarify). Consider two states labelled by

l(ower) and u(pper), where u lies higher in energy. Their internal energies are E_u and so forth. Consider that each has hidden internal structure such that there are g_l or g_u independent ways to be in states l or u. These are called their statistical weights and they simply leave open the possibility that there is other knowledge whose details we can leave for later. We would say here that these independent possibilities are degenerate – they are treated as being coincident – but under special conditions (recall the Stern-Gerlach experiment?) they might be made to be distinguishable. The sub-levels could themselves have degenerate sublevels and so on.

Einstein showed that there must be three basic processes connecting two levels *via* electromagnetic radiation; their strengths are all proportional but at least one of them was utterly new to the world of physics when he discovered it. The one truly classically understandable process is absorption of radiation at wavelength $hc/(E_u - E_l)$ (frequency $\nu_{ul} = (E_u - E_l)/h$) and the strength of this process is denoted by an atomic coefficient denoted by B_{lu}. What Einstein realized was that there must also be an analogous inverse process $B_{ul} = (g_l/g_u)B_{lu}$ called stimulated emission. When radiation of the proper wavelength passes by a system in the excited state, there is some chance ($\propto B_{ul}$) that it will induce decay, and that the excited state will emit radiation in the same direction, and with the same energy and phase, as that which was incident. This is why we can make lasers and masers.

Einstein's remaining insight (into this matter) was the spontaneous decay of excited states. The u-state will decay with a constant probability per unit time, called A_u; the population of the u-state of an isolated atom declines as $e^{-A_u t}$. The state u may lie above and have the possibility of decaying into more than one lower state so that $A_u = \sum_l A_{ul}$. When a system in the u-state decays, it does not know ahead of time for which lower state it is destined. At each decay, the lower states are populated in proportion to their respective probabilities or rates A_{ul}. It also transpires that the finitude of A_u induces into all the transitions $u \to ?$ [1] an element of non-monochromaticity in the energy of the emitted radiation. Because of the uncertainty principle, the radiation is spread over a spectrum whose width in frequency is of order A_u. This so-called "natural" width is a property of the state u, not of any particular transition $u \to l$. If the state u decayed instantly, A_u and the spectral width of all its transitions would (unpleasantly) be infinite.

Spontaneous emission – unlike stimulated emission – is isotropic and of random phase. A region having n_u atoms (cm^{-3}) in state u emits energy in the $u \to l$ transition at a rate $n_u h \nu_{ul} A_{ul}/4\pi$ (erg cm^{-3} s^{-1} per steradian) over the sky. Thus, if we observe a certain amount of energy coming from a region, we can under the simplest circumstances infer how many excited state systems are responsible. Of course the number of systems in the excited state may be a small fraction of the total, but what is important is that this fraction may sometimes be *calculated*. However, life is rarely this straightforward.

So what are these Einstein A's and B's (as they are called)? The A-coefficients are calculable from atomic physics and the B's are proportional to them (hence, at least some Einstein B's would be infinite if the A's were, another

---

[1] Because, when the decay is initiated, the final state is, effectively, still in doubt.

unfortunate circumstance). There are different gauges possible for the definition of the B's but since the time of the first edition of Spitzer's (1978) book it has been most common to employ the radiation density $U_\nu(T) = (4\pi/c)B_\nu(T)$ where $B_\nu(T)$ is the Planck function at temperature T (this is an unfortunate but utterly standard overuse of the second letter of the alphabet). Thus we write

$$B_{ul} = (g_l/g_u)B_{lu} = \frac{c^3}{8\pi h\nu_{ul}^3} A_{ul} \,, \qquad (1a)$$

so that the quantities $U_\nu(T)B_{ul}$, $U_\nu(T)B_{lu}$ and A_{ul} are all rates (units of s^{-1}). When we formulate the problem of determining how many systems will be in the various states l and u, these will be the rates which enter calculations of the equilibrium population densities. To satisfy the optical astronomers, we note that A_{ul} is related to the oscillator strength f_{lu} by:

$$A_{ul} = \frac{8\pi^2 e^2 \nu^2}{m_e c^3}(g_l/g_u)f_{lu} \,. \qquad (1b)$$

There just is not a one-size-fits-all example of what the Einstein A-coefficients are, nor are the strengths of radiative transitions assessable in such a simple way. But in general, the lifetimes of the states observed at radio wavelengths are much much much much longer than those seen at optical wavelengths. For instance, the A-coefficient for the upper state of the Lyman-α transition of neutral atomic hydrogen (H I) is 5×10^8 s^{-1}. On the other hand, the famous λ21cm hyperfine line of H I has an A-coefficient of only 2.7×10^{-15}s^{-1}. Note however, that the Einstein B-coefficients are proportional not just to A_{ul}, but to $A_{ul}\lambda^3$. Since radio wavelengths are so much longer, even rather weak transitions with rather long lifetimes can be good absorbers. And if they are good absorbers, we can see them.

3. Doppler Shifts and Line Broadening Mechanisms

3.1. Prelude: Doppler shifts, redshifts and restframes

Because of the Doppler effect – the shift in frequency/wavelength which is manifested when electromagnetic radiation is observed from a platform which is in motion with respect to the rest frame in which the radiation was emitted – spectral lines are broadened about and/or shifted away from their restframe or laboratory frequency/wavelength. The broadening of spectral lines will occupy the rest of this Section; here we deal with some gross effects, external to an emission region, which move its spectral lines about in the electromagnetic spectrum. We ignore, for instance, the gravitational redshift and other exotic effects, such as the transverse Doppler shift. Only the component of motion along the line of sight to an object contributes to its Doppler shift under ordinary circumstances. We only sense that component of an object's space velocity which projects onto our line of sight to it, one of the three about which it might actually be moving. So if we look at an ensemble of bodies with truly random space velocities we would expect to see a zero mean and a dispersion in radial velocity equal to $1/\sqrt{3}$ of the three-dimensional velocity dispersion in the ensemble.

Relative motion of the source and observer can arise from, for instance (moving outward), the rotation of the Earth, the motion of the Earth about the Sun, and from various perturbations of the orbit by other bodies: motion of the emitting body itself will be considered separately. The various velocity vectors are added (vectorially) and projected onto the line of sight to the observed object. Left to themselves, optical astronomers employ a heliocentric rest frame while radio astronomers refer motions to the Local Standard of Rest (LSR), the mean motion of the stars in the Solar Neighborhood, about which the Sun has its own peculiar movement. The Solar Motion is approximately $V_{Sun} = 19.7$ km s^{-1} toward $\alpha_0, \delta_0 = 271°, +30°(1950)$ and the projection of this motion onto a line of sight in the direction (α, δ) of the same epoch is such that

$$V_{lsr} = V_{hel} + V_{Sun}(\cos(\alpha_0 - \alpha)\cos(\delta_0)\cos(\delta) + \sin(\delta_0)\sin(\delta)) . \qquad (2)$$

The radio definition often eliminates confusion as to why gas appears at a particular heliocentric velocity in some direction (*i.e.* it is at rest with respect to the LSR but not the Sun) but the optical definition is perhaps more immediate.

The redshift of an object is defined in the context of optical astronomy (there is no such thing as a radio redshift, really) as

$$z_{opt} \equiv \lambda/\lambda_0 - 1 , \qquad (3a)$$

$$v_{opt} \equiv cz_{opt} , \qquad (3b)$$

where λ_0 is the laboratory or rest wavelength of a transition and λ is the observed wavelength in the heliocentric rest frame. The optical velocity of an object is *defined* as $v_{opt} \equiv cz_{opt}$ and as such is simply immune to questions of correctness. But physically, the optical velocity so defined is equal to the actual physical velocity of the object only in the limit of very small redshift.

Radio astronomers (mercifully) have no definition of a radio redshift on their own, but they do use a different definition of *velocity* such that

$$\nu \equiv \nu_0(1 - v_{radio}/c) . \qquad (3c)$$

Thus for any given object,

$$v_{opt} = \frac{v_{radio}}{(1 - v_{radio}/c)} , \qquad (3d)$$

$$v_{radio} = \frac{v_{opt}}{(1 + v_{opt}/c)} . \qquad (3e)$$

When the redshift of an object is very small, the radio and optical definitions of its velocity coincide with each other and with the physical velocity. Then, to calculate the frequency or wavelength at which radiation will appear to our instruments, it is sufficient to add all velocity vectors (linearly, vectorially) and project them onto the line of sight to the object. When the redshift is larger than, say, a few thousand km s^{-1}, this is incorrect because v_{opt} and v_{radio} are unphysical and the usual relativistic formula for addition of velocities is inapplicable. Instead what is done is to redshift the rest frame wavelength using the actual z of the object and then to apply local corrections for orbital motion and so forth; the redshifted wavelength is treated as the rest wavelength at which radiation appears in our vicinity, which is then shifted to and from by local effects.

3.2. A helpful mnemonic device

The amount of Doppler shift (km s^{-1}) corresponding to 1 MHz is equal to the wavelength measured in mm. For the HI line at $\lambda = 21.1$ cm, 1 MHz subtends to 211 km s^{-1}.

3.3. One effect which is negligible at radio wavelengths

The natural line width of the u (upper) level of a transition $u \to l$ is responsible for a semi-classical phenomenon called damping which should be familiar from college-level E&M. The model is a harmonic oscillator at some line frequency and the spontaneous decay rate is treated as a drag or damping term. This produces an exponential time-decay of the amplitude which, via complementarity, becomes a finite width in frequency (of order $A_u/2\pi$) when the Fourier transform is taken. In detail, the frequency profile so formed varies as $(\nu - \nu_{ul})^2$ about the line center, and is dubbed (a) Lorentzian (profile).

The phenomenon of damping is quite important at optical wavelengths where the Einstein A's are large, but is of little significance in radio astronomy because the upper states of the transitions we observe generally do not have the possibility of decaying at optical wavelengths as well. In this case, there are two possibilities for making a finite linewidth: disordered (turbulent) and ordered motions.

3.4. Thermal motions and microturbulence

At any point in the medium, there will usually be some well-defined kinetic temperature characterizing the motions of individual particles from Maxwellian physics, or at least, this is often assumed to be the case. Keep in mind however, that the gases we study are rather dilute, conditions in the interstellar medium (ISM) are not that steady, and thermal equilibrium is not always a given: the kinetic temperature locally could still be different from that of the processes which are driving the physics. In any case, this temperature is called the kinetic temperature T_K and for particles of mass m, there will be a one-d dispersion to the particle velocities, of size kT_K/m. This component of the motion occurs on the smallest considered scales and varies with the mass of the particles.

Additionally, it is typical to allow for the possibility of a so-called microturbulent velocity contribution whose dispersion (yes, it is also assumed to be Maxwellian) is defined as $v_{\text{turb}}^2/2$. v_{turb} is the microturbulent velocity and it is independent of the particle masses; the physical assumption is that small gas parcels containing all kinds of particles are bodily moving, carrying their constituents along together, on small scales.

The thermal and microturbulent contributions are assumed to add in quadrature, as follows, giving rise to an equivalent temperature called the Doppler temperature T_D:

$$\sigma_v^2 = kT_D/m = kT_K/m + v_{\text{turb}}^2/2 \ . \tag{4}$$

The Doppler temperature is just the temperature the gas would need to have to give particles of mass m the motion arising from this addition of turbulent and thermal velocities.

Are we being consistent? What does small scales mean here? Imagine photons scattering in a gas; if they typically travel a physical distance λ (which will be a mean free path ...) before being absorbed in the gas, the microturbulence should be established on scales no larger than this or the details of individual turbulent elements will be dominant, rather than their ensemble average. This other situation is usually called mesoturbulent (when λ and the turbulent size scales are comparable) or macroturbulent (as the size scale of the turbulence increases). Only microturbulence is a tractable physical situtation and turbulence, of course, is the most complicated thing in all of physical science.

3.5. Line broadening through ordered motions

When there are ordered motions in a medium, this will profoundly affect both the observed width of lines and the transfer of radiation within the medium. Imagine a spherical stellar atmosphere expanding outward from a star as $v \propto r$. If we observe toward the center, we encounter at each point along the line of sight a different velocity. Where these differ by more than the local linewidth in Equation 4, there is no way for photons originating in one region to be absorbed by or cause stimulated emission in the other; these regions do not communicate radiatively and the photons we receive from them are independent. But imagine looking well off center, so that, where the line of sight passes closest to the center, the expansion motion is exactly across the line of sight. From that subcentral region (that is what it is called, do not blame me) there is no perceptible Doppler shift from the systematic motion. Thus the range of velocities seen by us along these off-center columns is smaller, which changes the line width visible from outside and makes it more likely that two gas parcels will communicate with each other in this direction.

We may as well continue this explanation further and explain the old phenomenon of P-Cygni line profiles. In the expanding atmosphere, the gas closer to the center is hotter. So gas behind the star is occulted generally by somewhat hotter gas, at least nearby where there is some possibility of radiative communication. This means that the red-shifted gas behind the star reaches us unabsorbed or perhaps slightly augmented by nearer, even warmer gas at about the same projected velocity. But material in front of the star, which we see blue-shifted, is occulted only by cooler gas and if there is any radiative interaction, the tendency will be for cooler gas to absorb and replace it. Thus the overall profile we see toward the star will have a hotter redshifted wing, and blue–shifted absorption both from gas absorbing the central star, and closer gas absorbing gas nearer the star. The line profile is dominated by a combination of the velocity field, projected onto the line of sight, and the internal structure. An entirely analogous phenomenom occurs in CO when line emission is observed from infalling gas in a molecular cloud; there is a red-shifted absorption notch.

The important applications of this sort of reasoning are quite varied. Profiles of atomic hydrogen seen near the galactic equator are typically shaped by the ordered galactic velocity field, but are heavily confused by the local turbulent motions which cause emission from well-separated regions to overlap in frequency. A typical velocity gradient along the line of sight in the inner galaxy is 10 km s^{-1} kpc^{-1} but the local cloud-to-cloud velocity dispersion is 6 km s^{-1}: indeed in this case we are imagining the H I gas to be made up of pc-size particles with a speed of sound given by their macroscopic velocity dispersion. So the width of the local turbulence in the ISM, 13 km s^{-1}, confuses regions which are separated by more than a kpc. This limits the precision with which we can locate a gas particle in the Galaxy, even in the absence of other effects like the streaming associated with spiral arms (Burton 1988).

4. Optical Depth and Radiative Transfer

Now we have to substantiate the claim made at the end of Section 2 that we can see good absorbers (recall however, that the basic principles of thermodynamics *dictate* that good absorbers are also good emitters ...). In practice we see ensembles of systems and observe average properties, so that we need to formulate some means of gauging how astronomical bodies produce radiation *en masse*. To do this we introduce the concept of a column density $N = \int n(r)dr$ (whose units are cm^{-2}), which is the integral over some length r within which there is an assumed volume density $n(r)$ (cm^{-3}) of particles.

This transformation from volume to column density is our *expression* of the transformation from microscopic (local, volume) to macroscopic (integrated, column) quantities, since the lengths over which we typically integrate – pc or kpc – are very big indeed. Think of it this way; from outside we are presented with some body seen in projection. This is astronomy, not medicine, so there is no obvious possibility of tomography, we cannot undo the effects of the projection. What to do? The body is irregular so we break it up and consider each square cm separately[2]. Just what is projected onto that tiny patch of area? The column density behind it. Carry the dimensional analysis further. The strength of an interaction is typically measured in terms of a cross-section (units of area) so that the product of a column density and cross-section is dimensionless, a pure number. It is the mean number of interactions which would occur in traversing the length of the column. It is the size of the column measured in mean interaction lengths.

The usual way of defining the problem physically is in terms of a related dimensionless quantity, the optical depth, which is the size of a medium as seen by scattered photons. If the medium is small compared to even one scattering length, it is said to be optically thin. If it is many scattering lengths across, it is optically thick. Because photons do not sense time or distance in any meaningful way except on cosmological scales or near black holes *etc.*, the important thing for them (or us, now) is not linear size, but the size measured in units of the mean free path Λ (defined below) for interacting with the ambient matter. This dimensionless ratio (size/mfp), which is essentially identical to the (column density) - (cross section) product in the previous paragraph, is the total optical depth of the medium. Recall that the mean free path is the typical linear distance traversed before an interaction occurs, or, alternatively, the inverse of

---

[2]Well, maybe only each sq pc.

the (constant) probability per unit length for interacting with the medium. So the probability of travelling a distance r *without* interacting is $\exp{-r/\Lambda}$, which is an expression of Poisson statistics – the n=1 case of the familiar Poisson distribution in the continuum limit. Study this equation carefully before playing Russian Roulette.

The Einstein B-coefficients do not exactly have units of area, but they are proportional to the cross-sections for absorption σ_{lu}. If there is a flux of photons F_ν (units of $cm^{-2}\ s^{-1}$) of energy E_{lu} past a system in the l state, the rate of absorption will be $F_\nu \sigma_{lu}$. If the flux is incident on an ensemble of systems with a local density n_l in the l state, the ensemble will absorb photons at a rate $F_\nu n_l \sigma_{lu}$ per unit volume. This is the local microscopic rate at which the medium takes photons out of the incident flux ($n_l \sigma_{lu}$ is recognizable as $1/\Lambda$, right?). Of course it will also return some, and the flux which is incident on an atom deep within a medium will not be the same as that near the edge; radiative transfer problems are microscopically complex and macroscopically devilishly non-linear.

The basic equation defining the optical depth τ_{lu} which arises over a length dr, crossing a medium with number density $n_l(r)$ and column density N_l etc. is

$$\int \tau_{lu} d\nu = dr \int \kappa_{lu} d\nu = \frac{h\nu_{lu}}{c}(N_l B_{lu} - N_u B_{ul}) , \qquad (5a)$$

where the frequency on the right hand side is that measured in the laboratory or rest-frame, and the integrals extend over the (local) frequency span of the transition. We complicated this relationship by inserting the absorption coefficient $\kappa_{lu} = d\tau_{lu}/dr$ not only to confound the student but because there is an important relationship between κ and the emission which occurs over the region, and this is usually used in deriving the equation of radiative transfer.

Earlier we noted that there is a natural width in frequency, but the more important cause of linewidth in radio astronomy is of other origin and will be discussed separately below. The first term on the RHS is the quantity of photons removed by absorption, but this is ameliorated by the second term, representing photons which are coerced back out of the ensemble by stimulated emission. If there is a big imbalance between N_l and N_u, the medium will absorb more strongly. But the correction for stimulated emission may reduce the optical depth considerably.

We rearrange Equation 5a by using the relationships between the Einstein coefficients in Equation 1. We change the integral from one over frequency to one over velocity ($dv = cd\nu/\nu_{lu}$). We assert $N_u/N_l = (g_u/g_l)\exp(-h\nu/kT_{\rm exc})$ in analogy with the Boltzmann distribution in thermodynamic equilibrium (this new thing $T_{\rm exc}$, the excitation temperature, will be discussed in the following Section). The result, expressed equivalently in terms of the emission A-coefficient or absorption f-value is

$$\int \tau_{lu} dv = \frac{g_u}{g_l} \frac{N_l \lambda_{lu}^3}{8\pi} A_{ul}(1 - e^{\frac{-h\nu}{kT_{\rm exc}}}) , \qquad (5b)$$

$$\int \tau_{lu} dv = (\frac{\pi e^2}{m_e c}) N_l f_{lu} \lambda_{lu} (1 - e^{\frac{-h\nu}{kT_{\rm exc}}}) . \qquad (5c)$$

Note that the wavelength enters both expressions. This dependence is crucial to our ability to see the "weak" transitions involved in radio astronomy. The A-rates may be small, and the f-values may be low, but this is a lot more tolerable at wavelengths measured in cm than wavelengths measured in fractions of a micron. Note also that $h\nu/k$ is 0.068 K at a wavelength of $\lambda 21$ cm and only about 5 K at 3 mm wavelength. This means that the term in parenthesis correcting the optical depth for stimulated emission will often be quite small at radio frequencies.

4.1. Velocity gradients – a bit of dimensional analysis

Formally, the integrated optical depth produced over a small length dr varies as $\int \tau d v \propto n(r) dr$ so, if a typical line width over the region is dV, the line-center optical depth varies as $\tau_0 \propto n(r) dr/dV = n(r)/(dV/dr)$. In the second form we can discern the effect of a velocity gradient in a medium (should one actually exist). Regions with a large velocity gradient along the line of sight contribute over a somewhat stretched-out velocity interval and so make a smaller contribution to the optical depth at any given velocity. This discussion ignores the intrinsic local linewidth, which smooths in frequency and so is equivalent to blending optical depth from spatially distinct regions (recall the discussion of low-latitude H I in Section 3.5).

5. Thermodynamic and Statistical Equilibrium

At this point we have a general view of how radiation and matter interact (the Einstein coefficients), we have an equation which defines the optical depth in the same terms and we have a general view of how gas motions might combine to form line profiles. But we do not know anything about any particular medium nor do we have any idea of what particles harbor the levels we have defined. We will tackle the first point first and explore how an ensemble of particles is distributed over the various possible states, of which our u and l were two, perhaps among many and perhaps not.

From undergraduate thermo, we recall that, in thermodynamic equilibrium, the states of any system are populated in proportion to their statistical weights and Boltzman factors $\exp(-E/kT)$, without regard to the mechanisms which distribute systems among various levels. In this case we would have $n_u/n_l = (g_u/g_l)\exp(-E_{ul}/kT) = (g_u/g_l)\exp(-h\nu_{ul}/kT)$ which is what prompted the definition in the previous section $n_u/n_l = (g_u/g_l)\exp(-E_{ul}/kT_{\rm exc})$. This is a one-to-one analytic transformation so there is nothing wrong with doing so. It offers the possibility that $T_{\rm exc}$ will be the same as the real temperature if equilibrium is truly attained, or that $T_{\rm exc} < T_{\rm K}$ when collisional excitation is weak (and the excitation is said to be sub-thermal) and it shows that, if $n_u/n_l > g_u/g_l$, the excitation temperature would have to be negative. So right away it clues us in to unusual conditions. Negative excitation temperatures are a property of lasers and masers.

Equilibrium does not imply stasis but, when there is balance in some ensemble, the number of systems which change from state l to state u over some "small" time interval must be equal to the number which make the opposite transition from u to l; alternatively, the instantaneous rates at which particles make the upward and downward transition must be equal.

If the upward rate at which a system in state l is induced to shift to state u is called C_{lu}, the condition of equilibrium or balance is just $n_l C_{lu} = n_u C_{ul}$. In order that $n_u/n_l = (g_u/g_l)\exp(-E_{ul}/kT)$ in true thermodynamic equilibrium at some temperature T, it follows that

$$C_{lu} = C_{ul}(g_u/g_l)\exp(-E_{ul}/kT) . \tag{6}$$

This equality holds for any *single* process occurring at some temperature T always even though it was derived under the particular aspect of equilibrium. If this equality did not hold, thermodynamic equilibrium could not be attained by this process, i.e. it would be a perpetual motion machine of some kind. Coupled with this is the assumption that the process can be characterized by some temperature.

Physically think of the following. To cause the $u \to l$ transition involves extracting energy; it is always allowed energetically. But the upward transition might involve more energy than is typically available to the particles as they interact, so the upward rate is smaller by a Boltzmann factor to account for this. The upward transition gains from the ratio of statistical weights, but loses (usually) in the exponential. In fact, this kind of argument is precisely what Einstein did 100 years ago to derive the relationships (1a) and (1b) between his A's and B's, balancing an upward rate $n_l C_{lu} = n_l B_{lu} U_\nu$ against the downward rate $n_u C_{ul} = n_u(B_{ul}U_\nu + A_{ul})$, insisting that the ratio of populations be given correctly when the radiation density took the Planckian form, in the absence of other influences.

Now generalize so that the rates are the sums over many different processes. If all these processes are characterized by the same temperature, they will individually obey the equality as will their sum. But if these processes are not in equilibrium with each other, there will be competing influences on the excitation. If this seems confusing, an example will be given now.

5.1. Statistical equilibrium via the radiation field

In detail, a closed system of particles will distribute itself over its internal states, tending toward and eventually attaining true thermodynamical equilibrium. On short time scales, the path to equilibrium proceeds according to how the particles interact with other particles (of which there may be many kinds) and with the ambient radiation field: over long times, the final state is independent of such details, and of the outside world. But if the system is not closed, if it communicates with the world outside itself, the radiation field will consist partly of what is imposed – that in free space or at the periphery of the medium – and partly of that which is generated as the system absorbs and reradiates energy. So here we have the possibility of competing temperatures influencing the system. Intuitively, we sense that the ambient radiation field in some medium will be like that outside if the medium is optically thin; outside photons penetrate freely and those generated internally escape. But if the medium is optically thick, the radiation field will be dictated by local conditions in the medium itself as contact with the outside world becomes more and more difficult.

This intuition is commonly used to solve difficult radiative transfer problems crudely using something called the escape probability or Sobolev approximation. Pretend that we know at some point within the medium, separated from the outside world by an optical depth τ, the probability $p(\tau)$ that a photon, once emitted, will escape without further interaction. Functionally it happens in a variety of situations that $p(\tau) = (1 - \exp(-a\tau))/a\tau$ where a is a number of order unity. This certainly has the correct functional dependence ($p \to 1$ as $\tau \to 0$) and it is almost an exact solution in situations like rapidly expanding stellar winds. In the escape probability approximation, we assert that the local radiation density is $U_\nu = pU_\nu(T_{\rm exc}) + (1-p)U_\nu(T_{\rm C})$ where $T_{\rm C}$ is the characteristic temperature of the radiation density outside. Typically, $T_{\rm C} = T_{\rm cmb}$ but there can also be a contribution from the galactic or extragalactic background. Because τ depends explicitly on $T_{\rm exc}$, and because $T_{\rm exc}$ is physically dependent on the ambient radiation field (which causes radiative transitions between states) the problem is highly non-linear, but often still tractable (see Goldreich & Kwan 1974).

If one writes the equations of statistical equilibrium, use of the escape probability formulation preserves the functional form of the rate equations, the *only* alteration being that the Einstein coefficients, wherever they appear, are replaced by their product with $p(\tau)$. So if you have a set of rate equations involving radiative terms in the absence of any idea of radiative transfer, you can make this change (without going through the bother of rederiving the equations) and include radiative transfer in the physics quite trivially (of course this is not the same thing as *solving* the equations and they can hardly be said to be exact). If you have manipulated the rate equations so that only the A_{ul} are present (using Equation 1), just replace A_{ul} by $A_{ul}p(\tau_{ul})$.

Include radiative transfer effects when you know the optical depths will be appreciable. If other physical processes are weak, then radiative excitation will be more important at low optical depths. Certainly think about optical depth effects when radiative rates are the dominant ones in the problem.

5.2. Thermalization and statistical equilibrium through particle interactions

Particles interact by changing places (an ambient free electron replaces one bound inside the H atom, changing the orientation of the electron and proton spins there), by banging on each other lightly or as hard spheres, or in ways which are mediated by their individual and or mutual electromagnetic potentials. Typically, there will be some cross-section for excitation or deexcitation, known as a function of the velocities of the particles, their impact parameter, and so forth. These are averaged over geometry and a thermal energy distribution (since even microturbulence occurs on scales larger than the inter-particle separation) and presented in the form of *rate constants* with units of cm^3 s^{-1}

$$\gamma_{ul} \equiv <\sigma_{ul} v>,$$

$$\gamma_{lu} = \gamma_{ul}(g_u/g_l)\exp(-E_{ul}/kT).$$

In the earlier notation, $C_{ul} = n_X \gamma_{ul}^X$ might be the deexcitation rate (check out the units) for interactions with particles of some species X of local number density n_X. Atomic systems have dimensions of Angstroms; particle interaction cross sections correspondingly have sizes about 10^{-15}cm^2. Note that it is the rate constants γ, not the crosssections which satisfy the reciprocity condition between upward and downward changes.

Once we have this idea in mind, we can write very simple but typical microscopic expressions for the rates at which our hypothetical u- and l-state systems exchange population

$$n_u C_{ul} = n_u (A_{ul} + B_{ul} U_\nu + \sum_X n_X \gamma_{ul}^X) ,$$

$$n_l C_{lu} = n_l (B_{lu} U_\nu + \sum_X n_X \gamma_{lu}^X) ,$$

where the sum is over all kinds of particles and their possible influences. For a two level atom, these rates can be equated and the population ratio derived without further rate equations. But when there are more than two levels, the rates into and out of each must be considered. In a real world example, the equations could be highly non-linear (if optical depth effects are important, for instance).

In general the particle interactions $\sum_X n_X \gamma_{ul}^X$ will occur at the local kinetic temperature T_K, while the radiation will be characterized by some other temperature (recall the discussion of the escape probability approximation). If there were no particle interactions, the relationship between the Einstein coefficients would cause all excitation temperatures to be identically equal to the temperature used in the Planck formula for the radiation density; typically, this would be T_{cmb}. In the opposite limit, as particle interactions dominate, $T_{\text{exc}} \to T_K$. We say that the transitions or level populations become increasingly thermalized; otherwise we might say that the excitation is strongly sub-thermal ($T_{\text{exc}} << T_K$).

In the simple case where a radiative decay is possible to only one lower level, the condition that a transition $u \to l$ or level u be thermalized is, *very approximately* that the collisional rate be large compared to the spontaneous emission rate $R_{ul} \equiv \sum_X n_X \gamma_{ul}^X)/A_{ul} >> 1$. But, for H I, because the energy separation is so small, the condition is actually $(h\nu/kT_K)R_{ul} >> 1$. This is impossible to satisfy in the real ISM under typical conditions if $T_K > 1000$ K, so H I is in fact only thermalized by collisions in cooler gas. At higher temperatures, the gas density is too small to thermalize H I, partly because the ISM is in (at least) rough pressure equilibrium ($n = P/kT$) and partly because the condition for thermalization becomes so stringent at higher temperature (Liszt 2001).

5.3. Isothermal radiative transfer

We now write an expression for the manner in which rays of light propagate through a medium. Let I_ν be a specific intensity (erg cm^{-2} Hz^{-1} s^{-1} sr^{-1})[3]. The equation of radiative transfer, which describes what happens when the radiation traverses a small distance dr is

---

[3] It is at this point that most astronomers decide not to make the study of radiative transfer more than a temporary activity.

$$dI_\nu/dr = -I_\nu \kappa_\nu + j_\nu \qquad (7a)$$

and it says merely that there is absorption of photons, compensated by insertion of some which arise locally. Note that κ is already reduced by the existence of stimulated emission, and the j_ν term represents generally 'new 'photons (it is called the source function in radiative transfer parlance). Because of Equation 1, it happens (you could look it up) that $j_\nu = \kappa_\nu B_\nu(T)$ (this is the origin of the statement that good (poor) radiators are good (poor) absorbers) and

$$I_\nu(\tau) = I_\nu(0)e^{-\tau_\nu} + B_\nu(T_{\text{exc}})(1 - e^{-\tau_\nu}), \qquad (7b)$$

$$I_\nu(\tau) - I_\nu(0) = (B_\nu(T_{\text{exc}}) - I_\nu(0))(1 - e^{-\tau_\nu}), \qquad (7c)$$

where the second form gives the emergent intensity above the background. When the background is pervasive – which is the case for the cosmic microwave background – only the second form of the emergent intensity is really observable.

The exponentials express the Poisson nature of the underlying statistics as noted earlier; they are the probability of traversing the medium unhindered ($e^{-\tau_\nu}$) and its complement. $I_\nu(\tau)$ is the radiation which streams out of the region, toward us. It has an attenuated component $I_\nu(0)\exp-\tau$ of whatever was incident from behind, and a component $B_\nu(T)(1 - e^{-\tau_\nu})$ corresponding to photons which were created inside and managed to escape. Note that, beyond one optical depth into a medium, our view rapidly becomes very dim indeed.

It is common to express $I_\nu = 2kT_B/\lambda^2$, the equivalent Rayleigh-Jeans ($h\nu/k \ll T_B$) brightness temperature (this is a formal definition, not an assertion that the limit actually obtains!). If we also replace the emergent intensity $I_\nu(\tau)$ by $B_\nu(T_B)$, and set $I_\nu(0) = B_\nu(T_C)$ to allow for a continuum radiation field in the background, we find for the brightness temperature T_B *above the background* (the second form just above)

$$T_B = (h\nu/k)(1 - e^{-\tau_\nu})(1/(e^{h\nu/kT_{\text{exc}}} - 1) - 1/(e^{h\nu/kT_C} - 1)). \qquad (8)$$

The quantity on the left hand side of this expression corresponds to the brightness temperatures we observe above the pervasive cosmic microwave background ($T_C = T_{\text{cmb}}$). When we really are in the Rayleigh-Jeans limit, this expression reduces to

$$T_B = (T_{\text{exc}} - T_C)(1 - e^{-\tau_\nu}). \qquad (9)$$

When $T_{\text{exc}} = T_C$ any amount of gas, of any optical depth, is quite literally invisible. Otherwise, the cosmic microwave background gives a correction to T_B which is actually larger than T_{cmb}, increasingly so at higher and higher frequency. This is just an artifact keeping a quantity on the left hand side which was defined in a certain limiting case which does not strictly apply. When T_{exc} is derived in this way, one must always consider that a correction for the background is needed, even for H I when the background is only T_{cmb}.

In the limit of large optical depth, Equation 9 says that the emitted radiation is characterized only by the excitation temperature achieved in the medium.

In the limit of low optical depth the emitted radiation is the product of τ and T_{exc}. To the extent that the optical depth formally (functionally) depends on a column density within the gas, optically thin emission can be used to derive N_l, if we have sufficient other information about the gas.

5.4. Callen & Welton (1951)

Callen is known to students of an older generation for his textbook. In 1951 he and Welton showed that there is noise associated with a Planck radiation field, a zero-point energy $h\nu/2$. At wavelengths below about 10μ, this term is larger than the noise of the devices used to detect radiation. The noise manifests itself as an added power in the photon field. Instead of discussing the Planck function and taking the Rayleigh-Jeans limit, we really should consider $T_{CW} = (h\nu/k)(1/(\exp(h\nu/kT) - 1) + 1/2)$. Photon noise also manifests itself in a floor on the noise temperature of any detector of radiation, which is a full photon's worth, $h\nu/k$.

6. Kinds of Systems

At this point we are really making great progress! If only we had some idea of what atoms or molecules we could study, or what their internal energy states were, and if we knew the conditions under which they existed in interstellar space, we could maybe actually do something. So let us take the time now, finally to specify some properties of the different kinds of atomic and molecular structure we might study.

6.1. The λ21cm line of H I

This year marks the 50th anniversary of the detection of this first, most important radioastronomical spectral line. It arises because, like Na and all the other alkali atoms, hydrogen has an unpaired outer electron and a nucleus also with spin 1/2. For an atom in the ground ($n = 1$) electronic state, when these spins are parallel ($g_u = 3$, a triplet state) the energy is higher by a very small amount than when they are anti-parallel ($g_l = 1$). The energy separation of 0.068171 K corresponds to a frequency of 1420.405752 MHz or a wavelength of 21.1 cm (so that 1 MHz subtends 211 km s^{-1}) and the f-value for such transitions is $f_{lu} = h\nu/(2m_e c^2)$. For the H I line this means that $A_{ul} = 2.68 \times 10^{-15}$ s^{-1} (by comparison, the A-coefficient for the Ly-α transition downward from the ^{2}p level is 4.7×10^8 s^{-1}).

For the λ21cm line, the excitation temperature is always referred to as the spin temperature, T_{sp}. The low Einstein A means that interactions with other particles will easily thermalize the hyperfine transition – $T_{\text{sp}} \approx T_{\text{K}}$, $n_u/n_l = 3\exp(-h\nu/kT_{\text{K}})$ whenever the gas is cool, say below 1000 K (see the previous Section).

The frequency of the analogous hyperfine transition in Na I is 1771 MHz, but the abundance of sodium is far too small for Na I to be seen in this way. Na I is less abundant even than deuterium, whose 327 MHz hyperfine line has been sensitively but unsuccessfully sought for 50 years. Instead, the hyperfine splitting in alkali atoms is manifested in an unavoidable splitting of the optical resonance lines; for the Na I D lines, this is about 1 km s^{-1}; each line of the famous optical pair has this bothersome structure; this is radioastronomy's revenge.

There are a variety of handy relationships relevant to H I. Because the energy is so small, $\exp(-h\nu/kT_{\text{exc}}) \approx 1 - h\nu/kT_{\text{exc}} \approx 1$ for all realizable $T_{\text{exc}} \geq T_{\text{cmb}}$ and the level populations will be in a near 3:1 ratio independent of the excitation temperature. In this limit $1 - e^{-h\nu/kT_{\text{exc}}} = h\nu/kT_{\text{exc}}$ so the expression for the integrated optical depth becomes

$$\int \tau(HI) dv = \frac{N_{\text{HI}}/1.823 \times 10^{18} \text{cm}^{-2}}{T_{\text{exc}}}, \qquad (10a)$$

where N_{HI} is the total amount of H I (which is just 4 times that in the lower level). The fact that the optical depth is just exactly inversely proportional to the excitation temperature has an interesting side by-product. With reference to Equation 9, integrating over the profile, we see that for small τ

$$\int T_{\text{B}}(HI) dv \approx (1 - T_{\text{C}}/T_{\text{exc}}) N(\text{HI})/1.823 \times 10^{18} \text{cm}^{-2}. \qquad (10b)$$

Thus, in this case, where the populations are in a 3:1 ratio, when the optical depth is small and either T_{C} is known or $T_{\text{C}}/T_{\text{exc}}$ is small (a lot of if's but still often the case), the amount of hydrogen may be obtained just by summing over the channels of a λ21 cm spectrum. This is an impressive capability. If there were more than two states with appreciable population we would have needed some means of extrapolating to them as well (perhaps by observing another line?) to find N(H I). But how do we know that an H I line is optically thin without measuring the optical depth? How would we measure the optical depth? What if we found out that the line was optically thick and did not know the optical depth exactly?

Much of the volume of the ISM in the disk of the Galaxy is conducive to the presence of neutral atomic hydrogen. This is good (we can use H I to probe more of the Galaxy) and bad (our telescopes can be made almost snowblind by the profusion of H I gas). Despite 50 years of observations, many important issues remain to be settled. We cannot tell from H I alone how much of the volume of the ISM is filled by neutral gas; the H I can be clumped in gas parcels with cold cores and warmer envelopes, leaving the majority of the volume free for ionized gas at any conceivable temperature from 10^4 K on up. We do not know the exact balance between cool gas below 200 or so K, and warm gas, above 2000 K or so. We do not know just how the mix changes with height above the galactic plane (although we do know that warmer gas has a higher scale height) or with galactocentric radius. The amount of H I in the midplane is uncertain by nearly as much as that of the molecular gas (say 50%).

Excellent, comprehensive reviews of H I in the galactic environment are given by Burton (1988) and Dickey & Lockman (1990).

6.2. Recombination lines

Radiative transitions between bound states of neutral hydrogen arise when protons and electrons (re)combine radiatively in ionized gas – gas in which the balance between ionization and recombination is statistical, not static. Although the lines in question arise because of the nature of the bound states of the hydrogen atom, when the region is heavily ionized it is still ionized (not neutral) gas which is being observed. If the ionization fraction in the region is $x = n(e)/(n(\text{H}^0)+n(\text{H}^+))$ (this is the usual definition) any given atom is ionized for a fraction x of the time.

The population in any given level is determined by a combination of direct recombinations into that level, the cascade of systems which recombine into higher states and gradually work their way down, and decay of that level itself to lower levels. The line strengths are highest for recombinations into or decays down to the closest and furthest available lower levels. Photons released when recombinations occur into the $n = 1$ level can ionize other atoms and do so immediately. When the physical properties of the region are determined, the only recombination rate that matters is that into levels above $n = 1$. Recombinations into the 2s level cause the temporary absence of a bit of energy, because (since angular momentum must change by 1 unit) that level has no place to decay into; it is metastable.

Bright recombination lines from discrete sources are the signposts of formation of massive stars in the Milky Way; only stars of type B0 and earlier have enough flux near the Lyman limit to ionize enough gas that the emission is readily detected across the Galaxy. Near such a star – which is always a young star because of the short main-sequence lifetime – the gas is photoionized, forming a so-called Strömgren sphere (after the famous Danish astronomer Bengt Strömgren) at roughly 8000 K. The one-dimensional velocity dispersion of H atoms in such a region, solely due to thermal motions, is about 10 km s^{-1}.

In most cases even a very hot star does not have enough flux to ionize all the surrounding material and the edge of the ionized region is just where the ionizing flux peters out, either absorbed by dust or degraded – downconverted after many recombinations to Lyman-series and Balmer-series photons longward of the ionization limit. Such a region, running out of flux, is called 'ionization bounded'. Because the rate of recombination in any volume is proportional to the product of the densities of ionized atoms and electrons (basically, the square of the density of protons), denser regions are harder to ionize and denser Strömgren spheres are much smaller. If a star has more flux than the surrounding material could handle, the sphere would be said to be density-bounded. For ionization bounded regions, by definition, all of the ionizing flux of the star is absorbed in the gas; so the luminosity of the star – or all the stars which are present, not all of which are always visible – should be ascertainable from study of the gas.

The bound electronic energy levels of hydrogen atoms follow the famous formula derived by Bohr; for an atom of nuclear charge Z and atomic mass m (including all electrons) the frequencies of recombination transitions between two levels $n_l = 1, 2 ...$ and $n_u = 2, 3...$ are given by

$$\nu = Z^2 R_\infty c (1 + m_e/m)^{-1} (\frac{1}{n_l^2} - \frac{1}{n_u^2}) , \qquad (11)$$

where $R_\infty = 109737.312$ cm^{-1} is the Rydberg constant in wavenumbers (optical spectroscopy! 1 wavenumber = 1.43883 K).

Recombination lines have been observed from hydrogen, helium (both isotopes), carbon and nitrogen at radio wavelengths ranging upward from 25 MHz. Recombination lines are seen all the way from 25 MHz to the Ly-α line of HeI, spanning a positively mind-boggling factor of more than 10^8 in wavelength. Because of the complicated manner in which the levels are populated, it is possible to create recombination line masers. Because there is some level of ionization even in dense, largely neutral cool material (due, for instance, to cosmic rays), and because recombination coefficients (rates) typically have a $T^{-0.5}$ dependence (i.e. the electrons are easier to capture when they are moving more slowly) there is a dilute component of recombination line radiation which is distributed over the entire galactic plane.

A comprehensive review of recombination line radiation is given by Gordon (1988) and tables of recombination line wavelengths are contained in Lilley & Palmer (1965)

6.3. Rotational transitions of diatomic and linear molecules

Molecules have several internal degrees of freedom in addition to their own analog of the electronic multiplet structure in atoms. Molecules can vibrate because the electronic forces between pairs of nuclei create approximately harmonic potentials (deep in the potential well); the quantum harmonic oscillator, with evenly spaced energy levels and a ground state having a half-quantum of (zero-point) energy above the bottom of the potential well, applies to the bonds between pairs of molecules. Additionally, the molecules can, almost like dumbbells, rotate about their center of mass.

The electronic, vibrational, and rotational energies are in the approximate ratio $1 : \sqrt{m_e/m_p} : m_e/m_p$. This fundamental insight, part of the famous Born-Oppenheimer approximation (done while O was B's postdoc) informs us that vibrational motions appear in the infra-red (IR) while rotational transitions occur in the microwave, at least for heavier molecules (basically those with at least two atoms heavier than hydrogen). For molecular hydrogen and diatomics (like OH) with a hydrogen, the rotational energies occur in the far IR and sub-mm regions.

Molecules vibrate much more rapidly than they rotate, so, during one rotation, the nuclei adjust at any moment to the rotation; for this reason, the bonds between nuclei can be treated as if they were very nearly rigid during rotation. We will begin by considering the rotation of diatomic and linear molecules. Such molecules have almost no moment of inertia about the line joining the nuclei (since only electrons exist off-axis) and cannot rotate about this axis. The remaining two moments of inertia (through the center of mass, about either axis perpendicular to the line joining the nuclei) are equal and the energy of rotation $L^2/2I$ is quantized as a (slightly imperfect) quantum mechanical rotating top

$$E(J) = B_v J(J+1) - D_v J^2(J+1)^2 , \qquad (12)$$

where J is the quantum number or the (hopefully familiar) angular momentum of rotation in units of $h/2\pi$. In this equation D_v is very much smaller than B_v; both quantities are subscripted by the vibrational quantum number v ($= 0,1...$) to remind the reader that they vary slightly with v, but we seldom observe molecules in any but the lowest vibrational state. D_v represents a very slight stretching out of the chemical bonds by centrifugal forces.

As written above, B_v has units of energy; physically, it is inversely proportional to the moment of inertia (for two atoms, the product of the reduced mass of the nuclei times the inter-nuclear separation) of the molecule. Inter-nuclear distances are always of order one Bohr radius a_0. They are determined by electronic forces and do not vary much with the mass of the nuclei, so that lighter molecules have smaller moments of inertia and higher rotation frequencies. When one of the atoms in a diatomic is hydrogen, the reduced mass is so small that the rotational lines do not appear at radio wavelengths at all, but only in the far-IR. Tables of molecular constants and line frequencies are given by (Pickett et al. 1998). For ^{12}CO, $B_0/k = 2.76$ K. Because the moments of inertia of molecules are strongly dependent on nuclear mass, isotopic species have distinctly separate transitions. For instance, the lowest transitions of ^{12}CO and ^{13}CO occur at 115.3 and 110.2 GHz. Electronic multiplet transitions of isotopically different atoms nearly coincide and are not useful to optical spectroscopy.

When the two atoms of a diatomic molecule are identical, the center of mass and the center of charge must coincide. Otherwise, they do not, and, as a molecule rotates about the former, the latter swings around about it. This means that the molecule has a permanent dipole moment μ and radiative dipole transitions $J \to J \pm 1$ are permitted. Thus

$$\nu_{J \to J-1} = 2B_v J/h - 4D_v J^3/h \tag{13}$$

and the frequencies increase in proportion to the rotational quantum number. If the distribution of population over the rotational energy ladder can be characterized by an excitation temperature T_{exc}, then

$$n_J \propto g_J \exp\frac{-BJ(J+1)}{kT_{\mathrm{exc}}} = (2J+1)\exp\frac{-BJ(J+1)}{kT_{\mathrm{exc}}}.$$

The statistical weights are $g_J = 2J + 1$, representing the (ordinarily) degenerate magnetic sub levels: in an external magnetic field the molecules will align themselves about the field, projecting their angular momenta along it in quantized amounts ranging from $-J$ to J. Taking dn_J/dJ we find that the state having the greatest population is (approximately) kT_{exc}/B. As T_{exc} increases, population shifts out of the lower levels into those above. This means that there are two possibilities for the weakness of any given transition. Either the species could be of low abundance generally, or we might be looking at a line which happens to be weak because the population is distributed over other levels. But it is seldom the case that the energy levels are thermalized at a single temperature. Most likely, the lower levels are more easily excited and the lower-lying transitions have higher T_{exc}.

The permanent dipole moment μ is typically of order $a_0 e$ where a_0 is the lowest Bohr radius (0.53 Angstroms) and e is the unit of charge. Permanent dipole moments are measured in units of 10^{-18} esu, which quantity is known as one *Debye* (after Pieter, the famous Dutch physicist). CO has an anomalously low permanent dipole moment 0.112 Debye, while those of CS and HCN and most other non-homonuclear molecules are of order 2-3 Debye.

The Einstein A-coefficients of the dipole transitions of linear molecules can be expressed in terms of the dipole moment μ and frequency as

$$A_{J,J-1} = \frac{64\pi^4 \nu^3 \mu^2 J}{3hc^3(J+1)} \ . \tag{14}$$

Recall that the frequencies are proportional to J, so the A-rates increase rapidly up the rotational energy ladder. For CO, $A_{1,0} \approx 7.5 \times 10^{-8}$ s^{-1} much higher than for the 21 cm line of H I, but still far below that of any permitted optical transition. For species with more typical dipole moments, decay rates of low-lying states are typically about 10^{-5} s^{-1}.

We can put Equation 14 into the definition of the optical depth, finding

$$N_J = \frac{(2J+1)}{(J+1)} \frac{8.0 \times 10^{12} \mathrm{cm}^{-2} \int \tau_{J,J+1} \mathrm{d}v}{\mu^2(1 - \exp(-h\nu/kT_{\mathrm{exc}}))} \ , \tag{15}$$

where the optical depth integral is in km s^{-1}.

Unlike H I, we cannot ordinarily take the low-frequency (Rayleigh-Jeans) limit here; there are more than two levels, and the line profile integral is not as closely identifiable with the column density, even in the levels observed. Plugging in the numbers, we see that the lowest transition of CO attains an optical depth of unity (over 1 km s^{-1}) for N(CO) $\approx 10^{15}$ cm^{-2}, while other species become optically thick for $N \approx 10^{12}$ cm^{-2}. It is something of an accident that CO, whose transitions are relatively weak, is more abundant than most other species by a factor 10^3 in diffuse gas and 10^4-10^5 in dense gas. So its optical depths are the highest.

It is something of a paradox that, except for hydrogen (atomic or molecular) the sensitivities of radio astronomical observations are about as good as those in the optical regime. For hydrogen, optical astronomers can detect much lower column densities in the Ly-α line than we can at 21 cm; roughly the difference is between 10^{12} cm^{-2} and 10^{18} cm^{-2}. The same is true for ionized hydrogen as well, because the Hα line is so much stronger than radio recombination lines. And only optical or IR astronomers can directly see H$_2$. But with the slight exception of CO, with its small permanent dipole moment, radio astronomy is at least as sensitive. Thus we readily see column densities of many molecules at levels 10^{12} cm^{-2} in fairly diffuse gas, and optical absorption line studies along similar lines of sight have trouble reaching such low levels.

6.4. Molecules which are not linear

Most molecules are not linear [4]; assuming for the moment that there is a backbone or obvious physical body axis to the structure, any off-axis structure will cause the molecule to have three non-zero moments of inertia. Symmetry principles then tell us the following. For a structure with 3-fold or higher rotational symmetry about the backbone, two of the moments of inertia will be equal but may be either larger or smaller than the third moment of inertia. When the third moment of inertia is smaller, this more nearly resembles the case for the linear molecule (take the limit; let the third moment of inertia be 0). Such molecules

---

[4]It is at this point that many astronomers decide not to study molecular spectroscopy in more than a cursory fashion.

are described as prolate. The opposite case, where the unequal moment of inertia is larger, is called oblate (like the slightly flattened Earth),

In order to deal with the existence of two distinct moments of inertia we define an additional rotational constant to complement B_v, and employ a new quantum number K which is the projection of the rotational angular momentum about the body axis of the molecule ($K \leq J$). Unfortunately, the symmetries which simplify the energy level structure place very particular constraints on the wave functions, because of the need to obey overall symmetry principles, such as that the electronic wave function must be anti-symmetric. This brings into play a multitude of very detailed rules, each of which profoundly influences the existence and or statistical weights of the various levels.

If the off-axis structure has either no symmetry or only two-fold symmetry, the molecule as a whole must have three distinct moments of inertia and there is no simple closed formula for the energy levels because there is only one good quantum number (K is not strictly defined any more). In most cases, two of the moments of inertia are much closer to each other than to the third, that is, the molecule is approximately prolate or approximately oblate, but the situation is very complex. Specification of the Hamiltonian for complicated cases is something of a black art, best left to the initiated (not me).

6.5. Non-rotational structure

Lastly we note that radio astronomers sometimes observe microwave transitions from levels which exist within individual rotational states. The best known of these are the inversion lines of NH_3 at K-band (roughly 23 GHz) and the lines of OH seen at 1612, 1665, 1667 and 1720 MHz. Sadly, much of the time our view of the 1612 line is obliterated by interference from one or more satellites. The OH lines arise in part because the ground molecular state is the molecular analog of an atomic p-state; there is one quantum of angular momentum arising from the electrons and this splits all levels according to a phenomenon known as Λ-doubling. On top of this, there is an interaction with the spin of the H atom and the result is a four-level ground state. OH is a very common molecule in the interstellar medium, forming in appreciable quantities in just about any region which has a substantial proportion of molecular hydrogen and its emission is very widely distributed throughout the Galaxy. An ADS search on 'OH Maser' returns 600 entries.

6.6. Resources

For microwave spectroscopy, the 'bible' is the book by Townes & Schawlow (1955). Various aspects of molecules seen in the galactic environment are covered in texts such as Rohlfs & Wilson (1996). Tables of rate constants for chemical reactions and dipole moments for many neutral molecules are given in the UMIST database (see Le Teuff et al. 1999). Line frequencies are tabulated in Herb Pickett's database (Pickett et al. 1998), available online at http://spec.jpl.nasa.gov/.

References

Burton, W. B. 1988, in Galactic and Extragalactic Astronomy, ed. G. Verschuur & K. Kellermann (New York: Springer), 295

Callen, H. & Welton T. A. 1951, Phys. Rev., 83, 1

Dickey, J. M. & Lockman, F. J. 1990, ARA&A, 28, 215

Goldreich, P. G. & Kwan, J. 1974, ApJ, 189, 441

Gordon, M. A. 1988, in Galactic and Extragalactic Astronomy, ed. G. Verschuur & K. Kellermann (New York: Springer), 37

Le Teuff, Y. H., Millar T., & Markwick A. J. 1999, A&AS, 146, 157

Lilley, A. & Palmer, P. 1965, A&AS, 16, 143

Liszt, H. S. 2001, A&A, 371, 698

Pickett, H. M., Poynter, R. L., Cohen, E. A., Delitsky, M. L., Pearson, J. C., & Muller, H. S. P. 1998, Journal of Quantum Spectroscopy & Radiative Transfer, 60, 883

Rohlfs, K. & Wilson, T. L. 1996, Tools of Radio Astronomy, (New York: Springer)

Spitzer, L. 1978, Physical Processes in the Interstellar Medium (New York: Wiley-Interscience)

Townes, C. H. & Schawlow, A. L. 1955, Microwave Spectroscopy (New York: McGraw-Hill)

John Dickey confronted the audience with "Any questions?"

Spectral Line Advanced Topics

John M. Dickey

*University of Minnesota, Department of Astronomy,
116 Church St. SE, Minneapolis, Minnesota 55455, USA*

Abstract. This lecture is in three parts. The first gives the fundamental equations of level populations and radiative transfer which govern spectral line emission and absorption. Lines in the cm-wave band have similarities which allow us to simplify the equations. The further simplification of a two level system (a good approximation for HI) gives the familiar formulae for the 21-cm line.

The second part of the lecture deals with mapping, and the combination of single-dish and interferometer data. We consider the effect of gridding single-dish data using a convolving function. Mapping speeds and techniques are discussed.

The third part discusses the spectral line cube and its moments. The use of the first moments to determine the dynamical structure of a disk is discussed. The need for windowing in the cube before computing moments is justified.

1. Spectral Line Basics: Radiative Transfer

Spectral lines below 10 GHz come from transitions between quantum levels with miniscule energy separations; this is necessary if the photon energy is to be less than 4×10^{-5} eV, so that its frequency will be in this range. Atomic hydrogen has two ranges where the energy levels are so closely spaced: at very high quantum numbers ($n \gtrsim 80$) and in the ground state itself due to the hyperfine splitting. Several molecules also have transitions in this range, generally caused by similar hyperfine splitting of energy levels due to the magnetic moment of the nucleus, as in OH and H_2CO. The more important spectral lines with frequencies below 10 GHz are listed on Table 1.

Recombination lines are generated as an electron cascades down the energy levels of an atom; at cm-waves the transitions between levels with high quantum numbers are important. These make a comb of lines whose frequencies are given by the Rydberg formula,

$$\nu_{ul} = Z^2 \frac{R_\infty}{1 + \frac{m_e}{M_{\text{amu}}}} \times \left(\frac{1}{l^2} - \frac{1}{u^2} \right), \quad (1)$$

where l and u are the quantum numbers of the lower and upper states of the transition, Z is the screened nuclear charge (i.e. the number of protons minus the number of other electrons), M_{amu} is the atomic mass in amu (1.007825 for

H), and m_e is the electron mass. The Rydberg constant, R_∞ is 3.289842×10^{15} Hz (Rohlfs & Wilson 1996, p. 316). Recombination lines always come in groups, with the hydrogen line slightly lower in energy than the corresponding He line, and the corresponding C (and heavier element) lines piling up at frequencies just slightly higher than He. A few examples of recombination lines are included on the lower half of Table 1.

Table 1. Some Spectral Lines with Frequencies Below 10 GHz.

Transition	Rest Frequency (GHz)	A Recent Reference
HI	1.4204058	Gibson et al. (2000)
OH	1.665402, 1.667385	Liszt & Lucas (1996)
	1.61223, 1.720559	Lewis, Oppenheimer, & Daubar (2001)
OH	4.765562, 4.660242	Szymczak, Kus, & Hrynek (2000)
CH	3.2638, 3.3355, 3.3492	Magnani & Onello (1993)
H_2CO	4.82966	Pauls, Johnston, & Wilson (1996)
OH	6.035092	Caswell (1997)
CH_3OH	6.668518	Phillips et al. (1998)
^3He	8.665	Bania et al. (1997)
	A Few Recombination Lines	
H92α	8.309382	Lang, Goss, & Morris (2001)
H109α	5.008923	Peck et al. (1997)
H271α	0.3285959	Roshi & Anantharamaiah (1997)
C271α	0.3287597	Roshi & Anantharamaiah (1997)

1.1. Level populations

Line intensities are determined by the populations of the two quantum levels of the transition, which set the amount of emission and absorption at each point along the line of sight. Whether or not the excitations are in thermal equilibrium with the kinetic temperature of the gas, we can describe the ratio of the populations of the two levels using the Boltzmann equation with some temperature $T_{\rm ex}$, as:

$$\frac{n_u}{n_l} = \frac{g_u}{g_l} e^{\frac{-h\nu}{kT_{\rm ex}}}, \qquad (2)$$

where n_u and n_l are the level populations of the upper and lower levels, g_u and g_l are the statistical weights of the two levels, $h\nu$ is the energy of the photon, k is Boltzmann's constant, and $T_{\rm ex}$ is a temperature which is equal to the kinetic temperature in the special case where the collision rate is high enough to thermalize the transition (Osterbrock 1989, section 3.5).

At cm-waves the photon energies are so low that we can usually make the approximation $\frac{h\nu}{kT_{\rm ex}} \ll 1$; in that case we can expand the Boltzmann equation in a Taylor series:

$$\frac{n_u}{n_l} = \frac{g_u}{g_l} \times \left(1 - \frac{h\nu}{kT_{\rm ex}} + \cdots\right). \qquad (3)$$

For a two level system, i.e. one in which all the atoms are in one or the other of two levels, so that $n_u + n_l = n$, the first order term of the expansion makes the number in the upper level, n_u, just proportional to the total density, i.e.

$$n_u = n \times \frac{n_u}{n_u + n_l} = n \times \frac{\frac{n_u}{n_l}}{1 + \frac{n_u}{n_l}} \simeq n \times \frac{g_u}{g_u + g_l}, \qquad (4)$$

this gives $n_u \simeq \frac{3}{4} \times n$ for the hyperfine-split levels of the HI ground state. The two level approximation is valid for atomic hydrogen when all the atoms are in the ground state, as in the cool phases of the interstellar medium.

1.2. Emission and absorption

The emission and absorption coefficients can be derived once we know the level populations. Generally the emission coefficient, j_ν, is given by:

$$\int j_\nu \, d\nu = \frac{n_u \, A_{ul} \, h\nu}{4\pi}, \qquad (5)$$

where A_{ul} is the Einstein coefficient for spontaneous emission for the transition from level u to l. The frequency integrals are taken over the line profile (the subscript ν indicates that this is a function of frequency). The absorption coefficient is given by

$$\int I_\nu \, \kappa_\nu \, d\nu = h\nu \, (n_l B_{lu} - n_u B_{ul}) \frac{I_\nu}{c}, \qquad (6)$$

where I_ν is the radiation intensity and B_{lu} and B_{ul} are the Einstein coefficients which describe the probability of absorption and stimulated emission of a photon, respectively. These are in the ratio

$$B_{lu} = \frac{g_u}{g_l} B_{ul} \qquad (7)$$

(Spitzer 1977, section 3.4).

Note that the units of the emission and absorption coefficients are different. For j_ν the units are erg cm^{-3} sec^{-1} Hz^{-1} sterad^{-1}, which is just the energy in the radiation coming from a unit volume of gas per second, going into a unit bandwidth of the emission spectrum and into a unit solid angle in direction. The units of κ_ν are the same, divided by I_ν, since the absorption is always proportional to the intensity of the incident radiation field itself; this works out to give κ_ν simply units of cm^{-1}. Thus the combined effect of the emission and absorption of a differential volume element, ds, somewhere along the line of sight is to change the radiation intensity by dI where

$$dI_\nu = j_\nu \, ds - \kappa_\nu \, I_\nu \, ds. \qquad (8)$$

The ratio of the emission and absorption coefficients is the Plank source term:

$$\mathcal{B}_\nu = \frac{j_\nu}{\kappa_\nu} \qquad (9)$$

$$= \frac{2h\nu^3}{c^2} \frac{1}{e^{h\nu/kT} - 1} \qquad (10)$$

$$\simeq \frac{2kT}{\lambda^2},\qquad(11)$$

where the approximation is the Rayleigh-Jeans law which defines the brightness temperature:

$$T_B = \frac{B_\nu\, c^2}{2\,k\,T\,\nu^2}.\qquad(12)$$

For a two level system with energy separation giving a cm-wave line, the emission coefficient is proportional to the density, using Equation 4 in Equation 5 gives:

$$\int j_\nu d\nu = \frac{g_u}{g_u + g_l}\,\frac{A_{ul}\,h\nu}{4\pi}\times n.\qquad(13)$$

For the absorption coefficient we must include the second term in the Taylor expansion in Equation 3, since to first order the number of photons absorbed is just cancelled by the extra number produced through stimulated emission. Thus we get

$$\int \kappa_\nu d\nu = \frac{h\nu\,B_{ul}}{c\,g_l}(\,g_u n_l - g_l n_u)\qquad(14)$$

$$= \frac{h\nu\,B_{ul}\,n_l}{c\,g_l}\left(g_u - g_l\frac{n_u}{n_l}\right)\qquad(15)$$

$$= \frac{h\nu\,B_{ul}\,n_l}{c\,g_l}\left\{g_u - g_l\left[\frac{g_u}{g_l}\left(1 - \frac{h\nu}{kT_{ex}}\right)\right]\right\}\qquad(16)$$

$$= \frac{(h\nu)^2\,B_{ul}\,n}{c\,k\,T_{ex}}\left(\frac{g_u}{g_u + g_l}\right).\qquad(17)$$

1.3. Radiative transfer

When we integrate along the line of sight to determine what the telescope sees, the emission integral gives

$$I_\nu = \int j_\nu\,dx = \frac{A_{ul}\,h\nu}{4\pi}\,\frac{g_u}{g_u + g_l}\int n\,dx\qquad(18)$$

or in brightness temperature units with Doppler velocity in place of frequency

$$\int T_B(v)\,dv = C_0 \times N,\qquad(19)$$

where $C_0 = 5.485 \times 10^{-19}$ K km s^{-1} cm^2 for the 21-cm line. Note that we have to take the velocity integral as well as the line of sight integral to determine the total column density, N, because the radial motions of the atoms give them slight Doppler shifts which generate a line profile in velocity or frequency. In the case of an unresolved object such as a distant galaxy, Equation 19 gives the HI mass by

$$\frac{M_H}{m_\odot} = 2.3 \times 10^5 \left(\frac{d}{\text{Mpc}}\right)^2 \frac{\int S_v\,dv}{\text{Jy km s}^{-1}}.\qquad(20)$$

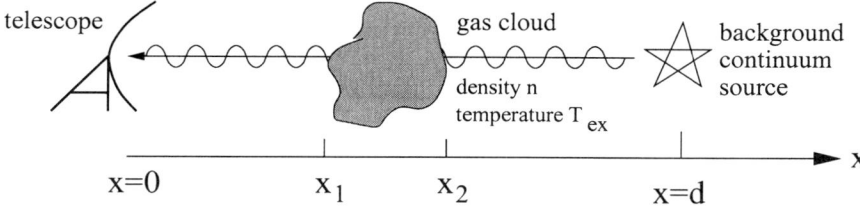

Figure 1. The geometry for the radiative transfer calculation in Equations 29 - 31.

The corresponding integral of the absorption coefficient along the line of sight gives the optical depth:

$$\tau_\nu = \int \kappa_\nu \, dx = \frac{(h\nu)^2}{c \, k} \frac{B_{ul} \, n}{T_{\text{ex}}} \frac{g_u}{g_u + g_l} \int \frac{n}{T} \, dx \quad (21)$$

or for the velocity integral

$$\int \tau_v dv = C_0 \frac{N}{T_{\text{harm}}}, \quad (22)$$

where again we replace the frequency integral with an integral over Doppler velocity, and C_0 for the 21-cm line has the same value as above. The T_{harm} in Equation 21 is the harmonic mean temperature,

$$T_{\text{harm}} \equiv \frac{\int n \, dx}{\int \frac{n}{T_{\text{ex}}} \, dx}. \quad (23)$$

This is only equal to the excitation temperature if the gas is isothermal along the line of sight.

Once we have determined the emission and absorption coefficients we can use radiative transfer to calculate how the radiation intensity builds up through propagation along the line of sight, as shown in Figure 1. Dividing Equation 8 by κ_ν gives

$$\frac{dI_\nu}{d\tau_\nu} = -\frac{j_\nu}{\kappa_\nu} + I_\nu, \quad (24)$$

since $d\tau = \kappa_\nu \, dx$ and $dx = -ds$, i.e. x increases away from the telescope, opposite to the direction of propagation, ds. We can integrate this along the line of sight, $x = 0$ to distance d which translates to $\tau = 0$ to $\tau(d)$, after multiplying both sides by $e^{-\tau}$, i.e.

$$e^{-\tau_\nu} \frac{dI_\nu}{d\tau_\nu} - I_\nu e^{-\tau_\nu} = -\frac{j_\nu}{\kappa_\nu} e^{-\tau_\nu} \quad (25)$$

or

$$\frac{d}{d\tau_\nu}\left(e^{-\tau_\nu} I_\nu\right) = -\frac{j_\nu}{\kappa_\nu} e^{-\tau_\nu}, \quad (26)$$

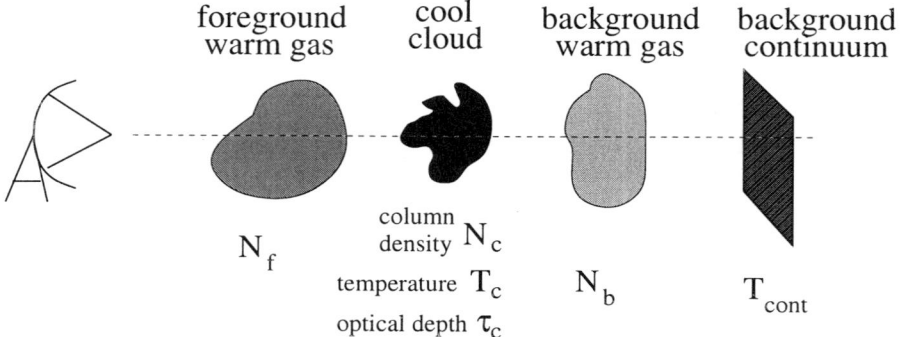

Figure 2. The geometry for the radiative transfer calculation in Equation 32.

$$\int_{I(0)}^{I(d)} d\left(e^{-\tau} I_\nu\right) = -\int_0^{\tau(d)} \frac{j_\nu}{\kappa_\nu} e^{-\tau} d\tau, \quad (27)$$

$$e^{-\tau(d)} I(d) - e^{-\tau(0)} I(0) = -\int_0^{\tau(d)} \frac{j_\nu}{\kappa_\nu} e^{-\tau} d\tau. \quad (28)$$

Since we are starting at the telescope, $\tau(0) = 0$, so

$$I(0) = e^{-\tau(d)} I(d) + \int_0^{\tau(d)} \frac{j_\nu}{\kappa_\nu} e^{-\tau} d\tau. \quad (29)$$

A simple case is that of a single cloud with a continuum background source, so that $\mathcal{B}_\nu(x) = \mathcal{B}_o$ inside the cloud (distance x_1 to x_2) and zero everywhere else, and $I(d) \equiv I_{\text{bkg}}$. Then

$$\tau = \int_{x_1}^{x_2} \kappa \, dx = \kappa(x_2 - x_1) = C_0 \times \frac{N}{T_{\text{ex}}}, \quad (30)$$

where $N = n(x_2 - x_1)$ is the column density through the cloud, and the velocity integral has been taken on both sides. Then the received intensity of the radiation is

$$I_0 = \mathcal{B}_c \left(1 - e^{-\tau}\right) + I_{\text{bkg}} \, e^{-\tau}, \quad (31)$$

where $\mathcal{B}_c$ is the Plank function for the excitation temperature of the gas in the cloud. In brightness temperature notation this is

$$T_B = T_{\text{ex}} \left(1 - e^{-\tau}\right) + T_{\text{bkg}} \, e^{-\tau}. \quad (32)$$

A more complicated case is shown on Figure 2, with optically thin foreground and background clouds of gas (column densities N_f and N_b) that are warm enough to have negligible optical depth, plus a cool, absorbing cloud with temperature T_c and column density N_c, plus background continuum, T_{cont}. This case gives

$$T_B(v) = \frac{C_0 \, N_f}{\sqrt{2\pi} \, \sigma_f} e^{\frac{(v-v_f)^2}{2\sigma_f^2}} + T_c \left(1 - e^{-\tau(v)}\right) +$$

$$+ \left(T_{\text{cont}} + \frac{C_0 \, N_b}{\sqrt{2\pi} \, \sigma_b} e^{\frac{(v-v_b)^2}{2\sigma_b^2}} \right) e^{-\tau(v)} . \tag{33}$$

Here we assume that the shapes of the profiles are Gaussians with center velocities and dispersions (v_f, σ_f), (v_c, σ_c), and (v_b, σ_b) for the foreground, cloud, and background gas, respectively.

1.4. Velocity profiles

In most areas of astronomy we use the optical version of the Doppler shift formula,

$$v_{\text{radial}} \equiv cz = c \times \frac{\Delta \lambda}{\lambda_r} , \tag{34}$$

where $\Delta \lambda$ is the difference between the observed wavelength and the rest wavelength, λ_r. This defines the redshift, z, without any relativistic corrections, so that z can be greater than one. A useful formula to remember is that **the velocity range corresponding to one MHz of bandwidth is just equal to the wavelength in mm.** For the $\lambda 21.1$ cm line of HI this means that 1 MHz corresponds to 211 km s^{-1}, or 1 km s^{-1} is $\frac{1000}{211}$ kHz or 4.73 kHz.

For a Maxwellian distribution of velocities corresponding to a thermal gas with temperature T_{kin}, the line profile is Gaussian, and so j_ν, κ_ν, and I_ν all have frequency or Doppler velocity profiles following

$$I_v = I_0 \times e^{\left[-\frac{(v-v_0)^2}{2\sigma_v^2} \right]} , \tag{35}$$

with

$$\sigma_v = \sqrt{\frac{2kT}{M_{\text{amu}}}} . \tag{36}$$

This reduces to

$$\sigma_v = 0.91 \text{ km s}^{-1} \times \sqrt{\frac{T}{100 \text{ K}} \times \frac{1}{M_{\text{amu}}}} \tag{37}$$

for hydrogen, $\sigma_v \simeq 1$ km s^{-1} for gas at 120 K temperature, $\sigma_v \simeq 10$ km s^{-1} for gas at 12,000 K, and $\sigma_v \simeq 100$ km s^{-1} for gas at 1.2 million K.

2. Mapping Basics: The uv Plane

Whether we are observing with a single-dish telescope, an aperture synthesis telescope, or a combination of the two, it is crucial to keep in mind the two Fourier conjugate representations of the telescope beam and the sky brightness. These are shown on Figure 3, with the uv plane functions on the left, and their conjugate functions on the plane of the sky on the right. Since the aperture plane is the transform of the image plane, the telescope beam is the Fourier transform of the aperture illumination in both cases. But there is a fundamental difference between a single-dish telescope (and/or an adding array interferometer) and a multiplying or correlation interferometer (aperture synthesis telescope). In the single-dish case (shaded on the lower right of Figure 3) it is the autocorrelation

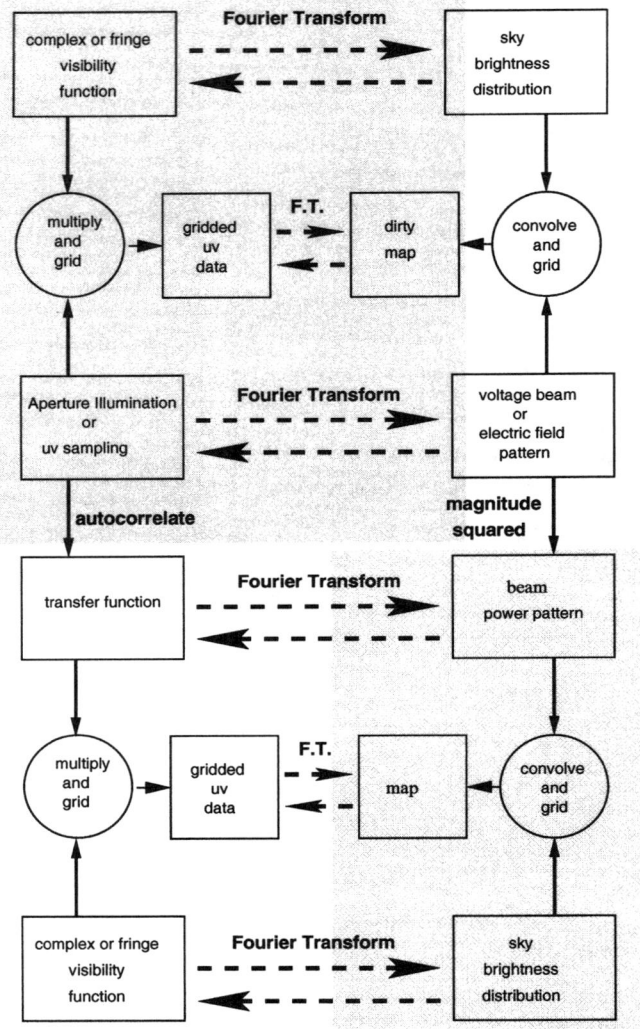

Figure 3. The relationship between the instrumental response and the sky brightness which lead to the measured map or image of the sky for two types of telescopes, aperture synthesis telescopes, i.e. correlation interferometers, (upper shaded area) and single-dish telescopes or adding interferometers (lower shaded area). The left side is the uv plane, the right side is the lm or sky plane.

function of the illumination pattern which matters; the transform of this function is the beam power pattern. For the multiplying interferometer (shaded on the upper left of Figure 3) the synthesized beam (dirty beam) is the transform of the illumination pattern itself. Thus there are negative sidelobes in the latter case, but not the former, since the autocorrelation function is always symmetric, so that its transform is real and positive definite.

2.1. Gridding

An important step in the processing of either interferometer or single-dish data is the gridding process. The Nyquist relation requires that the pixels in the map plane should be small enough that the FWHM of the beam is at least 2.4 pixels across, preferably 2.8 or more. If the pixels are larger than this relative to the beam width, then the higher spatial frequency information is lost due to undersampling. This is controlled in the aperture synthesis case by the gridding of the interferometer data on the uv plane before the transform back to the sky plane. When mapping with a single-dish a similar gridding step is needed to make a map out of data taken either in separate pointings or "on-the-fly" while the telescope is driving. Even if the single-dish observations are carefully taken in a lattice of pointings, it is still helpful to regrid the data using a convolving function which controls the behavior of the transform of the map. For example, noise in the spectra from the individual pointings can appear to have very high spatial frequencies, which alias on the uv plane back to low spatial frequencies unless they are suppressed by a convolving function in the gridding process. Thus the gridded map is:

$$T(x,y) = \frac{\Sigma_i T_i(l,m) f\left[(l-x),(m-y)\right]}{\Sigma_i U_i(l,m) f\left[(l-x),(m-y)\right]} , \qquad (38)$$

where the individual observations T_i are taken at positions (l, m) and the regular grid of the map pixels is given by (x, y). $f(x, y)$ is the convolving function, and $U(x, y)$ is a normalization function which is one at the positions of the observations. The division is suppressed for positions (x, y) so far from any observation that the denominator is less than some minimum threshold, typically between 0.2 and 0.05. A Gaussian is often used for f in the sky plane; in the uv plane it is better to use an exponential times sinc because of its aliasing suppression properties. If a Gaussian is used, the gridded map has resolution given by

$$\text{FWHM}^2_\text{gridded} = \sqrt{\text{FWHM}^2_\text{original} + \text{FWHM}^2_f} . \qquad (39)$$

The normalization illustrated in Equation 38 by U has many variations, especially in aperture synthesis applications. The easiest to understand are "uniform weighting", in which each uv cell has the same weight regardless of the number of samples which contribute to it, and "natural weighting", in which each sample has the same weight, so that regions which are heavily sampled are weighted more highly than sparsely sampled regions. "Robust weighting" achieves some of the best features of both strategies. In any event, the gridding should provide pixels in the sky plane spaced by less than 1/2.4 times the FWHM of the telescope beam.

2.2. Mapping speeds

For single-dish mapping the speed depends on the desired sensitivity of the spectra. Using the radiometer equation:

$$\sigma_T = \frac{\sqrt{2}\, T_{\text{sys}}}{\sqrt{\delta\nu\, T_{\text{int}}}}, \qquad (40)$$

with $\delta\nu$ the channel bandwidth, T_{int} the integration time, and T_{sys} the system temperature, we get the rms noise, σ_T. The factor of $\sqrt{2}$ in the numerator is usually needed to account for calibration, e.g. frequency switching. For the example of on-the-fly mapping, we should drive the telescope and read out the spectra so that one spectrum is read in the time it takes the telescope to move by less than 1/2.4 times the beamwidth, and so that this time provides σ_T less than 1/5 times the weakest features we hope to measure in the spectra after gridding. For Arecibo at 21-cm, this means we should read out spectra spaced by no more than about 1.2′. For channel bandwidth $\delta\nu$ of 0.5 km s^{-1} or 2.4 kHz we would need integration time per spectrum of 75 seconds to achieve rms noise of 100 mK, assuming system temperature of 30 K. Thus we can cover one square degree with 2600 spectra representing total integration time of 54 hours (plus overhead for telescope motion and calibration). So single-dish mapping can be a slow process, even at such a modest sensitivity as this. The most common response to this is to undersample the beam, spacing the spectra further apart than the Nyquist condition requires. This compromises the quality of the map, particularly if it is to be combined with interferometer data. It also causes the gridded map to have an uneven sensitivity function, which is given by the noise divided by the normalization in Equation 38, i.e.

$$S(x,y) = \frac{\sigma_T}{\sqrt{\Sigma_i U_i(l,m)\, f\,[(l-x),(m-y)]}}. \qquad (41)$$

2.3. Multibeam surveys and mosaicing

A very effective response to this problem is to build a multibeam receiver. An example is the Parkes 21-cm multibeam, which has been in use for more than four years on the 64-m telescope of the Australia Telescope National Facility. Using just the seven inner beams a team of astronomers led by Naomi McClure-Griffiths have used this instrument to map the Southern Milky Way at 21-cm (McClure-Griffiths et al. 2000, 2001). This project is similar to a Northern Milky Way survey, the Canadian Galactic Plane Survey (CGPS, English et al. 1998; Normandeau, Taylor, & Dewdney 1996). Both surveys combine single-dish and aperture synthesis data to achieve high dynamic range, good resolution, and most of all uniform sensitivity to all angular scales from $\sim 1'$ to many degrees.

Figure 4 indicates the telescope time required to survey a given area to a surface brightness sensitivity of 1 K (rms) in a velocity channel with width 0.8 km s^{-1}. The crosses mark single-dish telescopes, while interferometers are marked with crescents which illustrate how tapering (weighting down the longest baselines) can improve the brightness sensitivity. Only a little tapering helps; tapering to beamwidths larger than about two times the untapered value decreases the brightness sensitivity because so much data is weighted down by the

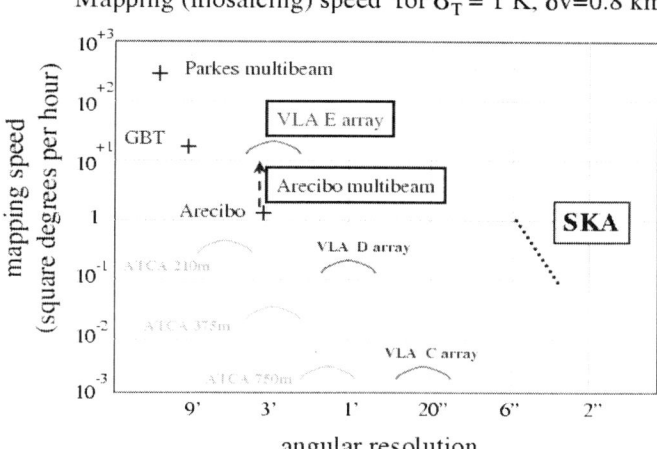

Figure 4. Mapping speeds with various telescopes. The resolution is indicated on the horizontal axis, while the number of square degrees per hour is on the vertical axis. The single-dish telescopes (marked with crosses) would never be driven so fast as this figure suggests. Driving them more slowly (say, at 1 degree per hour) would give rms noise, σ_T, smaller than 1 K in proportion to the inverse square root of the effective integration time.

tapering function. In boxes on Figure 4 are shown three future instruments. The Arecibo multibeam illustrates with the dashed arrow how much improvement in survey efficiency could be achieved if a seven beam system were installed on the Arecibo telescope. The E array is a possible enhancement to the VLA which would concentrate the antennas even more densely than in the D array. The SKA is the Square Kilometer Array. The design of this telescope is not settled yet, but it will provide several orders of magnitude improvement in brightness sensitivity and resolution over any existing cm-wave telescope.

Large area surveys like the Southern Galactic Plane Survey (SGPS) and the CGPS using interferometer telescopes must make use of the technique of mosaicing. This observing strategy moves the pointing center frequently (typically every 30 seconds) over a raster of positions spaced by roughly the half width (FWHM/2) of the primary beam of the interferometer. These "snap-shot" observations are repeated many times to build up good uv coverage. This is a more effective use of telescope time than longer integration times per snap-shot, since the Earth turns so slowly that most baselines do not change grid cells on the uv plane for many minutes. The more profound advantage of combining data from many nearby pointing centers of the interferometer is that it allows reconstruction of the shorter uv spacing information which cannot be directly measured because the antennas would shadow each other on such short baselines (Ekers & Rots 1979; Cornwell 1988; Sault, Staveley-Smith, & Brouw 1996). The effect of mosaicing on the uv plane is to deconstruct the antennas, and give information for all baselines corresponding to the nearest bit of one antenna to the nearest bit of the next.

2.4. Combining single-dish and aperture synthesis data

Combining the data from the two types of telescopes can be done in a variety of ways. These were compared quantitatively using 21-cm data on the Small Magellanic Cloud by Stanimirović (1999). The basic alternatives are to combine the uv data before mapping, to combine dirty maps and jointly deconvolve the beam shapes (implemented in the Miriad task MOSMEM), and to deconvolve separately and combine the cleaned maps (implemented in the Miriad task IMMERGE). The last has given consistent and trustworthy results in many tests.

However they are combined, the relative calibration of the two sets of data is crucial. Generally single-dish data is calibrated either using a standard brightness region (e.g. Weaver & Williams 1973) which gives units of K for T_B, or by observing unresolved continuum sources of known flux density, which gives units of Jy per beam. Interferometer maps are always calibrated using unresolved sources, thus they also generally have units of Jy per beam. The two are related by the gain of the synthesized beam, i.e.

$$G = \frac{A_e}{2\,k} = \frac{\lambda^2}{2\,\Omega_B\,k}, \qquad (42)$$

where A_e is the effective area of the dish and Ω_B is the solid angle of the synthesized (clean) beam, i.e.

$$\Omega_B = 1.13 \text{ FWHM}_1 \times \text{FWHM}_2, \qquad (43)$$

where FWHM$_1$ and FWHM$_2$ are the major and minor axes of the clean beam. For these beam widths in arc minutes, the gain is

$$G = 169 \text{ K Jy}^{-1} \, (\text{FWHM}_1 \times \text{FWHM}_2)^{-1}. \qquad (44)$$

This definition of the gain is $\frac{\lambda^2}{4\pi k}$ times the standard engineering quantity called the directive gain. The advantage of this definition is that it gives the conversion between units of Jy per beam and K of brightness temperature. For the Arecibo telescope at 21-cm $G \simeq 10$ K Jy^{-1}, for the GBT it may be about 1.5 K Jy^{-1}. For the VLA D-array with clean beam size of 45″ the gain is 300 K Jy^{-1}. Using the gain, we can convert from the observed antenna temperature to the true flux density of the source, S. In the spectral line case, the analog of the brightness temperature integral of Equation 19 becomes the flux integral ($\int S(v)\,dv = \frac{\int T_A(v)\,dv}{G}$). This gives the HI mass for the case of an unresolved 21-cm line source such as a distant galaxy,

$$\frac{M_H}{m_\odot} = 2.3 \times 10^5 \left(\frac{d}{\text{Mpc}}\right)^2 \frac{\int S(v)\,dv}{\text{Jy km s}^{-1}}, \qquad (45)$$

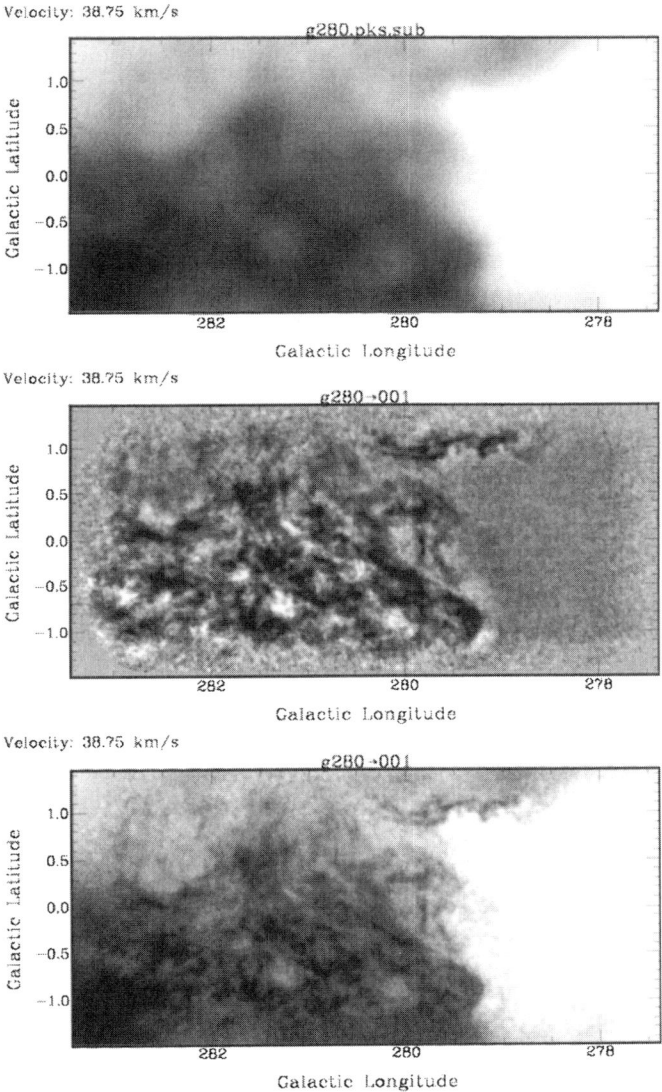

Figure 5. Images of part of a Galactic Supershell at longitude 277°, latitude 0°, velocity 38.75 km s^{-1}. The upper panel shows Parkes data only, the middle panel shows Australia Telescope Compact Array data only, and the lower panel shows the combined map.

222 Dickey

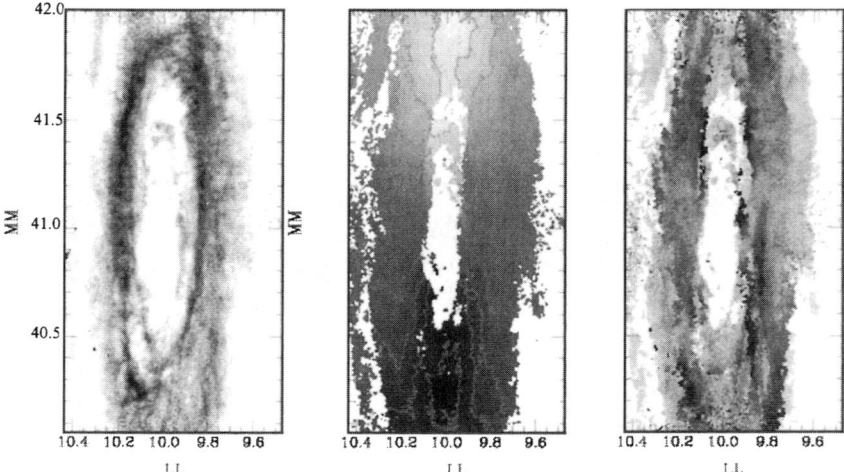

Figure 6. Moment maps of M31 made from the Westerbork Synthesis Radio Telescope data of Brinks & Shane (1984). The left figure is the zero moment, showing the HI column density, the center figure is the first moment, showing the radial velocity field, and the right figure is the second moment, showing the line width.

where d is the distance to the galaxy.

Before the data can be combined the calibration scales of the two instruments must be matched carefully. The only sure way to do this is by comparing data for the same source **in the overlap region on the uv plane**, which is the range of spatial frequencies for which both instruments have good sensitivity. Such an overlap calibration is implemented in the Miriad task IMMERGE.

Figure 5 shows the combination of single-dish and interferometer maps of the edge of a supershell in the outer Milky Way HI (McClure-Griffiths et al. 2000). The range of uv spacings contributing to each is manifest in its appearance. Only by combining them is the full range of structure clear.

3. Using Spectral Line Cubes

The spectral line cube is a three dimensional data structure made by stacking maps taken at different frequencies, i.e. Doppler velocities. The cube can be represented as a movie, either as a series of images of the sky at different frequencies, or a series of position-velocity diagrams taken along different lines on the sky. These and several other very useful ways of displaying spectral line cubes are implemented in the KARMA package (Gooch 1995). It is important not to confuse the spectral line cube with an image of the line emission in three spatial dimensions, but often there is a velocity gradient along the line of sight

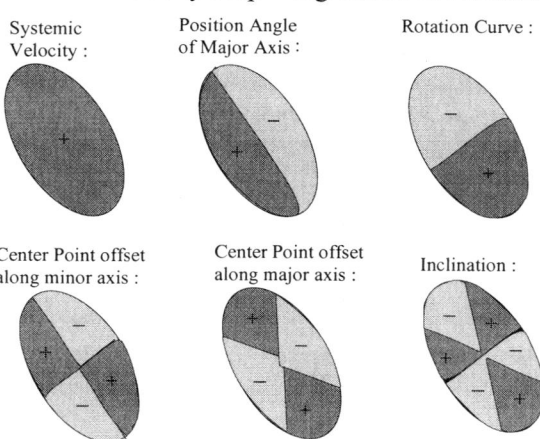

Figure 7. Errors in the fitted parameters give rise to patterns in the velocity residual map.

due to the dynamics of the system which allows us to link some velocities with distances.

Another way to represent the data in the cube is by computing the velocity moments of the brightness. The nth moment map, $V^n(x,y)$ is defined by

$$V^n(x,y) \equiv \int_v T_B(x,y,v) \times (v - v_0)^n \, dv \,, \qquad (46)$$

where the cube is $T_B(x,y,v)$, v_0 is the systemic velocity or some appropriate zero point for the velocity scale, and the integral is taken as a sum over some or all of the planes of the cube. The zero-th moment map is simply a map of the integral of the spectrum, as in Equation 19 but taken for every spatial pixel. For the 21-cm line this gives column density (as long as the optical depth is not high). The first moment map shows the mean velocity (weighted by intensity) in each pixel. Often this gives a good image of the velocity field of the source. The second moment map generally shows the linewidth as a function of position. Examples of moment maps are shown on Figure 6.

The first and higher order moment maps are very unstable to bias in the presence of noise. To be safe, the cube should be blanked to eliminate regions in both space and velocity which have no signal. This can be done by hand (the AIPS task BLANK is a good interactive tool for this), or automatically using a smoothed version of the cube as a template for determining where there is and is not signal. The AIPS task MOMNT is a good implementation of this strategy; it sets a threshold well outside the region with detected signal, so as to include the faint tails of strong lines, but it excludes spurious noise peaks in regions of the cube far from any emission.

The first moment map of a spectral line from the disk of a galaxy is generally used to fit the rotation velocity field. A good task for doing this is the GIPSY task ROCUR, which has been implemented on some AIPS systems. The objective is to determine several unknown parameters: the position of the center of the rotation pattern, the inclination (perhaps as a function of radius if the galaxy is warped) and the position angle of the major axis (i.e. the line of nodes, perhaps also as a function of radius), and the rotation curve, i.e. the circular velocity as a function of radius. The rotation curve is of critical importance for dynamical modelling since, with the assumption of circular rotation, it shows the distribution of gravitational force as a function of radius. Departures from circular rotation, e.g. in spiral arms, can often be analysed based on the first moment map as well.

Fitting the parameters of the disk rotation gives a model velocity field which can be subtracted from the first moment map, leaving the velocity residual map. A good fit shows only noise in the residual map, but if the fit is not perfect, the residual map shows distinct patterns. Figure 7 shows some examples, taken from Warner, Wright, & Baldwin (1973).

4. Conclusions

Single-dish telescopes are indispensable for observations of any spectral line source which is so extended that it cannot be imaged in a single interferometer field. The combination of single-dish and aperture synthesis data is becoming the standard technique for mapping Galactic sources, especially for the ubiquitous HI line. Single-dish telescopes like Arecibo and the GBT will be hugely valuable for this, particularly when equipped with multi-beam receivers to enhance their mapping speed. We may be on the threshold of a renaissance in single-dish, Galactic, spectral line astronomy.

References

Bania, T. M., Balser, D. S., Rood, R. T., Wilson, T. L., & Wilson, T. J. 1997, ApJS, 113, 353

Brinks, E. & Shane, W. W. 1984, A&AS, 55, 179

Caswell, J. L. 1997, MNRAS, 289, 203

Cornwell, T. J. 1988, A&A, 202, 316

Ekers, R. D. & Rots, A. 1979, in "Image Formation from Coherence Functions in Astronomy", ed. C. van Schooneveld (Dordrecht: Reidel), 61

English, J., Taylor, A. R., Irwin, J. A., Bougherty, S. M., Basu, S., et al. 1998, Proc. Astron. Soc. Austr., 15, 56

Gibson, S. J., Taylor, A. R., Higgs, L. A., & Dedwney, P. E. 2000, ApJ, 540, 851
Gooch R. 1995, in ASP Conf. Ser. Vol. 77, Astronomical Data Analysis and Software Systems V, ed. R. A. Shaw, H. E. Payne & J. J. E. Hayes (San Francisco: ASP), 144
Lang, C. C., Goss, W. M., & Morris, M. 2001, AJ, 121, 2681
Lewis, B. M., Oppenheimer, B. D., & Daubar, I. J. 2001, ApJ, 548, L77
Liszt, H. & Lucas, R. 1996, A&A, 314, 917
Magnani, L. & Onello, J. S. 1993, ApJ, 408, 559
McClure-Griffiths, N. M., Dickey, J. M., Gaensler, B. M., Green, A. J., Haynes, R. F., & Wieringa, M. H. 2000, AJ, 119, 2828
McClure-Griffiths, N. M., Green, A. J., Dickey, J. M., Gaensler, B. M., Haynes, R. F., & Wieringa, M. H. 2001, ApJ, 551, 394
Normandeau, M., Taylor, A. R., & Dewdney, P. E. 1996, Nature, 380, 687
Osterbrock, D. E. 1989, Astrophysics of Gaseous Nebulae and Active Galactic Nuclei (Sausalito: University Science Books)
Pauls, T., Johnston, K. J., & Wilson, T. L. 1996, ApJ, 461, 223
Peck, A. B., Goss, W. M., Dickel, H. R., Roelfsema, P. R., Kesteven, M., Dickel, J. R., Milne, D. K., & Points, S. D. 1997, ApJ, 486, 329
Phillips, C. J., Norris, R. P., Ellingsen, S. P., & McColloch, P. M. 1998, MNRAS, 300, 1131
Rohlfs, K. & Wilson, T. L. 1996, Tools of Radio Astronomy, (2nd edition; Berlin: Springer-Verlag)
Roshi, D. A. & Anantharamaiah, K. R. 1997, MNRAS, 292, 63
Sault, R. J., Staveley-Smith, L., & Brouw, W. N. 1996, A&AS, 120, 375
Spitzer, L. 1977, Physical Processes in the Interstellar Medium (New York: John Wiley)
Stanimirović, S. 1999, Ph.D. Thesis, University of Western Sydney Nepean, 157
Szymczak, M., Kus, A. J., & Hrynek, G. 2000, MNRAS, 312, 211
Warner, P. J., Wright, M. C. H., & Baldwin, J. E. 1973, MNRAS, 163, 163
Weaver, H. & Williams, D. R. W. 1973, A&AS, 8, 1

Visit to the bioluminescent bay in La Parguera on the south of Puerto Rico.

Pulsar Observations I. – Propagation Effects, Searching, Distance Estimates, Scintillations and VLBI

James M. Cordes

Astronomy Department, Cornell University, Ithaca, New York 14850, USA

Abstract. Propagation effects from the interstellar medium (ISM) are discussed along with their implications for pulsar observations and analysis. Then I discuss (1) dedispersion techniques, particularly that for postdetection systems; (2) pulsar search analysis and the role of instrumental, interstellar and orbital broadening of pulses; and (3) methods for determining distances to pulsars and a new Galactic electron density model now nearly completed. I end by discussing new observations including multibeam searches and VLBI astrometric observations.

1. Introduction

Radio pulsar signals consist of pulsed continuum noise that can be highly elliptically polarized. The signals have a very high degree of spatial coherence but very little temporal coherence. Pulses are modified significantly as they propagate through the ionized and magnetized interstellar medium. The signal complexity requires refined techniques for discovering new pulsars and for extracting all information contained in pulsar signals.

In this article, I discuss the phenomenology and measurement of pulsar signals. Methodology will center on 'postdetection' methods, i.e. those that involve recording the signal after it has been squared. Predetection methods are discussed by I. Stairs in a separate lecture.

2. Wave Propagation Basics

For the interstellar medium (ISM), the plasma frequency is
$\nu_p \approx 1.56\,\text{kHz}\,(n_e/0.03\,\text{cm}^{-3})^{1/2}$ and the gyrofrequency is $\nu_\text{B} \approx 2.8\,\text{Hz}\,B_{\mu G}$. Magnetic fields introduce birefringence that is most easily detected as Faraday rotation. The index of refraction in a cold magnetized plasma like the ISM for $\nu \gg \nu_p$ and $\nu \gg \nu_{\text{B}\parallel}$ is (e.g. Thomson, Moran, & Swenson 2001)

$$n_{l,r} \approx 1 - \nu_p^{\,2}/2\nu^2 \mp \nu_p^{\,2}\nu_{\text{B}\parallel}/2\nu^3, \qquad (1)$$

where $\nu_p = (n_e e^2/\pi m_e)^{1/2}$ is the plasma frequency and $\nu_{\text{B}\parallel} = eB\cos\theta/m_e c$ is the electron gyrofrequency calculated for the magnetic field component along the line of sight; the $\mp$ cases apply for LHCP and RHCP waves.

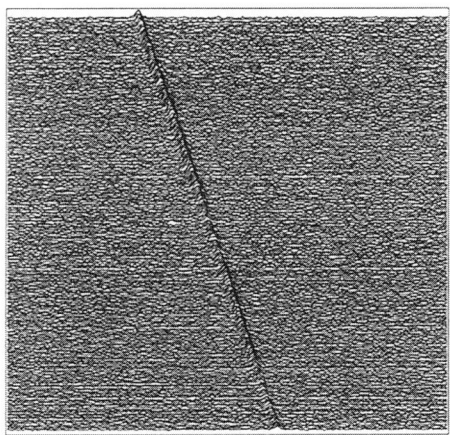

Figure 1. Frequency-pulse phase plot for PSR B1534+12 for 1024 frequency channels (vertical axis) at 0.43 GHz (Courtesy M. McLaughlin).

Dispersive arrival times & Faraday Rotation: The frequency dependent time delay (see Figure 1) is calculated by integrating (group velocity)$^{-1}$ = $dk/d\omega = c^{-1}\left(n_{l,r} + \nu dn_{l,r}/d\nu\right)$ along the line of sight, giving

$$t = t_{\rm DM} \pm t_{\rm RM}. \quad (2)$$

The dispersive time delay and the small correction due to birefringence are

$$t_{\rm DM} = \frac{e^2}{2\pi m_e c}\frac{\int_0^D ds\, n_e}{\nu^2} = 4.15\,{\rm ms\ DM}\ \nu_{\rm GHz}^{-2}, \quad (3)$$

$$t_{\rm RM} = 2{\rm RM}c^2\nu^{-3} = 0.18\,{\rm ns\ RM}\ \nu_{\rm GHz}^{-3}, \quad (4)$$

where the dispersion and rotation measures and their standard units (for D in pc, n_e in cm^{-3}, and $B_\parallel$ in μG) are

$$\rm DM = \int_0^D ds\, n_e(s) \quad (pc\,cm^{-3}), \quad (5)$$

$$\rm RM = \frac{e^3}{2\pi m_e^2 c^4}\int_0^D ds\, n_e B_\parallel = 0.81\int_0^D ds\, n_e B_\parallel \quad (rad\,m^{-2}). \quad (6)$$

The birefringent delay is too small to be measureable in practical situations but is manifested as Faraday rotation of the plane of polarization. The differential phase between LHCP and RHCP is

$$\Delta\phi = \frac{2\pi}{\lambda}\int_0^D ds\,(n_r - n_l) = 2{\rm RM}\lambda^2, \quad (7)$$

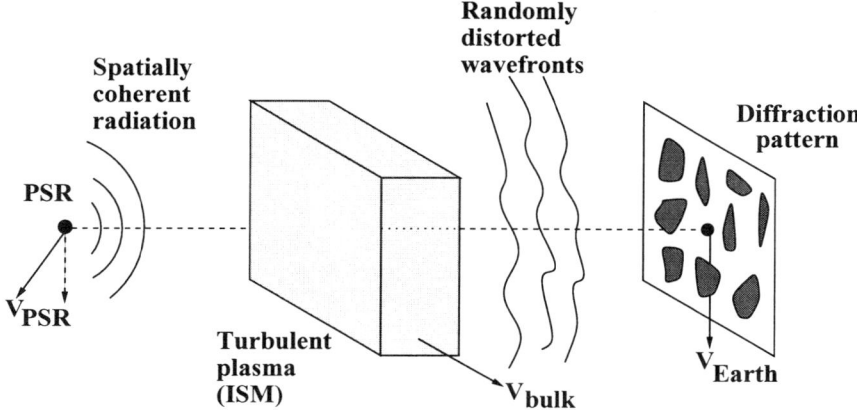

Figure 2. Geometry of scattering in the ISM that is responsible for a variety of seeing and scintillation effects.

where the factor of 2 accounts for the fact that the polarization position angle rotates over π when the phase rotates 2π.

Multipath Propagation: Interstellar Scattering & Scintillations: Phase perturbations from electron-density fluctuations are

$$\phi(\mathbf{x}_\perp, z) = \frac{2\pi}{\lambda} \int_0^D ds\, n_{l,r} \approx -\lambda r_e \int_0^D dz\, \delta n_e(\mathbf{x}_\perp, z), \qquad (8)$$

where the classical electron radius is $r_e = e^2/m_e c^2$. Geometrical optics (see Figure 2) gives the *refraction angle*, $\theta_r = k^{-1}\nabla_\perp \phi$, while physical optics gives the *diffraction angle*, $\theta_d \sim \lambda/\ell_\perp$, where $\ell_\perp$ is the length scale on which the phase changes by ~ 1 radian. For the ISM, θ_d ranges from < 1 mas to ~ 1 arc sec at 1 GHz for different lines of sight.

Effects from multipath propagation include (Rickett 1990; Cordes & Lazio 1991):

Angular broadening ("seeing"): Compact radio sources of many kinds, Galactic and extragalactic, show angular broadening by amounts ranging from $\lesssim 1$ mas to 1 arcsec at 1 GHz, depending on the distance and direction. The angular smearing scales approximately as ν^{-2}.

Time of arrival fluctuations: Changes in geometry induce DM variations. Also, variable scattering causes variable arrival times.

Pulse broadening: Multipath propagation causes a multiplicity of arrival times, usually seen as an exponential-like 'tail' to pulses from pulsars.

Diffractive intensity scintillations (DISS): Intensity variations in time and frequency on times scales ~ 100 sec and frequency scales ~ 1 MHz. These scales are very highly dependent on frequency, direction and source distance and velocity. These variations result from diffraction of radiation into an angle $\theta_d \sim \lambda/\ell_d$, where ℓ_d is the diffraction scale, the transverse length on which the phase perturbation from δn_e is one radian.

Refractive intensity scintillation (RISS): Intensity variations from large scale focusing and defocusing of the radiation. The refraction scale is $\ell_r \sim D\theta_d$, implying the well known relation, $\ell_r \ell_d \sim \lambda D \sim$ (Fresnel scale)^2.

Spectral Broadening: Broadening of narrow spectral lines from a combination of scattering and time-variable geometry. The effect is very small in the ISM ($\lesssim 1$ Hz) but has been measured from spacecraft viewed through the interplanetary medium.

Superresolution phenomena: The optical/atmospheric "Stars twinkle but planets do not" translates into "Pulsars twinkle, AGNs do not," at least for DISS. The (angular) isoplanatic scale for DISS is typically $\sim \ell_d/D \lesssim 10^{-6}$ arc sec at 1 GHz, or about $\times 10^3$ smaller than an AGN and $\times 10$ larger than a typical pulsar magnetosphere. Thus, we expect pulsars to show fully modulated scintillations in the strong scattering regime, whereas AGNs will not show DISS, typically. RISS is more forgiving of a large source size because it is caused by much larger turbulence scales than is DISS, with a critical angle $\sim \ell_r/D \sim \theta_d$. Thus, AGNs and masers, as well as pulsars, show RISS.

Scaling Laws for Pulse Broadening: Pulse broadening from multipath propagation is approximately

$$\tau_d \sim \frac{D\theta_d^2}{2c}. \qquad (9)$$

The scaling laws are approximately $\theta_d \propto \nu^{-2}$ and $\tau_d \propto \nu^{-4}$. Figure 3 shows a strongly-scattered pulse at three frequencies for a large-DM pulsar compared to a low-DM pulsar with negligible scattering.

Pulse broadening is correlated with DM, though not perfectly. Figure 4 shows τ_d at 1 GHz plotted against DM along with a parabolic fit (solid line) and $\pm 1.5\sigma$ lines (dashed). The fit is with $\log DM$ as the independent variable:

$$\log \tau_d(\mu s) = -3.59 + 0.129 \log \mathrm{DM} + 1.02 (\log \mathrm{DM})^2 - 4.4 \log \nu_{\mathrm{GHz}}. \qquad (10)$$

The scattering time for individual pulsars can deviate considerably from the fit, with $\sigma_{\log \tau} = 0.65$.

Scintillations: Figure 5 shows diffractive interstellar scintillation (DISS) as a dynamic spectrum $I(t,\nu)$ and the 2D autocorrelation function (ACF) of the dynamic spectrum, $\langle I(t,\nu)I(t+\delta t, \nu+\delta \nu)\rangle$ with time and frequency lags $\delta t, \delta \nu$, and where angular brackets denote time average. Slices along the ACF axes have features whose widths yield the characteristic scintillation (or diffraction) bandwidth of $\Delta \nu_d$ and time scale Δt_d of the DISS. Quoted values in the literature

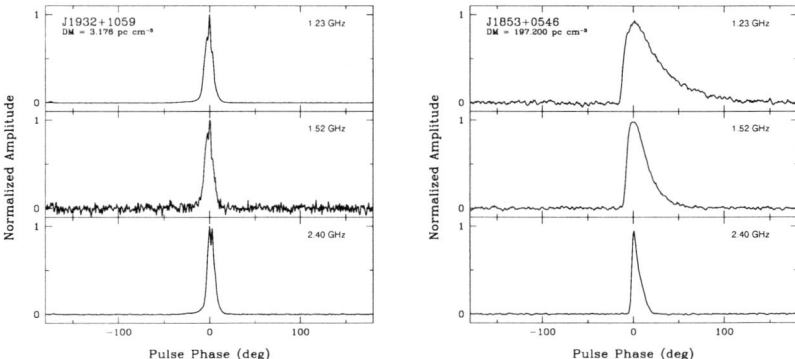

Figure 3. Left: Pulse shape at 3 frequencies for PSR J1932+1059 (aka B1929+10), a low DM pulsar whose profile evolution with frequency is all intrinsic. The low S/N at the middle frequency is due to DISS. Right: Pulse shape at 3 frequencies for PSR J1853+0546 showing the asymmetry caused by multipath propagation; the pulse-broadening time scales strongly with frequency, $\tau_d \propto \nu^{-4}$. Data are from R. Bhat et al. (unpublished).

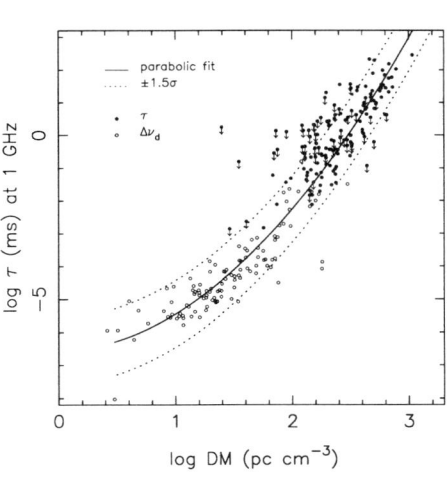

Figure 4. Pulse broadening time at 1 GHz plotted against dispersion measure. Filled circles indicate direct measurement of the scattering time; open circles represent scattering times estimated from scintillation bandwidth measurements. All measurements are scaled to 1 GHz using $\tau \propto \nu^{-4.4}$. The solid line is the parabolic fit given in Equation 10; dashed lines are at $\pm 1.5\sigma$.

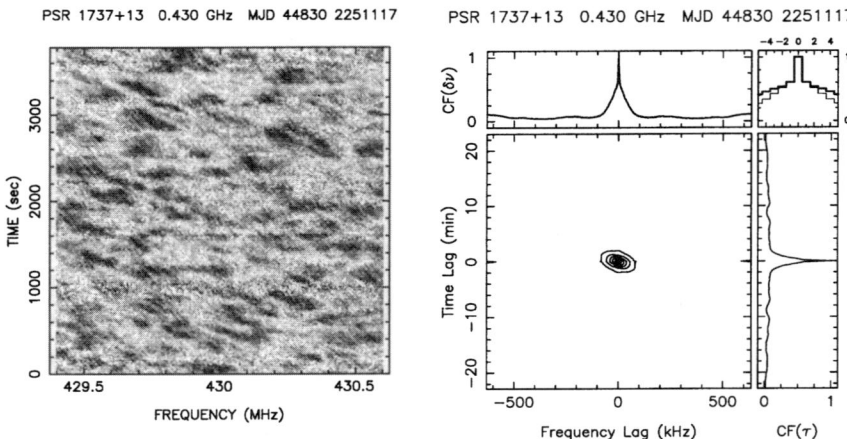

Figure 5. Left: Dynamic spectrum. Right: Autocorrelation function of the dynamic spectrum.

are usually taken to the halfwidth at half maximum for $\Delta\nu_d$ and the halfwidth at 1/e for Δt_d. These widths are strong functions of the strength of scattering. Scattering increases with wavelength and distance and is also much stronger in the inner Galaxy. For stronger scattering, the characteristic scales of the DISS get *smaller*.

DISS is important because it strongly modulates the pulsar intensity according to an *exponential* probability density function (PDF). For the more weakly scattered pulsars, an entire observation (i.e. the total bandwidth and duration) may comprise only one patch of constructive or destructive interference. Thus in a pulsar search, it is possible that a given object will be scintillated up or down. The exponential shape of the PDF implies that it is more likely to be scintillated downward. Therefore searches may profitably cover a given sky position more than once.

Scaling Laws from the Electron Density Wavenumber Spectrum:
Electron-density variations are commonly described by a power-law wavenumber spectrum of the form (e.g. Armstrong, Rickett, & Spangler 1995)

$$P_{\delta n_e}(q) = C_n^2 q^{-\alpha}, \quad \frac{2\pi}{\ell_0} \leq q \leq \frac{2\pi}{\ell_1}, \tag{11}$$

where C_n^2 is the spectral coefficient of the wavenumber spectrum of δn_e, ℓ_1, ℓ_0 are the inner and outer scales of the turbulence and the spectrum vanishes outside the specified interval. Equation 11 explicitly assumes isotropic irregularities. Evidence for anisotropies exists but it is not clear on what length scale these are present. Also, there is strong evidence (from refractive effects) that the spectrum departs from a power law on scales $\sim 1 - 10$ AU. The spectral coefficient varies by many orders of magnitude between lines of sight, as diagnosed from scattering and scintillation observations.

The *scattering measure* is defined as (e.g. Cordes et al. 1991)

$$\text{SM} = \int_0^D ds\, C_n^2(s). \tag{12}$$

Observables yield different effective values for SM because of different weight factors:

(1) $\text{SM}_{\theta,x} = \text{SM}$ for measurements of angular diameters of extragalactic sources;

(2) $\text{SM}_{\theta,g} = 3\int ds\,(1 - s/D)^2 C_n^2$ for angular diameters of galactic sources;

(3) $\text{SM}_\tau = 6\int ds\,(s/D)(1 - s/D) C_n^2$ for pulse broadening and scintillation measurements.

Units are usually m$^{-20/3}$ for C_n^2 and kpc m$^{-20/3}$ for SM. With ν in GHz, the seeing disk size (FWHM) is

$$\theta_d = \nu^{-11/5} \times \begin{cases} 128\,\text{mas}\,\text{SM}_{\theta,x}^{5/3} & \text{Extragalactic Sources} \\ \\ 71\,\text{mas}\,\text{SM}_{\theta,g}^{5/3} & \text{Galactic Sources.} \end{cases} \tag{13}$$

For pulse broadening and scintillation bandwidth,

$$\tau_d = 0.90\,\text{ms}\,\text{SM}_\tau^{6/5}\nu^{-22/5} D, \tag{14}$$

$$\Delta\nu_d = 0.171\,\text{MHz}\,\text{SM}_\tau^{-6/5}\nu^{22/5}(C_1/1.16) D^{-1}, \tag{15}$$

where D is in kpc and $C_1 = 1.16$ for a Kolmogorov spectrum. The DISS time scale is determined by the pulsar's transverse speed $V_{p\perp}$:

$$\Delta t_d = 2.53 \times 10^4\,\text{km s}^{-1}\,W_{\text{C}} W_{\text{D,PM}} \frac{\sqrt{D \Delta\nu_d}}{\nu V_{p\perp}}, \tag{16}$$

where W_C and $W_{\text{D,PM}}$ are correction factors equal to unity for a uniform, Kolmogorov medium. See Cordes & Rickett (1998) for further discussion about W_C. The factor $W_{\text{D,PM}}$ takes into account the spatial distribution of scattering material and is approximately

$$W_{\text{D,PM}}(D) \approx \left[\frac{\text{SM}_\tau}{3\text{SM} - (\text{SM}_{\theta,g} + \text{SM}_\tau)}\right]^{1/2}. \tag{17}$$

3. Dedispersion Techniques

Dispersion may be viewed as a delay of a pulse or as a phase rotation of its Fourier components. Correction for dispersion is done in two primary ways (e.g. Hankins & Rickett 1975). *Post-detection* schemes make use of a multichannel receiver, realized as an analog filter bank in the old days, or today as either an FFT

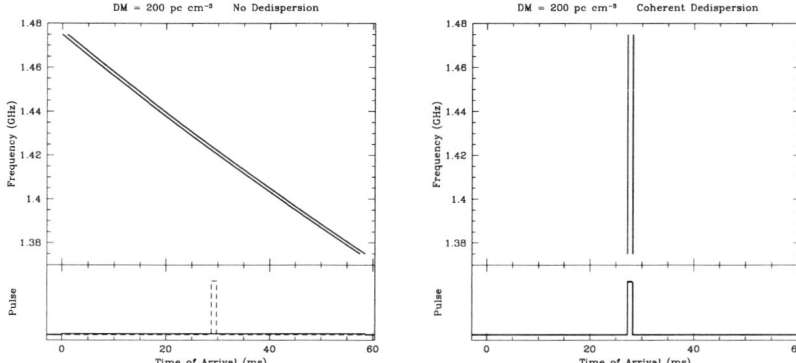

Figure 6. Schematic diagrams showing dispersive arrival times and postdetection dedispersion. Left: the top panel shows arrival time vs. frequency for a 1-ms wide pulse; the bottom panel shows the intrinsic pulse shape (dashed line) and the dispersed pulse (heavy line). Right: Coherent dedispersion corrects Fourier phases perfectly, yielding time resolution equal to the reciprocal of the total bandwidth.

system or correlation spectrometer. The detected (i.e. squared) outputs of each channel can be advanced to compensate for the delays from the dispersion law of Equation 3. *Predetection* methods correct the Fourier phases and are described further by I. Stairs. Post-detection devices achieve high-time resolution at high frequencies, but are limited by filter response times and by scattering at low frequencies. Predetection techniques use wider filters, so the filter response time can be much smaller than for the postdetection case. However, predetection methods are still limited by scattering at low frequencies and for large DM.

Figures 6-7 show the dispersion effect and its compensation in pre-and-post detection schemes schematically.

Minimum time resolution for postdetection Systems: The time resolution is determined by a combination of *dispersion over an individual channel* (with DM in pc cm^{-3}):

$$\Delta t_{\rm DM}(\Delta\nu) = 8.3\,\mu s\, DM\, \Delta\nu_{\rm MHz}\nu_{\rm GHz}^{-3}, \tag{18}$$

which is the differential form of Equation 3, and from the *impulse response* of an individual channel:

$$\Delta t_{\Delta\nu} \sim (\Delta\nu)^{-1} = 1\,\mu s (\Delta\nu_{\rm MHz})^{-1}. \tag{19}$$

For given DM and ν, the minimum dispersion time implied by Equations 18, 19 is

$$\Delta t_{DM,min}(\mu s) = \left(\frac{8.3 DM}{\nu_{GHz}^3}\right)^{1/2}. \tag{20}$$

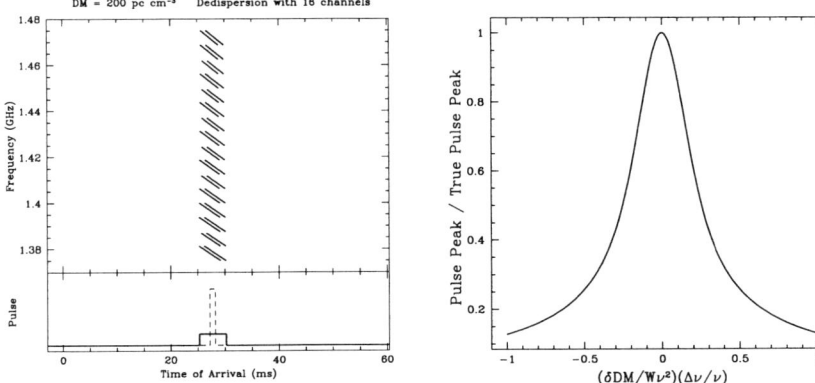

Figure 7. Left: the top panel shows how postdetection dispersion compensation yields residual dispersion across individual channels; the bottom panel shows that in this case the pulse is still significantly smeared. Smaller channel bandwidths yield less residual smearing but only to the limit where the dispersion smearing equals the impulse response time of the channel $\sim \Delta\nu_{ch}^{-1}$. Right: Pulse amplitude loss from imperfect dedispersion using a DM in an error by δDM. The abcissa is the composite quantity $(\delta DM/W\nu^2)(\Delta\nu/\nu)$ with δDM in pc cm^{-3}, pulse width W in ms, and frequency ν and bandwidth $\Delta\nu$ in GHz. The width of the function ~ 0.52 (FWHM).

The minimum resolution due to the combined effects of dispersion and the filter response, which are equal, is $\sqrt{2}\Delta t_{DM,min}$. (The achievable resolution, of course, may be determined by scattering rather than dispersion)

How accurately must DM be known? To dedisperse optimally, DM must be known to a precision that is frequency, bandwidth and pulse-width dependent. For a Gaussian shaped pulse, the amplitude of the dedispersed pulse (relative to perfect dedispersion) for an error δDM is

$$I_{\max}(\delta\mathrm{DM}) = \frac{\sqrt{\pi}}{2}\zeta^{-1}\mathrm{erf}\,\zeta, \qquad \zeta = \frac{6.91\,\mathrm{ms}}{W}\frac{\delta\mathrm{DM}}{\nu^2}\frac{\Delta\nu}{\nu}, \qquad (21)$$

where W is the intrinsic pulse width (FWHM) in ms and the units are pc cm^{-3}, GHz and GHz for δDM, ν and $\Delta\nu$, respectively. $I_{\max}$ is plotted in Figure 7 (right panel). The FWHM of the curve implies a DM tolerance

$$\delta\mathrm{DM} \approx 0.52\,\mathrm{pc\,cm^{-3}}\frac{W\nu^3}{\Delta\nu}. \qquad (22)$$

Achievable Time Resolution vs. DM & Frequency: For postdetection systems the achievable time resolution is approximately the root-quadratic sum of the filter-response, dispersion smearing and pulse-broadening times:

$$\Delta t_{\min} = \left[(\Delta t_{DM})^2 + (\Delta t_{\Delta\nu})^2 + \tau_d^2\right]_{\min}^{1/2} = \left[2(\Delta t_{DM})^2 + (\tau_d)^2\right]^{1/2}. \qquad (23)$$

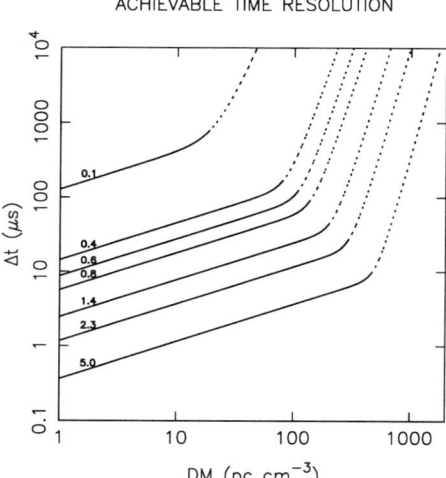

Figure 8. Best achievable time resolution of postdetection systems plotted against pulsar dispersion measure. The curves are labeled with radio frequency ranging from 0.1 to 5 GHz. (Solid lines) Time resolution is limited by dispersion; (Dashed lines) Time resolution limited by interstellar scattering. Note that predetection systems can achieve better time resolution than the post detection case if the latter is dispersion bound. But predetection systems will be limited by scattering according to the dashed curves shown here.

Figure 8 shows the minimum resolution plotted against DM for frequencies ranging from 0.1 to 5 GHz. The solid lines designate those DM-ν combinations that are *dispersion limited* while the dashed lines correspond to *scattering dominated* situations. Predetection dispersion removal over a bandwidth $B > (\Delta t)^{-1}$ can do better than postdetection schemes only if the latter are dispersion limited; that is, only for cases in Figure 8 where solid lines are plotted.

Viewed another way, there is a maximum DM for which some desired time resolution Δt can be achieved. Excluding scattering, the maximum DM is

$$\mathrm{DM}_{\max,\Delta\nu} = \frac{(\Delta t_{\mu s})^2 \nu_{\mathrm{GHz}}^3}{16.6} . \qquad (24)$$

However, at low frequencies scattering will limit even further the range of dispersion measures over which the specified time resolution can be obtained to a maximum:

$$\mathrm{DM}_{\max,\mathrm{scatt}} = 0.86\,\mathrm{pc\ cm^{-3}} 10^{1.88} (1 + 0.28 \log \Delta t_{\mu s} + 1.22 \log \nu_{\mathrm{GHz}})^{1/2} . \qquad (25)$$

4. Post-Detection Pulsar Observations

Table 1 summarizes typical observations that are made with postdetection systems. The data rate that is output from the filterbanks or correlator systems before any processing (such as realtime or offline 'folding' of data according to a pulsar emphemeris) is typically 1-20 Mbytes s^{-1}.

Table 1. Post-Detection Observations & Parameters

Type	Δt (μs)	T (s)	N_ν	Comments
Searching	~ 100	$\gtrsim 100$	256 to 1024	polarizations summed Trial DMs: ~ 50 to 500 (ℓ, b) FFTs of length M Megasamples
Timing	$10^{-3}P$	~ 100	~ 256	polarizations summed (calibrated!)
Polarization	$10^{-3}P$	~ 100	~ 256	full Stokes
ISS	$10^{-2}P$	~ 3600	~ 1024	polarizations summed
Single Pulse	$10^{-3}P$	~ 3600	$\gtrsim 256$	full Stokes

5. Pulsar Searches

Periodicity Searches: Figure 9 outlines the salient, generic features of periodicity searches. Alternative schemes have been used, including a 2D Fourier analysis of $I(t, \nu)$ rather than dedispersing with trial values of DM. A general discussion can be found in Lyne & Smith (1990).

The single-harmonic threshold, $S_{\min_1}$, is the minimum amplitude that a single FFT component must have to detected:

$$S_{\min_1} = \frac{mS_{\text{sys}}}{(n_{pol}\Delta\nu T)^{1/2}}, \qquad (26)$$

where m = number of σ corresponding to the detection threshold; $S_{\text{sys}} = T_{sys}/G$; T_{sys} = system temperature; G = gain (K Jy^{-1}); $n_{pol} = 2$ if two polarization channels are used in the search; $\Delta\nu$ = total bandwidth; and T = total integration time. Because many statistical tests are performed, $m \sim 8 - 10$.

Harmonic Summing: Survey DFTs are analyzed by constructing partial sums of harmonics (of the DFT magnitude or squared magnitude) for different trial periods. If multiple harmonics are above threshold, then the minimum detectable flux density is

$$S_{\min} = \frac{S_{\min_1}}{h(N_h)}, \qquad (27)$$

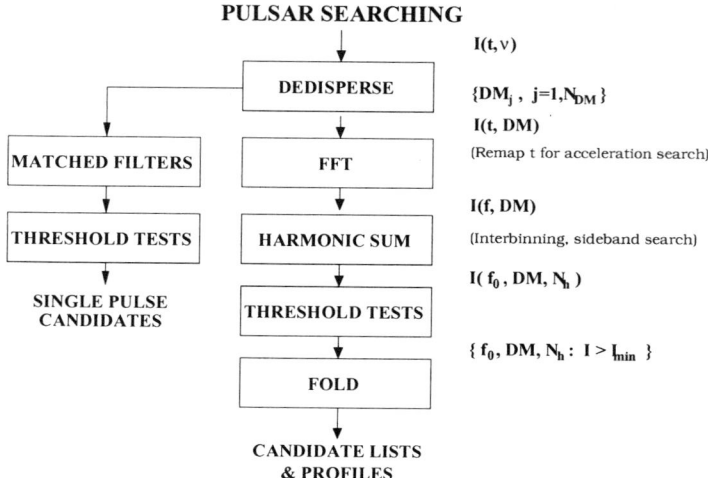

Figure 9. Generic pulsar search analysis including alternative paths for treating pulsars in binaries and for detection of individual pulses with matched filters corresponding to different pulse widths.

$$h(N_h) = N_h^{-1/2} \sum_{\ell=1}^{N_h} R_\ell, \qquad (28)$$

where $h(N_h)$ is the harmonic sum, normalized by $N_h^{-1/2}$ to take into account reduction of the noise in the harmonic sum, which lowers the threshold. For a given data set, the harmonic sum is maximized as a function of N_h. The quantity R_ℓ is related to the Fourier transform of the measured pulse shape. In actual practice, the harmonic sum is calculated up to some maximum N_h and Fourier amplitudes calculated from the DFT are usually interpolated to compensate for discreteness effects.

Distortions of the pulse shape may all be considered in the Fourier domain. These include: dispersion smearing, scattering broadening, rise time of spectrometer filters, postdetection smoothing, and smearing from orbital acceleration. Letting the DFTs of the intrinsic pulse shape, and the time constant, dispersion, scattering and orbital functions be $\tilde{s}_i$, $\tilde{s}_{tc}$, $\tilde{s}_d$, $\tilde{s}_s$ and $\tilde{s}_{orb}$, respectively, we may write the effective envelope function of harmonics, where $\ell =$ harmonic number, as

$$\tilde{s}_{\text{eff}}(\ell) = \tilde{s}_i(\ell)\tilde{s}_{tc}(\ell)\tilde{s}_d(\ell)\tilde{s}_s(\ell)\tilde{s}_{orb}(\ell). \qquad (29)$$

The quantity R_ℓ is the ratio

$$R_\ell \equiv \left| \frac{\tilde{s}_{\text{eff}}(\ell)}{\tilde{s}_{\text{eff}}(0)} \right|. \qquad (30)$$

For a pulsar with period-averaged flux density S at distance D, the maximum distance that the pulsar is detectable is

$$D_{\max} = \left(\frac{L_p\sqrt{N_h}}{S_{\min_1}}\right)^{1/2} = D\left(\frac{S}{S_{\min_1}}\right)^{1/2} N_h^{1/4}, \qquad (31)$$

where $L_p = D^2 S$ is the "pseudo luminosity" and N_h is the number of harmonics that maximizes the harmonic sum in a periodicity search.

Figure 10 shows $D_{\max}$ plotted against $S_{\min_1}$, calculated by using the electron density model of Taylor & Cordes (1993; hereafter TC93) to obtain dispersion and scattering as a function of distance.

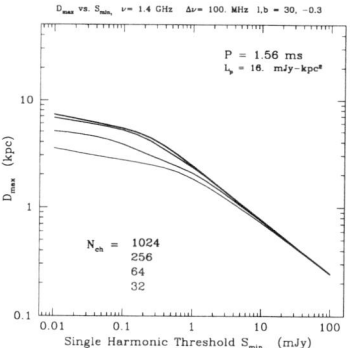
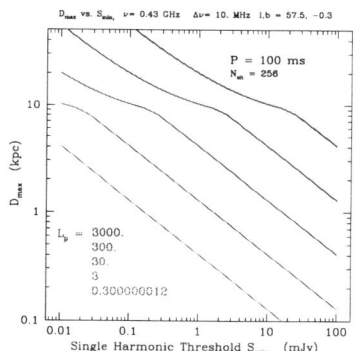

Figure 10. Left: $D_{\max}$ plotted against single harmonic threshold, $S_{\min_1}$, for a 100 MHz bandwidth at 1.4 GHz, a direction $\ell, b = 30°, -0.3°$, P=1.56 ms, and $L_p = 16$ mJy kpc^2. The different curves correspond to different numbers of spectrometer channels (heaviest line = 1024 channels, lightest line = 32 channels). At high thresholds, $S_{\min_1} \gtrsim 1$ mJy, $D_{\max}$ is *luminosity limited*, i.e. $D_{\max} \propto S_{\min_1}^{-1/2}$. For lower thresholds and small numbers of channels, it is *dispersion limited*. For large numbers of channels, dispersion smearing is negligible and $D_{\max}$ becomes *scattering limited* at low thresholds. These curves apply to *postdetection* search systems. For predetection ("coherent") dedispersing systems, the search can still be scattering limited. Right: $D_{\max}$ vs. $S_{\min_1}$ for P=100 ms and a series of pseudoluminosities, as labelled. The curves apply to a 10 MHz bandwidth at 0.43 GHz and a direction $\ell, b = 57.5°, -0.3°$.

In considering how $D_{\max}$ varies with $S_{\min_1}$, we define three regimes:

1. *Luminosity limited:* $D_{\max} \propto S_{\min_1}^{-1/2}$, i.e. the inverse square law.
2. *Dispersion limited:* $D_{\max}$ varies more weakly with $S_{\min_1}$ because at greater distances and hence larger DM, the pulse is smeared more, yielding fewer harmonics. In this case, $D_{\max}$ is a sensitive function of the number of spectrometer channels used in the dedispersion process. As N_{ch} increases,

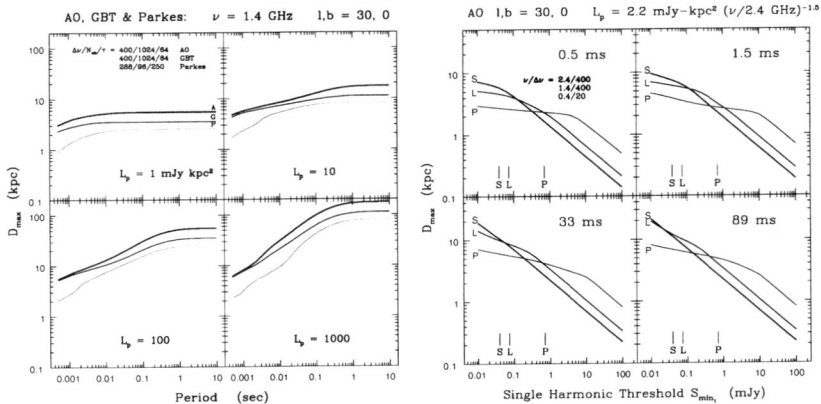

Figure 11. Left: D_{max} vs. P for Arecibo (AO), the Green Bank Telescope (GBT), and Parkes at 1.4 GHz and $\ell, b = 30°, 0°$. For AO and the GBT, spectrometers with 400 MHz total bandwidth, 1024 channels, and 64 μs time resolution are assumed while parameters of the multibeam survey are assumed for Parkes (288 MHz, 96 channels and 250 μs resolution). The four panels are for four different pseudoluminosities. For larger P and L_p, D_{max} extends beyond the nominal disk of the Galaxy. However, high velocity pulsars (e.g. $V > 1000$ km/s) can reach > 30 kpc in their radio emitting lifetimes. Right: D_{max} vs. S_{min_1} for Arecibo at 0.4, 1.4 and 2.4 GHz, labelled as P, L and S, respectively using the scaling $L_p \propto (\nu/2.4\,\mathrm{GHz})^{-1.5}$. The curves indicate that 1.4 GHz is the best of the three frequencies when surveying the Galactic plane.

D_{max} also increases until scattering dominates the pulse broadening. D_{max} is also a strong function of frequency in the dispersion-limited regime.
3. *Scattering limited:* D_{max} varies more weakly with S_{min_1} because scattering broadens the pulse and reduces the number of harmonics. In this case, D_{max} increases quickly with increasing frequency until either the dispersion or the luminosity limited regime is reached.

The *search volume* quantifies a telescope survey's performance:

$$V_s = \frac{1}{3}\Omega_s D_{max}^3, \qquad (32)$$

where Ω_s is the total solid angle covered and we assume (only for simplicity here) that D_{max} is the same in all directions. Note that some of the search volume can be empty if D_{max} extends past the pulsar population.

Figure 11 shows (left panel) D_{max} plotted against period for Arecibo, the GBT and the Parkes telescope for four luminosities. The same figure also shows (right panel) D_{max} vs. S_{min_1} for Arecibo at three frequencies. Except for short periods and very low thresholds, surveys at L band (1.4 GHz) are superior to those at S band (2.4 GHz) or P band (~ 400 MHz). Figure 12 shows contours of $S_{\min}$ plotted against P and DM for two surveys: 0.43 GHz using the Arecibo Telescope and 1.4 GHz using the Green Bank Telescope. Scalloping of the contours is caused by gridding in the set of DMs used to dedisperse. However, it should be emphasized that surveys with the GBT toward the Galactic Center will be influenced by much heavier scattering than is included in these figures. Consequently, frequencies as high as 10 GHz may be required to find pulsars in the Galactic Center (Cordes & Lazio 1997).

Guidelines for Optimizing Surveys: By considering the distribution of pulsars in the Galaxy along with the role of propagation effects in searches, a number of basic guidelines emerge:

1. For equal time-bandwidth products, aperture efficiencies, and system temperatures, the larger of two telescopes searches the greatest volume in a single beam.
2. For fixed total survey time, maximizing the number of independent pointings also maximizes the volume surveyed. It is more efficient to increase the volume by adding new pointings rather than increasing the integration time per pointing, which increases $D_{\max}$ fairly slowly.
3. When pulses are smeared by distance dependent propagation effects, building a better telescope and backend produces diminishing returns. The total volume surveyed then increases much more strongly with number of pointings (for fixed total time).
4. When searching by tracking a grid of positions, the slewing overhead time determines the optimal integration time per pointing and thus the solid-angle coverage of a survey with fixed total integration time. Note this conclusion holds only if all directions and locations are equally good for discovering objects.
5. When searching the galactic disk, the number of pointings is bounded, thus determining the survey depth as a function of the total survey time.
6. When searching a subpopulation located at some distance from the observer, the integration time is set by the need to reach this distance.
7. It is suboptimal to use telescope time to integrate longer in a given direction than the time it takes to reach a distance such that pulse broadening becomes important. It is better to move to another sky position unless a subpopulation's distance requires a longer integration time.
8. The number of spectrometer channels should be optimized so that pulse broadening from dispersion smearing is smaller than the pulse width for all spin periods of interest.
9. For a disk population with scale height H, it is optimal to (i) choose an integration time per pointing that reaches the edge of the distribution at distance $H/|\sin b|$; and (ii) If the entire volume cannot be searched, then the search should start at $|b| = 90°$ and work downwards in $|b|$.

Effects of Binary Motion: Figure 13 shows how binary motion smears the pulse and diminishes the Fourier amplitudes and harmonic sums. Such effects are

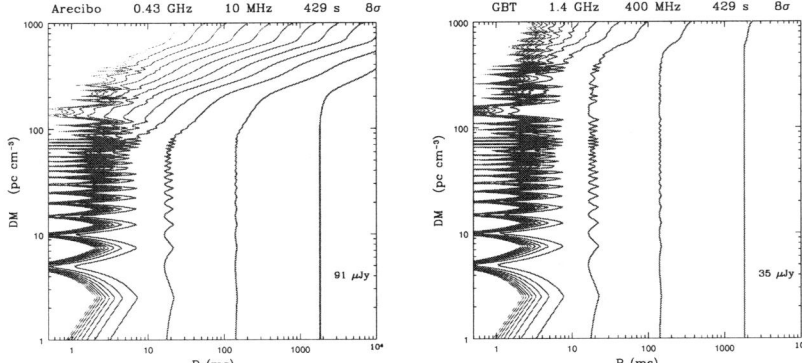

Figure 12. Contours of $S_{\min}$ vs. P, DM. The minimum detectable flux density is given on the bottom right of each plot and the contours are at $\times 2^n$ larger values. Contours take into account the change of pulse duty cycle with period ($\propto P^{-1/2}$), dispersion and pulse broadening according to the TC93 model, discreteness of the DM grid used for dedispersion, and harmonic summing up to 32 harmonics. Left: 0.43 GHz and 1024 channels for the Arecibo Telescope. Right: 1.4 GHz and 512 channels for the Green Bank Telesope.

important for integration times as short as a few minutes if the spin period is small and the orbital period is small.

Details about the effects of binary motion on pulsar searches can be found in Johnston & Kulkarni (1991).

Giant Pulse Searches: Giant pulses are defined as those from a distinct component of the pulse amplitude distribution having much larger than average amplitudes. Though only two pulsars are known to show giant pulses — the Crab pulsar (Staelin & Reifenstein 1968) and the millisecond pulsar B1937+21 (Cognard et al. 1996) (Figure 14)— only the Crab pulsar and any of its clones would be found with greater S/N in a search for single dispersed pulses than in a periodicity search. (In fact, the Crab pulsar was first detected in the radio via its giant pulses.) However, because individual giant pulses can be detected quite far away ($\sim$0.5 to 1 Mpc for the largest pulse seen in 1 hour) and because young pulsars like the Crab are important to survey, it is always sensible to search for isolated pulses while doing periodicity searches. As shown in Figure 9, an isolated pulse search may be conducted on the same dedispersed time series used in the periodicty searches. The computational cost is minimal.

For a single-pulse search of pulses with width W and amplitude I_m in a bandwidth $\Delta\nu$, we have $(S/N)_{\rm gp} = I_m (n_{\rm pol}\Delta\nu\, W)^{1/2}/S_{\rm sys}$, where $n_{\rm pol}$ is the number of polarizations summed. This can be compared with a periodicity search over a time $T = N_p P$. With $N_h \approx P/W$ harmonics detected and summed, we have $(S/N)_{\rm hs} = \bar{I} N_p^{1/2} (n_{\rm pol}\Delta\nu\, W)^{1/2}$, where $\bar{I}$ is the period-averaged intensity $\bar{I} \sim (W/P)\langle I \rangle$ and $\langle I \rangle$ is the mean intensity over the PDF of *peak* pulse

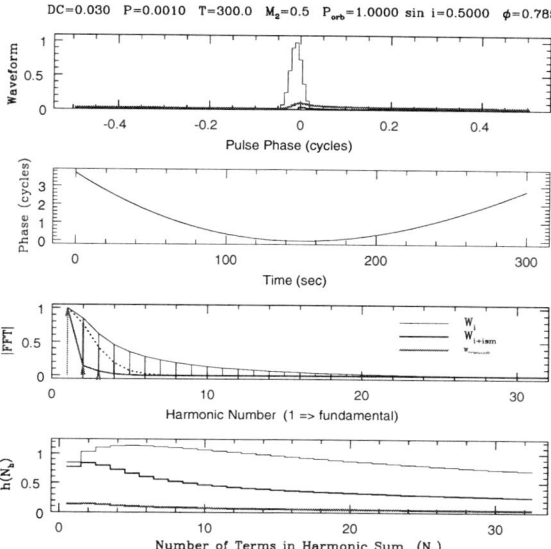

Figure 13. Pulse smearing by binary motion results in attenuation of harmonics and the harmonic sum. Shown is a case for a pulsar with P = 1 ms and an intrinsic duty cycle of 3%. The orbital period is 1 hr and the companion mass is 0.5 $M_\odot$. From top to bottom the panels are: 1. Intrinsic pulse shape (tallest), smeared pulse shape (next tallest) and smeared delta function. 2. Spin phase perturbation from orbital motion. 3. FFT magnitude for the intrinsic pulse shape (light line), pulse smeared by interstellar effects (dashed line), and pulse smeared by orbital and interstellar effects (heavy line). 4. Harmonic sum vs. number of terms in sum for intrinsic, ism-smeared, and ism+orbitally smeared. The harmonic sum maximizes at progressively smaller N_h as the smearing increases.

intensities. To compare the two kinds of seaches, we consider a simple bimodal distribution

$$f_I(I) = (1-g)\delta(I - I_1) + g\delta(I - I_2), \qquad 0 \leq g \leq 1, \tag{33}$$

where g is the probability of giant pulses with amplitudes I_2. We must have at least one giant pulse in a given interval of N_p pulse periods, requiring on average $g \geq g_{\min} \sim N_p^{-1}$. To have $S/N_{\rm gp} > S/N_{\rm hs}$, we require

$$g < g_{\max} \approx \left[N_p^{-1/2} - I_1/I_2\right] / (1 - I_1/I_2). \tag{34}$$

Figure 15 shows the domains of g and I_1/I_2 for which a giant pulse search is the superior method. The results depend on N_p in that larger N_p yield smaller domains.

244   Cordes

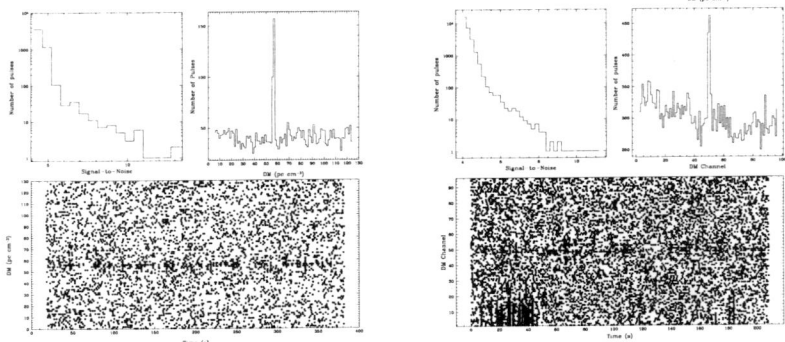

Figure 14. Single pulses from two pulsars. Left: Crab pulsar (P = 33 ms, DM = 56.8 pc cm^{-3}). Right: B1937+21 (P = 1.56 ms, DM = 71.04 pc cm^{-3}). For each pulsar, the top left panel is a histogram of signal to noise ratio, the right panel a histogram of candidate DMs, and the lower panel shows events plotted against time and DM channel. For B1937+21, radio frequency interference causes the events at low DMs. (Figures courtesy Maura McLaughlin).

Lundgren et al. (1995) inferred a bimodal distribution for the Crab pulsar with a ratio of typical giant pulse (in one hour) to mean pulse amplitude in the weaker component $\sim 10^{-3}$ and with a probability $\sim 10^{-4.4}$. The Crab pulsar has a power-law giant-pulse PDF, however.

6. Arrival-time Precision

Pulsar timing is discussed in the lecture by I. Stairs. Here we summarize briefly how some of the effects discussed in this paper affect the measurement of arrival times. These include

1. **radiometer noise:** contributes a Time of Arrival (TOA) error

$$\sigma_n \approx \left[\frac{21.9\mu s}{(S/N)_1}\right]\left(\frac{\theta_{\rm FWHM}}{1\ ms}\right)\left(\frac{\Delta}{\theta_{\rm FWHM}N_3}\right)^{1/2}, \qquad (35)$$

where Δ is the temporal resolution ($\sim$receiver time constant) and $(S/N)_1$ is the signal-to-noise ratio of a single pulse.

2. **pulse phase jitter:** like that seen in Figure 16 yields a TOA error

$$\sigma_J \approx 13.4\mu s\, f\, (1 + m_I^2)^{1/2} N_3^{-1/2} \left(\frac{\theta_{\rm FWHM}}{1\ ms}\right), \qquad (36)$$

where $\theta_{\rm FWHM}$ is the pulse width (full width at half maximum); $f \approx 1$ is the amount of jitter in units of the observed pulse width; $m_I \approx 1$ is the intensity modulation index (rms/mean); and $N_3 \equiv N/10^3$.

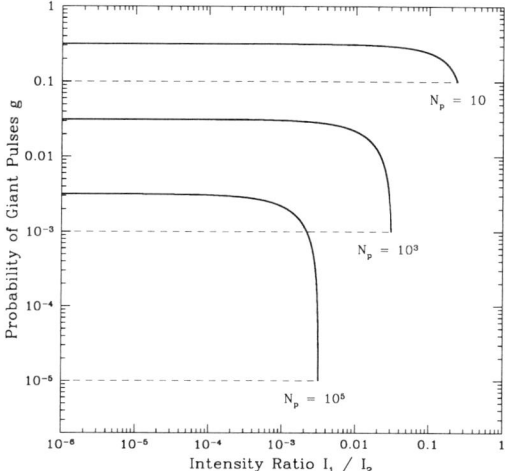

Figure 15. Domain for which a giant pulse search is more sensitive than a periodicity search. The pulse amplitude distribution is assumed to be bimodal. I_1 is the amplitude of ordinary pulses with probability $1-g$ and I_2 is the giant-pulse amplitude with probability g. The results depend on how many pulse periods, N_p, are analyzed. For each of the three cases shown, a giant pulse search is superior for values of g and I_1/I_2 between the dashed and solid lines.

An important conclusion follows from Eqs. 35-36: If $S/N_1 \gtrsim 1$, then the TOAs will be jitter dominated rather than noise dominated. When this is the case, the TOA error can be reduced only by integrating more pulses, not by building a better telescope!

3. **diffractive interstellar scintillation:** The finite number of bright patches in the frequency-time plane yield a $\sqrt{N}$ type error.
4. **pulse broadening:** produces a frequency-dependent and sometimes time-dependent error.
5. **variations in DM and angle-of-arrival from refraction:** DM variations occur because of the electron density variations (e.g. Phillips & Wolszczan (1992).
6. **instrumental polarization:** Total intensity profiles of pulsars are very stable (apart from cases where precession is seen). But pulsars can be highly linearly and circularly polarized. With imperfect calibration, a TOA error is induced on the order of the error in the Stokes parameter I.

7. Pulsar Distances

Table 2 summarizes types of distances estimates and approximate numbers. By far, most distances are estimated using DM and a model for the Galactic distri-

Table 2. Pulsar Distance Estimates

Type	Number	Comments
Parallax		
Interferometry	9	1 mas @ 1kpc, ionosphere
Timing	5	1.6 μs @ 1kpc, timing noise
Optical	1	HST point-spread function
Associations		
Supernova Remnants	8	ISM perturbations
Globular Clusters	16 clusters	Spectroscopic distances
LMC/SMC	8	
HI Absorption	74	bright pulsars, galactic rotation model
DM + Electron Density Model	all radio pulsars	ISM perturbations

bution of free electrons. The model by TC93 is currently being revised (Cordes & Lazio, in preparation). The new model (see Figure 17) consists of several components (thin and thick disks, spiral arms, Galactic Center, and individual regions) that are fitted for and calibrated by the independent measurements summarized in Table 2. The new model, like TC93, yields $n_e(\mathbf{x})$ and $C_n^2(\mathbf{x})$. It can be used with known pulsars by integrating n_e until DM is reached, thus yielding a distance estimate and an estimate for the scattering measure, SM. It can also be used in simulation studies where a pulsar distance is generated by Monte Carlo and then DM and SM are determined.

The new model differs from TC93 by including:

1. doubling of the number of lines of sight with DM and SM measurements;
2. redefinition of the spiral-arms.
3. improved treatment of the local ISM;
4. improved treatment of the outer Galaxy;
5. treatment of the Galactic Center region; and
6. improved fitting analysis using a likelihood function;

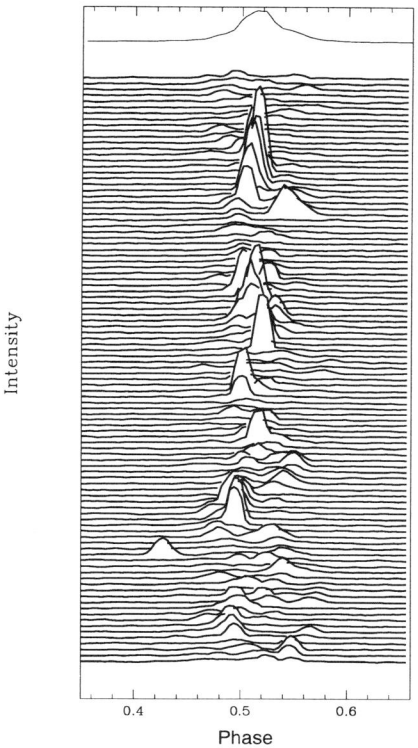

Figure 16. A sequence of pulses from the recently discovered pulsar J1740+1000 (McLaughlin et al. 2002).

8. VLBI Astrometry: If you want a good distance, measure a parallax!

Though the DM + electron-density model method for estimating distances is quite useful, it is also true that the model will never be good enough to estimate distances to all pulsars with reasonable precision (e.g. $\lesssim 10\%$).

Fortunately, the prospects for measuring parallaxes with VLBI are quite good, even for pulsars as far as 5 kpc or more if they are bright enough. Recent work (Brisken et al. 2000; Chatterjee et al. 2001) has demonstrated methods for removing deleterious ionospheric effects that plague astrometry efforts at cm wavelengths. A combination of self-calibration and usage of extragalactic background calibrators is needed, as discussed in these references. A crucial aspect of pulsar VLBI is *gating* of the correlator (e.g. Gwinn et al. 1986) so that only flux in a narrow window (appropriately shifted according to the DM

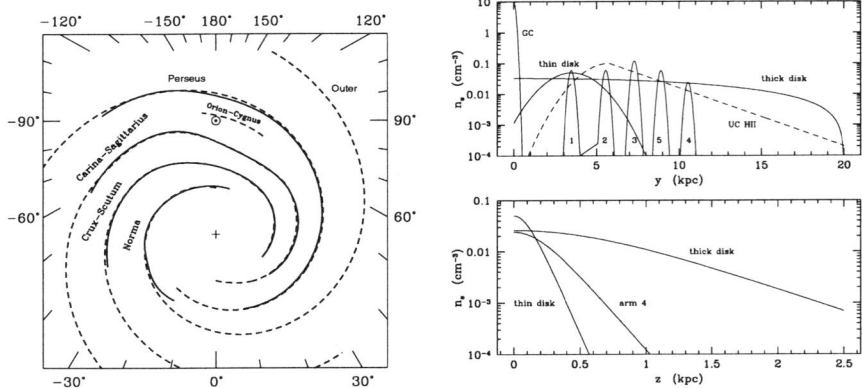

Figure 17. Features of the revised electron density model (Cordes & Lazio, in preparation). Left: Solid lines show the spiral arm definitions used in TC93 while dashed lines show the new spiral arms. The new arms are based largely on a 4-arm logarithmic spiral and yield a better fit. Right: Cross sections of the electron density model. The top panel shows n_e vs. y = coordinate with origin at the Galactic Center and increasing toward the Sun. The bottom panel shows $n_e(|z|)$ for different cuts at different Galactocentric radii ($r = 3.5$ kpc for the thin disk slice, $r = R_\odot$ for the thick disk, and the 'arm 4' cut is through the Perseus arm).

delay across the entire band) centered on the pulsar pulse is included. The boost in S/N is $\sim$(duty cycle)$^{-1/2} \sim 3$ to 5 for typical duty cycles of the *window*.

A powerful combination will result when Arecibo, the GBT and the phased VLA are included with the VLBA antennas and EVN antennas. With such an array, pulsars with flux densities less than 1 mJy can be reached in measuring parallaxes as well as proper motions. The large collecting areas on some of the baselines will also allow observations to be conducted at 5 or 8 GHz, frequencies where the ionosphere's effects are much smaller.

9. Pulsar Survey Examples

Galactic Plane Surveys: The Parkes Multibeam (PMB) survey (Camilo et al. 2000) has been extraordinarily successful in discovering new pulsars at 1.5 GHz, owing to the greater $D_{\max}$ afforded by the lessening of propagation effects. Results presented above confirm that 1.5 GHz is the best among 0.4, 1.5 and 2.4 GHz for searching the furthest. Arecibo can search ~ 480 deg^2 of the Galactic plane with $|b| \leq 5°$. Scaling from the PMB survey by solid angle only, we estimate that 240 pulsars could be discovered. However, Arecibo can search to significantly greater $D_{\max}$ and thus greater volume. For $L_p \lesssim 10$ mJy kpc^2, Arecibo can 'see' about 2.2 times further than Parkes for $P \gtrsim 10$ ms, corresponding to $\times 10$ volume. For millisecond pulsars (MSPs), Arecibo can see factors of 3 to 5 further than the PMB survey, owing to the anticipated narrower channel bandwidths and shorter dump times of Arecibo spectrometers. About 1000 pulsars could be found using Arecibo in a 7-beam multibeam survey, requiring a total bandwidth of 300 MHz with 1024 channels and 3000 hr of telescope time divided into 300-500 s per beam. The GBT can search a much larger solid angle in the Galactic plane though with less sensitivity than Arecibo. Nonetheless, the GBT will be a powerful instrument for discovering large samples and exotic pulsars.

High Latitude Searches for MSPs, NS-NS and NS-BH Binaries, and High-velocity Pulsars. Looking out of the Galactic plane, propagation effects are smaller and it is advantageous to search at lower frequencies in order to exploit the larger pulsed flux. MSPs have a scale height $H \sim 0.5$ kpc, so out of plane searches can be designed that search out to $H/|\sin b|$. The greatest volume available to be searched is at small $|b|$, where propagation effects are greatest. Thus the optimal latitude is roughly 10 to 20 deg. NS-NS binaries have sufficiently large space velocities that their scale height is expected to be about 5 kpc. This also favors a high-latitude search. NS-BH binaries also should have a sizable scale height though not as great as NS-NS binaries, since the BH formation process may not yield a kick and also because such binaries are more massive than NS-NS binaries. Isolated pulsars with high space velocities extend to very large $|z|$. About 50% of all NS will escape the Galaxy (Arzoumanian, Chernoff, & Cordes 2002, submitted; Lyne & Lorimer 1994) and can reach $> 10^3$ km s^{-1} × 10 Myr = 10 kpc in their typical radio-emitting lifetimes. A deep, low frequency search is clearly appropriate for discovering members of this population.

The Galactic Center(GC): The star cluster may contain as many as 10^7 NS; most will be radio quiet though there are probably many MSPs in the star cluster. A star burst within the last 10 Myr may have produced a significant population of active radio pulsars. Pulse broadening is notoriously large for pulsars near Sgr A*: $\tau_d \sim 300$ s at 1 GHz (Cordes & Lazio 1997). Exploiting the strong frequency dependence ($\tau_d \propto \nu^{-4}$), a search at 10 GHz is suggested. Some known pulsars placed at the GC are detectable at this frequency. Payoff includes study, through pulse timing, of the ISM in the GC and the gravitational potential in the star cluster and, possibly, near Sgr A*.

Extragalactic Giant Pulses: The Crab pulsar sporadically emits large amplitude ("giant") pulses that are hundreds to thousands of times the mean pulse amplitude (Hankins & Rickett 1975; Lundgren et al. 1995). The pulse amplitude distribution is a power law such that, at 0.4 GHz, the largest pulse seen at roughly one-hour intervals is visible to ~ 1 Mpc using an Arecibo-sized telescope. The giant-pulse phenomenon is poorly understood but it may be associated with pulsars having small light cylinders, i.e. young pulsars and MSPs. Searches for giant pulses from pulsars in nearby galaxies may be made either through blind searches or through targetting of known supernova remnants in those galaxies. The payoff is greatest if an ensemble of pulsars can be found in a given galaxy so that the contributions of the intergalactic medium and the ISMs of the Galaxy and the host galaxy can be disentangled.

References

Armstrong, J. W., Rickett, B. J., & Spangler, S. R. 1995, ApJ, 443, 209
Arzoumanian, Z., Chernoff, D. F., & Cordes, J. M. 2002, ApJ, 568, 289
Brisken, W. F., Benson, J. M., Beasley, A. J., Fomalont, E. B., Goss, W. M., & Thorsett, S. E. 2000, ApJ, 541, 959
Camilo, F. et al. 2000, in ASP Conf. Ser. Vol. 202, in IAU Colloq. 177, Pulsar Astronomy – 2000 and Beyond, ed. M. Kramer, N. Wex & R. Wielebinski (San Francisco: ASP), 3
Chatterjee, S., Cordes, J. M., Lazio, T. J. W., Goss, W. M., Fomalont, E. B., & Benson, J. M. 2001, ApJ, 550, 287
Cognard, I., Shrauner, J. A., Taylor, J. H., & Thorsett, S. E. 1996, A&A, 323, 211
Cordes, J. M. & Lazio, T. J. W. 1991, ApJ, 376, 123
Cordes, J. M. & Lazio, T. J. W. 1997, ApJ, 475, 557
Cordes, J. M. & Rickett, B. J. 1998, ApJ, 507, 846
Cordes, J. M., Weisberg, J. M., Frail, D. A., Spangler, S. R., & Ryan, M. 1991, Nature, 354, 121
Gwinn, C. R., Taylor, J. H., Weisberg, J. M., & Rawley, L. A. 1986, AJ, 91, 338
Hankins, T. H. & Rickett, B. J. 1975, in Methods in Computational Physics Volume 14 – Radio astronomy (New York: Academic Press), 55
Johnston, H. M. & Kulkarni, S. R. 1991, ApJ, 368, 504
Lundgren, S. C., Cordes, J.M., Ulmer, M., Matz, S., Lomatch, S., Foster, R. S., & Hankins, T. H. 1995, ApJ, 453, 433
Lyne, A. G. & Smith, F. G. 1990, Pulsar Astronomy (Cambridge: Cambridge University Press)
Lyne, A. G. & Lorimer, D. R. 1994, Nature, 369, 127
McLaughlin, M., Arzoumanian, Z., Cordes, J. M., Backer, D. C., Lommen, A. N., Lorimer, D. R., & Zepka, A. F. 2002, ApJ, 564, 333
Phillips, J. A. & Wolszczan, A. 1992, ApJ, 385, 273
Rickett, B. J. 1990, ARA&A, 28, 561
Staelin, D. H. & Reifenstein, E. C. III 1968, Science, 162, 1481
Taylor, J. H. & Cordes, J. M. 1993, ApJ, 411, 674
Thomson, R., Moran, J., & Swenson, G. W. 2001, Interferometry and Synthesis in Radio Astronomy, (2nd edition; New York: Wiley-Interscience)

Pulsar Observations II. – Coherent Dedispersion, Polarimetry, and Timing

I. H. Stairs

National Radio Astronomy Observatory, P.O. Box 2, Green Bank, West Virginia 24944, USA

Abstract. Pre-detection, or "coherent," dedispersion is a powerful technique which completely eliminates the effects of dispersive smearing on pulsar profiles, greatly increasing the precision of timing and polarization observations. Various software and hardware implementations of this method are described. Polarization observations with filterbanks and with coherent dedispersion are discussed, including calibration and compensation for instrumental effects. Finally, the technique of pulsar timing is presented, with descriptions of the method for obtaining pulse times of arrival and the software used to fit models of the pulsar's spin-down behavior, and a broad overview of scientific goals.

1. Introduction

Pulsar timing and polarization observations have a wide range of astrophysical applications, from studies of the internal structure of neutron stars, the pulsar radio emission mechanism, and the interstellar medium, to the exploration of binary evolution models, and stringent tests of general relativity and cosmological theories. The quality of all these experiments improves with increased signal-to-noise and, in the case of timing experiments, decreased pulse width. Conclusions regarding polarization are best drawn when the observed pulse profile most closely replicates the pulse shape emitted at the pulsar. This paper will discuss methods for obtaining the best possible timing and polarimetric signals and provide guidelines for their use.

2. Coherent Dedispersion

Pulsars are steep-spectrum objects, with signals typically strongest at observing frequencies of a few hundred MHz. At these frequencies, however, the pulses suffer significant dispersion as they propagate through the ionized interstellar medium (ISM). This phenomenon results in delays of lower-frequency components relative to the higher frequencies, and across a typical observing bandwidth can amount to hundreds or more times the pulse periods. The traditional method of dispersion removal has been a filterbank system, in which the bandpass is subdivided into a number of channels. The signal is detected in each channel, then shifted by the predicted dispersion delay in order to align the pulse peak (see Lecture by J. Cordes). However, far better timing and pulse-profile precision can be achieved with the use of a coherent dedispersion system, in which the data are sampled before detection and dispersive smearing is completely removed (Hankins & Rickett 1975).

2.1. Principles

For the purpose of describing its effects on a propagating electromagnetic wave, the interstellar medium can be modeled as a low-density electron plasma, with plasma frequency

$$\nu_p = \sqrt{\frac{n_e e^2}{\pi m_e}}, \qquad (1)$$

where n_e is the electron density, e the electron charge, and m_e the electron mass, all in cgs units. The wavenumber $k(\nu)$ for radiation with frequency ν can then be written as:

$$k(\nu) = \frac{2\pi}{c}\sqrt{\nu^2 - \nu_p^2}. \qquad (2)$$

As the wave propagates, its phase is rotated by the ISM transfer function $H(\nu) = e^{-ikz}$, where z is the distance to the pulsar. This phase rotation causes the observed dispersive retardation.

For a filterbank, the frequency-dependent group velocity ($v_g = 2\pi\, d\nu/dk$) can be used to determine the difference in pulse arrival times for two frequencies ν_1 and ν_2:

$$t_1 - t_2 = \frac{e^2}{2\pi\, m_e\, c}\, \mathrm{DM}\left(\frac{1}{\nu_1^2} - \frac{1}{\nu_2^2}\right), \qquad (3)$$

where the dispersion measure, given by

$$\mathrm{DM} = \int_0^d n_e dz, \qquad (4)$$

is the integrated electron density along the line of sight to the pulsar, usually measured in $\mathrm{cm}^{-3}\,\mathrm{pc}$. This delay is the required relative shift of the detected signals in the two filterbank channels.

With coherent dedispersion, the raw telescope voltages are mixed to complex baseband signals using combinations of quadrature local oscillators, then sampled prior to detection. Dispersion removal then involves calculating the inverse of the ISM "chirp" function and convolving the time series with this new function. Using an approximate electron density of $0.03\,\mathrm{cm}^{-3}$ (Weisberg, Rankin, & Boriakoff 1980), the plasma frequency is estimated to be roughly 2 kHz, much smaller than both the typical observing frequency of several hundred MHz and the typical observing bandpass of order 10 MHz, and permitting a Taylor expansion of the wavenumber $k(\nu)$. For bandwidth B, we take $\nu = \nu_0 + \nu_1$, where ν_0 is the central observing frequency and $|\nu_1| \leq B/2$. Keeping only the first two terms in the expansion, we have:

$$k(\nu_0 + \nu_1) = \frac{2\pi}{c}(\nu_0 + \nu_1)\left[1 - \frac{\nu_p^2}{2(\nu_0 + \nu_1)^2}\right], \qquad (5)$$

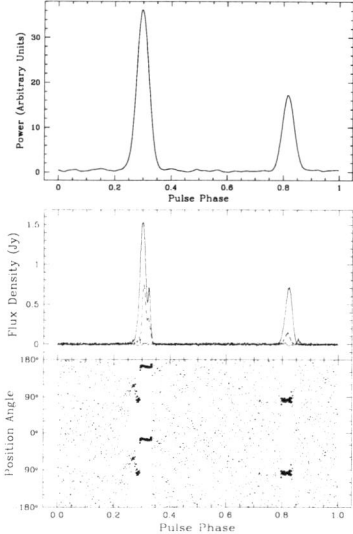

Figure 1. Pulse profile of the fastest rotating pulsar, PSR B1937+21, observed at 1400 MHz with the 76-m Lovell telescope at Jodrell Bank Observatory, U.K. The top panel shows the total-intensity profile derived from a filterbank observation; the true profile shape is convolved with the response of the channel filters. The lower panels shows the full-Stokes observation with the Princeton "Mark IV" coherent dedispersion instrument (Stairs et al. 2000; Stairs, Thorsett, & Camilo 1999). Here, and in the following figures, total intensity is indicated by a solid line, linear intensity by a dashed line, and circular power by a dotted line. The position angle is plotted twice for clarity. (From Stairs 2001.)

which leads to the transfer function

$$H(\nu_0 + \nu_1) = \exp\left[2\pi i \frac{\mathrm{DM}}{2.41 \times 10^{-10}} \frac{\nu_1^2}{\nu_0^2(\nu_0 + \nu_1)}\right], \qquad (6)$$

for frequencies in MHz (see Hankins & Rickett 1975). Once the voltage time series has been convolved with the inverse of this function, the pulses are perfectly aligned in the time domain, and the data streams can be detected (squared) to yield high-precision profiles. If two polarizations are observed, suitable crossproducts will yield the Stokes parameters, giving full polarimetric information. A comparison of filterbank and coherently dedispersed profiles for the 1.5-millisecond pulsar PSR B1937+21 is given in Figure 1.

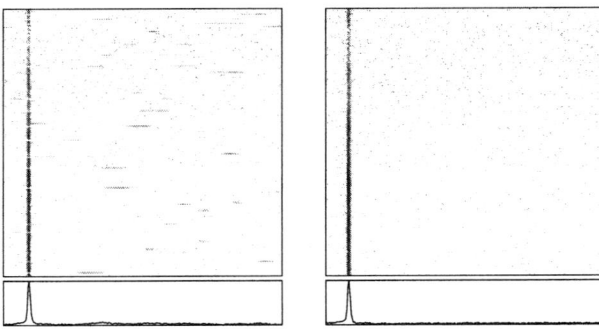

Figure 2. An example of broadband interference excision (Stairs et al. 2000). The left panel shows a grayscale plot of a contaminated 29-minute 610 MHz observation of PSR B1534+12 taken with the "Mark IV" instrument at Jodrell Bank. Each horizontal line represents ten seconds of data, and a cumulative pulse profile is displayed at the bottom. The broadband interference spikes have been "dedispersed" by roughly 6% of the pulse period, yielding the smeared-out bumps seen in the left-hand panel. The right panel presents the same observation, but with these broadband power spikes eliminated during processing.

2.2. Implementations

The coherent dedispersion method was pioneered more than 25 years ago (Hankins & Rickett 1975), but until recently, data storage and processing limitations resulted in mostly narrowband, hardware-based implementations with special-purpose chips performing the convolution of the data stream with the chirp function (e.g. Stinebring et al. 1992). These instruments and their more flexible, wider-bandwidth, modern digital counterparts (e.g. Backer et al. 1997; Kouwenhoven et al. 2000) operate by splitting the bandpass into narrow channels as in a filterbank, coherently dedispersing and detecting each channel, and then folding the data modulo the predicted topocentric pulse period. Such implementations allow the machines to cover large total bandwidths (100 MHz or more) and thereby produce high-signal-to-noise profiles. The primary disadvantage of a hardware implementation is that the data are recorded only after detection, making it impossible to reprocess with an improved dispersion measure, or to calculate and apply locally-relevant interference filters.

Software coherent dedispersion systems (e.g. Jenet et al. 1997; Weitfeldt et al. 1998; Stairs et al. 2000) trade wide total bandwidth for increased flexibility in processing. The instruments of the current generation do not subdivide the observing bandpass, but instead mix the full signal (typically 10 MHz) to baseband and Nyquist-sample the entire passband, storing the sampled voltages on disk or tape for offline analysis.

Data analysis in software allows useful corrections and filters to be applied to the time series before the final data products are obtained. Coarse quantization (typically 2 bits) is often selected to keep the data rate manageable. This quantization will necessarily affect the observed pulse shapes and signal-to-noise ratios in statistically predictable ways. Accurate pulse profile shapes are needed for the determination of precise times-of-arrival (see Section 4.1.). Various schemes have been suggested to compensate for 2-bit quantization artifacts (Jenet & Anderson 1998; Stairs et al. 2000); these involve setting the quantization thresholds carefully relative to the root-mean-square input voltage, and re-spacing the floating-point numbers assigned to the quantization levels. Corrections for coarse quantization may be applied retroactively to the folded output of hardware systems.

Radio frequency interference is an increasingly important problem facing radio astronomy. Software dedispersion systems permit the calculation and application of filters in both the frequency and time domains, often with striking improvements to the profile baseline, pulse shape, and quality of the derived TOA. Figure 2 displays an example of time-domain, broad-band RFI excision. Such techniques can easily be extended beyond pulsar observations and may prove useful in many subfields of radio astronomy.

The total data volumes and storage requirements for recorder-style software systems quickly become very large: for an observing bandpass of 10 MHz, sampled with only 2-bit quantization, the resulting data rate is 10 MB/s, or 35 GB/hour. As the convolutions are generally performed using Fast Fourier Transforms, the processing requirements are also heavy, but with typical desktop computer speeds now the same order of magnitude as radio observing frequencies, affordable processing power is rapidly becoming less of a limitation. Future instruments, such as the planned COBRA instrument at Jodrell Bank, will likely include wide-bandwidth, multichannel machines with real-time software-based, rather than hardware-based, dedispersion. This will allow flexibility in interference excision and output mode (e.g. folded profiles or single-pulse time series) while eliminating most or all of the data storage requirement.

3. Polarimetry

Pulsars tend to be highly polarized, showing both circular and linear polarization, and with polarized fractions of up to 100%. Although the exact radio emission mechanism is poorly understood, a model which has achieved a certain amount of success is that of a rotating magnetic dipole. For a misaligned dipole, there will be a set of field lines above the polar caps which cannot close without forcing the plasma on these field lines to rotate at a speed above that of light (Goldreich & Julian 1969). The basic picture of radio emission then involves coherent radiation from charged particles streaming along these open field lines (Sturrock 1971). It was soon realized (Radhakrishnan & Cooke 1969,) that the position angle of linearly polarized emission from such a rotating dipole should follow a well-defined "S"-shaped curve, in what is known as the "rotating vector

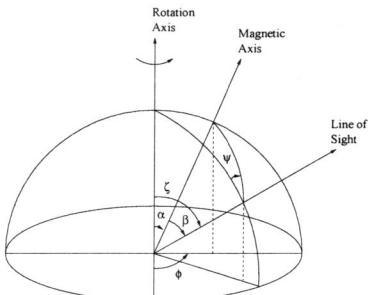

Figure 3. Definitions of angles in the Rotating Vector Model. The magnetic inclination is α and the impact parameter is β. (After Manchester & Taylor 1977.)

model" (RVM). The geometry of the RVM is illustrated in Figure 3. The exact shape of the predicted curve depends on the inclination α of the magnetic dipole axis relative to the spin axis and on the impact parameter β between the dipole axis and our line of sight. The position angle ψ can be written as a function of pulse phase ϕ:

$$\tan(\psi(\phi) - \psi_0) = \frac{\sin\alpha\sin(\phi - \phi_0)}{\cos\alpha\sin(\zeta) - \sin\alpha\cos(\zeta)\cos(\phi - \phi_0)}, \quad (7)$$

where $\zeta = \alpha + \beta$, ψ_0 is a constant offset and ϕ_0 is the phase of steepest position-angle swing. A number of pulsars, particularly slow "young" ones such as the pulsar depicted in Figure 4, fit the RVM fairly well.

Pulsar polarimetry, in combination with the RVM and statistics of profile morphology, has led to phenomenological models of pulsar emission (e.g. Rankin 1983, 1990, 1993; Lyne & Manchester 1988), and has shown that such models do not seem to apply to millisecond pulsars (e.g. Xilouris et al. 1998; Stairs, Thorsett, & Camilo 1999). Determination of pulsar rotation measures leads to information about the magnetic field of our Galaxy (e.g. Han, Manchester, & Qiao 1999) and, on occasion, the magnetic field of a pulsar's companion (Johnston et al. 1996). RVM fits can also be used to help constrain the orbital geometries of binary systems (e.g. Kramer 1998).

3.1. Data aquisition

Flux calibration of pulsar data is typically accomplished by pulsing an injected noise source, whose strength can be determined from comparison to a catalog flux calibration source. A typical procedure is to select a region of blank sky near the calibrator source, set the instrumental attenuation as for a normal observation, then pulse the noise calibrator signal and acquire data for a short time span. Subsequently, the telescope is moved to point at the calibrator, but the instrumental attenuation is not reset before acquiring more data. This allows

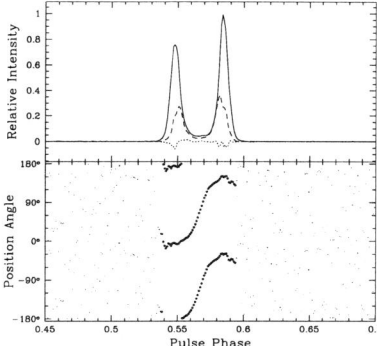

Figure 4. PSR B0525+21 observed at 1410 MHz with the 100-m telescope at Effelsberg, Germany (data from von Hoensbroech & Xilouris 1997). The position angle swing resembles the "S"-shaped curve expected from the Rotating Vector Model.

a direct comparison of the strength of the noise signal and the catalog source. If desired, small amounts of attenuation can be added or subtracted in order to test the overall linearity of the telescope/backend system. When observing a pulsar, it is only necessary to pulse the noise signal for a short time at the start or end of the observation in order to calculate the relative strengths of the pulsar signal and the noise source.

Assuming that left- and right-circular polarizations are observed and that some form of data cross-products are accumulated, it is straightforward to calculate the Stokes parameters:

$$I = |L|^2 + |R|^2, \tag{8}$$
$$Q = 2\operatorname{Re}(L^*R), \tag{9}$$
$$U = 2\operatorname{Im}(L^*R), \tag{10}$$
$$V = |L|^2 - |R|^2. \tag{11}$$

Hardware polarimeters are used with filterbank instruments. These fall into two categories: adding and multiplying polarimeters. Adding polarimeters are typically passive devices which sum the input voltages, in phase and rotated by $90°$, producing outputs such as R, L, $R+L$ and $R+iL$ for input right- and left-circular polarizations. As the signals are not detected in the process, an adding polarimeter may be used before the signals are passed to the filterbank. Stokes parameters may then be calculated once the outputs have been filtered and detected. Multiplying polarimeters perform signal multiplication in hardware, typically yielding $|R|^2$, $|L|^2$, $\operatorname{Re}(L^*R)$ and $\operatorname{Im}(L^*R)$; a separate polarimeter must therefore be used for each filterbank channel. See von Hoensbroech & Xilouris

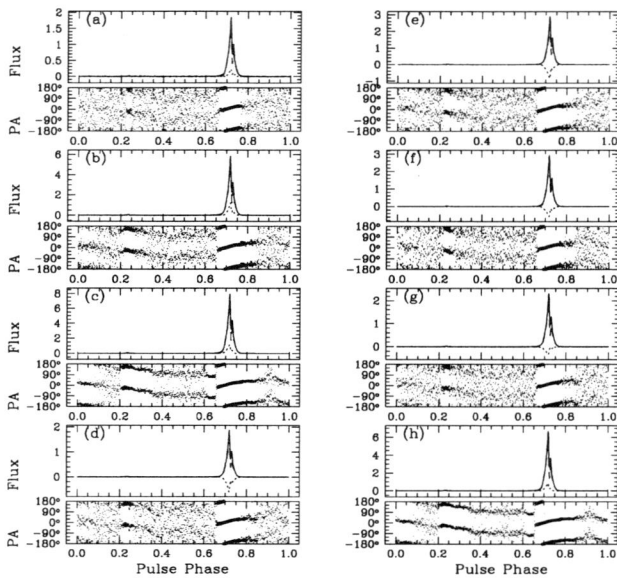

Figure 5. Arecibo observations of PSR B1929+10, uncorrected for the effects of antenna cross-coupling. Panels (a)–(g) are sequential 15- or 30-minute observations; panel (h) is cumulative. The circular polarization clearly changes shape and sign over the course of the observations, resulting in an incorrect cumulative profile.

(1997) for a thorough description of hardware polarimeter calibration at the Effelsberg 100-m telescope.

Correlator-style instruments produce self- and cross-product correlation functions after sampling the incoming voltages; the resultant data products are therefore similar to those produced by a filterbank plus multiplying polarimeter. See Navarro (1994) or the recently installed Wideband Arecibo Pulsar Processor (WAPP) instrument at Arecibo for examples.

Especially for fast, high-DM pulsars, coherent dedispersion offers a significant advantage over filterbank or correlator hardware systems. Coherent instruments produce profiles that, except for effects of interstellar scattering, exactly match those produced at the pulsar, and the Stokes parameters are quickly obtained by taking cross-products of the dedispersed time series.

3.2. Instrumental effects and corrections

For telescopes that are not equatorially mounted, the Stokes Q and U parameters are affected by the parallactic rotation of the feed during tracking. For a given hour angle, HA, telescope latitude ϕ and source declination δ, the parallactic

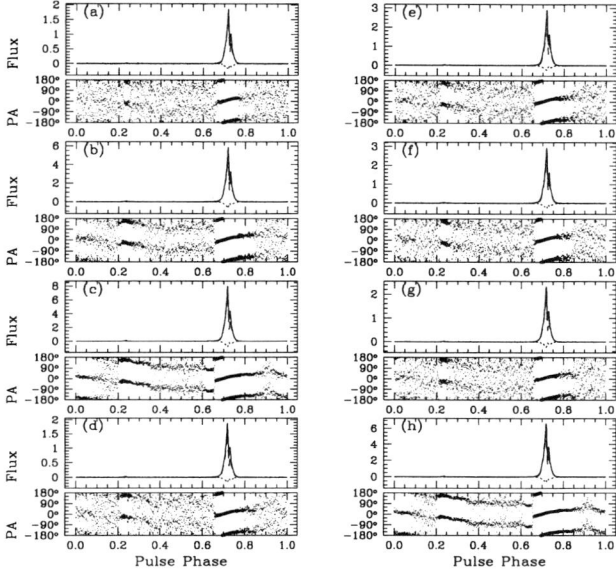

Figure 6. The same observations of PSR B1929+10 as in Figure 5, now corrected for the effects of antenna cross-coupling. Panels (a)–(g) are sequential 15- or 30-minute observations; panel (h) is cumulative. The circular polarization now maintains its intrinsic shape throughout the observing session.

angle χ may be calculated by (Rankin, Campbell, & Spangler 1975):

$$\tan \chi = \frac{\sin \mathrm{HA} \cos \phi}{\sin \phi \cos \delta - \cos \phi \sin \delta \cos \mathrm{HA}}. \tag{12}$$

The correct parameters can be recovered by rotating the observed (Q_{obs},U_{obs}) pair by the angle -2χ:

$$Q = Q_{obs} \cos(2\chi) - U_{obs} \sin(2\chi), \quad (13)$$
$$U = Q_{obs} \sin(2\chi) + U_{obs} \cos(2\chi). \quad (14)$$

The corrected Q, U values may then be used to derive the linearly polarized power L and source position angle ψ.

$$L = \sqrt{Q^2 + U^2}, \quad (15)$$
$$\psi = \frac{1}{2} \tan^{-1}(U/Q). \quad (16)$$

A further complication arises from the fact that, in general, receiver antennas are not perfectly orthogonal, nor, in the case of "circular" antennas, perfectly responsive to circularly polarized radiation. This results in a mixing of the Stokes parameters, so that the observed Stokes vector is the product of the true Stokes vector and the so-called Mueller matrix, and can result in significant distortion of the pulse polarimetry. An example can be seen in Figure 5, in which a strongly linearly polarized pulsar was observed with nominally circular feeds on the 305-m Arecibo telescope. The distortion of the circular polarization as a function of time is evident.

It is possible to correct for this cross-coupling; in fact pulsars can be used to calculate the Mueller matrix for a receiver (Xilouris 1991). A first-order correction can be found following the method of Stinebring (1982), in which it is assumed that the circular antennas are orthogonal but elliptical. Then the most significant mixing effects are between L and V:

$$I' = I, \quad (17)$$
$$Q' = Q - V\epsilon_0 \cos\phi_0, \quad (18)$$
$$U' = U - V\epsilon_0 \sin\phi_0, \quad (19)$$
$$V' = V + L\epsilon_0 \cos(\theta - \phi_0), \quad (20)$$

where θ is the incident position angle, and ϵ_0 and ϕ_0 are the cross-coupling parameters. For strongly linearly polarized pulsars such as PSR B1929+10, depicted in Figures 5 and 6, it is assumed that $L_{observed} \simeq L$. Then ϵ_0 and ϕ_0 can be determined by fitting a sinusoid to V'/L as a function of incident position angle; ϵ_0 and ϕ_0 may well vary somewhat across the observing bandpass. Once the cross-coupling parameters have been determined using a strong pulsar, they can be used to correct L and V for all pulsars observed with the same system. The corrected profiles from the Arecibo observations are displayed in Figure 6.

Stokes parameter mixing can also be induced by differences in path length from the antenna to the detector. This will mix Q and U for circular antennas, resulting in a measurable position angle offset, and U and V for linear antennas. In the latter case, the problem can be corrected by summing Stokes parameters taken with the antennas offset by a 90° rotation.

4. Pulsar Timing

Pulsars are among the most stable clocks in nature: some MSPs rival atomic clocks on timescales of several years (Rawley et al. 1987; Kaspi, Taylor, & Ryba 1994). High-precision timing requires two stages of analysis: determining the times of arrival (TOAs) of a set of pulses, and fitting a timing model to those TOAs.

4.1. Obtaining times of arrival (TOAs)

Although single pulses from any given pulsar fluctuate greatly in both intensity and shape, profiles representing the integrated waveform formed over several minutes are quite stable. This permits the accumulation of several hours' worth of data, at a given frequency and with a given observing system, to generate a "standard profile," against which any other profile may be cross-correlated.

During data acquisition and/or subsequent analysis, the data are folded modulo the predicted topocentric pulse period to form an integrated profile. The shape of this profile, $p(t)$, may be written as:

$$p(t) = a + b\,s(t-c) + g(t), \qquad (21)$$

where $s(t)$ is the standard profile, $g(t)$ is gaussian noise, a, b and c are constants, and $0 \leq t \leq P$, where P is the pulse period. Thus adding the time offset c to the time-stamp at the start of the integration yields the TOA. The constants a, b and c may be found by least-squares fitting in the Fourier domain (for instance, using the FFTFIT routine, Taylor 1992) or in the time domain. It may be desirable to use the pulsar's polarization information (for example, a steep position-angle swing) to help determine the time offset c. TOA precision depends on the pulse strength and width and pulse-to-pulse stability, and is affected by many of the interstellar propagation effects discussed in the lecture by J. Cordes, such as scintillation, scatter broadening, and DM variations.

The most stable MSPs have period derivatives on the order of $10^{-17}\,\mathrm{s\,s^{-1}}$ or smaller. Thus the best possible time standards must be used as references for the TOAs. The most commonly used time standard is Universal Coordinated Time (UTC), which is defined to be 1) an integral number of seconds from International Atomic Time (TAI), a weighted average of nearly 200 atomic clocks which is maintained by the Bureau International des Poids et Mesures (BIPM), and 2) never more than $0.9\,\mathrm{s}$ from UT1, the timescale defined by the rotation of the Earth. Leap seconds must occasionally be inserted into UTC in order to fulfill the second requirement, though it is possible that these may be eliminated in the next few years. A "real-time" version of UTC is readily accessible via broadcasts from the Global Positioning System (GPS) satellite array, and can be corrected retroactively. For the highest precision long-term timing (e.g. Kaspi et al. 1994), it is desirable to use TT ("Terrestrial Time"), which is specified relative to TAI, and is intended to represent the idealized geocentric timescale (Seidelmann, Guinot, & Doggett 1992). This retroactive time standard can be obtained from the BIPM as TT(BIPM). During observations, the data are given timestamps relative to the observatory's local time standard, often hydrogen masers. These timestamps are then corrected retroactively to UTC or TT(BIPM).

4.2. The spin-down model

The rotating magnetic dipole model is also vital to pulsar timing, providing the expectation that the pulsar will lose energy and therefore "spin down" over the long term. This effect is measurable via a slow decrease in the pulse spin frequency. To model this spin-down, the pulse phase ϕ may be written as a Taylor series in frequency:

$$\phi(T) = \phi(0) + \nu T + \frac{1}{2}\dot\nu T^2 + \frac{1}{6}\ddot\nu T^3 + \ldots, \tag{22}$$

where T is time in the pulsar rest frame and $\nu = 1/P$ is the pulse frequency. In practice, the frequency, ν, and its first derivative, $\dot\nu$, are measurable for all pulsars, but the second derivative, $\ddot\nu$, is measurable only for young, rapidly-decelerating pulsars.

4.3. Transformation to the solar system barycenter

The pulse times of arrival are measured in the observatory reference frame. Before any modeling of the pulsar parameters can be attempted, these topocentric TOAs must be transformed to an inertial reference frame. Though the approximation is not perfect, the most convenient inertial frame to use is that of the solar system barycenter. This transformation is sufficient for isolated pulsars; pulsars in binary systems require a further transformation to the pulsar frame, and more extensive parametrization.

A pulse arrival time t_b at the solar system barycenter is given by:

$$t_b = t + \frac{\mathbf{r}\cdot\hat{\mathbf{n}}}{c} + \frac{(\mathbf{r}\cdot\hat{\mathbf{n}})^2 - |\mathbf{r}|^2}{2cd} - \frac{D}{\nu^2} + \Delta_{E\odot} - \Delta_{S\odot}, \tag{23}$$

where t is the topocentric arrival time, $\mathbf{r}$ is the vector from the barycenter to the telescope, $\hat{\mathbf{n}}$ is a unit vector from the barycenter to the pulsar, and d is the barycenter-pulsar distance. The TOAs determined at observing frequency ν are corrected to infinite-frequency values by the dispersion term, D/ν^2. The dispersion constant, D, is related to the dispersion measure, DM, by $D \equiv \text{DM}/(2.41 \times 10^{-16})$ Hz. The Einstein delay term, $\Delta_{E\odot}$, consists of gravitational redshift and time dilation effects due to the Sun, planets and other objects in the solar system. The Shapiro delay, $\Delta_{S\odot}$, describes the bending of spacetime in the potential well of the Sun; its maximum magnitude is about $120\,\mu s$ at the limb of the Sun.

The vector $\hat{\mathbf{n}}$ is determined in part by the pulsar's position on the sky (right ascension α and declination δ) and its proper motion ($\mu_\alpha \equiv \dot\alpha\cos\delta$ and $\mu_\delta \equiv \dot\delta$). The third term on the right-hand side of Equation 23 represents parallax due to the Earth's orbit; it is measurable only for nearby millisecond pulsars (e.g. Kaspi et al. 1994; Toscano et al. 1999). The transformation presented in Equation 23 is accomplished by interpolating a solar system ephemeris; the Jet Propulsion Laboratory's DE200 ephemeris (Standish 1990) is the one commonly used.

4.4. Binary pulsars

In the case of a binary pulsar, additional parameters describing the binary orbit must be fit in order to translate a pulse from the center-of-mass frame of the

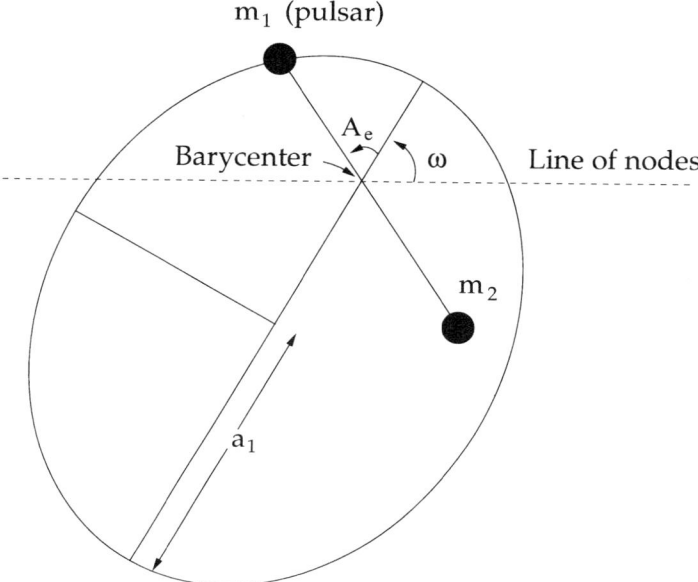

Figure 7. A binary pulsar orbit, with the line of nodes and definitions of the longitude of periastron, ω, and the true anomaly, A_e, indicated.

binary (considered to be moving at constant velocity relative to the solar system barycenter) to the rest frame of the pulsar. For non-relativistic orbits, it is often sufficient to determine the five accessible Keplerian orbital parameters: x, the projected semi-major axis of the orbit; ω, the longitude of periastron; T_0, the epoch of periastron; P_b, the orbital period; and e, the eccentricity, and to carry out the transformation on the basis of these values. For other systems, however, relativistic effects must be taken into account.

A complete "post-Newtonian" parametrization for binary systems was introduced by Damour & Deruelle (1986) and extended to include angular observables by Damour & Taylor (1992). It rewrites the orbital delay effects in terms of the five Keplerian parameters mentioned above and also a set of "post-Keplerian" parameters: $\dot{\omega}$, the advance of periastron; $\dot{P}_b$, the orbital period derivative; γ, the combined gravitational redshift and time dilation parameter; r and s, the "range" and "shape" of the Shapiro delay; A and B, the aberration parameters; and δ_r and δ_θ, which represent orbital shape corrections. These parameters are defined in a theory-independent manner; the measured values may then be compared with the predictions of gravitational theories (e.g. Taylor & Weisberg 1989; Taylor et al. 1992; Stairs et al. 1998).

The resulting equation describing the transformation between the barycentric arrival time t_b and the corresponding time T in the comoving pulsar frame may be written as

$$T = t_b - \Delta_R - \Delta_E - \Delta_S - \Delta_A, \qquad (24)$$

where Δ_R, Δ_E, Δ_S and Δ_A are the Roemer, Einstein, Shapiro and aberration delays across the pulsar orbit, defined by:

$$\Delta_R = x \sin\omega(\cos u - e(1+\delta_r)) + x(1-e^2(1+\delta_\theta)^2)^{1/2}\cos\omega\sin u, \quad (25)$$
$$\Delta_E = \gamma \sin u, \quad (26)$$
$$\Delta_S = -2r \ln\left\{1 - e\cos u - s\left[\sin\omega(\cos u - e) + (1-e^2)^{1/2}\cos\omega\sin u\right]\right\} (27)$$
$$\Delta_A = A[\sin(\omega + A_e(u)) + e\sin\omega] + B[\cos(\omega + A_e(u)) + e\cos\omega]. \quad (28)$$

These delays are written in terms of the eccentric anomaly u and true anomaly $A_e(u)$, and the time dependence of ω, which are related by:

$$u - e\sin u = 2\pi\left[\left(\frac{T-T_0}{P_b}\right) - \frac{\dot{P_b}}{2}\left(\frac{T-T_0}{P_b}\right)^2\right], \quad (29)$$

$$A_e(u) = 2\arctan\left[\left(\frac{1+e}{1-e}\right)^{1/2}\tan\frac{u}{2}\right], \quad (30)$$

$$\omega = \omega_0 + \left(\frac{P_b\dot{\omega}}{2\pi}\right)A_e(u). \quad (31)$$

Figure 7 depicts the geometry of the binary system. The longitude of periastron, ω, is defined relative to the line of nodes, where the pulsar's orbital plane intersects the plane of the sky. The orbital semi-major axis is a_1; it is related to the projected semi-major axis x by:

$$x \equiv \frac{a_1}{c}\sin i, \quad (32)$$

where i is the angle between the line of sight and the orbital angular momentum. (An inclination of 90° implies that the the orbit is viewed edge-on.) It is apparent that the pulsar and companion masses are not among the measurable Keplerian parameters; however, using Kepler's third law, a "mass function" can be derived from the Keplerian observables:

$$f_1(m_1, m_2, i) \equiv \frac{(m_2 \sin i)^3}{(m_1+m_2)^2} = x^3\left(\frac{2\pi}{P_b}\right)^2\left(\frac{1}{T_\odot}\right)M_\odot, \quad (33)$$

where $T_\odot \equiv GM_\odot/c^3 = 4.925490947\,\mu\text{s}$, $M_\odot$ is the mass of the Sun, and m_1 and m_2 are in solar masses. For a typical pulsar mass of $1.35\,M_\odot$ (Thorsett & Chakrabarty 1999), the fact that $\sin i \leq 1$ allows the mass function to be used to calculate a minimum possible companion mass.

If post-Keplerian parameters are measured, they may help to describe the geometry of the system more completely: for instance, in the theory of general relativity, the PK parameter $s \equiv \sin i$. Only for about 10 binary pulsar systems have any PK parameters been measurable.

Given the two-dimensional nature of astronomical measurements, there will always be an undetermined longitudinal velocity component to any timing solution. This amounts to a Doppler shift in the pulse period and, if applicable, the orbital period. This latter shift will change the system masses by a small fraction; however, as Damour & Deruelle (1986) point out, all instances of this Doppler-shift factor cancel out when calculating any expected orbital period derivative, thus it has no interesting impact on observable parameters.

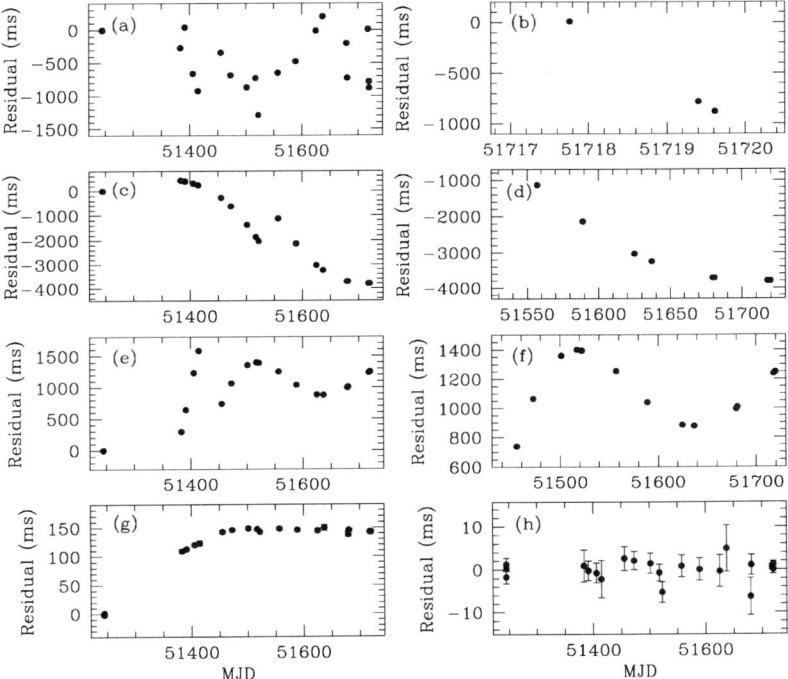

Figure 8. Obtaining a phase-connected solution for an isolated pulsar. (a) The scatter of residuals around the nominal initial solution consisting of the discovery parameters. The error bars are smaller than the plotted points. (b) Three closely-spaced points with no phase ambiguities are used to fit the frequency; the residuals relative to this new solution are in (c). (d) Adding more points allows a fit for frequency and frequency first derivative; residuals relative to such a fit are in (e). (f) A still-larger set of points containing a clear cubic term is selected, and the position terms are included in the fit. This solution leaves no phase ambiguities in the set of residuals in (g), and the fit may be extended to all points. (h) The final set of residuals; as the observations extend over more than a year, the position and spin parameters may be considered accurate.

4.5. TEMPO

To fit the pulsar timing model to the data, many observers use the TEMPO program (Taylor & Weisberg 1989; see also http://pulsar.princeton.edu/tempo). This software package was written by J. Taylor in the 1970s, and has been continually evolving since, with major contributions from R. Manchester, D. Nice, J. Weisberg, A. Irwin and N. Wex. In brief, the program reads in a set of TOAs and an initial estimate of the timing parameters, then performs an iterative least-squares fit to the data in order to determine the improved timing parameters. In the process, it calculates the residual of each TOA relative to the timing model and the standard goodness-of-fit statistic,

$$\chi^2 = \sum_{i=1}^{N} \left(\frac{\phi(t_i) - n_i}{\sigma_i} \right)^2, \tag{34}$$

where $\phi(t_i)$ is the measured pulse phase at time t_i, n_i is the nearest integer pulse number and the uncertainty σ_i is in units of pulse phase. The success of the fitting algorithm depends on correctly numbering each individual pulse; once this is accomplished the solution is said to be "phase-connected." Figure 8 demonstrates the fitting of a phase-connected solution for an isolated pulsar.

In practice, the art of pulsar timing requires careful examination of the post-fit residuals in order to check for any lingering systematics. The signatures of some problems are readily identifiable: for example, an incorrect proper motion will leave an annual sinusoidal curve that grows with time, while badly-fit orbital parameters will leave traces when the residuals are plotted against orbital phase. Ideally, the post-fit residuals should resemble random, gaussian noise and the χ^2 statistic should be close to the number of degrees of freedom. Finally, the accuracy with which the solution predicts arrival times should be tested against further observations; once all of these tests have been passed, the timing solution may be adopted with confidence. The TEMPO program can then be used to generate Chebyshev polynomial coefficients which will predict the pulsar's spin phase over a limited range of time. This feature is valuable for creating the folded timing-observation profiles needed to refine the spin-down model.

4.6. Starting points for science

With the assumption of purely dipolar radiation from an orthogonal rotator, measurement of a pulsar's ν and $\dot{\nu}$ leads directly to an estimate of the surface magnetic field B:

$$B = \left(\frac{-3I\dot{\nu}c^3}{8\pi^2 R^6 \nu^3} \right)^{1/2} \approx 3.2 \times 10^{19} \left(\frac{-\dot{\nu}}{\nu^3} \right)^{1/2} G, \tag{35}$$

where R is the radius of the neutron star, about 10^6 cm, and the moment of inertia $I = (2/5)1.4\,M_\odot\,R^2 \approx 10^{45}\,\mathrm{g\,cm^2}$.

It is often assumed that the loss of rotational energy will cause the pulsar spin frequency to evolve according to $\dot{\nu} \propto -\nu^n$, where the "braking index", n, will be exactly 3 for magnetic dipolar radiation. The braking index can be calculated if the frequency second derivative is measured; the few successful such

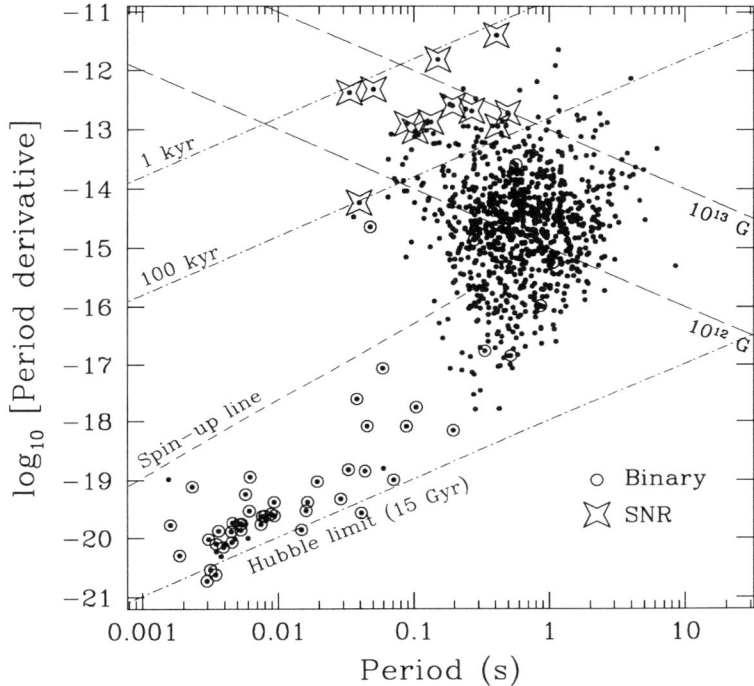

Figure 9. The $P - \dot{P}$ diagram for Galactic field pulsars. Pulsars associated with supernova remnants are indicated by stars; those in binary systems by circles.

measurements have yielded braking indices less than 3 (e.g. Lyne, Pritchard, & Smith 1988), indicating some level of departure from the simple dipole approximation.

The spin-down model also allows an estimate of the age of the pulsar. For constant n and an initial spin frequency ν_0, the age τ is:

$$\tau = \frac{-1}{(n-1)} \frac{\nu}{\dot{\nu}} \left[1 - \left(\frac{\nu}{\nu_0}\right)^{n-1} \right]. \tag{36}$$

With the assumption of a very large initial spin frequency and constant magnetic dipolar braking ($n = 3$), τ_c becomes an upper limit to the pulsar's age:

$$\tau_c = -\frac{\nu}{2\dot{\nu}}. \tag{37}$$

The age and magnetic field of an individual pulsar are most useful when looked at in the context of the entire known pulsar population. A qualitative understanding of the population may be achieved by plotting $\dot{P}$ against P on a logarithmic scale. Figure 9 does this for all known pulsars in the field of the Galaxy. (Though pulsars abound in globular clusters, the measured period derivatives of these pulsars are subject to bias from acceleration in the cluster potential.) Lines of constant age and constant magnetic field are also shown. The bulk of the population has spin periods of roughly 1 s and magnetic fields of the order of 10^{12} G. Several of the youngest pulsars have solid or plausible associations with supernova remnants, consistent with accepted ideas about the formation of pulsars in supernova explosions. These young pulsars age, passing through the densest region of the diagram, and eventually reach the point at which the radio emission mechanism no longer functions. A "dead" pulsar lucky enough to have a main-sequence binary companion will be "spun up" through accretion of matter and angular momentum as the companion evolves, and may be resurrected with a smaller magnetic field and much faster spin period. The recycled and millisecond pulsars in the lower left corner of the diagram were produced in this manner. It is clear from the distribution on this diagram that the pulsar population displays a great diversity, and our understanding of the population and evolutionary processes will be much enhanced by the timing of new pulsars from current and future generations of surveys.

Acknowledgments. The data for the profile in Figure 4 were obtained from the European Pulsar Network data archive. Figure 2 is copyright R.A.S. 2000. The author is supported by an NRAO Jansky fellowship. She thanks Michael Kramer and Jim Cordes for reading the manuscript.

References

Backer, D. C., Dexter, M. R., Zepka, A., Ng, D., Werthimer, D. J., Ray, P. S., & Foster, R. S. 1997, PASP, 109, 61
Damour, T. & Deruelle, N. 1986, Ann. Inst. H. Poincaré (Physique Théorique), 44, 263
Damour, T. & Taylor, J. H. 1992, Phys. Rev. D, 45, 1840
Goldreich, P. & Julian, W. H. 1969, ApJ, 157, 869
Han, J. L., Manchester, R. N., & Qiao, G. J. 1999, MNRAS, 306, 371
Hankins, T. H. & Rickett, B. J. 1975, in Methods in Computational Physics Volume 14 — Radio Astronomy (New York: Academic Press), 55
Jenet, F. A. & Anderson, S. B. 1998, PASP, 110, 1467
Jenet, F. A., Cook, W. R., Prince, T. A., & Unwin, S. C. 1997, PASP, 109, 707
Johnston, S., Manchester, R. N., Lyne, A. G., D'Amico, N., Bailes, M., Gaensler, B. M., & Nicastro, L. 1996, MNRAS, 279, 1026
Kaspi, V. M., Taylor, J. H., & Ryba, M. 1994, ApJ, 428, 713
Kouwenhoven, M. L. A., van Haren, P. C., Driesens, D., Langerak, J. J., Beijaard, T. D., Voûte, J. L. L., Stappers, B. W., & Ramachandran, R. 2000, in IAU Colloq. 177, Pulsar Astronomy – 2000 and Beyond, ed. M. Kramer, N. Wex & R. Wielebinski (San Francisco: ASP), 279

Kramer, M. 1998, ApJ, 509, 856
Lyne, A. G. & Manchester, R. N. 1988, MNRAS, 234, 477
Lyne, A. G., Pritchard, R. S., & Smith, F. G. 1988, MNRAS, 233, 667
Manchester, R. N. & Taylor, J. H. 1977, Pulsars (San Francisco: Freeman)
Navarro, J. 1994, PhD thesis, California Institute of Technology
Radhakrishnan, V. & Cooke, D. J. 1969, Astrophys. Lett., 3, 225
Rankin, J. M. 1983, ApJ, 274, 333
Rankin, J. M. 1990, ApJ, 352, 247
Rankin, J. M. 1993, ApJ, 405, 285
Rankin, J. M., Campbell, D. B., & Spangler, S. R. 1975, NAIC Report 46
Rawley, L. A., Taylor, J. H., Davis, M. M., & Allan, D. W. 1987, Science, 238, 761
Seidelmann, P. K., Guinot, B., & Doggett, L. E. 1992, in Explanatory Supplement to the Astronomical Almanac, ed. P. K. Seidelmann (Mill Valley, California: University Science Books), 39
Stairs, I. H. 2001 in ASP Conf. Ser., 248, Magnetic Fields Across the Hertzsprung-Russell Diagram, ed. G. Mathys, S. K. Solanki & D. T. Wickramasinghe (San Francisco: ASP), 587
Stairs, I. H., Arzoumanian, Z., Camilo, F., Lyne, A. G., Nice, D. J., Taylor, J. H., Thorsett, S. E., & Wolszczan, A. 1998, ApJ, 505, 352
Stairs, I. H., Splaver, E. M., Thorsett, S. E., Nice, D. J., & Taylor, J. H. 2000, MNRAS, 314, 459
Stairs, I. H., Thorsett, S. E., & Camilo, F. 1999, ApJS, 123, 627
Standish, E. M. 1990, A&A, 233, 252
Stinebring, D. R. 1982, PhD thesis, Cornell University
Stinebring, D. R., Kaspi, V. M., Nice, D. J., Ryba, M. F., Taylor, J. H., Thorsett, S. E., & Hankins, T. H. 1992, Rev. Sci. Instrum., 63, 3551
Sturrock, P. A. 1971, ApJ, 164, 529
Taylor, J. H. 1992, Phil. Trans. Roy. Soc. A, 341, 117
Taylor, J. H. & Weisberg, J. M. 1989, ApJ, 345, 434
Taylor, J. H., Wolszczan, A., Damour, T., & Weisberg, J. M. 1992, Nature, 355, 132
Thorsett, S. E. & Chakrabarty, D. 1999, ApJ, 512, 288
Toscano, M., Britton, M. C., Manchester, R. N., Bailes, M., Sandhu, J. S., Kulkarni, S. R., & Anderson, S. B. 1999, ApJ, 523, L171
von Hoensbroech, A. & Xilouris, K. M. 1997, A&AS, 126, 121
Weisberg, J. M., Rankin, J. M., & Boriakoff, V. 1980, A&A, 88, 84
Wietfeldt, R., Straten, W. V., Rizzo, D. D., Bartel, N., Cannon, W., & Novikov, A. 1998, A&AS, 131, 549
Xilouris, K. M. 1991, A&A, 248, 323
Xilouris, K. M., Kramer, M., Jessner, A., von Hoensbroech, A., Lorimer, D., Wielebinski, R., Wolszczan, A., & Camilo, F. 1998, ApJ, 501, 286

Dion Lewis, Darrel Emerson and Karin Sandstrom.

Single-Dish Radio Astronomy: Techniques and Applications
ASP Conference Series, Vol. 278, 2002
S. Stanimirović, D. R. Altschuler, P. F. Goldsmith, and C. J. Salter

Planetary Radar Astronomy

Gregory J. Black

National Radio Astronomy Observatory, P.O. Box 2, Green Bank, West Virginia 24944, USA

Abstract. Radar is a powerful tool for studying the Solar System, with its reach limited in theory only by the transmitter power available. It has been used to observe targets ranging in size from the rings of Saturn down to house-sized asteroids. An observer has control of the illumination source, so a radar experiment provides information not available from passive observing methods. On centimeter to meter scales it is a sensitive probe of surface characteristics such as dielectric constant and roughness, and on larger scales can map topography and determine shapes of irregular objects at resolutions finer than other ground-based methods. This lecture will cover the basic techniques of planetary radar astronomy, give an overview of the scientific questions that can be addressed, and survey some recent results. Key points of the lecture will be: what can be learned from radar experiments; types of radar experiments; observable quantities; the radar equation; and an outline of current radar systems.

1. Introduction

Unlike the majority of ground-based astronomy, planetary radar astronomy has the benefit of providing its own illumination of the target so that the observer has more control over the experiment than in passive techniques. By finely controlling the frequency, polarization, and temporal character of the transmitted signal, information on the target's scattering properties can be obtained at a very fine resolution compared with other ground-based observations. Planetary radar has applications in a wide range of areas, from measuring the scale of the Solar System, to mapping the shapes of small asteroids. A primary use of planetary radar is to probe the surface structure of Solar System objects. How a target scatters the impinging signal depends on its surface composition and structure. To first order the dielectric constant (set by composition and density) determines the overall reflectivity and the roughness of the surface determines the scattering function and polarization properties. Roughness is a generic term for the structure on scales comparable to the wavelength used; some examples of very rough surfaces are certain lava flows or impact ejecta.

On a larger scale the radar can measure global properties of the target, such as its size, shape, and rotation state. These types of measurements are routinely done for asteroids whose shapes and to a certain extent rotation states are difficult to measure via other ground-based methods (cf. review by Ostro 1993). Historically, such work has also been important in major planet studies, for example radar first measured the rotation of Mercury (Pettengill & Dyce 1965) and Venus (Carpenter 1964).

At an even larger scale the high precision radar measurements of a target's distance and velocity provide tight constraints on orbital parameters. Such data has played an important role in measuring the size of the Solar System (Muhleman, Holdridge, & Block 1962). Ranging to the inner planets to refine orbits and guide spacecraft continues to the present. Radar data is especially crucial in the orbit determination of asteroids, for which optical data alone is sometimes not even sufficient to recover newly discovered ones on subsequent orbits. Because of the high precision of radar measurements even tests of gravitational theories are possible (Shapiro 1964; Shapiro et al. 1972).

Most present day radar work involves either the S-Band transmitter on the Arecibo 305-m antenna which operates at 12.6 cm (2.38 GHz), or the X-Band transmitter on the JPL/Goldstone 70 m antenna which operates at 3.5 cm (8.4 GHz). The 70 cm (430 MHz) radar at Arecibo is mainly for ionospheric work but has also been used for some planetary targets.

In essence a radar experiment involves transmitting a signal for the round-trip light travel time to the target, then receiving the echo for an identical period of time. This two-way light time can range from a few seconds for near Earth asteroids to over two hours for targets in the Saturn system. Often the same system that transmits the signal is the one that receives the echo, and such an experiment is known as *monostatic*. It is also possible and often desirable to have a separate antenna (or several) receive the echo. An experiment with different transmit and receive locations is known as *bistatic*, and allows for a continuous transmission and reception for longer than the two-way light time to give more integration time. Alternatively, the echo can be received at multiple locations for an interferometric experiment.

Virtually all planetary radar experiments observe signals reflecting from targets in the exact backscatter direction, i.e. at zero phase angle. Observing the scattering at other angles away from exact backscatter requires a transmitter (or receiver) not located on the Earth and has only been done in a few cases, for example most recently with the Magellan spacecraft at Venus (Pettengill, Ford, & Simpson 1996), with the Clementine spacecraft at the Moon (Nozette et al. 1996), and with Mars Global Surveyor at Mars (Simpson & Tyler 2001).

2. Transmitters

The radar systems at Arecibo and Goldstone use klystron vacuum tubes to produce the emitted power. Inside the tube a magnetically confined beam of electrons passes through a potential drop of several tens of kV. An exciter signal of only a few watts is injected into the tube which modifies the electron velocity distribution entering the beam, alternatively boosting or retarding them. This results in an oscillation in the number density of electrons further along the beam. Resonance cavities along the tube amplify this oscillation, converting the input power into radio-frequency power which can then escape through a non-conductive window and out the feed. A simple cartoon of a klystron tube is shown in Figure 1. The transmission frequency can be varied slightly; the

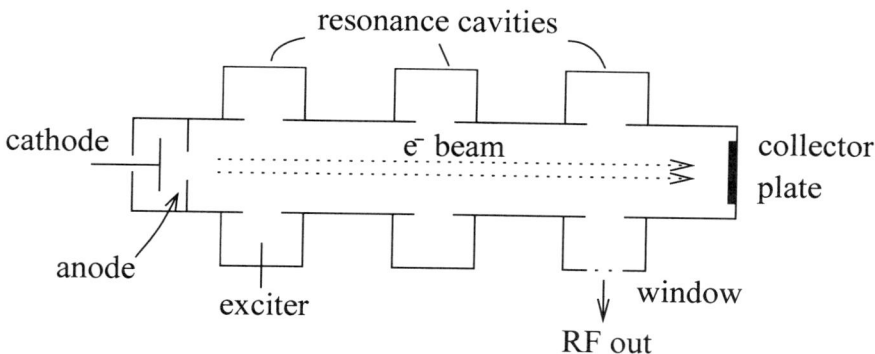

Figure 1. Cartoon of the basic details of the klystrons used to generate the RF transmissions.

Arecibo S-Band klystrons can be tuned to within about 25 MHz centered on the nominal 2.38 GHz frequency.

Typically these tubes are ∼ 50% efficient, so to obtain a megawatt of emitted power two megawatts are generated. The Arecibo S-Band and Goldstone X-Band transmitters are small enough to be located in their respective feed structures. The Arecibo 430 MHz radar transmitter is much larger and is located on the ground with a waveguide moving the power up to the platform and out the line feed.

3. Observables

3.1. Cross sections

The amount of energy reflected by a target back to the observer is known as the target's cross section, σ, and has the units of a physical area usually measured in km^2. This quantity is equivalent to the cross section that an isotropically, perfectly reflecting sphere would have at the same distance as the target. Often this cross section is normalized by the projected area of the target, $\hat{\sigma} = \sigma/A$, which is then called the *specific radar cross section* or the *radar albedo*. Since the cross sections are defined relative to an isotropically scattering sphere, strongly backscattering surfaces may even have cross sections greater than their projected area, i.e. $\hat{\sigma} > 1$.

The transmitted signal is always highly polarized, usually circular but linear polarization is also used. The echo can be received in both the same polarization sense at that transmitted as well as in the orthogonal sense. These two channels are generally referred to as the SC (same circular) and OC (opposite circular) echoes, or similarly SL and OL in the linearly polarized case. Other names for these can be found in the literature. In reference to a single scattering mechanism (discussed below), the OC and SL echoes are sometimes called the expected or polarized channel, while SC and OL echoes may be called unexpected or depolarized.

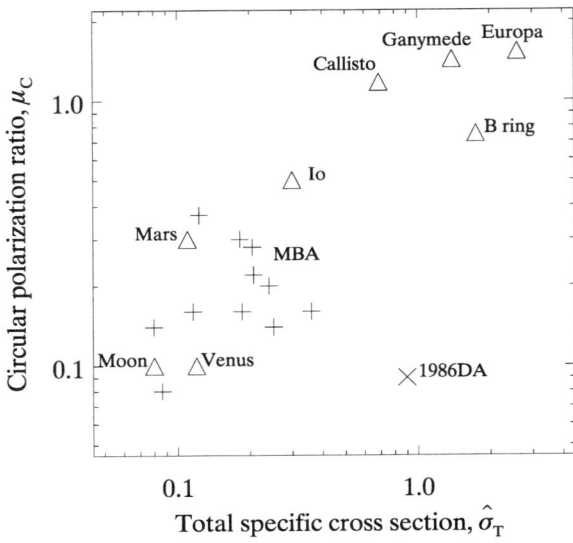

Figure 2. Typical parameters of Solar System objects. Crosses are a selection of main belt asteroids. Europa, Ganymede, Callisto, and Saturn's B ring are mainly water ice for which multiple scattering mechanisms dominate (= high μ_C), while most other targets are rock covered and have a strong single scattering component (lower μ_C). Io is intermediate and 1986 DA is a highly reflective metal-enriched asteroid.

3.2. Polarization

A single reflection from a surface which is smooth on scales larger than the wavelength will reverse the helicity of a circularly polarized signal. Thus for relatively smooth surfaces, which are the most common for terrestrial planets, most of the returned signal is in the OC sense. Structures with a scale similar to the wavelength can couple power into the orthogonal polarization. A parameter widely used is the *circular polarization ratio*, which is defined as the ratio of the echo strength in the two polarizations:

$$\mu_C = \frac{\sigma_{SC}}{\sigma_{OC}}. \tag{1}$$

With this definition a very smooth surface exhibiting only mirror-like reflections would have a ratio of zero, an extremely rough surface would have a ratio on the order of unity, and certain polarization preserving surfaces could have even larger ratios. If a reflection mechanism tends to preserve the incident circular polarization then more power can be returned in the same sense polarization than in the opposite sense.

If linear polarizations are used the linear polarization ratio is defined inversely as:

$$\mu_L = \frac{\sigma_{\rm OL}}{\sigma_{\rm SL}}, \qquad (2)$$

since now the linear polarization sense is preserved on a single reflection. Linear polarization measurements must take into account Faraday rotation in the Earth's ionosphere, rotation of the antenna's feed as it tracks the target (for an AZ/EL mount), and the dependence of the target's scattering properties on the plane of polarization relative to the local plane of incidence. Use of circular polarization avoids these issues and so is more often used. The exceptions are experiments where the latter effect provides some desired information about the target's surface.

A selection of cross sections and ratios for various Solar System targets is shown in Figure 2. Plotted are the circular polarization ratios versus the total specific cross sections,

$$\hat{\sigma}_T = \hat{\sigma}_{\rm OC} + \hat{\sigma}_{\rm SC} \qquad (3)$$

or

$$= \hat{\sigma}_{\rm OL} + \hat{\sigma}_{\rm SL}. \qquad (4)$$

Most objects in the inner Solar System whose surfaces are mainly rock have low cross sections and ratios. Some small asteroids can be very rough giving high ratios (Benner et al. 1997) and/or very reflective giving high cross sections due to high metallic content (Ostro et al. 1991a). In the outer Solar System, icy surfaces predominate which exhibit more exotic subsurface scattering mechanisms, and hence can have very high cross sections and ratios (Campbell et al. 1977; Ostro et al. 1992).

4. Scattering

The strength of an echo and its polarization properties are determined by the scattering characteristics of the reflecting surface. The penetration depth of the radio wave depends on the surface's dielectric constant. After traveling through a distance l of material the wave is attenuated by an amount $e^{-\alpha l}$. The absorption coefficient is given by (for low conductivity materials)

$$\alpha \simeq \frac{2\pi\,\epsilon_i}{\lambda\,\sqrt{\epsilon_r}}, \qquad (5)$$

where $\epsilon = \epsilon_r - i\epsilon_i$ is the complex dielectric constant. Sometimes the absorption is reported in terms of the *loss tangent* defined as $\tan\delta = \epsilon_i/\epsilon_r$, or the absorption depth, $1/\alpha$, which is the depth at which the signal has been reduced to $1/e$ of its initial strength. For solid and fragmented rock as on the Moon and asteroids, penetration depths can vary from 1 to 10 wavelengths depending on porosity and composition. For the Moon, the porosity of the surface reduces the solid basalt dielectric constant which is of order $7.7 - 0.2i$ (Campbell & Ulrichs 1969) to $\sim 3 - 0.02i$, yielding an absorption depth at 12.6 cm wavelength of ~ 1.5 m.

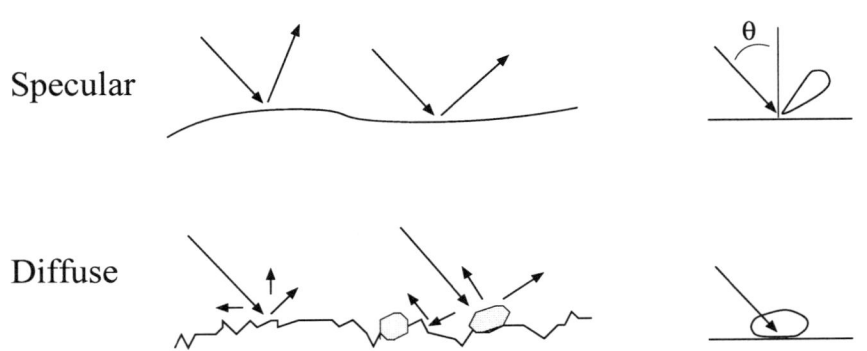

Figure 3. Simplified diagram scattering mechanisms. The scattering from a smooth surface is mirror-like (specular) with most energy in a scattering lobe directed at an angle equal to the incidence angle θ. Diffuse scattering from wavelength scale roughness and scatterers sends energy into more directions giving a more uniform scattering lobe in angle. Diagram after Simpson et al. (1992).

Keep in mind that the radar echo is a reflection so the wave will be attenuated by twice these factors as it penetrates in and then has to travel back out through the same medium. The ice component of surfaces in the outer Solar System may be much more transparent. At the low temperatures found in the outer Solar System, solid ice with even a fairly high impurity fraction of 10% has a dielectric constant around $3.1 - 3 \times 10^{-3}i$ (Thompson & Squyres 1990), giving absorption depths of order 10 m at 12.6 cm wavelength. Reducing the impurities to 1% increases the depth to ~ 100 m. Thus the 'surface' sensed by the radar must be considered to include these near subsurface layers, and embedded structures and rocks will play a part in any scattering mechanism.

In a general sense the scattering is due to structures on roughly the same scale as the wavelength used. To a large extent the terrestrial planets are smooth on the wavelength scale, presenting large or gently sloped surfaces. These smooth surfaces exhibit mirror-like *specular* reflections where the angle of reflection equals the angle of incidence (as in Figure 3), so they return significant power back to the observer only if the incidence angle is low. The echoes from such targets are extremely limb-darkened as they are dominated by a specular spike localized around the subradar point, with less echo near the limbs where the incidence angles become more glancing. The width and strength of the specular spike depend on the heights of large scale (relative to the wavelength) variations which can be thought of as a collection of large flat facets. Since this mechanism only involves a single scattering event, the polarization sense of a circularly polarized signal is reversed and the specular reflection is seen only in the OC received echo (or SL if linear polarization is used). The term *quasi-specular* is often used to describe this scattering for planetary targets since they are never ideally smooth spheres having a pure specular echo. Models of quasi-specular scattering assume some statistical distribution of surface slopes. A model that represents well the Moon's specular echo as a function of incidence angle for angles less than $\sim 40°$ is known as the Hagfors scattering law (Hagfors 1964):

$$\hat{\sigma}_{\text{spec}}(\theta) = \frac{C\rho}{2}\left(\cos^4\theta + C\sin^2\theta\right)^{-3/2}. \qquad (6)$$

Here C is related to the meter-scale rms surface slopes roughly as $\alpha \sim 1/C^{1/2}$, which for the Moon gives rms slopes of $\sim 7°$. ρ is the normal incidence reflectivity, related to the dielectric constant ϵ of the surface:

$$\rho = \left[\frac{1-\sqrt{\epsilon}}{1+\sqrt{\epsilon}}\right]^2. \qquad (7)$$

For Lunar soils $\epsilon =\sim 3$ so $\rho =\sim 0.07$. In cases where the polarization ratio is very low, the total OC cross section can be expressed as:

$$\hat{\sigma}_{\text{OC}} = g\rho, \qquad (8)$$

where g is a gain factor taking into account surface roughness and slopes. For the Hagfor's model the gain is $g = 1 + \alpha^2$, which is ~ 1 for fairly smooth, low μ_C surfaces. For the dry powdered surfaces of asteroids it is often assumed that g is unity, then the density of the surface can be determined from the reflectivity (Garvin et al. 1985):

$$d = 3.2 \ln\left[\frac{1+\sqrt{\rho}}{1-\sqrt{\rho}}\right] \text{ g/cm}^3. \qquad (9)$$

Another similar model by Muhleman (1964) for the OC scattering but covering all incidence angles is also used for the major planets:

$$\hat{\sigma}_{\text{spec}}(\theta) = \left(\frac{\alpha}{\alpha\cos\theta + \sin\theta}\right)^3 \cos\theta. \qquad (10)$$

Variations on the wavelength scale create more scattering events which depolarize the echo by distributing power in both polarizations. The coupling is accomplished by either single scattering from wavelength scale roughness or multiple scatterings (Figure 3). This scattering component is generally known as *diffuse*, and dominates SC echoes at all incidence angles and the portion of OC echoes away from the main specular contribution. It is always modeled empirically by a cosine law with incidence angle:

$$\hat{\sigma}_{\text{diff}}(\theta) \propto \cos^n\theta. \qquad (11)$$

A target which appears uniformly bright would have $n = 1$. The proportionality constant just provides a match to the total cross section. For the Moon and most other surfaces this diffuse component in both receive polarizations is well described by $n = 1 - 2$.

The statements regarding the use of the specular versus diffuse scattering laws is true only for surfaces made of silicate material that predominate the inner Solar System. In the outer Solar System ices are more common and due to the very low absorption at ~ 100 K at radio wavelengths they exhibit stronger multiple scattering effects. In particular, the icy Galilean satellites have scattering laws which show no specular component at all and instead show only strong diffuse scattering (Campbell et al. 1977; Ostro et al. 1992; and Black et al. 2001). In addition, as seen in Figure 2, they are extremely strong backscatters with high cross sections and polarizations, which indicates a polarization-preserving multiple- scattering mechanism. Similar strong backscattering behavior is seen in the polar deposits of Mars (Muhleman et al. 1991) and Mercury (Slade et al. 1992).

5. Techniques

5.1. Doppler

The first type of radar experiment to discuss involves transmitting a very narrow-band, continuous tone. In this *continuous wave* (CW) experiment the returned signal will be shifted in frequency due to velocity differences between the reflecting surfaces and the transmitter. The main frequency shift arises from the motion of the center-of-mass of the target relative to the transmitter. This is the standard Doppler shift and the returned frequency is shifted by

$$f = 2 f_0 \frac{v}{c}, \qquad (12)$$

where f_0 is the transmitted frequency and v is the relative velocity (positive for motion toward the observer). The factor of two accounts for the surface seeing a Doppler shifted incident signal, and the receiver then seeing the reflection Doppler shifted once again. For velocities typical of Solar System objects relative to the Earth this bulk Doppler shift at S-Band can be of order several hundred kilohertz, but will cross through zero during close approach. Determination of the bulk frequency shift gives an accurate measurement of the line-of-sight velocity of the target.

To compensate for most of the Doppler shift between the Earth-bound radar and the target either the transmitter frequency or the LO of the receiving system is drifted according to an *a priori* ephemeris so that the recorded echoes have a fixed center frequency. Any residual frequency offset of the echo in the data is of interest and can then be used to refine the target's ephemeris in an iterative manner (Ostro et al. 1991b).

In addition to this overall frequency shift, the target will likely have some rotation such that as seen by the observer different portions of the surface will be approaching or receding at varying velocities relative to the center-of-mass. It is straightforward to show that for a point on the surface this Doppler shift is proportional to its distance from the apparent spin axis projected into the plane of the sky. Denoting that projected distance by y, the shift for a point on a target with radius r and rotation rate $\omega = 2\pi/P$ is

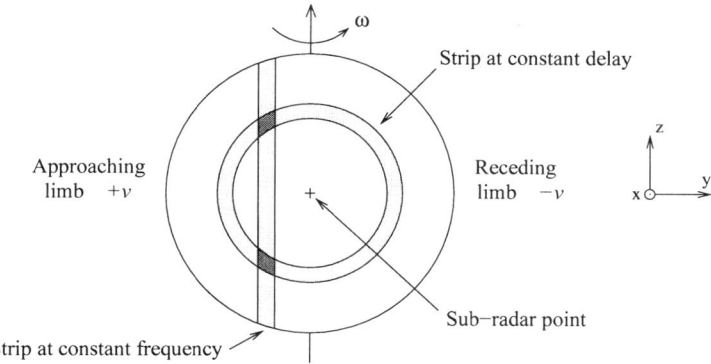

Figure 4. Delay-Doppler coordinates. The delay coordinate is proportional to the x axis out of the page, so all points on the ring centered on the sub-radar point have the same delay. The Doppler coordinate is proportional to the distance from the apparent spin vector, so all points on the strip (a half circle out of the page) have the same frequency shift. The darkest shaded points have the same delay and Doppler coordinates and so will be mapped into the same pixel in a radar image; this is the North-South ambiguity.

$$f = \frac{2 f_0 \, v_{\text{proj}}}{c} = \frac{-2 \, y \, \omega \, \cos \phi}{\lambda}, \qquad (13)$$

where ϕ is introduced as the angle between the spin vector and the plane of the sky. (Note the negative sign just arises from the definition of the coordinate system and direction of rotation.) Viewing an object pole-on results in no rotational Doppler shifts, while maximum shifts result from viewing an object directly over its equator. Figure 4 illustrates how for a sphere a strip of points equidistant in projection from the spin vector map into the same frequency, so a Doppler echo spectrum can be envisioned as the result of moving an observing slit across the target. For a non-spherical object the strip will be irregular on the surface but still at a constant projected distance from the spin vector.

The Doppler shift is maximum for points on the limb where $y = r$. Points on the receding limb will be shifted to lower frequencies while points on the approaching limb will be shifted higher, so that the total Doppler spread of the echo is twice that for one limb. Substituting r for y in the previous equation, taking twice the result and using the diameter $D = 2r$ gives the total echo bandwidth

$$B = \frac{4\pi D}{\lambda P} \cos \phi. \qquad (14)$$

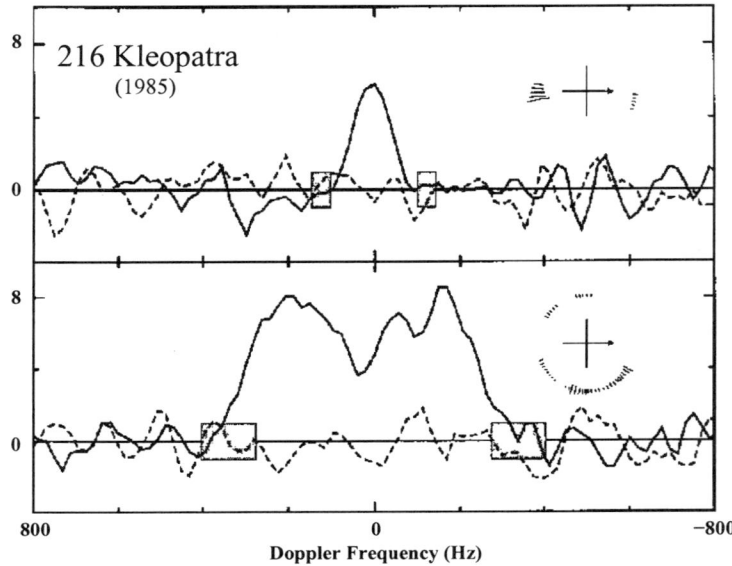

Figure 5. Doppler spectrum of main belt asteroid 216 Kleopatra from different orientations (Mitchell et al. 1995). Solid curves are OC echoes and dashed curves are SC echoes. The total echo bandwidth varies considerably since this asteroid is very elongated and presents a varying profile as it rotates. Boxes are estimates of where the echo edges should be based on an ellipsoid model.

This is also sometimes referred to as the *limb-to-limb bandwidth* or *Doppler dispersion*. Although a sphere is shown in Figure 4, this bandwidth expression holds for any object. For non-spherical shapes the maximum Doppler dispersion is given by the maximum projected dimension of the target, which for very irregular objects means the dispersion may change as the aspect and viewing geometry (ϕ) of the target change. Such additional variation provides a powerful estimate of the object's shape and rotation state. Example echo spectra from main belt asteroid 216 Kleopatra is shown in Figure 5 (Mitchell et al. 1995). This object has a rotation period of 5.4 hrs and a mean diameter of 135 km, which together give a Doppler bandwidth at S-Band of 700 Hz. In fact this asteroid is quite elongated and its projected size can vary considerably. Since the bandwidth is proportional to the longest projected dimension, this variation can be seen in the width of the spectra in Figure 5.

In some cases the distinction between a target's true rotation and its apparent rotation is very important. The most extreme case is the Moon, which because of its synchronous rotation has an apparent rotation vector dominated by libration and the changing perspective of an Earthly observer. The plane perpendicular to the apparent spin vector is known as the *Doppler equator*.

The frequency resolution of an echo spectrum is determined by the time span of voltage samples used in the Fourier transform, for example a 100 second span of data will yield a spectrum with 0.01 Hz resolution. For a fixed frequency resolution, the actual resolution on the surface will vary depending on the orientation of the surface within the corresponding Doppler strip, e.g. near the equatorial limbs the surface curves away from the observer so that more surface area falls within a frequency resolution cell than does near the subradar point. The coherence time of the frequency standards used for generating the transmitted tone as well as triggering the sampling during receive are much longer than any receive interval in practice, so frequency resolution is limited by other constraints. For example, with monostatic runs the short round trip light travel time to nearby objects rigidly fixes the maximum receive window per cycle. On the other hand, the strength of the echo can also limit the useful resolution. Dividing a finite echo into smaller channels increases the signal-to-noise ratio per channel, so a useful resolution should be chosen to maintain an adequate detection across the target.

5.2. Delay

Consider now a transmitted pulse where the energy emitted is a step function, only non-zero for a short time δt. The time it takes this pulse to travel a distance R to reach the target and then return is to a first approximation

$$t = \frac{2R}{c}. \quad (15)$$

Strictly speaking the distance on the out and back legs of the journey are not equal, and the distinction is necessary for precision astrometry. By timing the moment the echo returns an accurate distance to the target can be obtained. This technique, *ranging*, has played an important role over the past 40 years in measuring the scale of the Solar System and defining orbits of the planets, moons, and asteroids.

If short enough, the transmitted pulse will illuminate only a small area on the target at any instant as it sweeps over the surface. This results in an echo dispersed in time. All points on the target at the same distance from the observer will see and reflect the pulse at the same moment, so the pulse resolves the target in depth with a range resolution set by the pulse width, $\delta r = c\,\delta t/2$. The factor of 2 is included to account for the signal having to traverse the same distance once before impinging on the surface and again after reflection. Delay resolution cells are sometimes referred to as *range gates*. As with the Doppler resolution, the actual surface resolution on the target will vary depending on the local orientation of the surface within that range cell. For a spherical object this delay element encompasses an annulus centered on the sub-radar point as shown in Figure 4. In that figure, the delay coordinate is proportion to the x coordinate directed out of the page. The total delay depth of a target of radius r is $\tau_D = 2r/c$. Similar to the Doppler discussion above, for an irregular object the locus of points with equal delay will also be irregular, and the total delay depth may change with aspect.

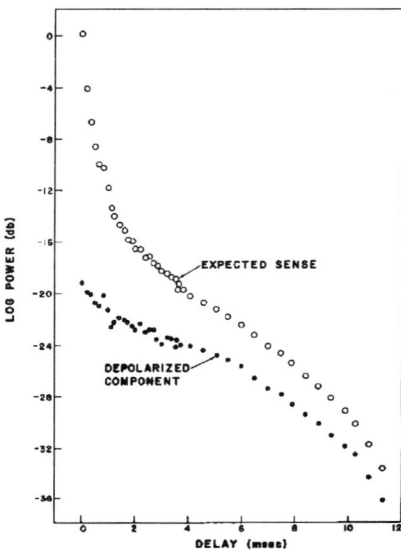

Figure 6. Delay profile of the Moon at 23 cm wavelength in OC (open circles) and SC (filled circles) polarizations (Evans & Hagfors 1966). Total delay depth of the Moon is 11.5 ms. The sharp rise is the quasi-specular echo from around the sub-radar point on the surface.

The delay resolution achievable is a function of the hardware, both how narrow a pulse can be transmitted and what sampling rates are available for receiving the echo. These constraints follow from the sampling theorem where in order to correctly measure (emit) a pulse of duration $\delta\tau$ a bandwidth of $1/\delta\tau$ must be sampled (transmitted). Typical pulses used today can be as narrow as 0.1 μs which provides a delay resolution of 15 m and requires a sampling rate of 10 MHz.

Delay profiles of the Moon made in both receive circular polarizations at 23 cm wavelength are shown in Figure 6 (Evans & Hagfors 1966). The first and strongest echo is returned from the sub-radar point, and as the pulse sweeps over the target at increasing delay depth the echo strength decreases as the local scattering angle on the surface increases. For a spherical target, a benefit of the delay profile is that it can easily be converted into a measurement of the surface's scattering function since the local incidence angle is proportional to the delay. In other words, each delay measurement is not only a ring on the surface at constant distance but also contains only a single incidence angle. The conversion is

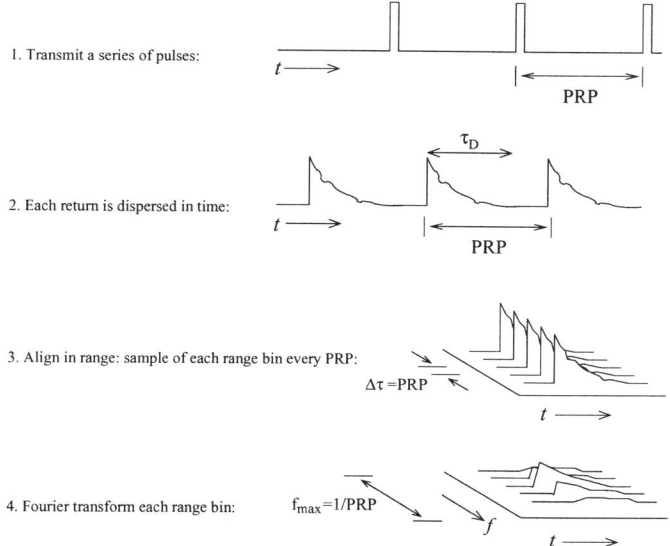

Figure 7. Delay-Doppler experiment schematic. A train of narrow pulses (possibly a coded CW transmission) produces a series of time-dispersed echoes spaced by the pulse repetition period (PRP). Aligning the echoes in range permits a spectrum to be made of each range bin, forming an image with time delay as one axis and frequency shift as the other. The frequency bandwidth imaged is set by the PRP.

$$\cos\theta = 1 - \frac{|\tau - \tau_0|}{\tau_D}, \qquad (16)$$

where τ_0 is the delay to the sub-radar point.

5.3. Delay-Doppler mapping

If a series of pulses is transmitted, the returned series of delay profiles can be aligned in delay and Fourier transformed to give a spectrum of each delay bin. If the pulses repeat with some period, called the *pulse repetition period* (PRP), then there is a regular measurement of the echo power in any given delay bin. Since the time series of each delay bin is sampled at the PRP, the spectra will sample a bandwidth of 1/PRP. In this way a two dimensional image can be constructed where the echo power from the target is mapped into time delay and frequency space. The full mapping from physical coordinates into delay-Doppler space is a combination of the individual delay and frequency expressions in the previous sections.

From the intersection of the strip of constant Doppler shift with the strip of constant delay in Figure 4 one can see that two areas on the surface will be

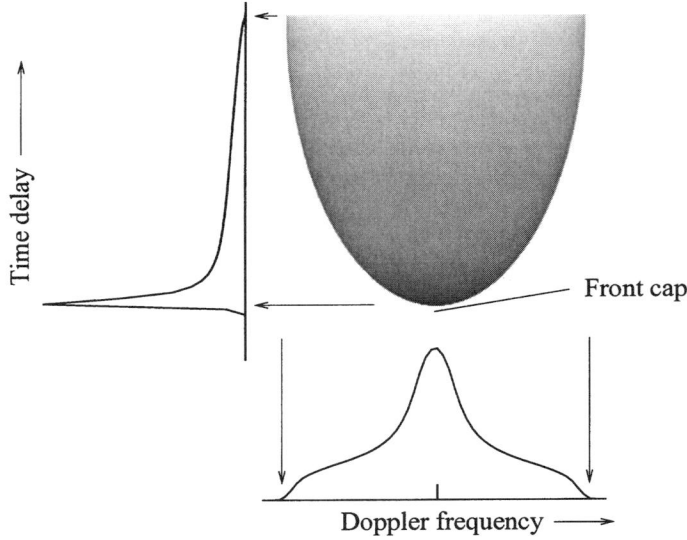

Figure 8. Model delay-Doppler image of a spherical target, such as a planet or large moon. Brighter regions are darker. The strongest echo is from the front cap at the smallest delay. Collapsing the image in time (shown at the side) or frequency (shown at the bottom) gives the profiles that would be obtained by a pulsed-only or CW run respectively.

mapped into the same time-frequency bin. Generally referred to as the *north-south ambiguity*, it effectively merges the two hemispheres together about the Doppler equator. For irregular objects the mapping can be even more severe, with more than two locations mapped into the same delay-Doppler cell. In general this ambiguity is unavoidable except for the Moon and Venus for which the antenna beam may be on the order of or smaller than the target and the hemispheres can be observed independently. Otherwise the ambiguity can be removed by either viewing the target over many orientations, or with interferometric techniques.

In practice, the pulses need not be formed by actually turning the transmitter power on and off. Instead a continuous wave transmission can be modulated by a pseudo-random code which switches the phase of the signal by 180° on regular intervals known as the *baud*. The code is chosen such that the correlation of the code with itself produces a narrow pulse and low sidelobes. Thus correlation of the received signal with a replica of the transmitted code effectively compresses the echoes into a train of single pulse echoes. The width of the compressed pulse is equal to the baud. Typically used are codes with lengths of $2^n - 1$, which are repeated every PRP. As an example, a 32767 length code ($n = 15$) with a 1 μs baud would give a PRP of 0.033 s. That combination would give 150 m range resolution on a object smaller than 0.033 seconds (~5000 km)

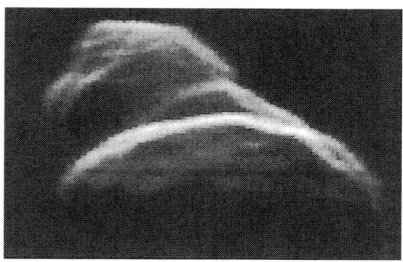 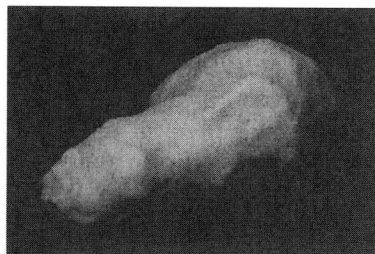

Radar map   Shape model

Figure 9. Radar delay-Doppler image and shape model of asteroid 4179 Toutatis. Image on left is one of many radar maps in delay-Doppler coordinates made at Goldstone (Ostro et al. 1995). Right image is a shape model created by modeling the set of radar images (Hudson & Ostro 1995). Toutatis has dimensions 4.6 km long and ~ 2 km wide; resolution in the radar map is ~ 20 m.

in radius. A schematic of the steps involved in a delay-Doppler experiment is shown in Figure 7.

A difficulty arises if the PRP is less than the delay depth of the planet, in which case individual echoes will start to overlap. But the PRP also sets the Doppler processing bandwidth, and if the period is too long the bandwidth sampled may be less than the Doppler dispersion of the target, causing aliasing in frequency. For example, the delay depth of Mars is 23 msec. To prevent aliasing in delay the PRP must be greater than this, which means the frequency bandwidth which can be imaged must be less than 1/23 ms$\simeq$ 44 Hz. Unfortunately, due to Mars' rotation rate of ~ 25 hours its Doppler dispersion is almost 3800 Hz at S-band and 14000 Hz at X-band, which will result in severe aliasing in frequency if sampled at only 44 Hz. Making the PRP short enough to prevent frequency aliasing makes an equal aliasing problem in delay. A target with this problem is said to be *overspread*, and if severe enough one is left with Doppler or delay only alternatives. This condition can be expressed as the product of the target's Doppler dispersion and delay depth, such that an overspread target has $B\tau_D > 1$.

A model of a delay-Doppler image from a spherical planet or large moon is shown in Figure 8. The mapping produces a parabolic shape in time-frequency space. The bright leading edge corresponds to the Doppler equator, and measurements of any deviations on this edge probes the topography along it. The front cap is brightest due since the radar is looking normal to the surface and receives a strong specular reflection. The echo power diminishes toward the limbs where the incidence angles increase and diffuse scattering becomes more important. This model uses a Hagfors scattering law with $C = 65$ (rms slopes $\sim 7°$) and a diffuse law with $n = 1.5$. Surface features such as crater, ridges, or mountains on an actual target would be apparent at coordinates corresponding to their positions relative to the front cap and apparent spin vector. Next to the image is a delay profile obtained by summing the image over frequency, and is what would be obtained using a pulse-only system. Similarly, below the image is a Doppler spectrum obtained by summing the image over delay and is what would be obtained in a CW experiment.

For non-spherical objects the delay-Doppler mapping is more complicated. An example is shown in the left panel of Figure 9 which is a radar image of near Earth asteroid 4179 Toutatis made at the Goldstone facility by Ostro et al. (1995). The asteroid has an irregular shape and as it rotates the images show various bumps and indentations and the total Doppler bandwidth and delay depth change significantly. Brighter ridges are radar facing areas and darker areas are at higher incidence angles or perhaps in shadow. Given a sufficient number of images obtained at different rotation phases and viewing geometries it is possible to invert the radar maps into a physical shape model. The right panel of Figure 9 shows perspective rendering of the shape model for Toutatis obtained by inverting the series of radar maps (Hudson & Ostro 1995). Of course, the resolution of the shape model depends on the quality of the radar data, and the ability to resolve any ambiguity in the radar maps depends on the addition of data from multiple views with respect to the rotation pole.

5.4. Astrometry

Since a radar operates in the time-frequency domain, it provides data complementary to standard plane-of-sky observations. In particular, the bulk delay and Doppler shift of the target directly measure the distance and relative velocity of the object along the line of sight. These measurements are orthogonal to plane-of-sky position measurements, and greatly constrain the position and motion of an object. Precision radar astrometry experiments have absolute uncertainties of < 100 m in range and ~ 1 mm/s in velocity (cf. Ostro et al. 1991b). Compare these to typical optical ground-based position measurements with, at best, $\sim 0.1''$ accuracy which is ~ 100 km at 1 AU. Combining optical and radar data effectively measures the full three-dimensional position and velocity of the target, albeit at varying precision (Yeomans et al. 1992).

Not only is such precision used to refine positions of planets and their satellites, but in extrapolating the orbits of newly discovered asteroids to future dates they can mean the difference between recovering or losing them on future orbits (Yeomans et al. 1987). As an example, optical observations of newly discovered asteroid 6489 Golevka during its 1991 close approach predicted an orbit from which the minimum distance during the next close approach in the year 2019 was uncertain by 1260 Earth radii. Combining radar data taken during the same 1991 apparition reduced the year 2019 uncertainty to only 4 Earth radii (Table 4 of Ostro 1997). Clearly, this magnitude of improvement can be vital to the future recovery of such objects as well as assessing any impact hazard.

6. Radar Equation

The radar equation relates the signal-to-noise ratio of an echo to known observing parameters. A primary use of this equation is to convert an observed signal, which is measured in standard-deviations of the noise level, into a physical cross section (e.g. km^2). It is also used before an experiment to predict echo strengths to assess the feasibility of an experiment and set observing parameters such as the necessary integration time.

A transmitter that radiates P_t watts into a solid angle Ω_A can be considered to be the same as an isotropic transmitter which radiates $P_t 4\pi/\Omega_A$ watts. The antenna gain is the factor $G = 4\pi/\Omega_A = 4\pi A_e/\lambda^2$, with A_e the effective collecting area of the dish. After traveling a distance R to the target, the power is spread over more space to a reduced power density of

$$\frac{P_t G}{4\pi R^2}. \tag{17}$$

The target reflects an amount equal to its radar cross section, σ. The echo propagates back to the antenna again spreading over a sphere of area $4\pi R^2$ and a portion is intercepted by the receiving dish with an effective area A_e. So the total received power is

$$S = \frac{P_t G}{4\pi R^2} \frac{\sigma A_e}{4\pi R^2}. \tag{18}$$

This power is returned over some Doppler bandwidth as discussed above. The noise power in the bandwidth B of the echo is $N = kT_{\text{sys}}B$, where T_{sys} is the system temperature of the receiving system. The absolute level of this noise may be large but the important issue is the size of the fluctuations in this background from which the echo must be distinguished. The fluctuations will be distributed with standard deviation $\Delta N = N/(\tau B)^{1/2}$, where τ is the integration time. Combining these expressions gives the signal-to-noise ratio:

$$\text{SNR} = \frac{S}{\Delta N} = \frac{P_t G A_e \sigma \sqrt{\tau}}{(4\pi R^2)^2 kT_{\text{sys}} \sqrt{B}}. \tag{19}$$

Note that the echo strength depends on the *fourth* power of the distance, making distant targets much more challenging. Conversely, an asteroid approaching very close to the Earth (to several lunar distances) can provide enormously strong echoes. For this reason radar experiments tend to be done at the target's close approach to the Earth.

The factor $G A_e$ may be rewritten to $G_t G_r \lambda^2/4\pi$, where the subscripted gains are for the transmitting and receiving antennas respectively. For a monostatic experiment these two gains may be identical; an exception that for distant targets where different portions of the Arecibo dish having slightly different gains may be used for the transmit and receive cycles.

Typical system parameters for the Arecibo and Goldstone radar systems are given in Table 1. Using the expression for a target's Doppler bandwidth from the previous section and writing the cross section as $\sigma = \hat{\sigma}\pi D^2/4$, the monostatic radar equation can be factored into a system dependent term and a target dependent term,

$$\text{SNR} \propto \left(\frac{P_t G^2 \lambda^{5/2}}{T_{\text{sys}}}\right)_{\text{system}} \left(\frac{\hat{\sigma} D^{3/2} P^{1/2}}{R^4}\right)_{\text{target}} \sqrt{\tau}, \tag{20}$$

allowing comparison of SNR attainable with different systems on the same target. The relative system factor (less numerous constants) is also given in Table 1.

Although Arecibo has a more powerful system by this measure, it sees a limited view of the sky which can also limit the integration time.

Table 1. Radar System Parameters as of Mid-2001.

	Arecibo S-band	Goldstone X-band	Arecibo 430 MHz
Power, kW	1000	500	1000
Wavelength, cm	12.6	3.5	70
Gain, dB	72.5	74.0	60.0
Collector diameter, m	305	70	305
Aperture efficiency	0.30	0.64	0.50
T_{sys}, K	25	20	60
Relative System Factor	20	1	2

7. Observing Considerations

Observations today typically record the raw voltage samples after the signal is filtered and mixed to baseband. Sampling rates can vary from several kHz for CW runs, to several MHz for coded runs. In the latter case a wide bandwidth is necessary to adequately sample the coded transmission; recall a signal modulated on a $1\mu s$ timescale has frequency components of 1 MHz. Complex samples are made of both the in-phase (I) and quadrature (Q) components so that full phase information can be recovered. If only frequency information is to be obtained, i.e. in a CW run, a spectrometer could be used provided it had suitable resolution.

Parameters for the data taking in a coded delay-Doppler imaging run might be a 10 MHz sampling rate, 2 bits per sample, complex sampling, and both polarization channels. The total data rate for this experiment would be 10 MB/sec (= 10 Msamples/sec times 2 bits/sample times 2 numbers/complex sample times 2 polarizations divided by 8 bits/byte) or 36 GB in a one hour run.

To control the telescope pointing during a session two separate ephemerides are required, one for the transmit portion and one for the receive portion. The receive ephemeris is what one would normally use to observe any object, but the transmit ephemeris will be different to account for motion of the observer and target during the round trip light travel time. In essence, to aim the transmission one must use a position based on the current Earth position but on a target position advanced in time by the round trip light time. The ephemerides should also include the estimate of the line-of-sight apparent velocity of the target so that the transmitter or receiving LO can be drifted accordingly. For a bistatic experiment a separate ephemeris for the receiving station (or stations) will be necessary to account for position and velocity parallax. In practice one typically uses services through either Arecibo in conjunction with the Harvard CfA or through Goldstone in conjunction with JPL.

8. Conclusion

Radar is capable of providing unique insight into the surface properties of Solar System objects, their shape and rotation states, and their orbit parameters. Since the observer has control of the illumination source, the resolution of radar data depends mainly on temporal and frequency discrimination rather than target angular size, and so provides information complementary to and not available from other ground-based observing methods. The range of planetary targets studied by radar has been from the smallest known asteroids to the rings of Saturn, so only a very few samples of actual results could fit into this summary. The previous sections give an introduction to basic types of radar experiments and the target properties which can be studied, and hopefully provide enough information for one to decide if and what type of experiment might address some scientific problem. A much more extensive review can be found in Ostro (1993) and for details of fundamental principles of radar measurements and scattering mechanisms the reader is referred to Evans & Hagfors (1968). This lecture has been an intended overview of various concepts but has not fully explored the techniques and science of radar astronomy.

References

Benner, L. A. M., et al. 1997, Icarus, 130, 296
Black, G. J., Campbell, D. B., & Ostro, S. J. 2001, Icarus, 151, 160
Campbell, D. B., Chandler, J. F., Pettengill, G. H., & Shapiro, I. I. 1977, Science, 34, 254
Campbell, M. J. & Ulrichs, J. 1969, JGR, 74, 5867
Carpenter, R. L. 1964, AJ, 69, 2C
Evans, J. V. & Hagfors, T. 1966, JGR, 71, 4871
Evans, J. V. & Hagfors, T. 1968, Radar Astronomy (New York: McGraw-Hill)
Hagfors, T. 1964, JGR, 69, 3779
Hudson, R. S. & Ostro, S. J. 1995, Science, 270, 84
Garvin, J. B., et al. 1985, JGR, 90, 6859
Mitchell, D. L., et al. 1995, Icarus, 118, 105
Muhleman, D. O. 1964, AJ, 69, 34
Muhleman, D. O., Butler, B. J., Grossman, A. W., & Slade, M. A. 1991, Science, 253, 1508
Muhleman, D. O., Holdridge, D. B., & Block, N. 1962, AJ, 67, 191
Nozette, S., et al. 1996, Science, 274, 1495
Ostro, S. J. 1993, Rev. Modern Phys., 65, 1235
Ostro, S. J. 1997, in IAU Colloq. 165, Dynamics and astrometry of natural and artificial celestial bodies, ed. I. M. Wytrzyszczak, J. H. Lieske & R. A. Feldman (Dordrecht; Boston: Kluwer Academic Publishers), 87
Ostro, S. J., et al. 1991a, Science, 252, 1399
Ostro, S. J., et al. 1991b, AJ, 102, 1490

Ostro, S. J., et al. 1992, JGR, 97, 18227
Ostro, S. J., et al. 1995, Science, 270, 80
Pettengill, G. H. & Dyce, R. B. 1965, Nature, 206, 1240
Pettengill, G. H., Ford, P. G., & Simpson, R. A. 1996, Science, 272, 1628
Shapiro, I. I. 1964, Phys.Rev.Lett, 13, 789
Shapiro, I. I., et al. 1972, Phys.Rev.Lett, 28, 1594
Simpson, R. A. & Tyler, G. L. 2001, Icarus, 152, 70
Simpson, R. A., et al. 1992, in Mars, ed. H. H. Kieffer, et al. (Tucson: Univ. Arizona Press), 652
Slade, M. A., Butler, B. J., & Muhleman, D. O. 1992, Science, 258, 635
Thompson, W. R. & Squyres, S. W. 1990, Icarus, 86, 336
Yeomans, D. K., Ostro, S. J., & Chodas, P. W. 1987, AJ, 94, 189
Yeomans, D. K., et al. 1992, AJ, 103, 303

Part 3
Calibration and Data Reduction

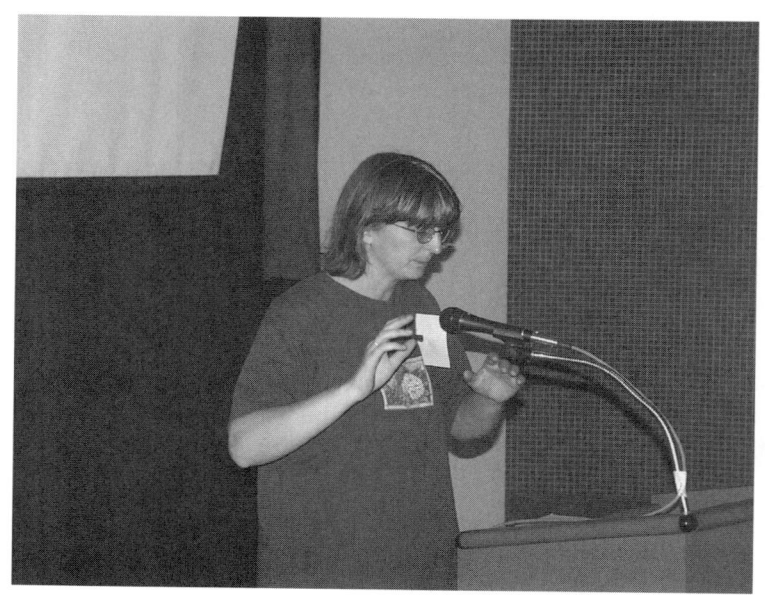

Karen O'Neil.

Single-Dish Calibration Techniques at Radio Wavelengths

K. O'Neil

National Astronomy and Ionosphere Center, Arecibo Observatory, HC 3 Box 53995, Arecibo, Puerto Rico 00612, USA

Abstract. Calibrating telescope data is one of the most important issues an observer faces. In this chapter we describe a number of the methods which are commonly used to calibrate radio telescope data in the centimeter wavelength regime. This includes a discussion of the various methods often used in determining the temperature and gain of a telescope, as well as some of the more common difficulties which can be encountered.

1. Introduction – The Importance of Calibration

As you likely know, every telescope is unique. One result of the uniqueness of individual telescopes is the difficulty of directly comparing measurements from one telescope with those from another. That is, Telescope A may record 280 counts for the peak of a given spectral line, while Telescope B may record only 100 counts. This is further complicated by the fact that even measurements taken on a given telescope, at a given frequency, can change over time. These changes can be the result of changes in, e.g. the telescope system temperature, the telescope response, and/or the atmospheric conditions. This means that even if you only observe your object with Telescope A, but measure it repeatedly over a year, you may discover that your object's peak count varies from 280 one day to 250 or 300 on another day, and so forth. You then need to understand whether the source emission is truly varying with time or if the differences are due to changes within the telescope and equipment.

In order to compare measurements between two telescopes, or even between one telescope taken at different times, we need a universal measurement system. That is, we need to be able to state that 280 counts from Telescope A is equivalent to X counts from Telescope B or equivalent to Y counts from Telescope A at a different epoch. This is the process of data (or telescope) calibration, and the rest of this chapter will be devoted to presenting the various methods available, both observationally and theoretically, to calibrate data.

2. A Brief Review

At this point it is useful to go over the more important equations and concepts that are used in the calibration process. All of these topics are also covered in other lectures, where more detailed discussions can be found.

2.1. The Rayleigh-Jeans approximation

Recall that the Planck Law for blackbody radiation is

$$B = \frac{2h\nu^3}{c^2} \frac{1}{e^{h\nu/kT} - 1}, \qquad (1)$$

where B is the brightness (specific intensity) measured in W m^{-2} Hz^{-1} rad^{-2} (or Jy rad^{-2}); h is Planck's constant (6.63×10^{-34} J s); ν is the frequency in Hz; k is Boltzman's constant (1.38×10^{-23} J K^{-1}); and T is the temperature, in K. In the centimeter wavelength regime, it is often true that $h\nu \ll kT$. In this case, you can use the Taylor Series to expand $e^{h\nu/kT} - 1$. This provides the Rayleigh-Jeans approximation for blackbody radiation at radio wavelengths:

$$B = \frac{2kT\nu^2}{c^2} = \frac{2kT}{\lambda^2}. \qquad (2)$$

For a discrete radio source of temperature T and subtending a solid angle Ω_S the source flux density (S) in the Rayleigh-Jeans limit is obtained by integrating over the source solid angle:

$$S = \frac{2k}{\lambda^2} \int_{\Omega_S} T(\theta, \phi) d\Omega. \qquad (3)$$

If the brightness temperature of the source is uniform across Ω_S, this reduces to

$$S = \frac{2kT}{\lambda^2} \Omega_S. \qquad (4)$$

2.2. Antenna temperature

The antenna temperature (T_A) can be defined as the temperature of the antenna radiation resistance. That is, let the telescope observe a point source (i.e. a source which is considerably smaller than the beam size) which has a flux density S. Then, replace the feed of the telescope with a matched resistor (or load). If you now adjust the temperature of the resistor until the power received is the same as it was for the point source (observed with the antenna or feed horn), the antenna temperature is equal to the resistor temperature. That is, the measured spectral power is simply $w = kT_A$.

If absorbing matter is present, or the source does not completely fill the beam, the measured antenna temperature will be less than the source temperature. In this case you will measure an antenna temperature T_A and you have

$$w = kT_A, \qquad (5)$$

$$S = \frac{2kT_A}{\lambda^2} \Omega_A = \frac{2kT_A}{A_e}, \qquad (6)$$

$$w = \frac{1}{2} A_e S. \qquad (7)$$

Here, A_e is the effective aperture of the antenna, and Ω_A is the solid angle of the telescope power pattern. Recall that the antenna theorem gives $A_e \Omega_A = \lambda^2$

Single-Dish Calibration Techniques at Radio Wavelengths   295

or, if ohmic losses are considered, $A_e \Omega_A \epsilon_r = \lambda^2$, where ϵ_r is the fractional power transmission of the antenna, typically close to unity. (See lectures by, i.e. Hagen or Goldsmith for further derivation of this quantity.)

In practice, the antenna temperature is given by

$$T_A = \frac{A_e}{\lambda^2} \int\int T_{\text{source}}(\theta,\phi) P_n(\theta,\phi) d\Omega = \frac{\epsilon_r}{\Omega_A} \int\int T_{\text{source}}(\theta,\phi) P_n(\theta,\phi) d\Omega \,. \quad (8)$$

Here Ω_A is again the antenna solid angle and $P_n(\theta,\phi)$ is the antenna power pattern (normalized to unity at maximum). If the source is a true "point source", i.e. it is small compared to the beam size, $P_n(\theta,\phi) \sim 1$ over the source solid angle and

$$T_A \approx \frac{\epsilon_r}{\Omega_A} \int\int_{\Omega_S} T_{\text{source}} d\Omega = \epsilon_r \frac{\Omega_S}{\Omega_A} T_{\text{avg}} \,, \quad (9)$$

where T_{avg} is the brightness temperature averaged over the source. If, on the other hand, the source is large compared to the beam size and has a constant brightness temperature T_{const}, the antenna temperature is

$$T_A = \frac{\epsilon_r T_{\text{const}}}{\Omega_A} \int\int_{\text{source}} P_n(\theta,\phi) d\Omega = \frac{T_{\text{const}}}{\Omega_A} \Omega'_b \epsilon_r \,. \quad (10)$$

Here, Ω'_b is the solid angle subtended by both the main beam and the side lobes falling on the source. (Note that if the source just fills the main beam, $\Omega'_b = \Omega'_m$, the main beam solid angle.) Finally, if the source fills the sky ($\Omega_S \gg$ beam size), $T_A = \epsilon_r T_{\text{const}}$.

2.3. Minimum detectable temperature and flux density

The minimum detectable antenna temperature is set by the fluctuations in the receiver output caused by the system noise. As has been discussed in other lectures, this noise is directly proportional to the system temperature (T_{sys}). T_{sys} can be broken down into three parts for analysis – the antenna contribution (T_A), the receiver contribution (T_R), and the power loss between the two. More specifically, the system temperature can be written as

$$T_{\text{sys}} = T_A + T_{\text{LP}}[1/\epsilon - 1] + (1/\epsilon) T_R \,, \quad (11)$$

where T_{LP} is the physical temperature of the transmission line between the antenna and the receiver, and ϵ is the efficiency of the transmission ($0 \leq \epsilon \leq 1$). The sensitivity of a radio telescope is then equal to the rms noise fluctuations of the system:

$$\Delta T_{\text{rms}} = \frac{K_S T_{\text{sys}}}{\sqrt{\Delta \nu \, t \, n}}. \quad (12)$$

Here, K_S is the sensitivity constant of the telescope (dimensionless and of order unity), $\Delta\nu$ is the pre-detection bandwidth (in Hz), t is the integration time for one record (in s), and n is the number of records averaged (dimensionless). The minimum detectable temperature is typically considered to be 3–5 times the rms noise temperature (e.g. $\Delta T_{\text{min}} = 3 T_{\text{rms}}$).

The minimum detectable temperature can be converted to a minimum brightness or flux density by applying the Rayleigh-Jeans approximation (Equations 2 and 7):

$$\Delta B_{\rm rms} = \frac{2k}{\lambda^2} \frac{K_S T_{\rm sys}}{\sqrt{\Delta \nu \, t \, n}}, \tag{13}$$

$$\Delta S_{\rm rms} = \frac{2k}{A_e} \frac{K_S T_{\rm sys}}{\sqrt{\Delta \nu \, t \, n}}. \tag{14}$$

3. System Temperature

System temperature, for the purpose of this lecture, will be defined by breaking it into two parts:

$$T_{\rm sys}(\alpha, \delta, az, za) = T_{\rm OFF}(\alpha, \delta, az, za) + T_{\rm source}(\alpha, \delta, az, za), \tag{15}$$

where α, δ, az, za are the source right ascension, source declination, and telescope position in azimuth and zenith angle, respectively. The "off source" temperature is simply the temperature measured if the telescope were pointed at a nearby region of blank sky:

$$T_{\rm OFF}(\alpha, \delta, az, za) = T_{\rm RX} + T_{\rm gr}(za, az) + T_{\rm atm}(za) + T_{\rm CMB} + T_{\rm BG}(\alpha, \delta). \tag{16}$$

$T_{\rm RX}$ is the receiver temperature; $T_{\rm gr}$ the temperature contributed from the ground; $T_{\rm atm}$ the contribution from the atmosphere; $T_{\rm CMB}$ the contribution from the cosmic microwave background; and $T_{\rm BG}$ the contribution from background and foreground celestial sources, including Galactic emission.

Theoretically, the system temperature of a telescope changes with telescope elevation due to atmospheric emission following:

$$T_{\rm atm} = T\left(1 - e^{-\tau A}\right), \tag{17}$$

where τ is the atmospheric opacity at zenith and A is related to the zenith angle at which the telescope is pointing ($A = \sec(za)$). A well-built telescope with an unblocked (or partially blocked) aperture and a raised, movable dish can come close to achieving this theoretical model (Figure 1). In reality, though, a number of factors must be added to this theoretical model. In particular, most telescopes experience temperature changes due to changes in the ground radiation with elevation, reflection of radiation (both atmospheric and celestial) off the telescope structure, and changes in the system itself.

In practice, the source temperature can be separated from the system temperature by making two observations – one which includes the source in the beam (the ON), and one which does not (the OFF). The source temperature is then just:

$$T_{\rm source} = T_{\rm ON} - T_{\rm OFF}, \tag{18}$$

$$\frac{T_{\rm source}}{T_{\rm OFF}} = \frac{T_{\rm ON} - T_{\rm OFF}}{T_{\rm OFF}}, \tag{19}$$

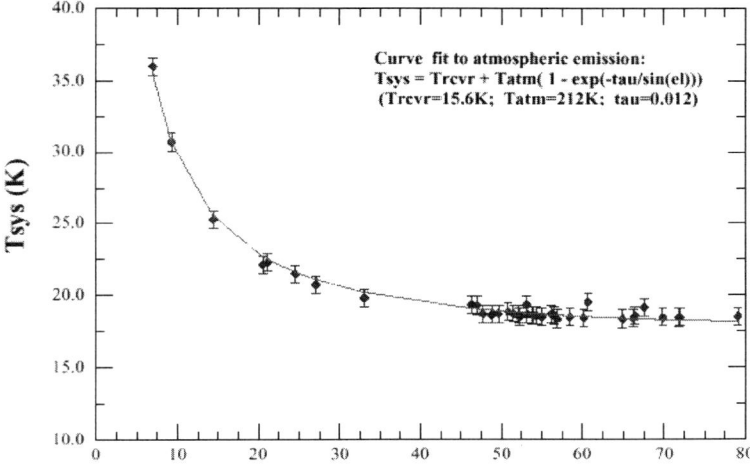

Figure 1. System temperature for the Green Bank Telescope as measured at 2 GHz on 21-22 March, 2001. (From Ghigo et al. 2001.)

$$T_{\text{source}} = \left[\frac{T_{\text{ON}} - T_{\text{OFF}}}{T_{\text{OFF}}}\right] T_{\text{OFF}} , \qquad (20)$$

where T_{ON} and T_{OFF} are the power measurements for the two celestial positions.

At this point it should be clear that one of the most important measurements to be made in order to understand, and calibrate, your data is the measurement of system temperature. In the regime of centimeter wavelength astronomy there are two techniques commonly used to make this determination. These will be discussed in the next section.

3.1. Switched noise diode

The first technique for measuring system temperature uses a "switched noise diode". With this method, a noise diode with known effective temperature at the desired frequency is coupled to the telescope system (Figure 2). The telescope is then pointed to the blank sky and two measurements are made – one with the diode turned on (ON_{CAL}) and one with the diode turned off (OFF_{CAL}). These measurements are then used to determine the off-source system temperature, T_{OFF}.

The basis behind the switched noise diode technique is quite simple. The ON_{CAL} measurement is simply the sum of the system temperature plus the effective temperature of the diode, while the OFF_{CAL} observation, in theory, measures only the system temperature. The system temperature can then be derived by comparing the temperature of the noise diode, measured as a fraction of the system temperature, with the known temperature of the noise diode. In

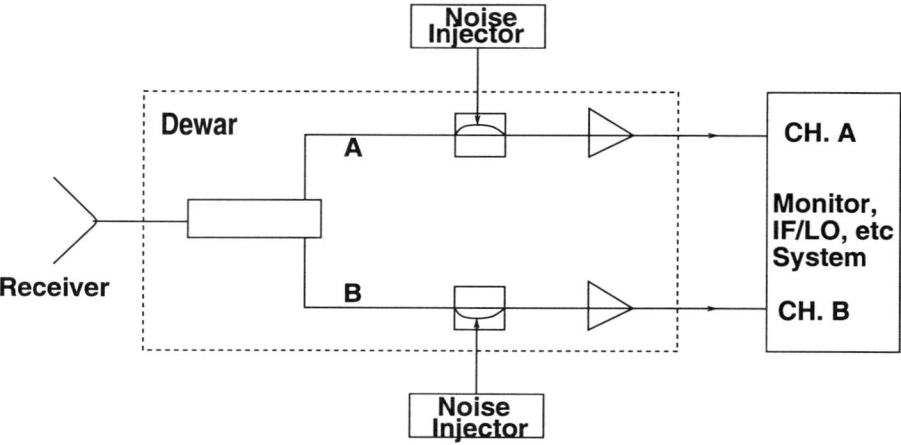

Figure 2. Schematic drawing of the noise diode injection into a receiver system. This schematic is modeled after the L-Narrow receiver system at Arecibo Observatory.

other words, system temperature is

$$\frac{T_{\rm OFF}}{T_{\rm CAL}} = \frac{\rm OFF_{CAL}}{\rm ON_{CAL} - OFF_{CAL}}, \qquad (21)$$

$$T_{\rm OFF}({\rm K}) = \frac{\rm OFF_{CAL}}{\rm ON_{CAL} - OFF_{CAL}} \times [T_{\rm CAL}({\rm K})]_{\rm known}. \qquad (22)$$

Although in theory switched noise diodes provide an excellent tool for measuring the system temperature, in practice a number of difficulties exist which limit the accuracy of the noise diode method. The first complication is that the effective temperatures of most diodes have a frequency dependence. This is not an overwhelming complication, as one can either obtain temperature measurements at a variety of frequencies or use some other multi-frequency calibration methods (i.e. van Zee et al. 1997). It is merely an added issue to be aware of when determining or applying noise diode measurements to your data.

A second issue to consider when using noise diodes to determine the system temperature is the time stability of the diodes. A good (well-built) noise diode, in a well controlled environment, should remain stable for a number of years. Yet the performance of a diode should be checked by frequent (bi-annual or more often) measurements of the diode response across the frequency band of interest, to insure that the diode is performing according to its specifications.

The final issue to consider when using the noise diode system is the accuracy of the measurement of the noise diode response. This is a non-trivial issue, as noise diode measurements are typically determined through bootstrapping off another ('standard') noise diode, which is believed to have a well-known, accurately measured value. Noise diode measurements therefore start off with an initial error from the measurement of the standard diode. As all other measurements are done through bootstrapping off the first diode measurement, this initial error is then propagated throughout all measurements. Added to the initial measurement errors are, of course, the errors inherent in the measurement process itself. Finally, it should be remembered that the initial noise diode measurements (that is, the measurements coming from the standard diode comparison) are not temperature measurements, but instead most be converted from, i.e. a voltage measurement, and in that process yet more error is introduced.

In sum, the errors introduced just by the measurements of the noise diode value are:

$$\sigma_{\text{total}} = \sqrt{\sigma^2_{\text{freq. dependence}} + \sigma^2_{\text{stability}} + \sigma^2_{\text{measured value}} + \sigma^2_{\text{conversion error}}}, \quad (23)$$

where

$$\sigma_{\text{measure value}} = \sqrt{\sigma^2_{\text{standard cal}} + \sigma^2_{\text{instrumental error}} + \sigma^2_{\text{loss uncertainties}}}. \quad (24)$$

In spite of these difficulties, though, noise diodes can give a fairly accurate indication of a telescope's system temperature, often to within 2% or less.

Note – System temperature is often measured at telescopes using a method known as the "Y-Factor" method. This method employs two diodes (or similar sources) of known effective temperatures and does not include any effects of the antenna. If diode one has a known effective temperature T_1 and diode two has a known effective temperature T_2, the ratio of the measured power of the two diodes will be

$$Y = \frac{T_1 + T_{\text{OFF}}}{T_2 + T_{\text{OFF}}}, \quad (25)$$

where recall T_{OFF} for these purposes is the system temperature without any effect of the antenna. This ratio (the Y-Factor) can then be used to obtain T_{OFF} via

$$T_{\text{OFF}} = \frac{T_1 - Y\,T_2}{Y - 1}. \quad (26)$$

3.2. Hot & cold loads

Another method for obtaining the system temperature is through the use of hot and cold loads. A cold load is typically an absorbing system placed inside some gas or liquid at a known temperature (often liquid Nitrogen). This load is then placed over, or otherwise coupled with, the receiver of choice, and the power level is measured. The hot load is often a similar measurement of a load at ambient temperature. Alternatively, the 'cold' load can be a blank sky measurement and the hot load can be a measurement of a load placed inside a liquid of known temperature. These measurements can then be used in the same fashion as a noise diode to obtain the system temperature.

Measuring the system temperature using hot and cold loads can be considerably more reliable than the noise diode method. The primary reasons for the superiority of the hot/cold load method are that the temperature of the loads can be measured directly, and the measurements are all temperature measurements – no conversions are necessary. However, the hot/cold load system is not used at most centimeter wavelength radio telescopes as building loads large enough to encompass a $\lambda > 5\text{--}6$ cm feed, and which have the easy availability necessary for on-the-fly measurements are highly impractical. As a result, the use of hot/cold loads to measure system temperature on-the-fly is generally restricted to the $\lambda < 5\text{--}6$ cm regime. In the $\lambda > 5\text{--}6$ cm regime, hot/cold loads are often used to check the noise diode measurements from the switched noise diode technique (and can be used in place of a standard diode for calibration, as is the case at the Green Bank Telescope). A more detailed discussion of this, and similar, calibration methods at short wavelengths is given in the lecture by Jewell.

4. Off-Source Observations

In Section 3 we showed that the temperature of the object of interest can be found by measuring the system temperature and using that measurement in combination with a blank-sky measurement of T_{sys}, to calibrate the observations. That is, Equation 20 gave:

$$T_{\text{source}}(\text{K}) = \left[\frac{T_{\text{ON}} - T_{\text{OFF}}}{T_{\text{OFF}}}\right] T_{\text{OFF}}(\text{K}). \tag{27}$$

As Section 3 discussed the various means for determining the system temperature, the next step is to determine the best method for obtaining an appropriate off-source, or blank sky, observation.

4.1. Baseline fitting

Baseline fitting to just the on-source records is potentially the simplest and most efficient of the methods for obtaining blank sky information for spectroscopic observations. The idea is straightforward – a baseline is fit to those spectral regions of an observation which are known to contain no signal from the object of interest. This baseline is then treated as the off source (blank sky) observation (Figure 3a).

Although this procedure is extremely efficient, it has several drawbacks. In particular, it is not a feasible option if the spectral line of interest is large compared to the bandpass (that is, if there are too few channels from which to fit a reasonable baseline), or if there are standing waves present which have a frequency higher than, or of order of, the bandpass of interest. Additionally, if the baseline is not reasonably flat for any reason this option should not be used, as a poor fit can artificially add or subtract considerable emission to the line of interest (Figure 3b).

Errors induced from the baseline fitting method come primarily from the quality of fit.

4.2. Frequency switching

The frequency switching technique obtains blank sky information by keeping the telescope pointed at the object of interest, but switching the center frequency of the measurements (changing the frequency of the first local oscillator). As this mode of observing does not require any movement of the telescope, it can

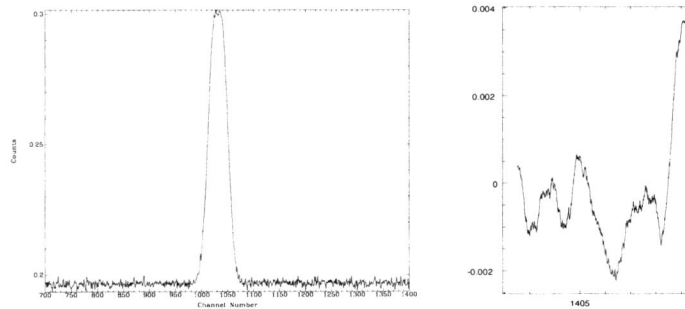

Figure 3. Examples showing successful fitting of a flat baseline to an observation (left), and an observation in which baseline fitting is not a viable option (right – courtesy of C. Conselice). Both images include the spectral line of interest and a number of 'blank' channels.

be done very quickly and efficiently. When the data from the two settings are subtracted, the quasi-stationary effects introduced into the data after the first mixer, such as spectral ripples, are cancelled. Additionally, if the frequency is shifted such that the frequency range of interest remains within the bandpass, but does not overlap its original "unswitched" range, frequency switching can be made extremely efficient (Figure 4). Liszt (1997) describes a deconvolution method for the case in which the two ranges do overlap.

Frequency switching has a number of advantages over the baseline fitting method. First, frequency switching reduces the chance of error induced through poor fits to the baselines in the regions of interest. Second, frequency switching can allow for higher resolution spectra as considerable bandpass does not need to be 'wasted' to accommodate a large number of blank channels, as is typically necessary for baseline fitting. Finally, because frequency switching can occur at a very rapid rate (on the order of a second, or less), frequency switching can cancel any post-mixer variations on this, or longer, time scales.

The disadvantages to frequency switching at centimeter wavelengths are few. The primary difficulties are that (a) the redshift of the line of interest must be accurately known *a priori*; (b) the system must be stable enough that the baselines of the primary observation and the frequency switched observation are virtually identical; and (c) as with baseline fitting, if there are significant standing waves in the baseline due to reflections off the telescope structure for a partially blocked aperture or the presence of strong continuum source within the beam, frequency switching may be unable to eliminate the standing pattern (provided, of course, the characteristic frequency is not considerably wider than the frequency shift).

4.3. Position switching

Position switching involves observing the object of interest for a fixed period of time, and then moving the telescope to a blank sky region to obtain the blank sky observations necessary for baseline subtraction. Although costly in terms

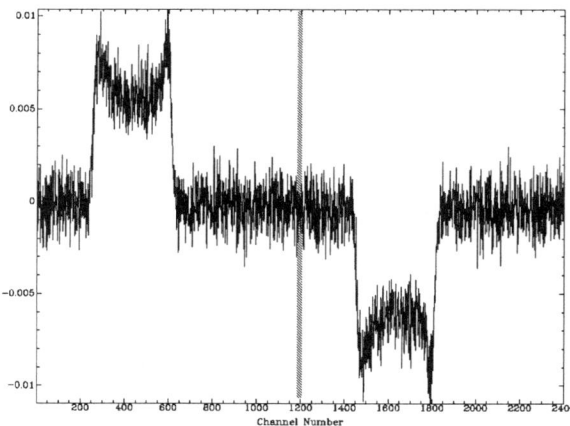

Figure 4. Data taken in frequency switching mode. The center line denotes the separation between the two images. Here the x-axis is channels and the y-axis in raw counts. As the spectral line of interest is available in both images, it would be possible to fold the data back together to achieve a higher signal to noise (and better observing efficiency).

of telescope time, this observing method has the advantages that no *a priori* information (other than position and a redshift known well enough that the line of interest will lie within the spectrometer bandpass) need to be known about the object and continuum emission within the beam can be less of a problem (although see Section 4.4, below).

There are a number of issues to consider when deciding to use position switching. The first consideration is the time stability of the telescope and baselines. As position switching typically requires re-pointing the telescope, it is not feasible to switch between the on-source and blank-sky observations at a rapid rate (although see the sections on wobbler switching and chopper wheel techniques in the lecture by Jewell). This means the telescope baselines must remain stable over a reasonable period of time, where this period of time is defined by the amount of time it takes to observe both the on and off-source positions. (That is, if five minute on- and off-source observations are being taken, with one minute between the observations, the baselines must be stable for *at least* 11-12 minutes, if not considerably longer.)

A second issue which needs to be considered when performing position switched observations is the dish illumination and aperture blockage. If the same portion of the dish is not illuminated during the on- and off-source observations, large differences in the baselines can be present. This is particularly true if there are significant standing waves in the baseline due to a partially blocked aperture. These difficulties are extreme for a telescope like the Arecibo 305-m which has a fixed primarily reflector. In this case, observations of different sky positions may not only illuminate different areas of the reflector but can also have markedly different contributions from reflections off the ground and telescope structure.

In this case it is prudent to obtain the off-source observation by tracking a blank sky region chosen such that the illumination pattern of the feed tracks across the same part of the primary reflector as for the on-source observation. In this case, if a five minute on-source observation of an object was taken, starting when the object was at an AZ of 283° and ZA of 13°, the off-source observation should also last for five minutes and start at (AZ , ZA) = (283°, 13°). Although this method is time consuming, it can offer considerably flatter baselines and therefore a considerable reduction in the spectral deviations introduced by the system.

The final issue to consider with position switching is that, if the size of the source is considerably more extended in angular distance than the difference between the on and off-source sky positions (as is the case for the ubiquitous Galactic HI), position switching becomes an impractical option. In these cases other alternatives such as frequency switching, baseline fitting, etc. must be considered.

4.4. Variations

To get around some of the difficulties inherent in the above 'standard' procedures for obtaining off-source observations, a number of variations of these methods have been (and are constantly being) devised. In this section we enumerate a few of these methods which have proven to be useful.

Baseline Fitting with an Average Fit One method for reducing random noise which can cause difficulties when fitting a function to the baselines (Section 4.1) is to average, or median average, all the observations together and use that average to determine the baseline fit for the data. This has the advantage of providing a fairly accurate baseline, but at the expense of losing detailed information which may be important in individual fits, particularly if the time over which the average is taken is longer than the stability time of the telescope system.

Position Switching on an Extended Source As was mentioned in Section 4.3, the position switching method is not easily applied when observing a source which is extended when compared with the offset distance between the on-source and off-source observations. Although frequency switching is typically the preferred observing method in these cases, occasionally it is not a viable option. In these cases, an alternative observing method must be considered. One option which can be used when mapping an extended, but finite, source is to extend the map beyond the edges of the source. The desired blank sky information is then obtained by averaging the off-source observations or by fitting baselines across the map using the off-source observations (Figure 5).

If the telescope being used does not have a constant telescope illumination as it points to different sky positions, such as is the case for the Arecibo 305-m telescope, it is often useful to use only those observations at a given Right Ascension (if the map is stepping in Declination – at a given Declination otherwise) to obtain the best blank sky information. Even flatter baselines can be obtained in this case through drift-scan mapping, as in this method the off- and on-source observations are taken at the same (az, za).

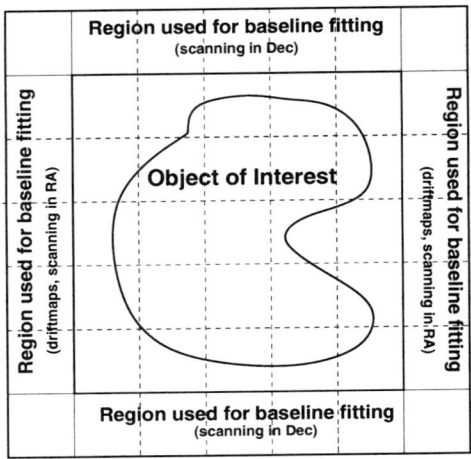

Figure 5. Diagram showing the off-source regions typically used to obtain blank sky information, using position switching, for an extended source.

Position Switching in the Presence of Continuum Emission When position switching is attempted on a source with considerable continuum emission (T_{source} is a significant fraction of the value of T_{sys}), a markedly different standing wave amplitude can be observed in the baselines between the on- and off-source observations. As a result, the components of the standing wave pattern produced by the strong continuum emitter is not cancelled by the off-source observation. This problem is particularly noticeable in telescopes with partially blocked apertures and which have significant standing waves even in the blank sky observations.

One method for dealing with this problem is to observe another continuum source, of similar strength to the object of interest, and then to divide the source difference (ON − OFF) spectrum by the reference difference spectrum to eliminate the residual standing waves. The result is a spectrum with a magnitude proportional to the ratio of the target and reference flux densities, and includes any spectral-line component (emission or absorption) that may be present in the target:

$$R = \frac{(\text{ON}(\nu) - \text{OFF}(\nu))_{\text{source1}}}{(\text{ON}(\nu) - \text{OFF}(\nu))_{\text{source2}}}. \tag{28}$$

In this case, the rms noise on the observed ratio is:

$$\sigma(R) = \frac{\sqrt{2}R}{\sqrt{\beta\tau}} \times \left[\frac{\text{SEFD}_{\text{S1,ON}}^2 + \text{SEFD}_{\text{S1,OFF}}^2}{(\text{SEFD}_{\text{S1,ON}} - \text{SEFD}_{\text{S1,OFF}})^2} + \frac{\text{SEFD}_{\text{S2,ON}}^2 + \text{SEFD}_{\text{S2,OFF}}^2}{(\text{SEFD}_{\text{S2,ON}} - \text{SEFD}_{\text{S2,OFF}})^2} \right]^{1/2}, \tag{29}$$

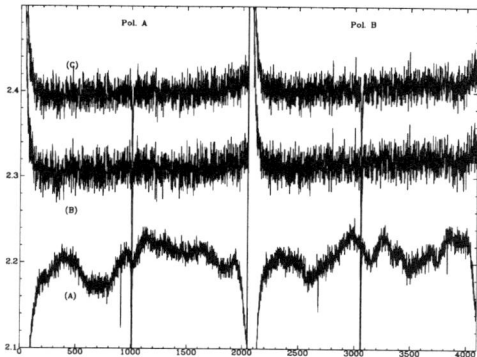

Figure 6. The average of three typical observing cycles with a strong continuum source; (A) is for the standard position switching technique, (B) is double position switching by dividing the [ON − OFF]/OFF of the source with that of the reference spectra, and (C) is double position switching using the double position switching method discussed in Section 4.4. (Figure from Salter & Ghosh 2001.)

where S_1 and S_2 are continuum source one and two, respectively, and SEFD is the system equivalent flux density of the telescope, which is equal to the system temperature divided by the telescope gain at the frequency of interest. This method has been carefully tested at Arecibo Observatory, where the unique design of the 305-m telescope makes eliminating standing waves from the baselines a challenge. The results show that, although this method of 'double position switching' and obtaining the ratio of target flux densities initially increases the rms noise when compared to standard position switched observations, the noise decreases as $1/\sqrt{time}$ with continued observation, while the noise of the standard position switched observations ceases to decrease after only four observation cycles have been averaged (Figure 6). Further information on this method can be found in Salter & Ghosh (2001), and references therein.

5. Flux Density Conversion

At this point we have discussed both steps necessary for converting observed (raw) data into units of antenna temperature. Although this is a significant step towards making data comparable between telescopes, it is not the complete picture. The reason for this is that every telescope has a different response, or gain. To complicate matters further, individual telescopes also have different gains at different frequencies or even different elevation angles. As a result, the temperature of a source must be converted from observed antenna temperature to a gain-corrected measurement. In radio astronomy this final, gain-corrected measurement is typically in units of flux density per beam or main beam brightness. Main beam brightness is simply the temperature measurement corrected for telescope efficiency ($T_{\text{MB}} = T_{\text{measured}}/\eta_{\text{beam}}$). As defined in Section 2, flux density per beam, on the other hand, is the integral of the source brightness over the telescope beam

$$S = \frac{2k}{\lambda^2} \frac{\int\int T(\theta,\phi)\, P_n(\theta,\phi)\, d\Omega}{\int\int P_n(\theta,\phi)\, d\Omega}. \tag{30}$$

The unit of flux density is W m^{-2} Hz^{-1}, or Janskies (Jy) (1 Jy = 10^{-26} W m^{-2} Hz^{-1}).

As flux density per beam is the commonly used brightness units for most centimeter wavelength radio astronomy, the rest of the discussion in this section will concentrate on calibrating data into flux density units. If conversion into main beam brightness is instead desired, the overall methods are the same.

5.1. The ideal telescope

In an ideal telescope – one with a completely unblocked aperture, no ground reflection, lossless instrumentation, cables, etc., and a transparent atmosphere – the telescope gain can readily be theoretically modeled. Even in the case of a fairly simple and well understood system with minimum aperture blockage, an accurate theoretical prediction of the telescope's gain can be obtained. (See discussions by Lockman and Condon in this volume).

As discussed by both Condon and Salter (this volume), if a telescope's response can be well modeled, the telescope can be used to obtain absolute flux densities of continuum sources. Then, if the absolute flux densities of a reasonable number of continuum sources are determined at a variety of frequencies, monitored, and recorded, a catalog of 'standard' continuum sources can be developed. These sources can then be used to monitor telescope performance and look for any degradation in the telescope gain due to deterioration in the system components, distortion of the reflector shape, etc.

5.2. Bootstrapping

For many telescopes, determining a telescope's gain theoretically is extremely complicated due to considerable or irregular blockage of the aperture, uncertain losses in the cabling or electronics, uneven reflection from the ground, etc. In these cases, rather than relying upon what could be fairly inaccurate models of the telescope gain, it is often good to take advantage of pre-existing catalogs of standard continuum source flux densities and "bootstrap" off those values to determine the gain of the telescope of interest.

When choosing which sources to observe from source catalogs, a few issues should be considered. First, the size of the chosen source should be small when compared with the size of telescope beam. If the source is of significant size when compared to the beam (a diameter more than $\sim 1/10$ the telescope beam at the frequency of interest), issues such as the detailed beam pattern and potential spillover of the source onto the ground need to be considered. These problems can be eliminated by insuring that the source is essentially point-like when compared to the beam.

The second item to consider when choosing a source for telescope gain calibration is that the source should be strong enough to supply a fairly high signal-to-noise ratio but not so strong as to cause difficulties with baseline ripples or the telescope's dynamic range.

Finally, a good calibrator source should have a non-variable flux density which has been well determined at the frequency of interest. For a well defined telescope, errors in the previously determined flux density measurements typically dominate all other errors inherent in determining a telescope's gain using the bootstrap technique. This means that reducing the errors in the pre-determined flux density values results in an essentially linear reduction in the errors in the telescope gain determination. One of the results of this is that the error in determining a telescope's gain (when obtained through bootstrapping) can be greatly reduced by observing a number of sources at a range of positions in the sky.

The primary difficulty with using the bootstrap technique to determine telescope gain is that if a telescope's gain changes significantly at different sky positions, or if the telescope's beam illumination or aperture blockage changes as the telescope points to different azimuth and/or zenith angles, it can readily become impractical for an observer to continuously monitor the telescope gain. It may take more time to determine the gain than to obtain the desired observations of the observer's source. In this case, the observer often needs to rely on a telescope's pre-determined standard gain curves.

5.3. Gain curves

When not available through theoretical models, gain curves are obtained by observing a large number of standard flux density calibrators located across a telescope's visible sky region at frequencies which range across the telescope's usable bandwidth. Gain curves can be extremely useful if a telescope's response is fairly stable over time and the gain of a telescope varies across the sky.

A telescope with an evenly illuminated reflector typically has a gain curve which varies with telescope zenith angle in a fairly predictable manner (Figure 7). Telescope gains can, though, become fairly complicated as aperture blockage is increased, the illuminated portion of the main reflector changes (as is the case with the Arecibo telescope), and the contribution from ground reflection changes. In this case it is possible to obtain gain curves which vary as the telescope moves in azimuth as well as zenith angle (Figure 8).

Although the existence of pre-determined gain curves can save the observer a considerable amount of telescope time, gain curves should never be blindly accepted. Instead, a circumspect observer should recognize that although the shape of a telescope's gain curve should remain relatively constant, a scaling of the curve may be necessary in the case of poor telescope focus, pointing offsets, or degradation in the feed, electronics, cabling, etc. As a result, it is always useful to obtain observations of a number of flux-density standards at telescope positions near those of the objects of interest. These observations can then be used to check the telescope gain curve and, when necessary, to scale the curve accordingly.

6. Other Issues

There are a variety of other issues which can affect the gain and system temperature of a radio telescope. In this section we will describe the more common issues an observer might encounter.

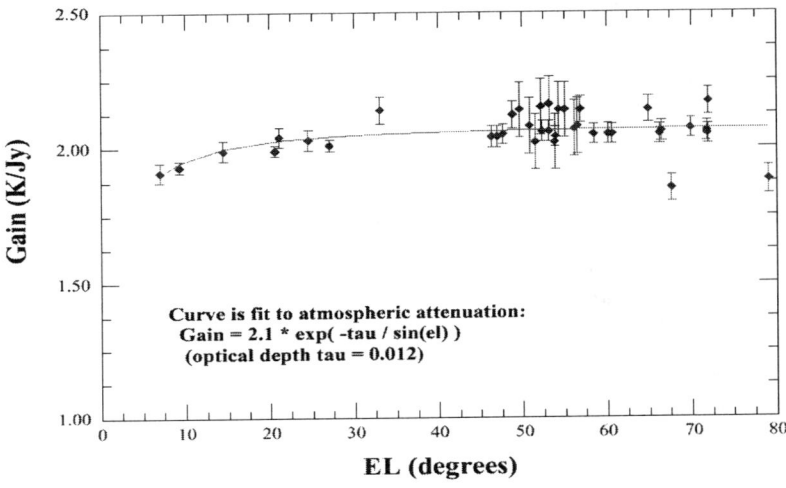

Figure 7. Gain curve for the Green Bank Telescope. (From Ghigo et al. 2001).

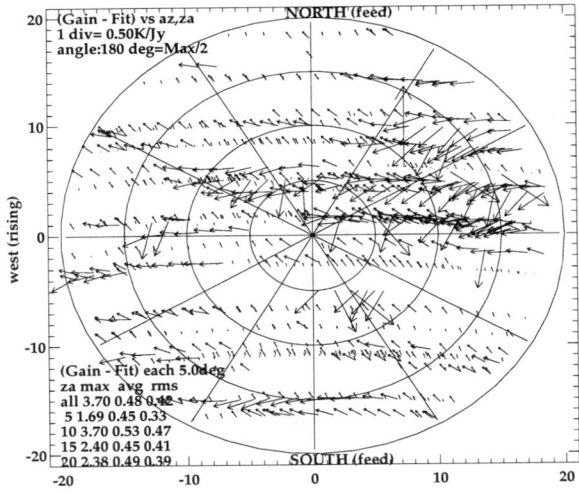

Figure 8. Gain distribution for the Arecibo 305-m telescope. Here both the length and angle of the arrow are proportional to the value of the gain, and the polar plot is in terms of zenith angle (radius) and azimuth angle (θ).

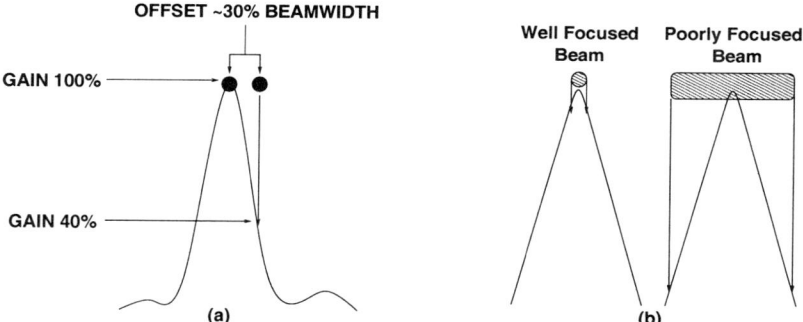

Figure 9. Figure showing the results of (a) pointing errors and (b) focus errors on the measured flux of an object.

6.1. Pointing and focus

Poor pointing or axial focus of a telescope can result in a reduction of the telescope gain. The reason for this is that a typical telescope main beam pattern is Gaussian in nature (to a fairly low level). As a result, a relatively offset in pointing can result in a significant reduction of the telescope gain at the position of the object (Figure 9a). In this case the signal-to-noise and calibration of the the observation can suffer severely. In a similar manner, poor telescope focus will artificially diffuse an object so that less of the object falls onto the center of the beam where the telescope gain is at the highest (Figure 9b). Again this will result in a degradation in the object signal-to-noise ratio.

The method for checking axial focus is telescope dependent. One reliable method for determining a telescope's pointing is by performing cross-scans across a strong, point-like continuum source with an accurately known position. If a telescope's pointing is accurate, the maxima of the scans will lie in the center of each cross scan. If the pointing is in error, an offset between the center of the cross scans and the peak intensity measured by the observation will be seen. In many cases, this offset can then be fed back into the telescope pointing model.

6.2. Side lobes

Issues of side lobes in a telescope's beam pattern are discussed extensively in the lecture by Lockman. However, as the presence of side lobes can affect the calibration of an observation, it is worthy of a brief mention here.

Because they can allow extraneous radiation to enter into an observation, the presence of side lobes in a telescope's response pattern can artificially increase the measured flux from a source. This can result both in inaccurate measurements of a source's flux density and, if a telescope's gain is determined through bootstrapping, can create avoidable errors in the telescope's gain determination. As a result, an observer should be aware of the presence and extent of a telescope's side lobes at the frequency of interest. If possible, an observer should simply avoid observing any sky position where the side lobe contributions will cause unacceptable errors in the flux density measurements. When this is not possible (much of the time, particularly when mapping an extended source), the observer should attempt to use models of the side lobes and beam pattern to deconvolve the side lobe contributions from the desired spectrum.

6.3. Coma & astigmatism

If a telescope's feed or secondary reflector is displaced or rotated from its principal axis it will cause asymmetries in the detailed telescope beam. If a subreflector is shifted perpendicular off the main reflector axis, a pointing error is generated. This is known as a comatic error. Astigmatism occurs due to deformations in the reflector(s) and results in irregularities (or lobes) on the beam pattern (Figure 10). As with side lobes, the presence of distortions in a telescope beam pattern can artificially increase the measured flux density of an extended source due to the addition of stray radiation, and decrease the flux density for a point source (due to decreased telescope efficiency).

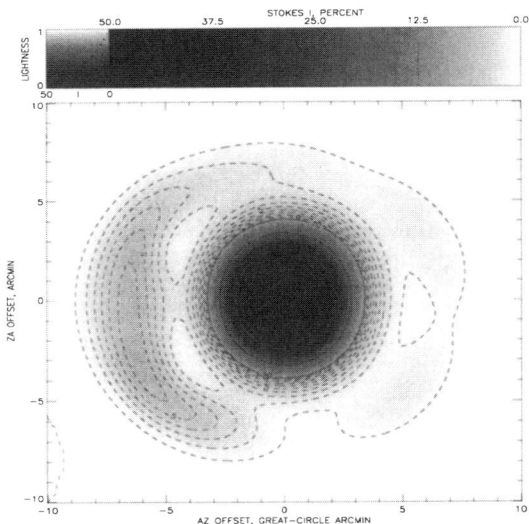

Figure 10. Beam response for the Arecibo 305-m telescope. The contours are labeled in units of percent peak intensity and are spaced by 0.4%. The dotted cross marks the position of intensity maximum. The asymmetry in the beam pattern at the left of the image is due in large part to the effects of the coma. (From Heiles et al. 2001).

7. Useful Resources

In this section we list a few of the resources that are useful when calibrating radio astronomy data.

- Baars et al. (1977)

- Baars (1973)
- Condon et al. (1998), available online at http://www.cv.nrao.edu/NVSS
- Kraus (1986)
- Kuehr et al. (1981)
- Ott et al. (1994)
- Tabara & Inoue (1980)

Acknowledgments. Thanks to Chris Salter and Ron Maddalena for their helpful comments on this paper, and of course many thanks to Paul Marganian for all his help.

References

Baars, J. W. M. 1973, IEEE Trans. Antennas Propagat., AP-21, 461–474
Baars, J. W. M., Genzel, R., Pauliny-Toth, I. I. K., & Witzel, A. 1977, A&A, 61, 99
Condon, J. J., Cotton, W. D., Greisen, E. W., Yin, Q. F., Perley, R. A., Taylor, G. B., & Broderick, J. J. 1998, AJ, 115, 1693
Ghigo, F., Maddalena, R., Balser, D., & Langston, G. 2001, GBT Commissioning Memo 10
Heiles, C., et al. 2001, PASP, 113, 1274
Kraus, J. D. 1986, Radio Astronomy (Ohio: Cygnus-Quasar Books)
Kuehr, H., Witzel, A., Pauliny-Toth, I. I. K., & Nauber, U. 1981, A&AS, 45, 367
Liszt, H. 1997, A&A, 124, 183
Ott, M., Witzel, A., Quirrenbach, A., Krichbaum, T. P., Standke, K. J., Schalinski, C. J., & Hummel, C. A. 1994, A&A, 284, 331
Salter, C. J. & Ghosh, T. 2001, AAS, 198, 7502
Tabara, H., & Inoue, M. 1980, A&AS, 39, 379
van Zee, L., et al. 1997, AJ 113, 1638

From left to right: Shih-Ping Lai, Kevin Healy, Jordi Gutiérrez, Esteban Araya, Wendy Lane, Mercedes Gomez, and Don Campbell.

Single-Dish Radio Astronomy: Techniques and Applications
ASP Conference Series, Vol. 278, 2002
S. Stanimirović, D. R. Altschuler, P. F. Goldsmith, and C. J. Salter

Millimeter Wave Calibration Techniques

P. R. Jewell

National Radio Astronomy Observatory, P.O. Box 2, Green Bank, West Virginia 24944, USA

Abstract. Calibration practices for millimeter wavelengths are somewhat different than those for meter and centimeter wavelengths. There are technological, atmospheric, and historical reasons for this. This lecture reviews the specific techniques used for millimeter wave calibration and highlights the differences between these and the techniques used at longer wavelengths. The importance of the atmosphere at centimeter, millimeter, and submillimeter wavelengths is discussed in detail. Topics include specific calibration techniques such as the hot/sky chopper wheel method, variations such as hot/cold/sky schemes, sky tipping calibration, and the possibilities of subreflector-based calibration sources. Calibration loss factors including rear and forward spillover and error beam losses are described and illustrated. The $T_A{}^*$, $T_R{}^*$, and T_{MB} temperature scales are defined. Techniques for absolute calibration are also discussed.

1. Introduction

The requirements of calibration of millimeter-wave observations are, in principle, no different from centimeter- or meter-wave observations. The objective of calibration is the same: convert an arbitrary instrumental response to an astronomically meaningful flux. Nonetheless, millimeter-wave calibration has traditionally employed different conventions and techniques. The origin of these differences dates from the earliest days of millimeter-wave astronomy (e.g. Penzias & Burrus 1973). The reasons for these differences relate both to available technology and to the importance of the atmosphere at millimeter wavelengths. These differences are best explained by reviewing the basics of calibration that pertain to observations at any wavelength, then describing the reasons why millimeter-wave calibration has diverged from the traditional practices at longer wavelengths.

When we observe "cold sky", i.e., a region of the sky with no detectable astronomical emission, we might have a bandpass such as in Figure 1 (left). The bandpass represents "system noise" which is the sum of the internal receiver noise and the noise that is coming from the sky. A region with narrow band, spectral line emission might show a bandpass as in Figure 1 (right).

In general, to distinguish the astronomical emission from the instrumental and background sky emission, we take a difference between the two bandpass responses as represented by Equation 1 and by Figure 2:

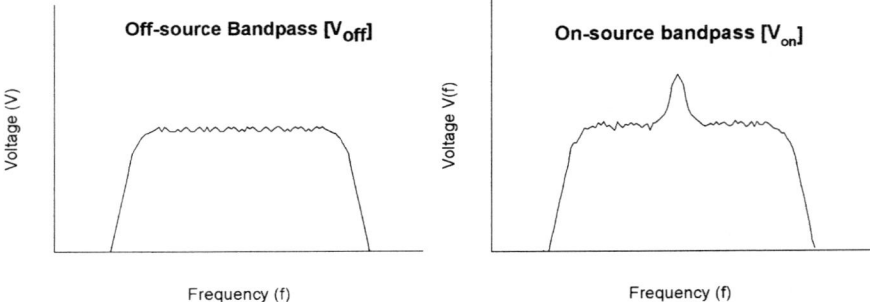

Figure 1. Off-source and on-source receiver bandpasses.

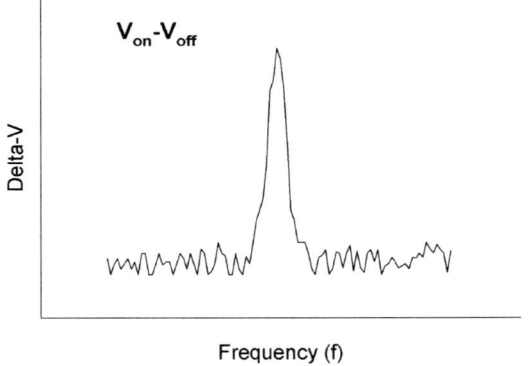

Figure 2. On-Off spectral bandpass.

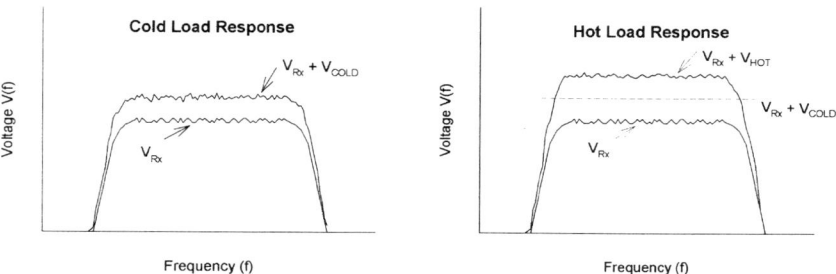

Figure 3. Cold-load and hot-load system responses.

$$\Delta V = V_{on} - V_{off}. \tag{1}$$

This resultant response contains the astronomical information we wish to measure, but will be wholly uncalibrated (raw voltage or computer counts). To turn this into something that is astronomically meaningful, we must first perform a temperature calibration. In some cases, this can be done by scaling against a source with known brightness temperature or flux density, but most often we go through the intermediate steps of establishing an antenna temperature scale. This is done by using a broadband source of known temperature and injecting its emission into the feed or waveguide in front of the first amplifier or mixer. Fundamental calibration is usually established by putting hot and cold loads of known temperature, such as ambient temperature and liquid nitrogen loads, in front of the feed. This produces bandpasses as shown in Figure 3.

A scale factor g that gives a "Kelvins per Volt" conversion scale is given by:

$$g = \frac{T_{hot} - T_{cold}}{[V_{Rx} + V_{hot}] - [V_{Rx} + V_{cold}]}. \tag{2}$$

The receiver noise temperature can also be computed as part of this same process, and is given by the "Y-Factor" equation

$$T_{Rx} = \frac{T_{hot} - Y T_{cold}}{Y - 1}, \tag{3}$$

where

$$Y = \frac{V_{Rx} + V_{hot}}{V_{Rx} + V_{cold}}, \tag{4}$$

i.e., Y is the ratio of the voltage response when the hot and cold loads are inserted in front of the feed. Once this conversion factor is established, a stable secondary calibration source such as a pulsed noise diode or other blackbody load can be calibrated once, and used thereafter to measure the system temperature of the receiver. When the secondary calibration source is on, the bandpass will change as shown in Figure 4.

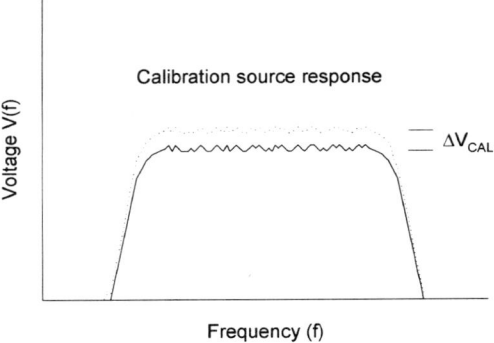

Figure 4. Calibration noise diode system response.

The system temperature is then given by the equation

$$V_{sys} = V_{Rx} + V_{sky}, \qquad (5)$$

$$\frac{V_{sys} + V_{cal}}{V_{sys}} = \frac{T_{sys} + T_{cal}}{T_{sys}} = \frac{V_{cal\_on}}{V_{cal\_off}} \qquad (6)$$

and

$$T_{sys} = T_{cal} \cdot \frac{V_{sys}}{(V_{sys} + V_{cal}) - V_{sys}} = V_{sys} \frac{T_{cal}}{\Delta V_{cal}}. \qquad (7)$$

Normally, signal processing is given by the equation

$$\Delta T_{on-off} \equiv T_{on} - T_{off} \equiv T_A \qquad (8)$$

or

$$T_A = \frac{V_{on} - V_{off}}{V_{off}} T_{sys}. \qquad (9)$$

where the ratio is used because T_{sys} is often measured separately. In the equations above, T_A is the antenna temperature of the source, V_{on} is the "on source" voltage and is the sum of source, sky, and system responses; V_{off} is the "off source" voltage and is the sum of sky and system responses ($V_{off} = V_{sys}$). For noise diode calibration, note that the term for the difference between "cal on" and "cal off" is just the calibration value itself, since the atmosphere and receiver noise are present in both phases:

$$V_{cal\_on} = V_{cal} + V_{Rx} + V_{sky} \qquad (10)$$

and

$$V_{cal\_off} = V_{Rx} + V_{sky}. \tag{11}$$

At millimeter wavelengths, waveguide noise devices are not available, at least at the shorter wavelengths. To overcome this problem, the earliest millimeter-wave receivers utilized rotating chopper wheels that alternately placed the sky and an ambient temperature blackbody load in front of the receiver feed. When the blackbody load is in front of the feed, it blocks the emission from the sky, and this changes the signal processing equations compared to those presented above; this is the essence of the difference between centimeter- and millimeter-wave practices:

$$\Delta T_{on-off} = (V_{on} - V_{off}) \frac{T_{cal}}{\Delta V_{cal}}, \tag{12}$$

where $T_{cal} = T_{amb}$, i.e., the temperature of the hot load. The difference between load and sky is

$$\Delta T_{cal} = g\Delta V_{cal} = T_{load} - T_{off} =$$
$$= [T_{Rx} + T_{hot}] - [T_{Rx} + \eta T_{atm}(1 - e^{-\tau}) + (1 - \eta)T_{spill}], \tag{13}$$

where η is the fraction of antenna power that reaches the sky. If

$$T_{hot} \cong T_{atm} \cong T_{spill} \cong T_{amb}, \tag{14}$$

then

$$\Delta T_{cal} = \eta T_{amb} e^{-\tau} \tag{15}$$

and

$$\Delta T_{on-off} = (T_{on} - T_{off}) \frac{1}{\eta e^{-\tau}} = \frac{T_A}{\eta e^{-\tau}} \cong T_A*, \tag{16}$$

where T_A* is an approximate, "corrected" antenna temperature.

Chopper wheel calibration is thus seen to have a very beneficial property: it makes a first order correction to signal attenuation by the atmosphere. We can see qualitatively why the chopper wheel method has this property by writing the signal processing equation as follows:

$$T_A* = (T_{on} - T_{off}) \frac{T_{cal}}{T_{load} - T_{sky}}. \tag{17}$$

Since the denominator of the cal term involves the difference (T_{load} - T_{sky}), when the sky opacity gets larger, T_{sky} increases and the difference gets smaller, thus scaling up the measured antenna temperature. The chopper wheel method is thus self-compensating for atmospheric attenuation, to first order.

Although these equations have great beauty in their simplicity, it was soon realized that the approximations made in this treatment introduced some significant calibration errors, and that there were also other antenna loss terms that should be included. The process of the refinement of the technique is quite interesting and can be followed in the sequence of papers from Penzias & Burrus (1973), Davis & Vanden Bout (1973), Ulich & Haas (1976), to Kutner & Ulich (1981). The method is, in fact, still being refined (e.g. Bock et al. 1998; Guilloteau et al. 2001). Although the techniques of millimeter wave calibration can be built up in historical sequence, it will be most useful to describe calibration techniques as we currently use them.

2. Antenna Losses

In centimeter-wave calibration, antenna efficiencies are usually lumped into global terms such as the overall beam and aperture efficiencies. These include all the losses outside the main diffraction beam of the telescope. Millimeter-wave calibration requires that efficiency terms be separated into components because they affect calibration in different ways that will become apparent later as we develop the methodology. Nevertheless, the efficiency terms used for centimeter and millimeter-wave calibration can ultimately be related to one another. Antenna losses occur in several categories – diffraction beam response, reflection losses, illumination losses, and finally, the way that the forward beam response couples with the astronomical source. These loss and coupling factors are discussed in the following sections. This treatment largely follows that of Kutner and Ulich (1981), with additional definitions from Baars (1973).

2.1. Diffraction losses

All antennas and feeds have diffraction patterns owing to their finite aperture within the incoming wavefront. Diffraction lobes arise from the edges of the aperture, and a circular aperture will give rise to a circular Airy diffraction pattern. In addition, diffraction arises from the feedlegs and support structures, unless the antenna has an unblocked aperture such as the Green Bank Telescope (GBT). Feedleg diffraction generally appears as a clover-leaf pattern. Diffraction can also arise from the gaps between the surface panels, and any other part of the reflective surface path where an edge or obstruction exists. In general, the amount of telescope gain or power in the sidelobes relative to the total forward gain is relatively low. However, if a sidelobe falls on strong emission from the sky, it can be detected and is a source of "stray radiation" response (see Lockman in this volume). This response is most damaging for 21 cm HI observations since strong HI is ubiquitous in the Galaxy, but can also be important for extended emission of carbon monoxide and a few other strong species at millimeter wavelengths.

For most millimeter wave observations, exact knowledge of the position and strength of the diffraction sidelobes is not essential, and the power lost through them is included in one of the source-beam coupling terms, η_M or η_{MB}, which will be discussed below.

In addition to diffraction effects, there are also antenna losses and noise contributions that arise from imperfect reflection of the signal. These are ohmic losses, blockage, scattering, and error beam response.

Ohmic loss. This arises from the fact that an antenna mirror is not a perfect reflector so that a fraction of the radiation is absorbed by the surface and dissipated through resistive heating. For a single reflecting surface, this term is typically much less than 1%. However, every reflection in the optical path will have an associated ohmic loss, and in systems using complicated, bent optics involving many reflections, the term can build up to significant levels. Ohmic loss can be minimized by polishing mirrors to very smooth surfaces. Ohmic loss terminates at the ambient temperature of the reflectors. Ohmic loss is formally defined as

$$\eta_r \equiv \frac{G}{4\pi} \iint_{4\pi} P_n(\Omega) d\Omega, \qquad (18)$$

where G is the gain of the antenna and P_n is the normalized power pattern of the antenna.

Blockage. Any blockage with the optical path will also lead to a loss of sky signal and emission at ambient temperature. The magnitude of these terms correspond to the fraction of area of the blocking structures relative to the total aperture. The most notable blocking structures are the feed legs, subreflector, and subreflector support structure in symmetrically illuminated antennas. The GBT does not suffer from this problem owing to its unblocked aperture. In addition to direct blockage, radiation from the ground and other parts of the telescope may scatter off telescope structures into the beam of the antenna. Ohmic and scattering losses are also sources of ambient temperature noise. *In general, any antenna power that falls away from the main beam is both a loss and a noise source.*

Error beam response. The final reflection loss, the error beam response, is quite important because it gives rise to a broad beam centered on the main diffraction beam of the antenna that may couple with emission outside the main beam. Practical antennas depart from the theoretical geometrical figure of the reflecting surface (spherical in the case of the Arecibo main reflector, and parabolic in the case of the GBT main reflector). The RMS deviations from a perfectly figured reflector take power away from the peak response of the main beam and give rise to the familiar aperture efficiency relationship

$$\eta_A = \eta_o \exp\left[-\left(\frac{4\pi\sigma}{\lambda}\right)^2\right], \qquad (19)$$

where η_o is the aperture efficiency of a perfect reflector with the given illumination properties, and σ is the RMS of surface deviations.

The gain of the antenna is closely related to this, and is known as the Ruze Equation:

$$G = \eta_o \left(\frac{\pi D}{\lambda}\right)^2 \exp\left[-\left(\frac{4\pi\sigma}{\lambda}\right)^2\right]. \qquad (20)$$

The forward power that results from surface irregularities goes into a broad error pattern, that was first characterized by Ruze (1966), and cast in the form presently used by Baars (1973). From both theoretical analysis and actual measurements, Ruze found that if the dish errors followed Gaussian statistics, the error beam was typically a broad, Gaussian beam of FWHM

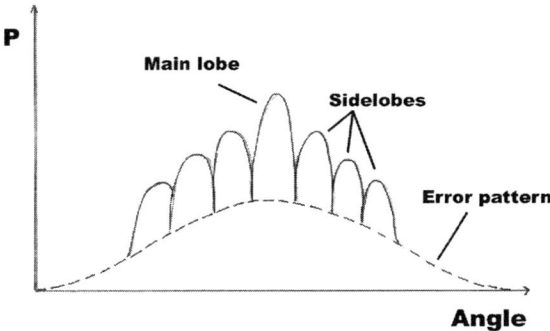

Figure 5. Schematic total forward beam pattern of an antenna.

$$\theta_E = 2(\ln 2)^{1/2} \frac{\lambda}{\pi c_\sigma} \qquad (21)$$

and relative amplitude

$$\frac{A_E}{A_M} = \frac{1}{\eta_o} \left[\frac{2c_\sigma}{D}\right]^2 \left\{\exp\left(\frac{4\pi\sigma}{\lambda}\right)^2 - 1\right\}, \qquad (22)$$

where c_σ is the correlation scale size of the errors, and A_M is the amplitude of the main beam. The correlation scale size arises from the fact that surface errors often have a characteristic scale size, which may correspond, for example to the size of the surface panels or some other structural element. The total forward beam of the antenna may be similar to that of Figure 5.

2.2. Illumination losses

In reflecting antenna systems, a receiver feed illuminates either the primary reflector (prime focus system) or a secondary mirror (Cassegrain, or Gregorian systems), or series of mirrors (e.g., bent Cassegrain and Nasmyth systems). Feeds can be designed with an illumination pattern or taper to provide nearly uniform illumination or a tapered illumination (typically 10-12 dB down at the edge). Uniform illumination gives the highest possible aperture efficiency, but at the cost of higher sidelobes. Most single dish radio telescopes use a tapered feed to reduce sidelobe response, whereas interferometers often opt for more uniform illumination since the beam response is usually deconvolved from the image anyway.

Regardless of the feed taper, there is almost always some radiation spillover past the edge of the dish (it is easiest to visualize this in a transmitting system, but by reciprocity, the same effect exists in a receiving system). Prime focus

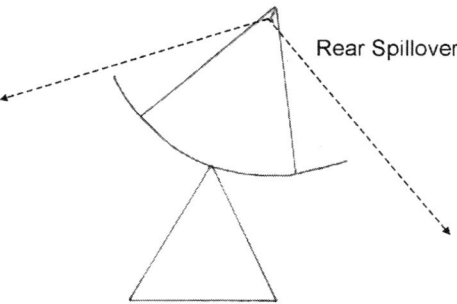

Figure 6. Prime focus antenna with rear spillover.

systems will have rear spillover, which is defined by the following equation and illustrated in Figure 6:

$$\eta_{rss} \equiv \frac{\int\int_{2\pi(forward)} P_n(\Omega)d\Omega}{\iint_{4\pi} P_n(\Omega)d\Omega} , \qquad (23)$$

η_{rss} is the rear spillover and scattering efficiency and the integral in the numerator is over the forward hemisphere. Systems in which the receiver is at a secondary focus will have both rear and forward spillover, as defined in the following equation and illustrated in Figure 7

$$\eta_{fss} \equiv \frac{\iint_{\Omega_d} P_n(\Omega)d\Omega}{\iint_{2\pi} P_n(\Omega)d\Omega} .$$

In this equation, Ω_d is the diffraction zone around the source, and the integral in the denominator is over the forward hemisphere. Rear spillover and scattering is usually terminated at ambient temperature, and is sometimes referred to as "warm spillover." Forward spillover and scattering typically terminates on the sky, and is sometimes called "cold spillover." Note that when the telescope is observing at low elevation angles, some rear spillover may actually terminate on the sky, and some forward spillover on the ground, as illustrated in Figure 8. This fact is often neglected, but might be important in projects seeking the best calibration accuracy.

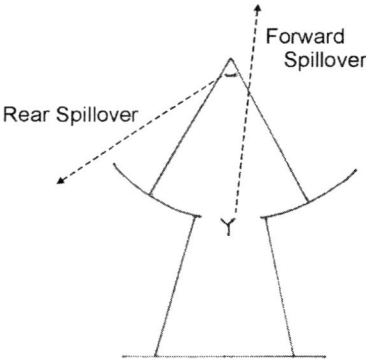

Figure 7. Secondary focus antenna with forward and rear spillover.

The two loss terms that terminate at ambient temperature, η_{rss} and η_r (ohmic loss) are usually grouped into a single efficiency, η_l, defined as

$$\eta_l = \eta_r \eta_{rss}. \tag{24}$$

To understand the relationship between the various efficiency and loss terms, the graphical representation in Figure 9 may be helpful.

2.3. Relationships between mm-wave and conventional efficiency and source coupling terms

For a source brightness temperature T_B

$$T_{A^*} = \eta_{fss} \eta_c T_B, \tag{25}$$

where η_c is a source coupling efficiency in the diffraction zone Ω_d around the pointing axis (boresight) which contains the main beam, error beam, and near sidelobes, and is given by:

$$\eta_c \equiv \frac{\int\int_{\Omega_{source}} P_n(\Psi - \Omega) B_n(\Psi) d\Psi}{\iint_{\Omega_d} P_n(\Omega) d\Omega}, \tag{26}$$

where B_n is the normalized source brightness distribution. For sources smaller than the main beam that do not couple with the error beam or sidelobes, it is useful to define a corrected main beam efficiency $\eta_M{}^*$. In terms of conventional beam efficiency η_B,

$$\eta_M{}^* = \frac{\eta_B}{\eta_l \eta_{fss}}. \tag{27}$$

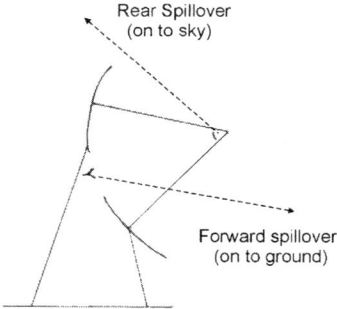

Figure 8. Low elevation operation in which rear spillover terminates on the sky and forward spillover on the ground.

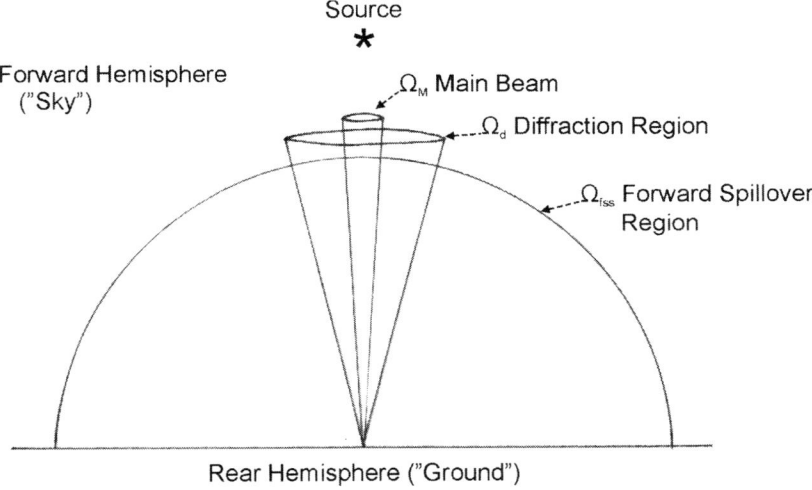

Figure 9. Schematic representation of beam areas.

The main beam brightness temperature is thus defined as

$$T_{MB} \equiv \frac{T_A*}{\eta_{fss}\eta_M*} = \frac{T_A}{\eta_B e^{-\tau}} \qquad (28)$$

and is the source brightness distribution convolved with the main diffraction beam of the antenna.

3. Atmospheric Effects on Calibration

3.1. Atmospheric absorption

Atmospheric water vapor (H_2O) and molecular oxygen (O_2) attenuate astronomical signals and emit noise. Water vapor emission is highly variable, depending on prevailing weather conditions. The amount of water vapor in the sky is often parameterized by the "precipitable water vapor" (PWV), which is the amount of water, usually expressed in millimeters, in a column from the surface of the Earth through the atmosphere. In good weather conditions at a high, dry site such as Mauna Kea or the Atacama Desert, the PWV can be 0.5 to 1 mm.

There are numerous absorption lines of H_2O and O_2 in the atmosphere. These lines are typically pressure-broadened, and the wings can have great extent in frequency. Extinction is expressed in nepers (τ), and actual attenuation is given by $e^{-\tau}$. Transmission is inversely related to absorption as $1 - e^{-\tau}$.

Assuming a plane parallel atmosphere, the absorption at an arbitrary zenith angle is given by the "secant z" term,

$$Attenuation = e^{-\tau} = e^{-\tau_o \sec(z)} = e^{-\tau_o/\sin(el)} = e^{-\tau_o A}, \qquad (29)$$

where τ_o is the zenith optical depth, A is the number of airmasses (=sec(z)), and z is the zenith distance, or (90° − elevation angle).

3.2. Sky noise equation

The antenna temperature from the sky is given by:

$$T_A(sky) = T_{Rx} + \eta_l T_{atm}(1 - e^{-\tau}) + (1 - \eta_l)T_{spill} + \eta_l T_{CBR} e^{-\tau}, \qquad (30)$$

where T_{atm} is the mean temperature of the atmosphere, T_{spill} is the effective spillover temperature, and T_{CBR} is the cosmic background radiation ($\sim$2.7 K). In general, the temperatures should be computed as Rayleigh-Jeans effective temperatures, as described below.

3.3. Millimeter-wave calibration formalism

Having defined antenna efficiency terms and the effects of the atmosphere, we can now return to the development of an accurate calibration scale factor for use with chopper wheel calibration. The treatment given here is that of Ulich and Haas (1976) as revised by Kutner and Ulich (1981).

To begin, we define a Rayleigh-Jeans-equivalent source radiation temperature based on the Planck blackbody equation

$$J(\nu,T) = \frac{h\nu/k}{\exp(h\nu/kT) - 1}. \tag{31}$$

This expression avoids the errors introduced by the Rayleigh-Jeans approximation, which can become significant at millimeter-wave frequencies.

Proper treatment of the chopper wheel calibration method requires that we take into account the response of the receiver and the sky in both the signal and image sidebands, and the fact that

$$T_{atm} n_e T_{spill} n_e T_{chop} n_e T_{amb}, \tag{32}$$

i.e., no approximations to the temperatures. The defining relations of the derivation are

$$\Delta T_{source} = T_{source} - T_{sky} = T_{on} - T_{off} \tag{33}$$

and

$$T_A^* \equiv \frac{\Delta T_{source}}{G_s \eta_l \exp(-\tau_s A)}, \tag{34}$$

where G_s and G_i are the signal and image sideband gains, respectively, and have a normalized sum

$$G_i + G_s = 1. \tag{35}$$

In terms of signal processing,

$$T_A^* = \Delta T_{source} \frac{T_c^*}{\Delta T_{cal}}, \tag{36}$$

so that

$$T_c^* = \frac{\Delta T_{cal}}{G_s \eta_l \exp(-\tau_s A)}, \tag{37}$$

where

$$\Delta T_{cal} = T_{load} - T_{sky}. \tag{38}$$

The load effective temperature is given by the sum of contributions from both sidebands:

$$T_{load} = G_s J(\nu_s, T_{chop}) + G_i J(\nu_i, T_{chop}). \tag{39}$$

Likewise, T_{sky} is the effective temperature of the sky with contributions from both sidebands:

$$T_{sky} = G_s \{\eta_l J(\nu_s, T_{atm})(1 - e^{-\tau_s A}) + (1 - \eta_l) J(\nu_s, T_{spill}) + \eta_l J(\nu_s, T_{CBR}) e^{-\tau_s A}\}$$
$$+ G_i \{\eta_l J(\nu_i, T_{atm})(1 - e^{-\tau_i A}) + (1 - \eta_l) J(\nu_i, T_{spill}) + \eta_l J(\nu_i, T_{CBR}) e^{-\tau_i A}\}. \tag{40}$$

Substituting the expanded expression for ΔT_{cal} into the equation

$$T_c* = \frac{\Delta T_{cal}}{G_s \eta_l \exp(-\tau_s A)}, \qquad (41)$$

we find that

$$\begin{aligned}T_c* =\ & \left(1 + \frac{G_i}{G_s}\right)[J(\nu_s, T_{atm}) - J(\nu_s, T_{CBR})] \\ & + \left(1 + \frac{G_i}{G_s}\right)[J(\nu_s, T_{spill}) - J(\nu_s, T_{atm})]\, e^{\tau_s A} \\ & + \frac{G_i}{G_s}[J(\nu_s, T_{atm}) - J(\nu_s, T_{CBR})]\left(e^{(\tau_s - \tau_i)A} - 1\right) \\ & + \frac{1}{\eta_l}\left(1 + \frac{G_i}{G_s}\right)[J(\nu_s, T_{chop}) - J(\nu_s, T_{spill})]\, e^{\tau_s A},\end{aligned} \qquad (42)$$

where we have assumed only that

$$J(\nu_s, T) = J(\nu_i, T), \qquad (43)$$

for all temperatures T, i.e., it is assumed that the frequency separation of the sidebands is small relative to the observing frequencies and that the equivalent radiation temperatures are approximately the same in both sidebands.

In addition to the T_A* scale, the T_R* scale is also frequently used, as defined by Kutner and Ulich (1981). The two scales are simply related by

$$T_R* \equiv \frac{T_A*}{\eta_{fss}}. \qquad (44)$$

The T_R* scale has the virtue that it contains all the correction factors one can make without knowledge of the source-beam coupling coefficients. For extended sources, $T_R* \approx T_R$, the source radiation (brightness) temperature. The forward spillover efficiency appears only as a final scaling factor, however, so the use of the scale is a matter of preference and convention. The NRAO 12-m telescope used the T_R* scale for many years, but most other millimeter-wave observatories use T_A*. It is important to understand the conventions and definitions in use at a particular observatory, since nomenclature is sometimes mixed.

4. Continuing Refinements to the Technique

The calibration accuracy of the chopper wheel technique is typically at the 5-10% level. Accuracy can be improved by using a two load system with known temperatures, and this is employed at a number of observatories (e.g., the JCMT). The use of two loads allows the receiver and sky noise to be separated, which can be used to calculate an atmospheric model and/or a refined value of the atmospheric optical depth. This can improve the relative calibration accuracy to be as good as $\sim 1\%$ (Mangum 2000).

Bock et al. (1998) have suggested an alternative calibration scheme based on a dual temperature chopping load in the center of the subreflector. This method is more akin to centimeter-wave direct calibration with a pulsed noise source in which sky noise is present in the calibration phases. Atmospheric attenuation is measured by tipping curves. Owing to the two-temperature loads, the calibration can, in principle, be accurate to the $\sim 1\%$ level.

Plambeck (2000) and Guilloteau et al. (2001) have suggested the use of a semi-transparent vane (or vanes) for millimeter-wave calibration. This technique utilizes chopper wheel calibration formalism. The advantage of the technique is that the semi-transparent vanes do not run the risk of saturating the amplifiers, which can happen when a ~ 300 K hot load is switched in front of a sensitive SIS mixer. This technique has not been tested with an actual receiver, but an analysis of the propagation of errors suggests that it also can produce accuracy at the $\sim 1\%$ level.

5. Measuring Efficiency Factors and Absolute Calibration

For chopper wheel calibration, absolute calibration is established by measuring the efficiency factors η_l and η_{fss}, and relevant source coupling efficiencies. η_l can be measured from fits of atmospheric tipping curves. Tipping measurements are made at several elevation angles, and the measured antenna temperature values are fit with the atmospheric emission equation

$$T_A(sky)_i = T_{Rx} + \eta_l J_{atm}(\nu, T_{atm}) \left[1 - e^{-\tau_o/\sin(El_i)}\right] +$$
$$(1 - \eta_l) J_{spill}(v, T_{spill}) + \eta_l J_{CBR}(v, T = 2.74) e^{-\tau_o/\sin(El_i)} , \qquad (45)$$

for elevation i (El_i). If the fit is made with both τ_o and η_l as free parameters, then η_l is determined. An accurate measurement of η_l requires good calibration of the antenna temperature at each elevation angle and a uniform atmosphere. Typical values of η_l might range from as high as 0.98 for a telescope with minimal blockage and spillover to 0.85 for a telescope with considerable blockage or rear spillover.

The forward spillover efficiency is usually determined by observing the Moon. The Moon is large enough to subtend the entire forward beam area, including the main diffraction beam, near sidelobes, and, usually, the error beam. Best calibration is achieved at new Moon in which the entire surface temperature of the Moon is uniform.

The corrected main beam efficiency $\eta_M{}^*$ can be measured by observations of a planet. The approximate brightness temperatures of the planets are given in Table 1 (Mangum 1999 and references therein).

Updated values for these brightness temperatures appear frequently in the literature, which should be consulted. For a disk-like source such as a planet, and a Gaussian antenna beam, the coupling efficiency is

$$\eta_c(disk) = 1 - \exp\left\{-4\ln 2 \left(\frac{r}{\theta_M}\right)^2\right\} , \qquad (46)$$

Table 1. Approximate planetary brightness temperatures at 90 and 227 GHz.

Planet	T_B (90 GHz) (K)	T_B (227 GHz) (K)	Planet Unit Semi-Diameter (arcsec)
Venus	367±10	317±30	8.34
Mars	207±6	207±7	4.68
Jupiter	172±1	171±4	95.20
Saturn	149±4	140±14	78.15
Uranus	135±4	98±5	35.02
Neptune	130±5	93±5	33.50

where r is the angular radius of the disk and θ_M is the FWHM of the main beam. The flux density is related to antenna temperature through the equations

$$S_v = \frac{2kT_A}{\eta_A A_p} , \qquad (47)$$

using the conventional definitions, and

$$S_v = \frac{\eta_l T_A*}{\eta_A} \frac{2k}{A_p} , \qquad (48)$$

using the millimeter-wave T_A* definition.

The flux density of a standard source such as a planet that is partially resolved by the main diffraction beam must be appropriately adjusted.

References

Baars, J. W. M. 1973, IEEE Antennas & Propagation, AP-21, 461
Bock, D., Welch, J., Fleming, M., & Thornton, D. 1998, ALMA Memo 225
Davis, J. H. & Vanden Bout, P. 1973, ApJ, 15, 43
Guilloteau, S. & Moreno, R. 2001, ALMA Memo 371, NRAO
Kutner, M. L. & Ulich, B. L. 1981, ApJ, 250, 341
Mangum, J. G. 1999, NRAO 12 Meter Users Manual,
    http://www.tuc.nrao.edu/12meter/obsinfo.html
Mangum, J. G. 2000, ALMA Memo 318, NRAO
Penzias, A. A. & Burrus, C. A. 1973, ARA&A, 11, 51
Plambeck, R. L. 2000, ALMA Memo 321, NRAO
Ruze, J. 1966, Proc. IEEE, 54, 633
Ulich, B. L. & Haas, R. W. 1976, ApJS, 30, 247

Reduction and Analysis Techniques

Ronald J. Maddalena

National Radio Astronomy Observatory, P. O. Box 2, Green Bank, West Virginia 24944, USA

Abstract. Single-dish observations can be made in myriad of ways with each observing technique almost always requiring different kinds of data analysis. This chapter will cover the basic continuum and spectral line analysis algorithms common to all single-dish data analysis packages. However, I will not cover the very specialized fields of polarimetry, pulsar, or radar data reduction. In the case of continuum observations the student will learn the steps used to derive the flux of a point source as well as the more common data analysis techniques for generating and analyzing maps of extended sources. For spectral line data, I will concentrate on the analysis algorithms usually applied to single spectra (bandpass and velocity calibration, data averaging and smoothing, baseline fitting, component fitting,...) and how to produce and analyze spectral-line data cubes.

1. Introduction

Reducing and analyzing single-dish data is similar to reducing other kinds of astronomical data in that it is a time consuming process. Typically, one spends more time analyzing data than taking it. And, in the same way that the reduction of X-ray data differs from the reduction of optical spectra, each type of single-dish data requires different reduction algorithms.

A full description of the gamut of single-dish data reduction techniques would require a textbook a few times the size of the manuals that accompany single-dish software packages. In this introductory lecture to single-dish analysis algorithms, I will only have space to concentrate on the common algorithms found in most packages. I will provide the basics so that the reader will be able to effectively start using any analysis system. No attempt will be made to describe the use of any individual software package nor will I give any description of pulsar, polarimetry, or radar analysis techniques.

In the following, I will first list some of the more common data analysis systems for single-dish telescopes (Section 2). I will then discuss the analysis of continuum (Section 3) and spectral line (Section 4) observations. Many of the spectral-line algorithms explained in Section 4 require an understanding of the continuum algorithms presented in Section 3. Section 5 gives my thoughts on the future of single-dish data analysis. An appendix provides definitions of the symbols used throughout.

330 Maddalena

2. Single-Dish Data Analysis Packages

Each telescope or observatory typically has its own single-dish data reduction system. Users of a telescope should be prepared to learn a new analysis system whenever they observe on a different telescope.

The following table is an incomplete list of single-dish analysis packages and telescopes where they are used.

Table 1. Common single-dish analysis packages

Package	Telescopes	Web Address
Aips	NRAO 12m	www.cv.nrao.edu/aips
Aips++	NRAO 100m	www.aips2.nrao.edu/docs/aips++.html
Analyz	NAIC 305m	www.naic.edu/menuimag/astronomy.htm
CLASS	IRAM 30m & others	www.iram.fr/doc/class/class.html
IDL	NAIC 305m & others	www.idlastro.gsfc.nasa.gov/homepage.html
SPECX	JCMT	www.jach.hawaii.edu/JACpublic/JCMT/ User_documentation/SPECX/part6/node2.html
UniPops	NRAO 43m & 12m	www.cv.nrao.edu/unipops

Each of the above systems differs in the powers and algorithms provided. Since the same class of analysis algorithms in various packages will make different simplifying assumptions, the user cannot expect fully rigorous algorithms in all packages. Some packages have a steep learning curve while others will be easy to master. Discussions of the merits and flaws of different analysis systems, like discussions on religion and politics, are usually very fervent.

2.1. Common traits of different analysis packages

Although differences between analysis systems are major, one can still find some commonalities. For example, almost all still use command line interfaces with graphical user interfaces still relegated to future development. Most analysis systems have some scripting ability to package often-used commands for batch processing.

The structure of the data that each telescope produces, although different in its details, consists not only of a vector or array of data values but also information on how the data were taken, the telescope environment (e.g., weather conditions), and a record of the time when the data were taken. Some analysis algorithms often will use some of this extra information.

Most analysis systems have rudimentary data base management systems that allow the user to peruse the collected data and specify selection criteria that the program will use to define a data set to be processed. Some analysis systems are tied into telescope control systems so that it is possible to look at and analyze data in near real time.

All of the most recent systems both read and write FITS files. FITS files should make it simple to export and import data between systems; however, to this day, the exchange of data between some systems still remains painful if at all possible.

Packages that reduce continuum or line data but not both are no longer prevalent. Many systems no longer just analyze one-dimensional data vectors

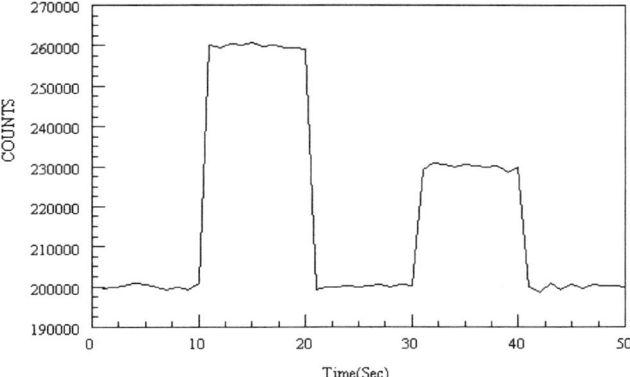

Figure 1. A typical On-Off observation before any data reduction or calibration. The first rise in power is due to a source; the second is due to the turning on and of the noise diode or receiver calibration source. From such an observation, one can determine the source flux relative to the calibration source as well as the system temperature.

but can analyze and visualize two-dimensional matrices and spectral-line data cubes. The palette of tools presented to the user is almost the same from system to system, only the names of the commands and details about the algorithms they use are different.

3. Continuum Data Reduction Techniques

The analysis of continuum observations depends somewhat on whether the object observed is a point or extended source. Many of the techniques used in the analysis of point sources will come in handy when dealing with extended sources. Understanding continuum data reduction will facilitate learning how to reduce spectral-line data.

3.1. Continuum observations of point sources

The analysis of observations of point sources will provide us with the rudiments of how one should reduce more complicated types of data. Continuum observations of point sources can be made in many ways of which I will discuss only three.

On-Off Continuum Observations: In On-Off observations, the telescope records data when the telescope is on the source's position, and then a few beam widths off the source. No data are recorded when the telescope is moving between the signal and reference positions. Such observations are useful for measuring the flux of a source when the pointing of the telescope is well known and the frequency low enough that gain or atmospheric variations are of no concern. In essence, it is only successfully used by cm-wave telescope.

The difference in power detected by the telescope between the signal and reference position can then be compared to the difference in power between when the noise diode or receiver calibration source is on and off (Figure 1). In many analysis systems, one will find a command to automatically reduce On-Off observations.

The user should be warned that most algorithms make some simplifying assumptions about the observing method and the object observed. If one assumes:

- narrow bandwidths,

- linear power detector,

- source intensity that is a small fraction of the system temperature,

- noise diode temperature that is a small fraction of the system temperature,

- equal time on the reference and source position,

- equal time with the noise diode on and off, and

- insignificant backend blanking time between reference and signal phases and between calibration on and off phases,

then the antenna temperature of the source is given by:

$$T_{\text{src}} = \frac{T_{\text{cal}}}{P_{\text{cal\_on}}^{\text{ref}} - P_{\text{cal\_off}}^{\text{ref}}} \cdot \frac{\left(P_{\text{cal\_on}}^{\text{sig}} + P_{\text{cal\_off}}^{\text{sig}} - P_{\text{cal\_on}}^{\text{ref}} - P_{\text{cal\_off}}^{\text{ref}}\right)}{2} \quad . \quad (1)$$

The resulting data has a theoretical noise of:

$$\sigma_{T_{\text{src}}} = \left(T_{\text{sys}}/\sqrt{\text{BW}}\right) \cdot \sqrt{\frac{1}{t^{\text{ref}}} + \frac{1}{t^{\text{sig}}}} , \quad (2)$$

where

$$T_{\text{sys}} = \frac{T_{\text{cal}}}{2} \cdot \left(\frac{\left(P_{\text{cal\_off}}^{\text{ref}} + P_{\text{cal\_on}}^{\text{ref}}\right)}{\left(P_{\text{cal\_on}}^{\text{ref}} - P_{\text{cal\_off}}^{\text{ref}}\right)} - 1\right) = \frac{T_{\text{cal}} P_{\text{cal\_off}}^{\text{ref}}}{\left(P_{\text{cal\_on}}^{\text{ref}} - P_{\text{cal\_off}}^{\text{ref}}\right)} \quad . \quad (3)$$

If any of the assumptions is not true, then the above equations need to be modified. A robust formalization of the above for the general case can be found at: http://www.gb.nrao.edu/~rmaddale/GBT/continuumCal.html.

T_{src} must be corrected for atmospheric attenuation and antenna efficiency, and then converted to flux density. The details on how to apply the corrections are covered in other lectures in this volume.

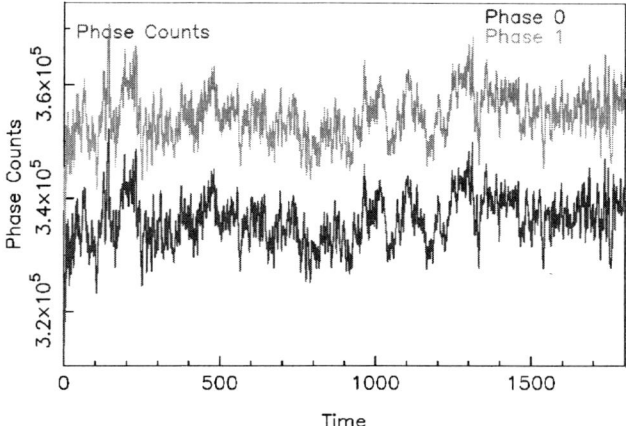

Figure 2. Typical beam-switched observation. Phase 0 data (darker color) are those with the beam of the telescope on blank sky while Phase 1 (lighter color) are those with the beam on the source. The fluctuations in both phases are extremely similar and typically due to atmospheric fluctuations. The difference between the phases will be proportional to the strength of the source and will have less fluctuation than either phase's data.

Beam-Switched Continuum Observations: On many single-dish telescopes that use Cassegrain or Gregorian optics, the secondary or other small mirror in the system will chop at a rate of a few Hz between two positions on the sky. The positions are usually labeled 'signal' and 'reference' (Figure 2). Or, a telescope receiver might have two feeds that point at slightly separated positions on the sky and a switch that toggles the detected signal between the signal and reference feeds.

The data are collected synchronously in that the data for the signal and reference positions are accumulated separately. Such beam-switched observations are useful for removing fluctuation in power due to the atmosphere at frequencies typically above 10 GHz.

The data are usually represented as a vector of sometimes hundreds of values. The x-axis of the vector is the time at which data were sampled. The y-axis will need to be converted from sampled powers into antenna temperatures (T_{ant}) for both the signal and reference phases of the observations. To derive T_{ant}, the power of each data sample is compared to the difference in power between when the noise diode or receiver calibration source is on and off. If we make similar assumptions as for On-Off observations, then for each data sample i:

$$T_{\mathrm{ant}}^{\mathrm{ref}}(i) = \left\langle \frac{T_{\mathrm{cal}}}{P_{\mathrm{cal\_on}}^{\mathrm{ref}}(i) - P_{\mathrm{cal\_off}}^{\mathrm{ref}}(i)} \right\rangle \cdot \frac{\left(P_{\mathrm{cal\_on}}^{\mathrm{ref}}(i) + P_{\mathrm{cal\_off}}^{\mathrm{ref}}(i)\right)}{2}, \quad (4)$$

$$T_{\text{ant}}^{\text{sig}}(i) = \left\langle \frac{T_{\text{cal}}}{P_{\text{cal\_on}}^{\text{sig}}(i) - P_{\text{cal\_off}}^{\text{sig}}(i)} \right\rangle \cdot \frac{\left(P_{\text{cal\_on}}^{\text{sig}}(i) + P_{\text{cal\_off}}^{\text{sig}}(i)\right)}{2}, \quad (5)$$

where the angle brackets represents an average over all data samples. Then,

$$T_{\text{src}} = \left\langle T_{\text{ant}}^{\text{sig}}(i) - T_{\text{ant}}^{\text{ref}}(i) \right\rangle. \quad (6)$$

T_{src} must next be corrected for atmospheric attenuation and antenna efficiency, and then converted to flux density.

On-the-Fly Continuum Observations: In on-the-fly (OTF) observing, the telescope slews across a source and simultaneously records data at a rate of up to a few samples per second (Figure 3). The extent of the slew is typically a few beamwidths. Such an observing technique is useful when one wants to determine the flux of the source or the telescope pointing and beam shape along a certain direction.

The data are usually represented as a vector of sometimes hundreds of points. The x-axis of the vector can be considered either the times or telescope positions at which data were sampled. The y-axis will need to be converted from sampled powers into antenna temperatures (T_{ant}) before one can derive source temperature, beam parameters, and telescope pointing offset.

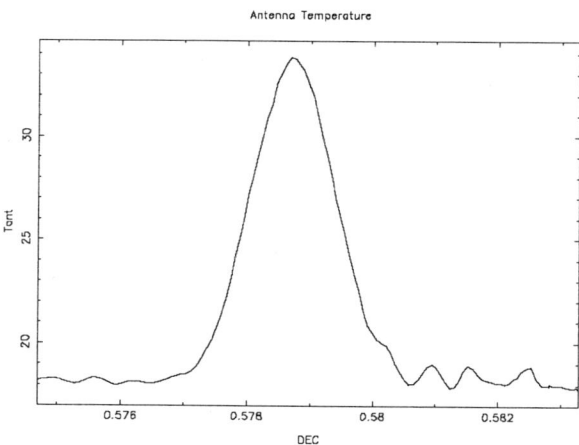

Figure 3. An on-the-fly observation through a source. The data have already been calibrated and converted to T_{ant}. The slew direction was in Declination, as implied by the x-axis of the plot.

To derive T_{ant}, the power of each data sample is compared to the difference in power between when the noise diode or receiver calibration source is on and off. If we make similar assumptions as for On-Off observations, then for each data sample i:

$$T_{\text{ant}}(i) = \left\langle \frac{T_{\text{cal}}}{P_{\text{cal\_on}}(i) - P_{\text{cal\_off}}(i)} \right\rangle \cdot \frac{(P_{\text{cal\_on}}(i) + P_{\text{cal\_off}}(i))}{2}. \tag{7}$$

Some telescopes combine on-the-fly observations with beam switching in which case we would use Equations 4 and 5 directly while Equation 7 becomes:

$$T_{\text{ant}}(i) = T_{\text{ant}}^{\text{sig}}(i) - T_{\text{ant}}^{\text{ref}}(i). \tag{8}$$

The following analysis steps can then be performed on $T_{\text{ant}}(i)$.

Averaging: If the same observation is repeated a number of times, you will want to average the data to reduce the noise. In some cases the observations could have been taken under various conditions like different integration times, different bandwidths, and different weather conditions. To produce an optimum average with the highest signal to noise, the data reduction system must perform a weighted average. The weights should be $1/\sigma^2$ where σ is either the theoretical rms (Equation 2) or measured rms of the data. The student should examine any textbook on statistics for details on weighted averages (e.g. Mandel 1964; Bevington & Robinson 1992).

However, there are problems with using either the theoretical or measured σ. The theoretical σ may be lower than the actual σ (due to such factors as receiver 1/f noise, atmospheric fluctuations) and the measured σ may be higher (due to slow drifts from things like changing ground pickup). It is usually up to the observer to determine which weights to use.

All analysis systems have statistical tools for determining the measured σ. The only thing the user needs to define is the source-free data samples to use in the calculation of σ.

After averaging, what should one then use as the σ of the observation? If you use the measured σ, then you should probably use the same statistical tools to remeasure the σ of the average. If you use the theoretical σ, the σ of the average will be:

$$\sigma_{\text{avrg}} = \frac{1}{\sqrt{\sum_j \frac{BW_j \cdot t_j}{Tsys_j^2}}} = \frac{1}{\sqrt{\sum_j \frac{1}{\sigma_j^2}}}. \tag{9}$$

Baseline Fitting: The derived $T_{\text{ant}}(i)$ include not only variations due to sources but any drifts in power due to changes in the atmosphere or ground pickup. The system temperature also introduces a DC offset to the data. Fitting a low-order polynomial to areas of the data thought to be devoid of emission eliminates most of these unwanted factors (Figure 4).

Baseline fitting involves defining in the analysis software the samples which one thinks are devoid of emission and the order of the polynomial to fit. After

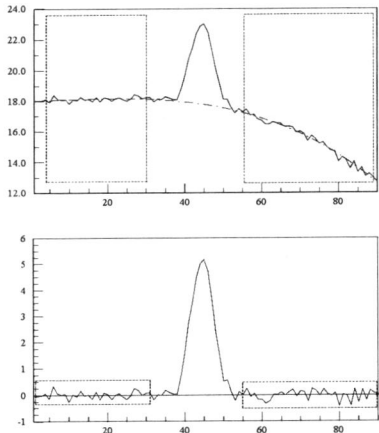

Figure 4. Illustrates baseline fitting to an on-the-fly observation. The top plot is the data before baseline removal and the bottom plot is the same data afterwards. Thin rectangles illustrate the regions thought to be devoid of emission that were used to fit a third order polynomial. The dashed line in the top plot is the fitted polynomial which, when subtracted from the data, produces the bottom plot.

fitting, the polynomial is subtracted from the data. See Press et al. (1992) or Bevington & Robinson (1992) for discussion on the details of polynomial linear least squares fit algorithms.

Unless the data have very good signal-to-noise or the baselines start out being exceptionally flat, fitting an arbitrary functional form such as a polynomial to your data will introduce biases in any estimates of source parameters or create false features (Rood, Bania, and Wilson 1984). Two observers fitting baselines to the same data can produce differences in measured source intensities that are larger than the rms calculated from Equation 2. Thus, baseline fitting introduces a random error to an observation, due to the uncertainty of the baseline determination (Lockman et al. 1986).

Gaussian Fitting: Once a baseline has been removed, one can then try to determine $T_{\rm src}$ by fitting a Gaussian to the residuals of the baseline fit (Figure 5). Note that most telescope beams are approximately Gaussian only between about the beam's half power points. Using data outside of the half power points will introduce systematic errors to the calculated source fluxes and estimated beam widths.

Most analysis systems will allow you to perform the following steps.

- Restrict the data points it will use for Gaussian fitting to those between the half power points.

- Define initial guess for the Gaussian half width, center, and height.

- Set a flag to fit or hold constant the half-width of the Gaussian (or any other Gaussian parameter). This flag should be used if one already knows well the width of the telescope beam.

- Number of iterations to use in the non-linear, least squares fitting routine before the fit will be claimed to be unsuccessful.

- Set the criteria that will be used to tell if the fit has converged. Usually this is the percentage decrease in χ-square between loops in the non-linear fitting routine.

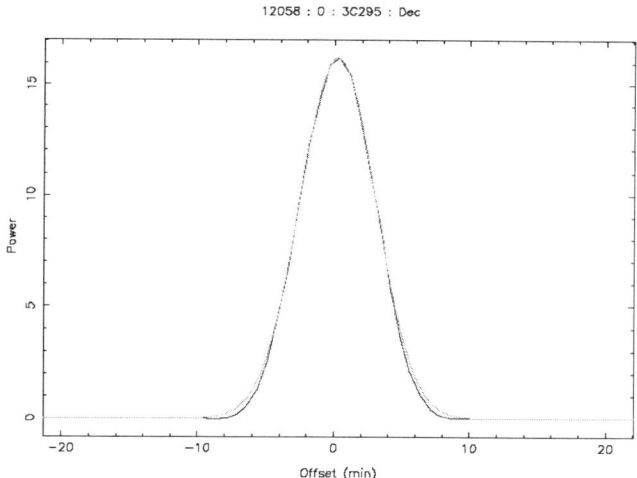

Figure 5. An on-the-fly observation taken with the Green Bank Telescope of the source 3C295. The excellent signal-to-noise of this observation illustrates how the data (dark line) and a Gaussian fitted to the data between the half-power points (thin line) differ in the wings of the Gaussian. Note that a baseline was subtracted from the data before fitting.

In addition to returning the values of the fitted parameters, most algorithms also return the χ-square of the fit and formal standard deviations of the fitted parameters. Quang-Rieu (1969) provides a discussion of the statistical errors inherent in Gaussian fitting. See Press et al. (1992) or Bevington & Robinson (1992) for discussion on the details of non-linear least squares fitting algorithms as is typically used for Gaussian fitting.

Template Fitting: Gaussian fitting is flawed in that beams are sometimes not symmetric due to optical aberrations like coma. And, Gaussian fitting can only be performed over the subset of the data collected between the half-power points of the beam. Template fitting is one way to avoid these flaws. It harks back to the old philosophy of fitting a physical model to your data instead of arbitrary functions.

If one has sufficient knowledge of the telescope, the beam of the telescope can be determined a priori. Or, an observer can average a large number of observations to produce a high signal-to-noise representation of the telescope beam. Some analysis systems allow you to fit a template or model of the beam to the collected data.

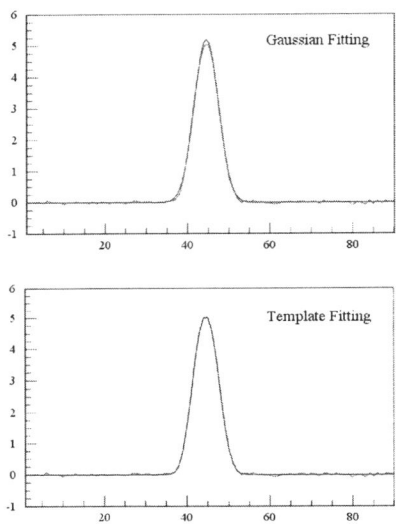

Figure 6. Comparison of Gaussian fitting (top) to template fitting for the same on-the-fly observation. Unlike the fit in Figure 5, the Gaussian fitting was across all of the data, not just between the half-power points. Note how the Gaussian fitting underestimates the intensity of the source and does not fit well the wings of the profile.

The standard algorithm usually convolves the template with the data to determine the x-axis offset of the data. The data are shifted by the offset usually by taking the Fourier transform of the data, adding a phase shift corresponding to the offset, and inverse Fourier transforming the data (Press et al. 1992). Then, the algorithm performs a linear least-squares fit of the template to the data: $T(i) = A \cdot \text{Template}(i) + B$ where A corresponds to the fitted scaling of template to the data and B to the fitted DC-offset between the template and the data. One implementation of template fitting can be found at http://aips2.nrao.edu/docs/user/Dish; see Van Zee et al. (1997) for a typical use of template fitting.

If the template is an adequate representation of the data, the results of a template fit will have fewer systematic problems. The results will be more accurate than Gaussian fitting since all of the data will be used in the fit (see Figure 6).

3.2. Continuum observations of extended sources

The previous section on point sources has provided us with the algorithms and techniques for simple analysis of continuum mapping observations of extended sources. All mapping observations of extended sources can be divided into the following two major categories.

Continuum Grid Mapping: In grid mapping, the telescope steps from pixel to pixel in the map only collecting data when the telescope has settled onto a pixel. While mapping, the observing technique could be beam switching to reduce atmospheric fluctuations.

The reduction of grid mapping requires converting the detected powers of each pixel to $T_{\rm ant}$ using the appropriate analysis described in Section 3.1. The resulting $T_{\rm ant}$ are then regridded into a matrix. Since grid mapping is going out of vogue, we will not go into any further details.

Continuum On-the-Fly (OTF) Mapping: OTF mapping has been used for decades by many cm-wave telescopes for strong sources. It is now the preferred technique at mm-waves since at those wavelengths OTF reduces the ill effects of atmospheric fluctuations. In OTF mapping, the telescope slews through a region of the sky and simultaneously records data at a rate of up to a few samples per second. The slewing is usually performed in one cardinal direction to produce a row or column in the map that is most often highly oversampled. Between rows or columns, the telescope is offset in the direction orthogonal to the slew direction by somewhat less than half the beam width and a new row or column is started. While mapping, the observing technique could be beam switching to further reduce atmospheric fluctuations.

As in grid mapping, the first step is to use the appropriate analysis described in Section 3.1 to convert from detected power to $T_{\rm ant}$. If warranted, a baseline might be fitted and removed from the data for each row or column in the map separately. The data are then gridded into a matrix using techniques that depend upon the analysis package. Each gridding algorithm has its pluses and minuses. Usually a gridding algorithm performs a weighted average of all data points that fall within some radius of a matrix pixel. The weights usually depend upon the distance a data point lies from the matrix pixel with those further away getting less weight.

For example, a 'top-hat' gridding algorithm uses uniform weighting of all data points that are within a specified radius of a matrix pixel. A top-hat algorithm is useful for when you expect little structure in your map. Other gridding functions use a weight function that is representative of the telescope's beam and are better suited to noisy maps or maps full of small-scale structure. Although Briggs, Schwab, and Sramek (1998) describe gridding algorithms in the context of array imaging, much of the general details of their discussion apply to single-dish gridding algorithms as well. An example of the OTF continuum map is shown in Figure 7.

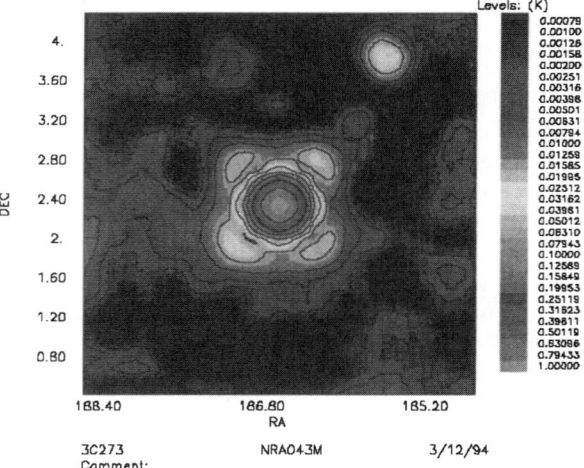

Figure 7. On-the-fly continuum map of the point source 3C273 taken with the NRAO 140-ft telescope at 1400 MHz. There is a second radio source located to the northwest of 3C273. Contours are spaced 1 dB apart and show the beam pattern or response of the telescope to a point source.

Advanced Techniques for Continuum Mapping: Once the data in a continuum map have been gridded, some systematic problems are often evident in the data, see Figure 8. The powers available to you will depend on the analysis system you are using and you should consult the program's manual for suggestions.

One common problem is striping along rows or columns in the matrix due to instrumental or atmospheric drifts from one row or column to the next. Fourier transforming the matrix, removing the low-order frequencies, and inverse transforming the matrix, can reduce stripping (Emerson 1995; also Klein & Mack 1995 for using the CLEAN algorithm on single-dish data).

If the data have been beam-switched, the separation of the beams on the sky might be smaller than the size of the mapped source. In the later case, certain algorithms, such as that of Emerson, Klein, and Haslam (1979) can be used to reconstruct the image.

For creating maps with the highest fidelity, one might consider making multiple maps maybe with each map taken with the telescope slewing in a different cardinal direction. Each map should be taken very quickly. The resulting maps should then be averaged using some measure of each maps σ for weights.

After a matrix is created, and any artifacts removed, many analysis programs have tools to calculate source strengths and sizes, take slices through the matrix, and remove drifts from varying ground pickup. Since different analysis programs possess different functionalities, the user must peruse the software's documentation for a list of available algorithms.

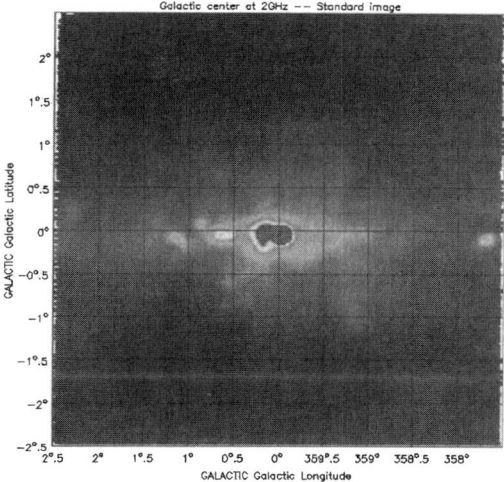

Figure 8. On-the-fly continuum map of the galactic center taken with the Green Bank Telescope at a frequency of 2 GHz. Note the stripping at galactic latitude of $-1.6°$ that will need to be fixed either with more observations or one of the techniques described in the text.

4. Spectral-Line Data Reduction Techniques

Most of the principles behind continuum data reduction also apply to spectral-line data reduction. Thus, this section will build upon what we have already discussed in the previous section.

4.1. Spectral-line observations of point sources

As with continuum observations, we will start by discussing observations of point sources before tackling spectral-line mapping. There are three different categories of spectral-line observations, each depending upon how a reference spectrum is created. Two of these schemes, beam switching and position switching, are very similar to the beam switching and On-Off techniques described above for continuum observations. Frequency switching is commonly used only for spectral line observing.

Position-Switched Observations: Position-switched observations are the spectral line equivalent of On-Off continuum observations. All analysis programs have some command to create a difference spectrum from the signal and reference observations. The difference spectrum is derived by comparing the difference in power between the signal and reference spectra with the difference in power between when the noise diode or calibrator is on and off (Figure 9). With the same assumptions we made in Section 3.1, the difference spectrum is:

$$T_{\text{ant}}(\nu) = \frac{T_{\text{cal}}(\nu)}{2} \cdot \left[\frac{\left(P_{\text{cal\_on}}^{\text{sig}}(\nu) + P_{\text{cal\_off}}^{\text{sig}}(\nu)\right) - \left(P_{\text{cal\_on}}^{\text{ref}}(\nu) + P_{\text{cal\_off}}^{\text{ref}}(\nu)\right)}{\left(P_{\text{cal\_on}}^{\text{ref}}(\nu) - P_{\text{cal\_off}}^{\text{ref}}(\nu)\right)} \right]$$

$$= \left(T_{\text{sys}}(\nu) + \frac{T_{\text{cal}}(\nu)}{2}\right) \cdot \left[\frac{\left(P_{\text{cal\_on}}^{\text{sig}}(\nu) + P_{\text{cal\_off}}^{\text{sig}}(\nu)\right) - \left(P_{\text{cal\_on}}^{\text{ref}}(\nu) + P_{\text{cal\_off}}^{\text{ref}}(\nu)\right)}{\left(P_{\text{cal\_on}}^{\text{ref}}(\nu) + P_{\text{cal\_off}}^{\text{ref}}(\nu)\right)} \right] . \quad (10)$$

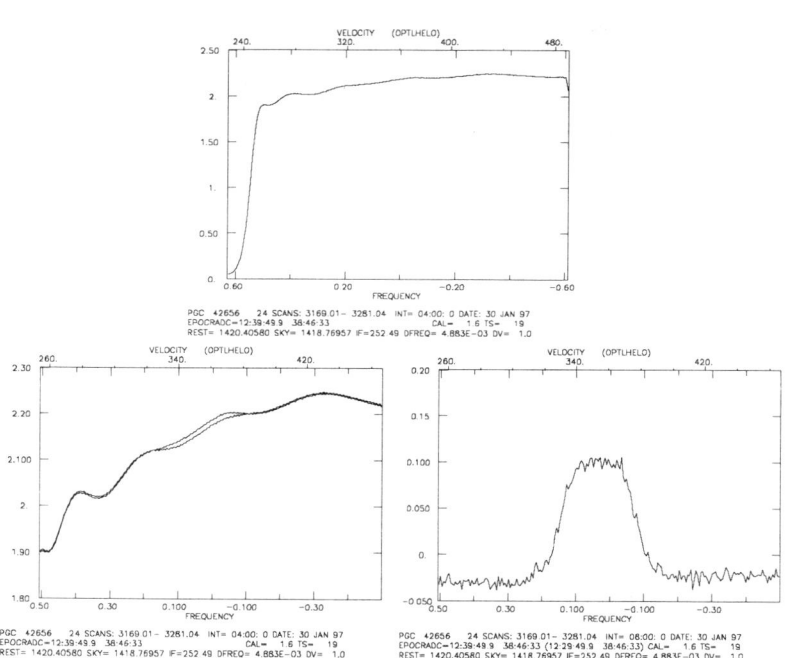

Figure 9. A position switched, atomic hydrogen observation taken with the NRAO 140-ft telescope toward an extragalactic source. At top is the bandpass of the 'off' position observation. At lower left is the superposition of the bandpasses of the 'on' and 'off' observation. Note how the two bandpasses are almost identical except between 320 and 380 km s^{-1}. The difference spectra, as defined by Equation 10, is at lower right and clearly shows the hydrogen spectral line profile from the galaxy.

The per-channel noise of the difference spectrum is:

$$\sigma_{T_{\text{ant}}} = \text{Constant} \cdot \left(T_{\text{sys}} / \sqrt{\text{BW/N}}\right) \cdot \sqrt{\frac{1}{t^{\text{ref}}} + \frac{1}{t^{\text{sig}}}} , \quad (11)$$

where the Constant depends upon the details of the spectrometer (e.g. channel spacing and number of sampling levels for an autocorrelation spectrometer). Note that the noise is higher in the difference spectra than in either the signal or reference spectra.

$T_{\text{ant}}(\nu)$ must be corrected for atmospheric attenuation and antenna efficiency, and, if one is observing a point source, converted to flux density.

Beam-Switched Observations: Beam switched spectral-line observations are very similar to their continuum cousin and their spectral line position-switched brother. In beam switched observations, both the signal and reference spectra are accumulated simultaneously and synchronously. For some control systems, the difference spectrum (Equation 10) is calculated for the observer and he or she never gets to see the individual signal and reference spectra. With other telescopes, the user will have to use the built-in analysis command to create the difference spectrum.

Frequency-Switched Observations: Frequency switched observations have some similarities to beam switched observations in that the signal and reference spectra are accumulated simultaneously and synchronously. Frequency-switched observations differ in that it is the center frequency of the observation that is toggled between two values. Depending upon the telescope's control system, the user might be presented with a difference spectrum or the individual signal and reference spectra. Frequency-switched observations come in two flavors:

(a) Out-of-band frequency switching:
In out-of-band frequency switching, the amount that the frequency is changed is larger than the bandpass of the observation. The spectral line of interest will lie only within the signal spectrum and not the reference. With this type of observing, the user or telescope control system need only apply Equation 12 to form a difference spectrum.

(b) In-band frequency switching:
If the frequency change is less than the bandwidth of the observation, it is possible that the spectral line of interest will lie in both the signal and reference spectra, see Figure 10. In fact, the frequency change might be less than the width of the line, a specialized observing mode discussed by Liszt (1997). The difference spectrum produced by Equation 12 will have a representation of the spectral line and an inverted image of the line separated by the amount of the frequency switch.

One is tempted to take the difference spectrum, shift it by the amount of the frequency switch, invert the shifted version, and average the shifted/inverted spectra with the original. However, this simplistic algorithm would produce a systematic lowering of the detected line strength if the line strength is even a small fraction of T_{sys}. The correct algorithm involves creating a second difference spectrum but with the roles of signal and reference reversed. The second difference spectrum is then shifted by the amount of the frequency switch ($\Delta\nu$) and averaged with the original (Figure 11). Luckily, most analysis programs have a command that will 'fold' frequency switched observations.

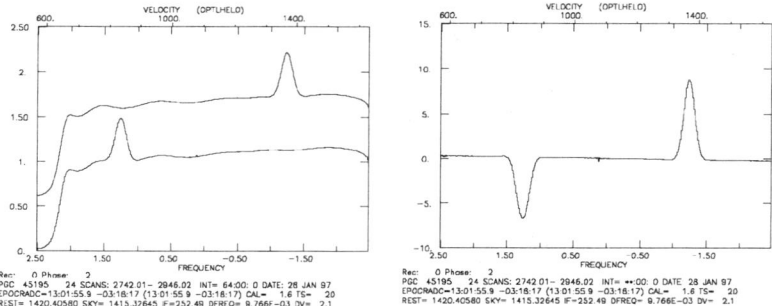

Figure 10. An example of in-band frequency switching. The left panel is the two bandpasses of the signal and reference spectra. The right panel is the difference spectra. Note the spectral line also appears as a negative image of itself but displaced by the amount of the frequency switch, the signature of in-band frequency switched observations.

$$T_{\text{ant}}(\nu) = \frac{T_{\text{cal}}(\nu)}{2} \cdot \left[\frac{\left(P_{\text{cal\_on}}^{\text{sig}}(\nu) + P_{\text{cal\_off}}^{\text{sig}}(\nu)\right) - \left(P_{\text{cal\_on}}^{\text{ref}}(\nu) + P_{\text{cal\_off}}^{\text{ref}}(\nu)\right)}{\left(P_{\text{cal\_on}}^{\text{ref}}(\nu) - P_{\text{cal\_off}}^{\text{ref}}(\nu)\right)} \right] +$$
$$\frac{T_{\text{cal}}(\nu)}{2} \cdot \left[\frac{\left(P_{\text{cal\_on}}^{\text{ref}}(\nu + \Delta\nu) + P_{\text{cal\_off}}^{\text{ref}}(\nu + \Delta\nu)\right) - \left(P_{\text{cal\_on}}^{\text{sig}}(\nu + \Delta\nu) + P_{\text{cal\_off}}^{\text{sig}}(\nu + \Delta\nu)\right)}{\left(P_{\text{cal\_on}}^{\text{sig}}(\nu + \Delta\nu) - P_{\text{cal\_off}}^{\text{sig}}(\nu + \Delta\nu)\right)} \right].$$
(12)

Frequency and Velocity Calibration and Data Shifting: Once a difference spectrum is formed, it might be necessary to calibrate the frequency and velocity axes of the resulting data. For example, some backends, such as with Acoustical Optical Spectrometers (AOS), the spacing of channels is non-uniform. Each observatory will recommend its own techniques and analysis algorithms for calibrating the frequency axis of their AOS data. Another issue neophyte observers should be aware of is that, since spectrometers sample in frequency, the velocity widths of channels change across the bandpass.

A subtler problem afflicts all spectral-line observations at some level. Since the Earth rotates and revolves, and the Sun moves relative to the local standard of rest, observations made at two different epochs will need different Doppler shifts since the projected motion of the telescope relative to the direction of the observation will change with time. Usually, the telescope control software will alter the observed frequency so as to align one epoch's data with all other epochs in some chosen rest frame. This observing technique has been sufficient in the days when observing bandwidths were a small fraction of the observing frequency or when only a single spectral line is observed. Today, these hardware limitations are no longer true so the assumptions many analysis systems make

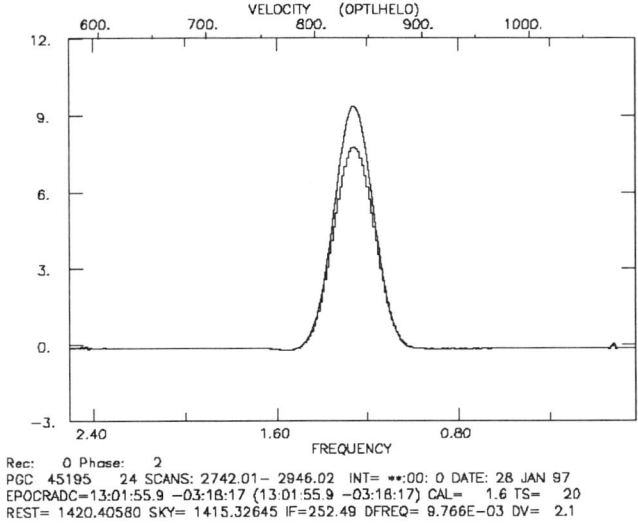

Figure 11. The comparison of the resulting spectra using the correct (upper curve) and incorrect (lower curve) algorithms for 'folding' an in-band frequency-switched observation.

are no longer adequate. For wide bandwidths, one can expect the velocity for all but the central channel of a spectrum to shift slowly with time. Additionally, Doppler tracking will only work for a single spectral line; any other simultaneously observed line will be mis-tracked especially if the second line has a rest frequency very different from the first.

All of these issues will require the occasional regridding and shifting of spectra in velocity or frequency. Some analysis programs have automated tools that will align the velocity of two spectra taken at different times. Typically these algorithms take the Fourier transform of the spectrum, apply a phase shift corresponding to the shift in velocity or frequency, and inverse Fourier transform the data. Unfortunately, these algorithms also have the undesired artifact of slightly altering the noise in your data and require well- or over-sampled data (Brigham 1974; Press et al. 1992).

Averaging Spectra: Our discussion in Section 3.1 on averaging data applies to spectral-line data as well. That is, you create an optimum average with the lowest noise only by proper weighting of the individual spectra.

The only caveat of concern for spectra is that one should velocity shift or regrid spectra before averaging so that all spectra will have the same velocity sampling. Failure to perform proper shifting will smear the line and make it have a greater width and lower intensity than would otherwise happen.

Baseline Fitting: Fitting a polynomial baseline to a spectrum is identical to the fitting of a baseline to a time sequence of continuum data as discussed in Section 3.1. All of the statistical issues associated with continuum baseline fitting apply to spectral line fitting as well. Most single-dish analysis packages also offer an algorithm that will fit a sinusoid to emission-free regions of a spectrum. A sinusoidal baseline shape in spectral-line data will often occur when the telescope optics produce a standing wave with a frequency that is lower than a few times the spectrum's bandwidths.

Gaussian and Component Fitting: The steps to fit a Gaussian to spectra are identical to those for continuum observations. It is up to the science at hand to dictate whether or not a Gaussian is a reasonable model for the expected line profile.

In the case where a spectrum has one or more features, you might want to fit multiple Gaussians. If the features do not overlap, then you can fit each feature separately. However, if the features are blended, or if you have extra information about the features (e.g. the spectrum contains multiple hyperfine transitions for which you know the frequency separations), then you will want to simultaneously fit multiple Gaussians. Many but not all packages allow for multiple Gaussian fits. Fewer systems allow you to use already-known information (like our example of known hyperfine separations, or transitions with known ratios of line strengths) to better constrain the Gaussian fit. Thus, if you plan on fitting multiple Gaussians, you should familiarize yourself with the software's documentation.

Spectral-Line Moments: In many cases you will be interested in the integrated intensity or the temperature-weighted velocity (centroid) of a spectral-line feature. All of the standard packages have some command that will perform these calculations of moments for you. The input to the algorithms usually involves defining the region of the spectrum over which spectral-line moments should be taken.

Either the analysis system or the user of a system should be able to generate the statistical uncertainty in the results of a moment calculation. For example, the uncertainty in the integrated intensity is $\sigma\sqrt{n}$ where σ is the per-channel noise and n is the number of channels across which the integrated intensity is taken.

Smoothing: There are a few reasons why one would want to smooth spectral-line data. If you need to average spectra that were taken at two different velocity or frequency resolutions, first you will need to smooth the data so that all will have the same velocity or frequency resolution as that possessed by the spectrum with the coarsest resolution. Smoothing also effectively removes "ringing" inherent with cross-correlation/FFT-style correlation spectrometer when observing a narrow, strong spectral feature like an RFI spike. Brigham (1974) discusses how leakage, inherent in the discrete Fourier transform these spectrometers use, manifests itself as ringing. Brigham also discusses how the correct solution for reducing ringing requires specifying a windowing function that the spectrometer should apply before it takes the Fourier transform.

Smoothing algorithms in single-dish analysis packages are of two types: those that decimate data (e.g. the smoothed data have less spectrometer channels than the unsmoothed), and those that do not (smoothed and unsmoothed data have the same number of spectrometer channels).

Algorithms that do not decimate data, like the prolific Hanning or Boxcar smoothing functions, are overused and give a statistically false impression of the noise level of your data. Individual data points are no longer independent of each other. Thus, we suggest you limit your use of these functions for those cases you want a quick look at smoothed data.

Better and more statistically correct algorithms are those that decimate data. These algorithms usually convolve the data with some function, usually Gaussian of user-specified half width, before decimating the data. Even with decimating functions, the data points in the resulting spectra are still no longer independent. However, the results from these algorithms, in comparison to algorithms that do not decimate, will produce results that better match what the observations would have been like if it were made at the coarser resolution. Many general texts (Press et al. 1992; Bevington & Robinson 1992) describe the details of smoothing functions.

RFI Excision: At some point during your career of using radio telescopes you will encounter radio interference. In some cases the observations can be altered to mitigate the problems, in others you must deal with the consequences in the data analysis stage.

We have discussed under the previous section on smoothing how a single RFI spike in a spectrum can cause ringing across a bandpass. In other cases, if the RFI comes in bursts, it is sometime possible to excise that fraction of the data that are affected. In such a case, the observer should set up the backend for the fastest sampling times available in the hope that only a small fraction of the resulting spectra will experience RFI. Some data analysis programs have the ability to display the incoming spectra as a 2-dimensional waterfall image of intensity as a function of frequency and time (Figure 12). The user would use the tool to flag which spectra, and maybe which channels, are to be excised. The software will then average all unflagged data to produce a spectrum of only those data samples deemed to be free of RFI.

4.2. Spectral-line observations of extended sources

We now have under our belts the necessary background to discuss spectral line mapping of extended sources. Like continuum maps, spectral-line maps can be made by grid mapping or on-the-fly mapping.

Spectral-Line Grid Mapping: In grid mapping, the telescope steps from pixel to pixel in a rectangular region of the sky taking spectra only when the telescope has settled onto a pixel. While mapping, the observing technique could use beam or frequency switching. If the observations are position-switched, then the observing technique frequently should pause to take reference spectra.

The reduction of grid mapping requires converting the raw data into difference spectra as described in Section 4.1. The resulting $T_{ant}(\nu)$ are then regridded into an appropriate 3-dimensional image commonly called a data cube. The axes

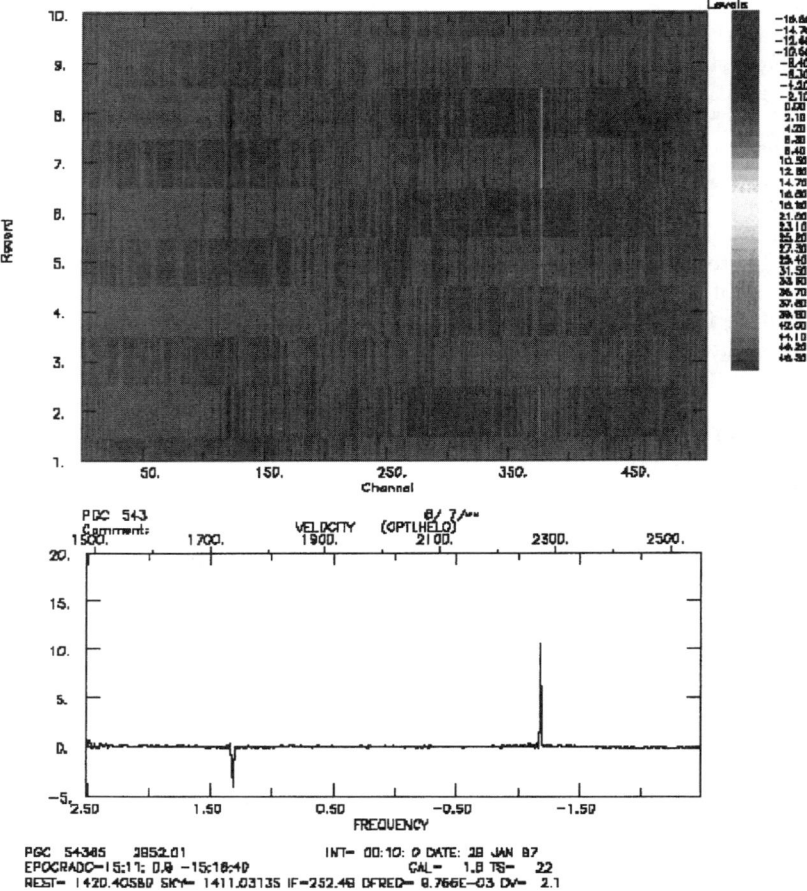

Figure 12. At top, a waterfall display showing ten sequential frequency-switched spectra in gray scale. Note the RFI in four of the spectra numbered 5-8 near spectral channel 370 (and a weaker RFI frequency-switched into the bandpass near channel 120). The spectrum at bottom is the weighted average of all ten spectra. By flagging the few channels or spectra with RFI, one could produce a cleaner average.

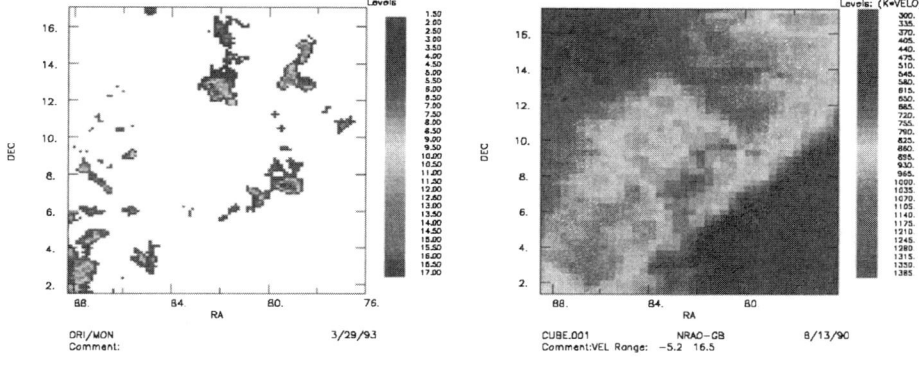

Figure 13. Examples of a grid-mapped (left) and on-the-fly (right) spectral-line observations. At left is the CO emission surrounding the λ Ori HII region as observed with the Columbia 1.2-m mm-wave telescope. At right, the HI emission from the same region as observed with the NRAO 140-ft telescope. Both images were created by slicing a 3-dimensional data cube.

of the cube are the two orthogonal sky coordinates in which the map was made and frequency or velocity.

Spectral-Line On-the-Fly (OTF) Mapping: OTF spectral-line mapping is no longer just the venue of cm-wave telescopes but has also become the preferred technique at mm-wave telescopes. In OTF mapping, the telescope slews through a rectangular region of the sky and simultaneously records data sometimes as fast as a few spectra per second. As in continuum OTF maps, the slewing is usually performed in one cardinal direction to produce a row or column in the map. After finishing a row or column, the telescope is offset in the direction orthogonal to the slew direction by somewhat less than half the beam width and a new row or column is started. While mapping, the observing technique could be beam or frequency switching. If the observations are position-switched, then the observing technique frequently will pause to take reference spectra.

The first step in the data analysis is to use the appropriate analysis described in Section 4.1 to create difference spectra. The data may then be smoothed, baselined, etc. before they are gridded into a data cube (see previous Section and Section 3.2). Figure 13 shows examples of a grid-mapped and an OTF observation.

Rudimentary Analysis of Spectral-Line Maps: The simpler single-dish packages allow one to produce a two-dimensional image (position-position or position-velocity maps) by taking a slice out of a data cube or integrating along a face of a data cube. Then, one uses the packages two-dimensional analysis tools to further analyze the map. In the single-dish packages, visualization tools usually are no more sophisticated than contour plots and the occasional grayscale or color-scale image. For more sophisticated visualization and analysis one usually needs to turn to systems like AIPS, AIPS++, and IRAF.

5. The Future of Single-Dish Analysis

Single-dish observing is becoming more and more sophisticated. Receivers are getting better, astronomers are asking for more sensitive observations or for wide-field maps. Backends are increasing in complexity and the number of channels they produce. Many telescopes are paving the way with multi-feed receivers. New spectrometers will have dozens of inputs, extremely wide bandwidths, and many hundreds of thousands of channels. As with other branches of astronomy, we will continue to see an escalating increase in the data rates coming out of single-dish radio telescopes.

We are also seeing users of single-dish telescopes whose previous observing experience has not been at radio wavelengths. Astronomers are asking for more robust and sophisticated analysis algorithms and better visualization tools. We are seeing a decrease in the amount of real-time, interactive data analysis as many observatories now allow for remote, queue, and service observing.

Single-dish observing has not stood still nor should single-dish analysis techniques. I predict that we will see in future single-dish analysis systems:

- An increase in the use of relational data base management systems to help control and organize the huge data sets we will be collecting.

- A unilateral archive of data that, after a reasonable time has passed since observations are completed is perusable by any astronomer, not just the observer who took the data.

- A rise in the sophistication of the visualization tools single-dish astronomers will use.

- More sophisticated, robust algorithms, especially those dealing with large-scale mapping.

- Further exploration of data pipelining for the general user.

- Automatic data calibration as sophisticated models of the telescope are incorporated and used by analysis programs.

- Development of algorithms that deal with data sets instead of individual spectra or continuum slices.

Individual observatories will no longer have the resources needed to provide the sophisticated packages users will request. Instead, within a decade only a few major systems maintained by cross-observatory groups will survive. Some observers and class of observers (e.g. pulsar or radar observers) might still have their specialized packages, but this will be the exception while just ten years ago it was the rule. Only time will tell whether these changes will benefit or hurt the science we love and cherish.

Appendix: Symbols and Definitions

Table 2. Symbols used throughout the paper.

BW	Bandwidth (Hz)
ν	Frequency
i	Sample number
N	Number of spectral-line channels
P	Detected power in A/D counts or Volts
σ_T	Standard deviation of detected source temperature (K)
t	Integration time (s)
T_{ant}	Antenna temperature due to all sources: sky, atmosphere, receiver,... (K)
T_{cal}	Temperature of noise diode or calibration source (K)
T_{src}	Source temperature (K)
T_{sys}	System temperature due to all sources but the object of interest (K)

Acknowledgments. I thank Harvey Liszt for his very helpful comments on many points in this paper.

References

Bevington, P. R. & Robinson, D. K. 1992, Data Reduction and Error Analysis for the Physical Sciences (Boston: McGraw-Hill)

Briggs, D. S., Schwab, F. R., & Sramek, R. A. 1998, in Synthesis Imaging in Radio Astronomy II, ed. G. B. Taylor, C. L. Carilli & R. A. Perley (San Francisco: ASP), 134-145

Brigham, E. B 1974, The Fast Fourier Transform, (Englewood Cliffs: Prentice Hall)

Emerson, D. T 1995, in Multi-Feed systems for Radio Telescopes, ed. D. T. Emerson & J. M. Payne (San Francisco: ASP), 309-317

Emerson, D. T., Klein, U., & Haslam, C. G. T 1979, A&A, 76, 92

Klein, U. & Mack, K. H. 1995, in Multi-Feed systems for Radio Telescopes, ed. D. T. Emerson & J. M. Payne (San Francisco: ASP), 318-326

Liszt, H. 1997, A&AS, 124, 183

Lockman, F. J., Jahoda, K., & McCammon, D. 1986, ApJ, 302, 432

Mandel, J. 1964, The Statistical Analysis of Experimental Data (New York: Dover)

Press W. H., Teukolsky, S. A., Vetterling, W. T., & Flannery, B. P. 1992, Numerical Recipes in Fortran (New York: Cambridge)

Rood, R. T., Bania, T. M., & Wilson, T. L. 1984, ApJ, 280, 629

Quang-Rieu, N. 1969, A&A, 1, 128

Van Zee, L., Maddalena, R. J., Haynes, M. P., Hogg, D. H., & Roberts, M. S. 1997, AJ, 113, 1638

Single-Dish Analysis within AIPS++

Joseph P. McMullin, Robert W. Garwood

National Radio Astronomy Observatory, 520 Edgemont Road, Charlottesville, Virginia 222903, USA

Tim Cornwell, Athol Kemball

National Radio Astronomy Observatory, P.O. Box O, Socorro, New Mexico 87801, USA

Jim Braatz

National Radio Astronomy Observatory, P.O. Box 2, Green Bank, West Virginia 24944, USA

Abstract. We present a short description and update of the single-dish analysis environment in AIPS++, called DISH. The focus on DISH has been to provide traditional single-dish spectral line data analysis tools with support for extensibility and customization of the package. It currently supports end-to-end processing of data, from data flagging and calibration, through averaging, fitting, statistics and imaging. DISH and AIPS++ are the primary analysis tools for data from the Green Bank Telescope and are also in use at several other observatories. We provide examples of its use and outline the general features. DISH is available now with the sixth public release of AIPS++ (V1.7) (*http://aips2.nrao.edu*).

1. Introduction

Single-dish analysis in AIPS++ is distilled into a flexible application called DISH which serves as both an interface and a toolkit for data reduction. As an AIPS++ package, DISH inherits its approach to working with data, its underlying Glish command line interface and its global access to every tool in other packages of AIPS++. We therefore first review the background and content of the AIPS++ toolkit. We provide brief instructions on initiating work within AIPS++ and the various means of getting data into and out of the package. We then provide a more detailed census of the DISH package capabilities and components. Finally, we provide a demonstration data analysis session with DISH, culminating in producing a publication quality graph.

Complete documentation on AIPS++ and DISH is available online at: *http://aips2.nrao.edu*.

2. AIPS++ Background

In this section, we review the AIPS++ software package and its current implementation (v.1.7).

2.1. AIPS++ overview

AIPS++ was tasked to provide facilities for general radio astronomical processing, including tools for calibration, editing, image formation, image enhancement, and analysis of images and other astronomical data. The focus has been on reduction of single-dish (single or multi-beam) and aperture synthesis data (including VLBA). The package is also useful for processing other types of astronomical data and images.

AIPS++ adopts a tool-based approach to provide the analysis routines for astronomers. This means that in general, the terminology and the look and feel is somewhat different from other astronomical processing packages. The user typically creates a tool which contains both the data to be operated upon and all of the relevant functions necessary to understand the data.

An overriding ideal in AIPS++ is to provide the user with direct access to all portions of a data set or image. All aspects of the data can be easily manipulated using existing applications, through user developed applications or from simple calculator-type commands from the command line.

AIPS++ is also designed to take advantage of modern computer software and hardware. Mouse-driven Graphical User Interfaces (GUIs) are used extensively for handling various applications, computations, and image display.

Glish is the command language used in AIPS++ and operates similarly to a public domain IDL. It is a very capable interactive programming language which allows high level operations on vectors and arrays. All data can be accessed and manipulated from Glish and placed back into AIPS++. Tasks in AIPS++ are written in C++ for efficiency and bound to Glish, therefore, all capabilities of interest to the user are controllable from the Glish command line window.

Glish can be used in an interactive or non-interactive mode (batch) and it also excels at process control, allowing event-driven programming. Clients can be easily bound (e.g. the basic Tk widgets are included in Glish); a sophisticated example of all of these features is the Green Bank Telescope (GBT) Observer Interface built by Rick Fisher (see the article in the Nov 2000 issue of the AIPS++ newsletter: *aips2.nrao.edu/daily/docs/newsletters/index.htm*). AIPS++ can be completely controlled from the command line window. For most applications, control using a combination of both command line window and GUI windows is probably most appropriate.

In order to access the extensive AIPS++ reference material and other documentation, the Help menu that appears on most GUIs can directly steer a Web browser to the appropriate section. The existing documentation to AIPS++ is vast but the main elements are: 1) Getting Started in AIPS++, Getting Started in Glish (an introduction to the package and the command line interface, CLI); 2) Getting Results in AIPS++ (a series of recipes and detailed examples using various tools); 3) User Reference Manual (which is a census of all of the functionality).

2.2. Data and images in AIPS++

In AIPS++, all forms of data (i.e. visibility, images, spectra, history, header information, logs) are stored in tables. Tables consist of information arranged in rows and columns. The columns have names (e.g. UVW, TIME, ANTENNA1, SIGMA) and the rows are numbered. Keywords may be attached to both the table as a whole or to specific columns and are used to specify additional information. Examples of keywords are revision number of a table, author of the table, units in a column, coordinate information for a column and so on. Keywords may also be used to store other tables.

Tables A Table and its contents are stored on disk as a directory with the Table name (e.g. 3C273XC1.ms) and may contain a number of subdirectories and files. Although there can be many different types of Tables in AIPS++, representing different types of data, there are two important types of Tables in common use: MeasurementSets (MS) (which represents data from a telescope, e.g. visibility data) and Images (which represents the distribution of intensities on the sky). An Image may have two or more dimensions, e.g. position, velocity, Stokes.

MeasurementSet (MS) A MeasurementSet is simply a Table obeying certain conventions that allow it to represent observations from many different types of radio telescopes. In addition to a number of data columns, it also contains sub-Tables stored as Table keywords (as discussed above). These keywords contain information on the antennas, feeds, fields, and spectral windows. An MS can be dynamic in that new columns can be written into it as data are processed. For example, the original measured UV data appear in the data column and subsequent processing can produce additional columns containing corrected data (e.g. after a Selfcal) or Model data (e.g. a set of clean or other components).

Image An Image is also a Table. The keywords in this Table contain information on units and coordinates. The pixels are stored as a single cell in a column. Other information about the image (e.g. its co-ordinates, origin, history, characteristics) are stored under separate keywords.

2.3. AIPS++ contents and capabilities

In this section, we briefly highlight the most commonly used applications in AIPS++. A toolmanager has been created to help organize and initiate these applications. They are organized in four levels: 1) package – this is the broadest group and delineates a general area of emphasis, for example, Display or Synthesis; 2) module – this narrows the focus within a package to a few subsets of utilities, for example the synthesis package has a simulator module (for creating test data sets) and an imager module (for manipulating images); 3) tools – this is the level of application which is directly tied to a data set; and 4) global functions which are operations or tasks on a data set. Once a tool is created, all global functions are accessible through it by typing:

- `toolname.functionname([args]);`

The following sections outline the highlights of AIPS++ functionality.

General
deconvolver
Tool to deconvolve a known point spread function; includes Hogbom and Clark clean, MEM, and multi-scale techniques plus more.
images
Tool that creates, manipulates and analyzes images; allows concatenation, scaling, smoothing, moments, convolutions, etc.
measures
Allows versatile operations on measured quantities with units and coordinate systems.

Synthesis
calibrater and imager
These form the core of the synthesis analysis. Calibrater has solvers for electronic and atmospheric gain, bandpass and polarization. Imager has support for multi-field processing and wide field imaging.
componentmodels
Allows manipulation of the sky brightness as a function of position on the sky and observing frequency.
flagger
Provides synthesis flagging capabilities for various selections.
map
Is a high-level synthesis calibration and imaging tool which leads the user through the various steps and associated options.
simulator
Allows simulation of telescope observations (either synthesis or single-dish) and corruption of data through Gaussian errors or other specific errors.

DISH
The single-dish analysis package. Supports general single-dish operations (flagging, averaging, baselining, smoothing, etc.).

Display
plotter
A plotting tool based on the CalTech PGPLOT subroutine library with all of its associated commands available through Glish. Allows easy plotting of Glish variables.
viewer
The image visualization tool which supports raster, contour, vector and 3-D slice displays.

Utility
catalog
Allows access and manipulation of files on disk.
fitting

Provides numerous tools for linear and non-linear fits (either real or complex).
mathematics
Provides tools for statistics (median, moments etc), FFTs, polynomial fits, 1-d interpolation, 1-d gaussian fits and evaluation, random number generation, matrix algebra, and least squares solutions of simultaneous equations.
table
This tool allows a sophisticated selection on table information, along with browsing, editing, and plotting of table data.

2.4. Starting AIPS++

AIPS++ requires setting of some environment variables before running. The setup commands are packaged into a single shell script (csh or bash).

```
. /aips++/release/aipsinit.sh       #Bourne/Korn/Bash shell;
or
source /aips++/release/aipsinit.csh  #csh and similar shells.
```

To run AIPS++, then simply type:

```
aips++
```

This starts Glish, opens a logger window where messages are displayed, and opens the Tool Manager GUI. Starting AIPS++ the first time is generally slow as various initialization files are created. Subsequent starts are faster.

2.5. Essential commands

```
- exit     # Exit from Glish
- bug()    # Report a bug
- help()   # Get help
```

2.6. Getting help

```
- help('general')# modules in the general package
- help('mom')    # help uses minimum matching-all items with
                 # the string 'mom' are revealed:
There are 3 matches. Please choose from:
  general.images.image.moments.function
  general.images.image.momentsgui.function
  utility.mathematics.moments
- help('general.images.image.moments.function')

 moments -- Function -- general.images.image
   Compute moments from an image
 Usage:  moments(moments, axis, region, mask, method, smoothaxes,
smoothtypes, smoothwidths, includepix, excludepix, peaksnr, stddev,
doppler, outfile, smoothout, plotter, nx, ny, yind, async)

You may find more information in the on-line documentation available
via your web browser.  Type the command
```

```
web()
```

to view more about general.images.image.moments.function.

```
- web()              # Drives to last requested help section
- help('Refman:')    # Drives to the Refman
- help('Glish:')     # Drives to the Glish User Manual
```

2.7. Import routines

Importing data into AIPS++ can be done in several ways. The relevant documentation is found in the 'ms' and 'images' modules.

- sdfitstoms – convert SDFITS to MS

- fitstoms – convert UVFITS to MS

- imagefromfits – convert FITS image to AIPS++ image

- imagefromforeign – convert foreign image (MIRIAD, Gipsy) to AIPS++ image

- imagefromarray – convert array to AIPS++ image

2.8. Export routines

Exporting data also has several options.

- sprintf – Glish capability: C style write ascii to file

- ms.tofits – convert MS to UVFITS

- image.tofits – convert AIPS++ image to FITS image

3. DISH

DISH is a collection of Glish scripts and clients which provide an environment within AIPS++ intended to be used for single-dish radio astronomy data analysis. For full documentation on DISH see its chapter in the AIPS++ "Getting Results..." document. DISH features all of the common analysis operations necessary for single-dish data reduction, with the added advantages of being able to use all AIPS++ utilities on the data (e.g. imaging/regridding capabilities) and with full extensibility through Glish scripting.

- Design goals. The primary goal has been to provide a graphical user interface that is intuitive, unsurprising and responsive. Users should feel that results go to obvious places and, whenever possible, are displayed as they occur.

- Data. Two types of data are typically handled by DISH

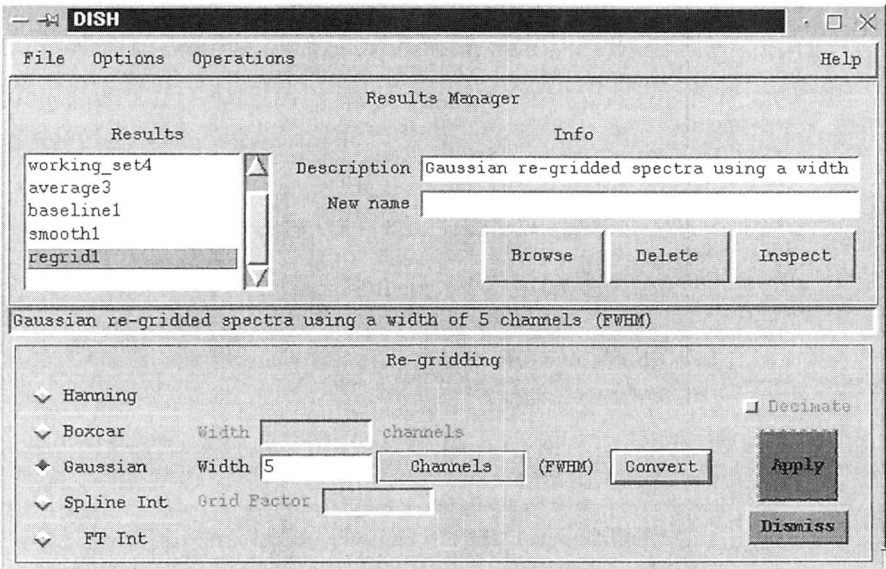

Figure 1. Example window showing basic DISH frame with Regrid Operation frame opened.

1. Individual scans. These are two-dimensional data arrays plus associated information including analysis history information. The first dimension can be anything but is most typically a frequency and/or velocity axis while the second dimension is a Stokes axis. These data units are referred to as *SDRecords*. DISH knows how to plot these. They can also be browsed in the sense that the associated information and analysis history can be examined in the SD Record Browser (i.e. a graphical way of examining the data).

2. Collection of individual scans typically representing a single observing run. Subset collections are easily obtained through selections on the data.

- Data flow. Operations may work on collections of scans or they may operate on individual scans. Most operations place their results in the DISH results manager. The sole exception to this rule is the Calculator which has its own stack. If the result can be displayed, it is. Operations on individual scans always operate on the most recently displayed scan.

- Data format. The primary archival format for single-dish data within AIPS++ is Single-Dish FITS. Single-Dish FITS is a convention for storing single-dish radio astronomy data in a FITS binary table (Garwood 2000). All astronomical data within AIPS++ resides in a MeasurementSet. MeasurementSets can be converted to and from Single-Dish FITS.

- Saving State. Each time you exit DISH its state is saved to disk. This includes all of the values known to the results manager as well as the various settings for each GUI element and operation. When you restart DISH, this state is restored. You can also choose to save and restore the DISH state while using DISH as a way of bookmarking key steps in the data analysis.

- Scripting. Underlying each GUI operation are Glish commands. Users who wish to do complex tasks not available through the DISH GUI or who want to capture existing sets of operations into a Glish script will need to use the command line interface to DISH. The AIPS++ scripting tool can be used to show the underlying Glish commands as each GUI operation occurs. These can be used to learn how the GUI does things, or the text can be copied to an editor and used as the basis for a Glish script.

3.1. DISH components

- Results Manager – the main DISH GUI; explained in Section 3.2.

- Browsers – show the contents of an SDRecord or group of records.

- Plotter – the display for all 1-d data; explained in Section 3.3.

- Scripter – the AIPS++ scripting tool which prints out CLI equivalents to GUI actions.

- Logger – logs all CLI commands; displays messages resulting from actions.

- Operation GUIs – each operation has its own frame which expands off of the main GUI (see Figure 1).
- Tools – higher level operations. Currently holds the flagging and imaging capabilities.
- CLI – simple command line interface.

3.2. A breakdown of the Results Manager GUI

- Menu bar

  New data files are created and existing data files are opened through the *File* menu. The *Options* menu has three selections which can be toggled. The "Write to Script" option signals that DISH should echo the Glish behind each GUI action to the AIPS++ Scripter as each action is invoked by the user. The "Save when done" option indicates that the state of DISH should be automatically saved when the current session of DISH ends. The *Operations menu* allows you to toggle on the GUI panels for the available operations in DISH. The "Select from RM" option allows bulk processing of groups of scans. Normally DISH operates on the currently displayed spectrum; "Select from RM" overrides this behavior, and forces DISH to work on whatever is currently selected (highlighted) in the Results Manager.

- Results Manager (main frame)

  The *Results Manager* is the heart of DISH. All Glish results which DISH creates in response to user actions are stored in the *Results Manager* and listed in the "Results" window. These results are all available at the Glish command line as well (i.e. the variable name is a record with all of the data and associated information). DISH immediately plots the currently selected result if it can be plotted. A description is associated with each result. The user can change the description or the name of the result if they choose to do so. Certain results can also be browsed. Pressing the "Inspect" button launches a simple GUI useful for looking at the contents of results which can not be browsed or displayed.

- Status Line

  This is the line of text immediately below the *Results Manager* which generally describes the results of user actions or the current state of the GUI. These messages are echoed to the AIPS++ logger for longer term storage.

- Operation Windows

  Each operation has an associated panel which is toggled on and off through the *Operations* menu and appears at the bottom of the DISH window. All operations have an "Apply" button which, when pressed, actually performs the selected operation. Currently, there are eleven predefined operations, including: data selection, averaging, baseline fitting and removal, apply a user-specified function to data, regridding, saving, smoothing, obtain data statistics, spectral calculator, gaussian fitting, and write to file.

All of the results in the *Results Manager* are available at the Glish command line for further manipulation (e.g. user defined functions can be applied to data arrays). The results of operations done at the Glish command line can then be placed in the *Results Manager*. As mentioned, the 'Write to Script' option is also a valuable tool to see the underlying Glish behind each GUI operation and to assemble Glish scripts to do things which DISH does not yet do.

3.3. Plotter

The DISH plotter (Figure 2) is built upon the existing AIPS++ pgplotter tool. Any activity or task in DISH which produces a result which can be plotted will automatically have the result displayed in the DISH plotter. The plotter features tools for: 1) changing the X-axis units; 2) changing the reference frame for the X-axis; 3) toggling header information display; 4) toggling the overlay of subsequent plots; 5) options for a) altering the colors of the plot (reverse color), and b) allowing a draw facility (chalkboard); 6) toggling the X-axis autoscale; 7) unzoom (incrementally unzooms any zooms); 8) activate the gaussian fit tool; and 9) line identification (marks any molecular lines within the frequency ranges of the displayed plot). The cursor position is continuously displayed in the grooved box at the bottom of the plot whenever the cursor is within the DISH plotter frame.

File Menu Open: This operation will read a plot file on disk written by the Save command. Save: This operation will write a vector plot file of the displayed image to disk. Save will write to the file specified in Feature L (defaults to dish.plot). Print: This operation will print the current spectrum. Dismiss: This closes the plotter window while Done destroys the tool (not recommended!).

Edit Menu This brings up a frame which allows interactive editing (changing of existing commands) or addition of new commands. For example, additional labeling or fiducial lines may be easily added. The frame lists all of the available PGPLOT commands. Selecting one brings up a brief explanation along with the various arguments required. The frame also lists the PGPLOT commands which produced the displayed frame. Again selecting a command brings up a small explanation along with the arguments used in its execution. Cut, Copy, and Paste features are available to duplicate and add additional commands.

Help Menu Standard pgplotter help menu. The options are: 1) PGPlotter, will drive your Netscape browser to the help section on the PGPlotter; 2) Reference Manual, will drive the browser to the Reference Manual main page; 3) About Aips++, provides basic information about the AIPS++ version being used.

Plotfile name This is the filename that the Save option in the File menu will save to by default. The default plotfile name is dish.plot.

Clear Plotter Clears all currently displayed plots. Resets most plot parameters to their defaults. The color of displayed data resets to red, but line styles and plot styles previously selected will be retained.

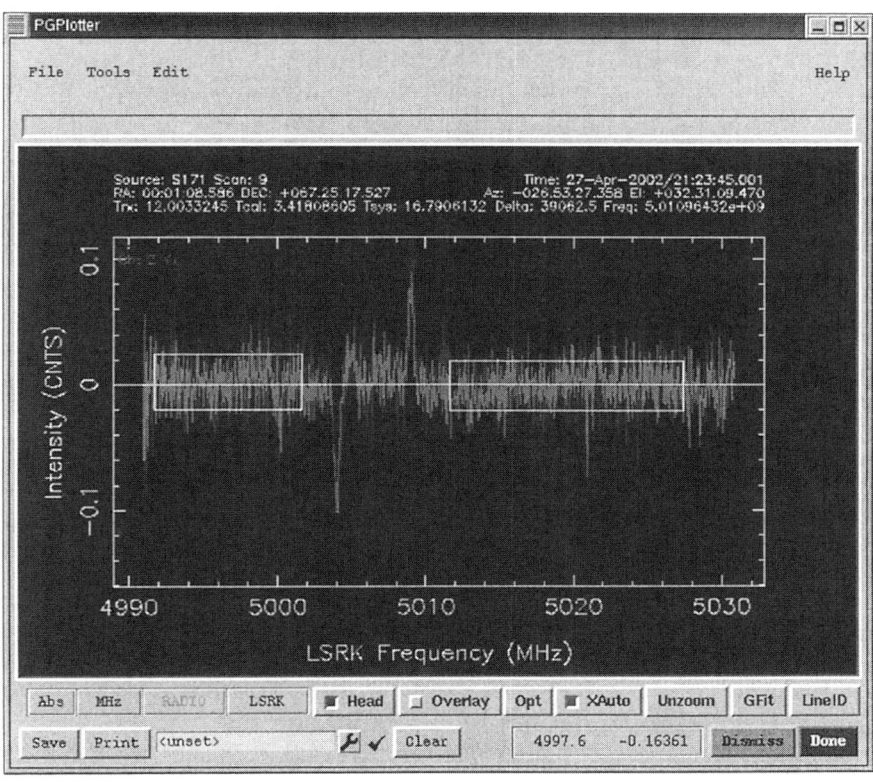

Figure 2. Dish Plotter Appearance. Displayed is a baselined GBT frequency switched spectrum. The baseline regions are marked by the white boxes whose height indicates the rms in that region. The different lines are different polarizations as labelled in the upper left.

Cursor Position Whenever the cursor is within the Dish Plotter frame, the cursor positions will be continuously read back. The coordinates shown are the X and Y position as defined by the X and Y axes' units.

Dismiss Plotter Dismisses or closes the Dish Plotter frame. The frame will be automatically re-opened if an action occurs which produces a plottable result, i.e., if a spectrum is selected, a new Dish Plotter frame will be initialized to display the selection.

3.4. CLI

The DISH command line interface layer provides simple scripting and general command line use. It is composed of a subset of functionality available through the GUI but with easier syntax for more facile manipulation of data. Most CLI commands return SDRecords which are basically spectra complete with all the associated header information. These records may also be stored in the Results Manager for further browsing and as a history mechanism. Any variables in the Results Manager are accessible at the command line by those names. Names of variables in the Results Manager may be changed by typing a string in the "New name" entry box of the Results Manager after selecting the variable.

3.5. Data reduction session

In this section we run through an example data reduction session with comments. For this example we use the CLI interface; the GUI should be intuitive, with selections made through mouse selections which highlight or check the relevant selections.

In the following examples ">" denotes a Linux command line prompt and "–" indicates a Glish prompt from within AIPS++. The logger window has also been closed which then ports all messages to the command line window.

- Create a DISH tool

```
> . /aips++/release/aipsinit.sh
> aips++ -l dish.g           # start AIPS++ and include
                             # the DISH tool
or

> aips++
- include "dish.g"           # you can include glish
                             # scripts once you've
                             # entered into AIPS++

- d := dish()                # this creates an instance
                             # of the DISH tool


- field_names(d)             # what functions are available

  addop addtool busy clearrm dologging files filein gauss wsname
```

```
find doselect done nogui getscan nsubscans setfeed getfeed gui
history dhistory info listscans logcommand message mult open
ops tools rm header plotcom plotscan plotxy rmadd plotter
restorestate savestate scanadd scansub scandiv scansrr
scanmult scanscale scanbias scanscale2D scanbias2D summary
statefile type uniput uniget view_sdrec gaussian boxcar eff
qdumps qscan PScal FScal sig ref temp plotr getr plotc getc
debug mscal imcal makeim calib import fixpnt tip1 tip gms
clip msr aver base save select smooth stat writetofile
fileout zline

- d.mult                          # typing in a function name
                                  # prints out the function.
function (val factor, val scanrec = F) {
{
if is_boolean(scanrec)
{{ (scanrec := sd.rm().getlastviewed().value); }}
if is_sdrecord(scanrec) {{ {
{
{
(temp := spaste(ARR*, factor));
(ok := sd.ops().function.setfn(temp));}

(ok := sd.ops().function.apply(scanrec));}

return private.getrecord()}
}}
else {{ return throw(FAIL: Bad SDRecord) }}}

return F}
```

- Get data in

  For this example, we use demonstration data from the AIPS++ repository (/aips++/data/demo/dishdemo/dishdemo2.fits).

  First we must convert the data from SDFITS to a MS.

```
- include 'ms.g'
- myms := sdfitstoms("dishdemo2_ms","dishdemo2.fits");
Starting server sdfits2ms
Server started: /aips++/weekly/linux/bin/sdfits2ms
(AIPS++ version: 1.6 (build #167))
Starting sdfits2ms::convert
Copying SDFITS file dishdemo2.fits to MeasurementSet
dishdemo2_ms
Finished sdfits2ms::convert
          8.57 real        1.85 user        0.36 system
Successfully closed empty server: sdfits2ms
```

```
Starting server ms
Server started: /aips++/weekly/linux/bin/ms
(AIPS++ version: 1.6 (build #167))
```

Next, we load the data into DISH; this creates an entry in the Results Manager "Variable" window which indicates that this is a Glish variable which can be manipulated through DISH or through user-written functions.

```
- d.open('dishdemo2_ms')        # this will appear in the
                                # results manager
```

HINT: for scripting, it is often useful to assign a function result to a value, for example:

```
- ok := d.open('dishdemo2_ms')
```

Functions which end successfully typically return a T, if the value of "ok" is something else, then a problem occurs which you can trap for in the script.

```
- field_names(dishdemo2_ms1)    # this shows those functions
                                # that are intrinsic to the
                                # data set after being
                                # loaded into DISH.

sel2string select resync reselect origin more next previous
setlocation location length iswritable get getempty put
appendrec deleterec getheader getother getdata getdesc
getvectors getheadervector getothervector getdescvector
name type id history appendHistory stringfields done

- d.listsummary()                  # give a summary of the data
   1    firstz 31-Dec-1999/00:05:37.500
   2   secondz 31-Dec-1999/00:11:15.000
   3    thirdz 31-Dec-1999/00:16:52.500
   4   fourthz 31-Dec-1999/00:22:30.000
   5    firstz 31-Dec-1999/00:28:07.500
   6   secondz 31-Dec-1999/00:33:45.000
   7    thirdz 31-Dec-1999/00:39:22.500
   8   fourthz 31-Dec-1999/00:45:00.000
...

- s.listscans()            # give a list of scan numbers
                           # in the data set
[1  2  3  4  5  6  7  8  9 10 11 12 13 14 15 16 17 18 19 20 21 22
 23 24 25 26 27 28 29 30 31 32 33 34 35 36 37 38 39 40 41
 42 43 44 45 46 47 48 49 50 51 52 53 54 55..
]
```

- Initial inspection

    The measurement set can be viewed by using the table browser. A glish variable is set to the data on disk by:

    ```
    - mytable := table('dishdemo2_ms');
    successful readonly open of default-locked table dishdemo2:
    44 columns, 256 rows
    - field_names(mytable)         # shows the various
                                   # functions available

    type ok open handle server toascii browse flush resync close
    done iswritable lock unlock datachanged haslock lockoptions
    ismultiused name selectrows query info putinfo addreadmeline
    summary setmaxcachesize rownumbers colnames isscalarcol
    coldatatype colarraytype ncols nrows addcols removecols
    addrows removerows getcell getcellslice getcol getvarcol
    getcolslice putcell putcellslice putcol putvarcol
    putcolslice getcolshapestring getkeyword getkeywords
    getcolkeyword getcolkeywords putkeyword putkeywords
    putcolkeyword putcolkeywords removekeyword removecolkeyword
    keywordnames colkeywordnames fieldnames colfieldnames
    getdminfo getdesc getcoldesc

    - mytable.browse()  # brings up a frame with the main tables
                        # (other subtables can be selected
                        # as well) from the table browser, data
                        # can be examined, edited, and plotted.
    ```

    In addition, once loaded into DISH, the data may be inspected by selecting it with the mouse (which highlights it in the Results Manager) and clicking the mouse on the "Browse" button of the Results Manager. A frame will then appear which lists an index (a row in the table), the scan number and the source. Selecting on any column in the browser window will automatically plot the spectrum.

- Flag data/RFI rejection

    Selecting on the "Tool" menu displays two options, msplot and imager. msplot allows an interactive visual look at a measurement set. Selecting on the tool pulls down an entry field. Type in the name of the measurement set on disk, select the "Edit" option, and hit 'go' to initiate. This brings up the msplot frame.

    The leftmost "Show" button is a drop down menu (default is "Plot X vs Y"). Change this to "Display data as an image". Next review your Spectral selection (select this button and examine the information) and Polarization selection. If all looks well, hit "Go". This brings up an image of the data arranged by row in the MS (corresponds to time) and channel; the color map or grey scale corresponds to the intensity at that channel.

Select on the relevant flagging parameters (e.g., channel, time) and where to flag (Spectral Window, Polarization). The left mouse button is used to click and drag out rectangular flagging regions. Hitting 'ESC' will eliminate a box. Hitting the "Flag" button will then flag that data (it will appear blanked). When you are finished with this, use the 'File' menu to delete the tool and make the flagging changes (Hit 'Done').

If you made some changes to the data, then it must be re-loaded into DISH (redo the "Getting it in" section).

- Calibrate data

  DISH analysis routines expect data is calibrated to one of the radio temperature scales (e.g., T_R^*). For instruments which do not naturally write data in this form, there are instrument specific Glish scripts which parse information to the AIPS++ calibrate module for calibration. An example interface for data taken with the Arecibo telescope looks as follows:

```
> aips++ -l naic.g
- ao.import('U3564.sdfits');
- aoname:='U3564_ms1';
- ao.gaincorr();
- ao.plot('average1');
```

The AIPS++ "Getting Results" document details some step-by-step calibration done with various types of data including Arecibo and GBT.

Here we illustrate a simple example function to reduce a GBT frequency switched scan.

```
- d.import('standards_apr27',outms='demo2',startscan=5,
stopscan=9,calflag=T);
T
- d.gms()
successful readonly open of default-locked table
demo_SP/NRAO_GBT_GLISH: 23 columns, 5 rows
Scan  Object Proctype  SWState    SWtchsig Procseqn  Procsize
   5  S171   OffOn     PSWITCHOFF TPWCAL       1         2
   7  S171   OffOn     PSWITCHOFF TPWCAL       1         2
   9  S171   Track     FSWITCH    FSW01        1         1
T
- d.calib(9); # calibrate scan 9
Scan 9 calibrated
T
- d.calib(9,T,order=5); # calibrate scan 9, baseline the
                        # data with a 5th order polynomial.
- d.plotc(9); # plot this data; if an integration number
              # isn't specified, it averages all
              # integrations together; if no polarization
              # is specified, it plots all polarizations.
```

- Average data

  Averaging data can be done based on various selections. A combination of the Select GUI and Average GUI will generally get you an averaged spectrum of a desired subset. For quick, scan/sub-scan based selections, DISH's CLI average tool is very efficient. It takes a vector of scan numbers, a vector of subscans, a weighting parameter (NONE, TSYS (TSYS and time), or RMS=default), and an optional alignment parameter. A simple example follows:

  ```
  myavg := d.aver([2,3,4],[1,3]); # myavg will contain the
                                  # average of subscans 1 and 3
                                  # in scans 2-4
  ```

- Baseline data

  A spectral line will often be offset due to a continuum background or from instrumental effects (e.g. standing waves from the telescope structure). A baseline fit is typically obtained, which is then subtracted to allow an accurate measure of the actual line intensity. This is done by:

  ```
  - myscan:=d.getscan(200,1)
  - d.base(myscan,order=2,action='subtract',
  range='[50:400],[1500:2000]', units='channels');
  ```

- Smooth data

  Smoothing data will improve signal-to-noise at the expense of resolution. This can be achieved through three different techniques: 1) Hanning smoothing; 2) Boxcar smoothing; and 3) Gaussian smoothing.

  ```
  - myscan:=d.getscan(200,1);
  - d.smooth(myscan, type="BOXCAR",width=5);
  ```

- Gaussian Fit data

  Typically spectral lines are broadened through thermal and non-thermal effects into a Gaussian profile. The 'gauss' routine allows you to fit a Gaussian profile to a spectrum.

  ```
  - d.gauss(1,prompt=T); # fit 1 gaussian, use cursor
                         # to set guesses.
  ```

- Working with images

  The process of making a single-dish image in AIPS++ is straightforward. The data are converted into the coordinate system of the image (via imager.setimage) and then added to the image using a gridding function. A variety of convolution functions can be used: BOX, SF or PB. The first is a simple nearest neighbor gridding, the second uses a prolate spheroidal wavefunction and the last uses a primary beam model appropriate to the telescope being used. The primary beam is optimal in the least squares sense but degrades the resolution, whereas the prolate spheroidal function can avoid the resolution degradation but the noise level will be higher.

To access an interface to the AIPS++ imager tool, select it from the list of Tools in the Results Manager. This opens a frame which has various entry fields for setting the parameters for making the image. An example set of fields would look as follows:

```
MeasurementSet: w49n_all_ms
                # Select the data on disk from
                # /aips++/data/demo/dishdemo/w49n_all.fits

setdata: fieldid=1,spwid=1,msselect='SCAN_NUMBER<423',
mode='channel',nchan=256, start=1,step=1
                # Set the field, spectral window, any
                # selections (in this case, all scans
                # less than scan 423), all channels
                # (256 channel spectrum, stepping along
                # by 1 channel).

setimage: spwid=1,nx=40,ny=40,cellx='0.79arcmin',
celly='0.79arcmin',stokes='I', doshift=T,mode='channel',
nchan=15,start=80,step=5
                # Make 15 planes, averaging 5 channels
                # per plane, starting at channel 80.

setoptions: gridfunction='PB'
                # Use the Primary Beam for the gridding.

weighting: 'natural'
                # Use natural weighting.

row of phase center: 25
                # Select the direction for the phase
                # center.
```

Streamlined tools for making data cubes from GBT and JCMT data are available. Again, see the Getting Results documentation for details.

3.6. Scripting – extending DISH functionality

DISH functionality can be extended in several ways. First, it can be extended simply by writing Glish scripts which perform other operations on data. Data are accessible through table functions, MS functions, and DISH functions. This example accesses scan data and then simply changes it.

```
- changesourcename := function(ref sdrecord,newsourcename) {
+ sdrecord.header.source_name:=newsourcename;
+ return T;
```

```
+     }
- testscan := d.getscan(425,1); #gets the first subscan of scan 425
- field_names(testscan)
data header hist other
- testscan.header.source_name
W49N
- changesourcename(testscan,"W49OH")
T
- testscan.header.source_name
W49OH
```

Plugins are another standard means of adding functionality to DISH. Any file called mydish_*.gp that lives in the current directory will be picked up when DISH is started. An example plugin file with a simple squaring function is shown here:

```
pragma include once;
note('mydish_example included');
mydish_example := [=];
mydish_example.attach := function (ref public) {
        #new function
#####User adds functions here
public.square := function(somevalue) {
wider public;
print 'the square of ',somevalue,' is: ',somevalue*somevalue;
}
# other functions go here
# public.otherfunction := function(input1,input2,etc) {
#           some code goes here to do something
#           return; # can return a value too if you like
#       }
# public.etc...
#
#####End user edit
return T;
}
```

If you then look at the functions contained in your DISH tool, d, it will now have a 'square' function.

Finally, any syntax can be roughly approximated by a combination of Glish and the existing tools in DISH. You can always set a new variable to equal an existing function. In this way you can modify and create your own command line reduction language. An example of changing the syntax to approximate a UniPops command would be:

```
UniPops: hanning xx             # This will hanning smooth a
                                # spectrum and redisplay it.

- hanning:=function() { d.smooth('HANNING');}
```

- hanning()
                              # Create a new function that
                              # automatically does a hanning
                              # smooth; CLI defaults to operate
                              # on the currently displayed
                              # spectrum. This now does the
                              # equivalent of the above
                              # UniPops command.

3.7. Getting publication quality graphs

The key to getting publication quality graphs is understanding that you have access to all functions in the PGPLOT library. In particular, the annotation commands (label, mtxt, etc) are useful, as are the CLI commands which allow plotting arbitrary points to the screen.

A sample creation/annotation of a plot:

- x := 1:100   # Create a vector of values from
                              # 1 to 100.
- y := sin(x)   # Create another vector which is
                              # the sine.
- d.plotxy(x,y)   # Plot the values.
- d.plotcom('sci',7)   # Set the color to yellow.
- d.plotcom('line',x,y/2)   # Plot a line of these values.
- d.plotcom('lab','my x-axis','my y-axis','my title')
                              # Label the title and axes.
- d.plotcom('text',80,1.0,'sample data');
                              # Move to the x=80, y=1.0 location
                              # and write text.
- d.plotcom('move',60,1.0)   # Move to the x=60, y=1.0 location.
- d.plotcom('draw',75,1.0)   # Draw a line from the last
                              # position to x=75, y=1.

References

Garwood, R.W. 2000, in ASP Conf. Ser. Vol. 216, Atronomical Data Analysis Software and Systems IX, ed. N. Manset, C. Veillet, & D. Crabtree (San Francisco: ASP), 243

Part 4
Special Topics

"You call this a pointer?!" – S. Stanimirović.

Short-Spacings Correction from the Single-Dish Perspective

Snežana Stanimirović

National Astronomy and Ionosphere Center, Arecibo Observatory, HC 3 Box 53995, Arecibo, Puerto Rico 00612, USA

Abstract. While, in general, interferometers provide high spatial resolution for imaging small-scale structure (corresponding to high spatial frequencies in the Fourier plane), single-dishes can be used to image the largest spatial scales (corresponding to the lowest spatial frequencies), including the total power (corresponding to zero spatial frequency). For many astrophysical studies, it is essential to bring 'both worlds' together by combining information over a wide range of spatial frequencies. This lecture demonstrates the effects of missing short-spacings, and discusses two main issues: (a) how to provide missing short-spacings to interferometric data, and (b) how to combine short-spacing single-dish data with those from an interferometer.

1. Introduction

All radio telescopes can be classified as either filled or unfilled-aperture antennas. The simplest filled-aperture antennas are single-dish telescopes. The desired astronomical object to be observed and the particular scientific goals determine which type of radio telescope to use. In general, single-dishes are considered as tools for low spatial resolution observations, while interferometers are used for high resolution observations. While compact objects are more suited for interferometric observations, extended objects are commonly observed with single-dishes as interferometers cannot faithfully recover information on the largest spatial scales.

However, in many scientific cases it is essential to obtain high spatial resolution observations of large objects, and to accurately represent emission present over a wide range of spatial scales. A simple recipe you may follow in such cases is:

- observe (mosaic) your object with an interferometer,
- observe your object with a single-dish,
- cross-calibrate the two data sets, and then
- combine the single-dish and interferometer data.

This combination of single-dish and interferometer data, when observing extended objects, is referred to as the short-spacings correction.

This 'simple' recipe may be considered as an artistic touch to the interferometric images as it makes them look much nicer but still preserves their high spatial resolution. This results from: (a) inclusion of more resolution elements, those seen by a single-dish; and (b) reconstruction of image artifacts. At the same time, these images contain information about the total power, and can be used to measure accurate flux densities, column densities, masses, etc. From a pure historical perspective, the short-spacings correction bridges the gap between the two classes of radio telescopes, essentially obtaining the best of 'both worlds', that is the high spatial resolution information provided by interferometers, and the low spatial resolution, including the total power, information provided by single-dishes.

This lecture will explain what the short-spacing problem is, how it is manifested, and how we can, both theoretically and practically, solve this problem. Section 2 depicts very briefly the fundamentals of interferometry, defines the spatial frequency domain, and draws an analogy between a single-dish and an interferometer. The effects of missing short-spacings are demonstrated in Section 3, as well as prospects for solving the problem. Section 4 considers the cross-calibration of interferometer and single-dish data which is a precursor to any combination method. Methods for data combination are discussed in Section 5 and Section 6, and compared in Section 7.

2. A Very Brief Introduction to Interferometry

A very brief review of the basics of interferometry is necessary right at the beginning of this lecture, in order to define and explain some terms that will be used further on. However, we do not want to go deeply into interferometry, as there is a vast literature available on this topic, starting with "Interferometry and Synthesis in Radio Astronomy" (Thompson, Moran, & Swenson 1986) and "Synthesis Imaging in Radio Astronomy" (Taylor, Carilli, & Perley 1999).

The fundamental idea behind interferometry is that a Fourier transform relation exists between the sky radio brightness distribution I and the response of a radio interferometer. If the distance between two antennas (the baseline) is $\mathbf{d}$, then the so-called visibility function, $V(\mathbf{d})$, is given by:

$$V(\mathbf{d}) = \int_{\text{source}} A(\boldsymbol{\sigma}) I(\boldsymbol{\sigma}) \exp\left[-2\pi i\, \mathbf{d} \cdot \boldsymbol{\sigma}/\lambda\right] d\Omega\,. \qquad (1)$$

Here, $A(\boldsymbol{\sigma})$ is an antenna reception pattern, or **primary beam**, and $\boldsymbol{\sigma}$ is the vector difference between a given celestial position and the central position of the field of view. The **aperture synthesis technique** is a method of solving Equation 1 for $I(\boldsymbol{\sigma})$ by measuring V at suitable values of $\mathbf{d}$.

To simplify Equation 1, a more convenient, right-hand rectilinear, coordinate system is introduced in Figure 1. Coordinates of vector $\mathbf{d}$ in this system are (u, v, w), where the direction to the source center $\mathbf{s}_0$ defines the w direction, and u and v are baseline projections onto the plane perpendicular to the $\mathbf{s}_0$ direction, towards the East and the North, respectively. A synthesized image in the $l - m$ plane represents a projection of the celestial sphere onto a tangential plane at the source center. In certain conditions, that is in the case of an Earth tracking, East-West interferometer array, with the w-axis lying in the direction

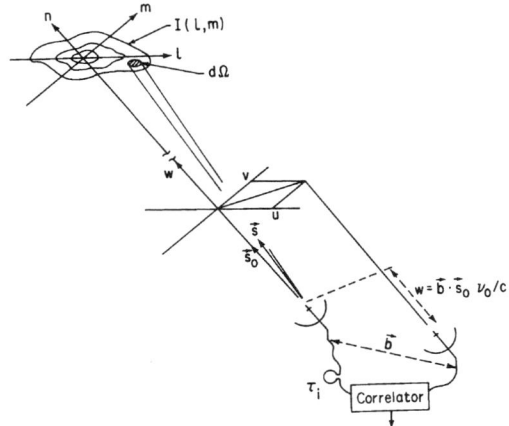

Figure 1. A schematic diagram of the spatial frequency and image coordinate systems. The spatial frequency domain, (u, v, w), is used to express the interferometer baseline, and the image domain, (l, m, n), is used to express the source brightness distribution. The direction to the center of the field of view is given by $\mathbf{s_0}$, and to any given position by $\mathbf{s}$. Hence, $\mathbf{s} = \mathbf{s_0} + \boldsymbol{\sigma}$. (From Thompson 1994.)

of the celestial pole, further simplifications of Equation 1 are possible:

$$V(u,v) = \int\int A(l,m)I(l,m)\exp[-2\pi i(ul+vm)]\frac{dldm}{\sqrt{1-l^2-m^2}}. \quad (2)$$

Therefore, the visibility function $V(u, v)$ can be expressed as the Fourier transform of a modified brightness distribution $A(l, m)I(l, m)$. Coordinates u and v ($w = 0$) are measured in units of wavelength and the $u - v$ plane is called **the spatial frequency domain**. These are effectively projections of a terrestrial baseline onto a plane perpendicular to the source direction. The $l - m$ plane is referred to as **the image domain**. To obtain $I(l, m)$, from Equation 2, an inverse Fourier transform of $V(u, v)$ is required, meaning that a complete sampling of the spatial frequency domain is essential. In practice however, a bit more than a simple inversion is needed as only limited sampling of the $u - v$ plane is available.

2.1. How do we 'fill in' the $u - v$ plane?

For a given configuration of antennas any interferometer array has a limited range of baselines, lying between a minimum, $d_{\min}$[1], and maximum, $d_{\max}$, baselines. As an example, 5 antennas of the Australia Telescope Compact Array

---

[1] The shortest baseline that can be achieved is constrained by the physical limitations in placing two dishes together and the shadowing effect of one dish by another one.

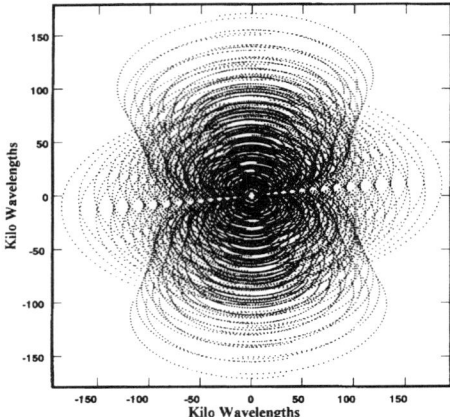

Figure 2. The $u - v$ coverage for an 8 hr tracking observation with the VLA, for an object at declination of 30° (from Burke & Smith 1997).

(ATCA) form 10 baselines, with $d_{min} = 31$ m and $d_{max} = 459$ m for a particularly compact configuration. The final resolution (θ_{int}) is inversely related to the maximum baseline by $\theta_{int} \approx \lambda/d_{max}$. In the case of an Earth tracking interferometer, as the Earth rotates the baseline projections on the $u - v$ plane trace a series of ellipses. The parameters of each ellipse depend upon the declination of the source, the length and orientation of the baseline, and the latitude of the center of the baseline (Thompson et al. 1986). The ellipses are concentric for a linear array (e.g. ATCA). For a 2-dimensional array (e.g. the Very Large Array, VLA), the ellipses are not concentric and so can intersect.

As each baseline traces a different ellipse, the ensemble of ellipses indicates the spatial frequencies that can be measured by the array (see Thompson et al. 1986). At each sampling interval, the correlator measures the visibility function for each baseline, thus resulting in a number of samples being measured over elliptical tracks in the $u - v$ plane. Hence, the resultant interferometer $u - v$ coverage will always be, more or less, incomplete, having a hole in the center of the $u - v$ plane whose diameter corresponds to the minimum baseline, gaps between measured elliptical tracks, and gaps between each adjacent samples on each elliptical track. The ensemble of ellipses (loci) is known as the **transfer** or **sampling function**, $b_{int}(u, v)$. An example of the sampling function obtained with the VLA is shown in Figure 2.

Hence, if $V(u, v)$ is a true (ideal) visibility function, the measured (observed) visibilities (V'_{int}) can be expressed as:

$$V'_{int}(u, v) = V(u, v) b_{int}(u, v) . \qquad (3)$$

b_{int} is usually representable by a set of δ-functions, between the lowest and the highest spatial frequency sampled by the interferometer (corresponding to the shortest and the longest baselines, respectively). The Fourier transform of Equation 3 gives the observed sky brightness distribution I_{int}^D (so called **'dirty' image**):

$$I_{int}^D(l,m) = I(l,m) * B_{int}(l',m') , \qquad (4)$$

where B_{int} is **the synthesized or 'dirty' beam**, which is the point source response of the interferometer. As usually, asterisks (∗) are used to denote convolution. When imaging, incomplete $u-v$ coverage leads to severe artifacts, such as negative 'bowls' around emission regions and negative and positive sidelobes (Cornwell, Braun, & Briggs 1999). We return to this in Section 3. The determination of I from I_{int}^D in the deconvolution process, requires beforehand interpolation and extrapolation of V'_{int} for missing data due to the discontinuous nature of b_{int} (Cornwell & Braun 1989). This process works well when a compact configuration of antennas is used and when the source is small enough, with angular size $\theta \leq 2\lambda/d_{min}$ (Bajaja & van Albada 1979).

For imaging larger objects, with angular size $\theta > \lambda/d_{min}$, a significant improvement in filling in the $u-v$ coverage can be achieved by using the **'mosaicing technique'**, where observations of many pointing centers are obtained and 'pasted' together (see Holdaway 1999). Mosaicing effectively reduces the shortest projected baseline to $d_{min} - D/2$, where D is the diameter of an individual antenna. Nevertheless, the center of the $u-v$ plane still suffers if significant large scale structure is present. This lack of information for very low spatial frequencies (around the center of the $u-v$ coverage) in an interferometric observation is usually referred to as the **'short-spacings problem'**.

2.2. A single-dish as an interferometer?

Extended objects with angular size $\theta > \lambda/d_{min}$ can be observed with a single-dish. Let us now think of a single-dish in a slightly unusual way, imagining filled apertures consisting of a large number of small panels packed closely together. Then all these panels can act as interferometer elements with their signals being combined together at the focus, making a so called phased or adding array (see lecture by D. Emerson in this volume). The distance between each two panels corresponds to a baseline, as shown in Figure 3. The baseline distribution then monotonically decreases from zero at the center up to the maximum baseline, determined by the single-dish diameter D_{sd}. This is also shown in Figure 3.

One observation with a single-dish provides a total flux density measurement, corresponding to the zero spacing, $(u,v) = (0,0)$. However, if a single-dish scans across an extended celestial object, it measures not only a single spatial frequency, but a whole range of continuous spatial frequencies all the way up to a maximum of D_{sd} (Ekers & Rots 1979). Hence, a single-dish behaves as an interferometer with an almost infinite number of antennas, and therefore has a continuous range of baselines, from zero up to D_{sd}. The nice thing about this representation is that we can now use the same mathematical notation to describe both single-dishes and interferometers.

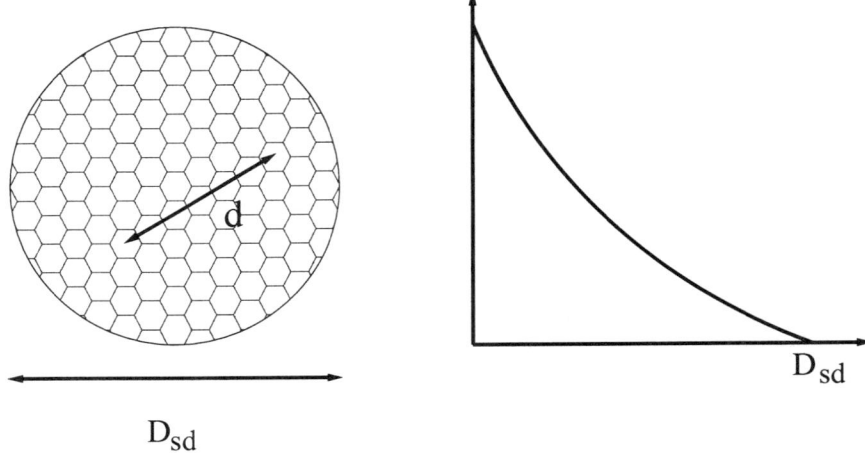

Figure 3. A schematic representation of a single-dish as an interferometer with a large number of elements and a monotonically decreasing distribution of baselines (d) from zero to D_{sd}.

The observed sky brightness distribution I_{sd}^D in the case of single-dish observations is then given by:

$$I_{sd}^D(l,m) = I(l,m) * B_{sd}(l',m'), \tag{5}$$

with B_{sd} being the **single-dish beam** pattern. The Fourier transform of Equation 5 gives the observed single-dish 'visibilities', V'_{sd}:

$$V'_{sd}(u,v) = V(u,v) \times b_{sd}(u,v), \tag{6}$$

where b_{sd} is the Fourier transform of the single-dish beam pattern which, unlike b_{int}, is a continuous function between zero and the highest spatial frequency sampled by the single-dish. Determination of I from I_{sd}^D requires deconvolution, but no interpolation of V'_{sd} is needed since this is a continuous function.

3. How is the Short-spacings Problem Manifested?

As shown in Equation 4, the sky brightness distribution can be reconstructed, in the case of interferometric observations, by deconvolving the 'dirty' image with the synthesized beam. As an example, Figure 4 shows the result of the HI 'mosaic' observations of the Small Magellanic Cloud (SMC) with the ATCA. More information about these observations and data reduction is available in Stanimirovic et al. (1999). The two adjacent panels on the right side show right ascension (RA) cuts through the image. Negative bowls (shown in white on the image) are seen around emission peaks (shown in black), as well as in RA cuts. These are typical interferometric artifacts resulting from an incomplete $u - v$ coverage.

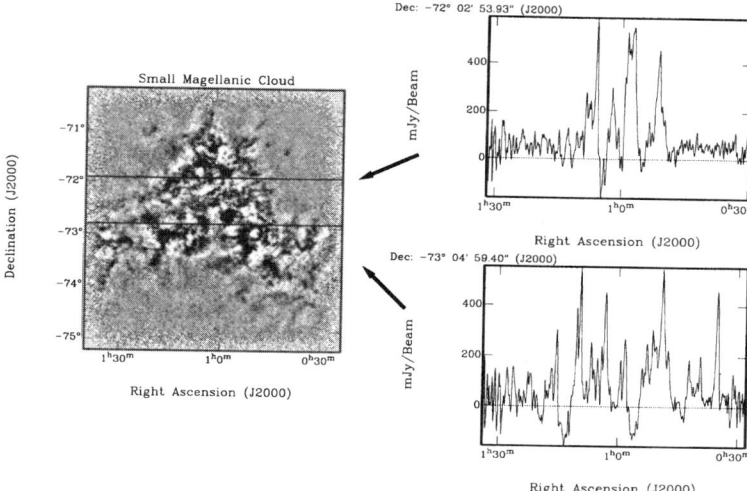

Figure 4. An HI image of the SMC from Stanimirovic et al. (1999) is shown on the left. The observations were obtained as a mosaic of 320 different pointing centers with the ATCA. Positive intensity values are shown in black, while white represents negative pixel values. RA cuts through the image, at Dec $-72°\ 02'\ 54''$ and Dec $-73°\ 05'$, are shown on the right.

A simple graphical explanation of why this happens, borrowed from Braun & Walterbos (1985), is shown in Figure 5 for the case of a point source. The solid vertical line in Figure 5 distinguishes the spatial frequency (a) from the image domain (b). The distribution of measured spatial frequencies, or what we have already defined as a transfer (or sampling) function, b_{int}, is given on the left side, while its Fourier transform, that is the synthesized beam, B_{int}, is shown in the right. An exclusion of the central values from the spatial frequency domain, is equivalent to a subtraction of a broad pedestal in the image domain, resulting in the presence of a deep negative 'bowl' around the observed object, as seen in Figure 4.

This demonstrates simply how severe the effects of missing short-spacings can be. The larger the object is relative to the reciprocal of the shortest measured baseline one tries to image, the more prominent the short-spacing problem becomes.

3.1. How can we solve the short-spacings problem?

There are two main questions concerning the short-spacings problem:

1. how to provide (observe) missing short-spacings to interferometric data;
2. how to combine short-spacing data with those from an interferometer.

In answering the first question, all solutions can be grouped into two array schemes: **homogeneous**, having all antennas of the same size, and **heteroge-**

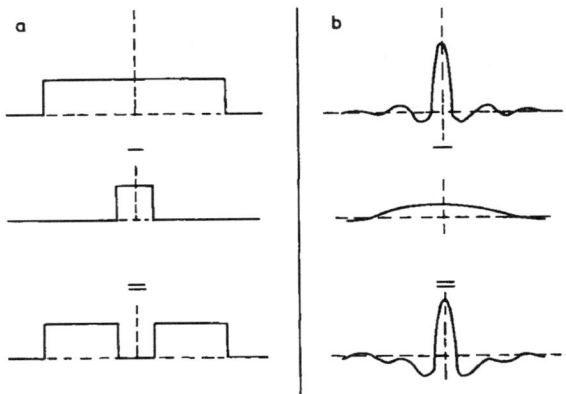

Figure 5. A 1-dimensional cut through $(u, v) = (0, 0)$ of the spatial frequency domain, (a), and its corresponding manifestation in the image domain, (b). An exclusion of the central values from the spatial frequency domain is equivalent to a subtraction of a broad pedestal in the image domain, resulting in the presence of a deep negative 'bowl' around the observed object. (From Braun & Walterbos 1985.)

neous, based on observations obtained with different-sized antennas. There are many possibilities concerning the heterogeneous arrays, such as using smaller arrays and even a hierarchy of smaller arrays. The simplest option, though, is a single-dish telescope with a diameter (D_{sd}) larger than the interferometer's minimum baseline.

We briefly touch here on some pros and cons for both array schemes. One of the difficulties in providing short-spacings with a single-dish is that it is hard to provide a large single-dish which would have the sensitivity equivalent to that of an interferometer (Holdaway 1999). Also, single-dish observations are complex (they require a lot of separate pointing centers to cover a large object) and very sensitive to systematic errors. Using theoretical analysis, numerical simulations and observational tests, Cornwell, Holdaway, & Uson (1993) show that a homogeneous array in which the short-spacings are obtained from single antennas of the array, allows high quality imaging. They find that a key advantage over the large single-dish scheme is pure simplicity, which is an important factor for the complex interferometric systems. As both interferometric and total power data are obtained with the same array elements, no cross-calibration is required in this case. Note that in this case total-power and interferometric observations have to be synchronized which is not a simple task because of different observing techniques involved (e.g. the single-dish observations require frequency or position switching modes!). This turns out to be an especially difficult task for the continuum observations.

However, to *fully* fill in the central gap in an interferometer $u-v$ coverage and *preserve sensitivity* at the same time, the heterogeneous array scheme appears more advantageous. This has been recently recognized in the planning of the future Atacama Large Millimeter/Submillimeter Array (ALMA). Imaging simulations have shown that antenna pointing errors of only a few percent of the primary beam width produce large errors in the visibilities in the central $u-v$ plane, causing a large degradation of image quality (see Morita 2001). To compensate for this problem, an additional, smaller array of 6 – 8 m dishes, the so called ALMA Compact Array (ACA), has been proposed to provide short baselines.

In answering the second question, methods for the combination of interferometer and single-dish data can be grouped into two classes: data combination in the spatial frequency domain (Bajaja & van Albada 1979; Vogel et al. 1984; Roger et al. 1984; Wilner & Welch 1994; Zhou, Evans, & Wang 1996), and data combination in the image domain (Ye & Turtle 1991; Stewart et al. 1993; Schwarz & Wakker 1991; Holdaway 1999). Each approach can be realized through a number of different methods. Both approaches are very common and are becoming a standard data processing step.

As the most common scheme of a heterogeneous array involves use of a large single-dish telescope, we proceed to consider this particular case further.

4. Cross-calibration of Interferometer and Single-dish Data

Before adding short-spacing data, it is necessary to be sure that both the interferometer and single-dish data sets have identical flux density scales. As calibration is never perfect, the calibration differences between the two sets of observations can be significant in some cases (e.g. observations spread over a long period of time, different data quality, use of different flux density scales for calibration, quality of calibrators, etc.). This results in a small but appreciable difference in the measured flux densities.

We define the calibration scaling factor, f, as the ratio of the flux densities of an unresolved source in the single-dish and interferometer maps:

$$f = \frac{S_{\text{int}}}{S_{\text{sd}}}. \qquad (7)$$

In the case of perfect calibration, $f = 1$. However, $f \neq 1$ otherwise, and needs to be determined very accurately. Unfortunately, it is hard to find suitable compact sources to directly determine f. Hence, the best way to estimate f is to compare the surface brightness of the observed object in the overlap region of the $u-v$ plane, see Figure 6. This region should correspond to angular sizes to which both telescopes are sensitive. For a source of brightness, I, both the interferometer and single-dish should measure within this region the same, $I_{\text{sd}} = I_{\text{int}}$, and calibration errors will appear as:

$$f = \frac{I_{\text{int}}}{I_{\text{sd}}}. \qquad (8)$$

For an extended source I_{int} and I_{sd} are often, for convenience, expressed in units of Jy beam^{-1} not Jy sr^{-1}, and so will be different numbers because of the

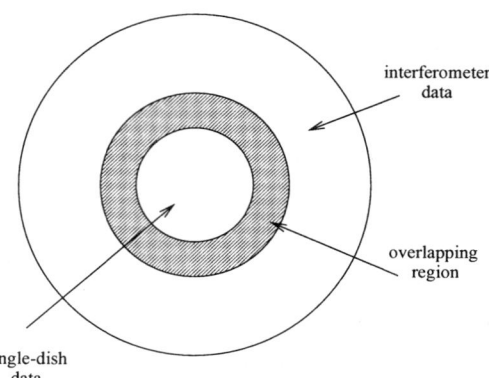

Figure 6. A schematic diagram of the spatial frequency domain of an observation in a heterogeneous scheme when using a single-dish for providing short-spacings: the inner spatial frequencies are sampled by the single-dish only, the outer ones are sampled only by the interferometer, while the overlapping region contains spacings to which both the single-dish and interferometer are sensitive.

different beams considered (with beam areas $\Omega_{\rm int}$ and $\Omega_{\rm sd}$, respectively). For this purpose, an estimate of the resolution difference between the two data sets ($\alpha = \Omega_{\rm int}/\Omega_{\rm sd}$) is also needed.

To determine f the following steps are required:

1. scale the single-dish data by α to account for the difference in brightness caused *only* by different resolutions,

2. Fourier transform the interferometer and scaled single-dish images,

3. deconvolve the single-dish data (by dividing them by the Fourier transform of the single-dish beam), and

4. compare 'visibilities' in the overlapping region of spatial frequencies.

Several important issues should be considered here:

- When Fourier transforming in Step 2 watch for edge-effects! To avoid nasty edge-effects in some cases apodizing of both interferometer and single-dish images may be required in order to make the image intensities smoothly decrease to zero near the edges.

- Step 3 requires a very good knowledge of the single-dish beam! To make things even harder, the FWHM of the single-dish beam and the calibration scaling factor are highly coupled (Sault & Killeen 1998). Therefore, an

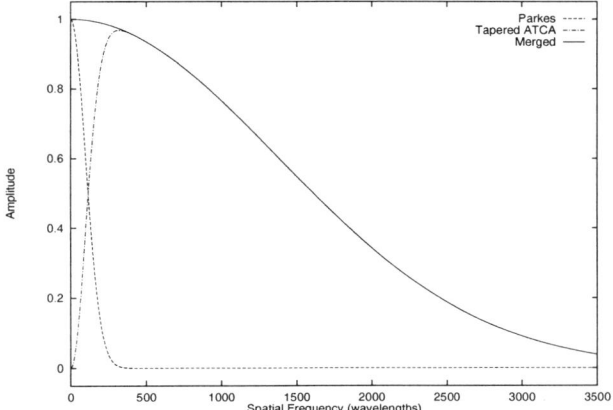

Figure 7. The tapering of spatial frequencies sampled by a single-dish (dashed line), by an interferometer (dot-dashed line) and as merged together by the MIRIAD task IMMERGE (solid line). The ATCA interferometer and the Parkes single-dish are shown as an example (from Sault & Killeen 1998).

error in the single-dish beam model has the same effect in the overlapping region as an error in the flux density scale. If the single-dish beam is poorly known, f will be a quadratic function of distance in the Fourier plane, in the first approximation (see Stanimirovic 1999).

- A sufficient overlap in spatial frequency is required for Step 4. Assuming a Gaussian-tapered illumination pattern for a single-dish, and considering a cut-off level of 0.2 for reliable data, we can estimate the minimum diameter, D, of a single-dish necessary to provide all spacings shorter than $d_{\min}$ for a given interferometer:

$$D > 1.5 \times d_{\min} . \qquad (9)$$

In order to have a reasonable overlap of spatial frequencies so that f can be derived, a slightly larger single-dish is required with $D > 2 \times d_{\min}$. For example, for the ATCA shortest baseline of 31 m, the single-dish providing short spacings should have diameter of $D \geq 62$ m. Therefore, the 64 m Parkes telescope can do a really great job. Also, while the 100-m Green Bank Telescope will be able to provide short-spacing data for the VLA C and D arrays (with $d_{\min} = 35$ m), only Arecibo could do so for the VLA B array (with $d_{\min} = 210$ m).

5. Data Combination in the Spatial Frequency Domain

5.1. Theoretically...

As shown by Bajaja & van Albada (1979), the true missing short-spacing visibilities can be provided from the function $V(u,v)$ in Equation 6, if the single-dish is large enough to cover the whole central gap in the interferometer $u-v$ coverage. The deconvolution of the single-dish data gives the true single-dish visibilities, where $b_{\rm sd}(u,v) \neq 0$, by:

$$V(u,v) = \frac{V'_{\rm sd}(u,v)}{b_{\rm sd}(u,v)} \ . \tag{10}$$

Function $V(u,v)$ can be then substituted in Equation 3, after rescaling by f, everywhere in the $u-v$ plane where Equation 10 holds. This would provide the resultant $u-v$ coverage having the inner visibilities from the single-dish data only (rescaled to match the interferometer flux-density scale due to the calibration differences), and the outer visibilities from just the interferometer data. This is effectively feathering or padding the interferometer visibilities with the single-dish data.

Depending on the type of input images used, and/or the type of weighting applied within the region of overlapping spatial frequencies, there are several applications of this technique (Roger et al. 1984; Vogel et al. 1984; Wilner & Welch 1994; Zhou et al. 1996). However, in all cases, for a good data combination the single-dish data set should fulfill the following two conditions.

- A sufficiently fine sampling of the single-dish data at the Nyquist rate (2.4 pixels across the beamwidth) is required to avoid aliasing during deconvolution (Vogel et al. 1984). The single-dish data must also have the same coordinate system as the interferometer data. Therefore, it is sometimes necessary to re-grid single-dish images.

- Visibilities derived from the single-dish data should have a signal-to-noise ratio comparable to those of the interferometer in the overlapping region in order not to degrade the combined map (Vogel et al. 1984).

5.2. Practically...

This linear data combination in the spatial frequency domain is very widely used and is implemented in several packages for radio data reduction, e.g. task IMERG in AIPS, IMAGE TOOL in AIPS++ and task IMMERGE in MIRIAD. As an example we will discuss IMMERGE here.

IMMERGE takes as input a clean (deconvolved) high-resolution image, and a non-deconvolved, low-resolution image. These images are Fourier transformed (labeled as $V_{\rm int}(k)$ and $V'_{\rm sd}(k)$) and combined in the Fourier domain applying tapering functions, $w'(k)$ and $w''(k)$, such that their sum is equal to the Gaussian function having a FWHM of the interferometer, $\theta_{\rm int}$:

$$V_{\rm comb}(k) = w'(k) V_{\rm int}(k) + f w''(k) V'_{\rm sd}(k) \ , \tag{11}$$

$$w'(k) + w''(k) = \frac{1}{\sqrt{2\pi}} \exp\left(-\frac{\theta_{\rm int}^2 k^2}{4 \ln 2}\right) \ . \tag{12}$$

Function $w''(k)$ is a Gaussian with the FWHM of the single-dish. The low-resolution visibilities are multiplied by f to account for the calibration differences. The final resolution is that of the interferometer image. The tapering functions w' and w'' are shown in Figure 7, together with the tapering function of the merged data set, for the case of the ATCA interferometer and the Parkes single-dish telescope. IMMERGE can also estimate the calibration scaling factor, f, by comparing single-dish and interferometer data in the region of overlapping spatial fequencies specified by the user.

As an example of how IMMERGE works in practice, Figure 8 shows a sequence of images at various stages of data processing: before merging, after Fourier transforming, and the final version.

6. Data Combination in the Image Domain

There are two distinct methods for data combination in the image domain. The first, the 'linear combination' method (Stanimirovic et al. 1999), merges data sets in a simple linear fashion before final deconvolution, while the second (Sault & Killeen 1998), the non-linear method, combines all data during the deconvolution process.

6.1. The 'Linear combination' approach

The theoretical basis for merging before deconvolution is the linear property of the Fourier transform: a Fourier transform of a sum of two functions is equal to the sum of the Fourier transforms of the individual functions. Therefore, instead of adding two maps in the Fourier domain and Fourier transforming the combined map to the image domain, one can produce the same effect (fill in missing short-spacings in an interferometer $u - v$ coverage) by adding maps in the image domain. This method was first applied by Ye & Turtle (1991), and Stewart et al. (1993).

As we have seen from Equations 4 and 5, the interferometer and single-dish data obey the convolution relationship. The dirty images and beams can be combined to form a composite dirty image (I^D_{comb}) and a composite beam (B_{comb}) with the following weighting:

$$I^D_{\text{comb}} = (I^D_{\text{int}} + \alpha f I^D_{\text{sd}})/(1 + \alpha), \tag{13}$$
$$B_{\text{comb}} = (B_{\text{int}} + \alpha B_{\text{sd}})/(1 + \alpha), \tag{14}$$

where $\alpha = \Omega_{\text{int}}/\Omega_{\text{sd}}$ is an estimate of the resolution difference between the two data sets. The convolution relationship, $I^D_{\text{comb}} = B_{\text{comb}} * I$, still exists between the composite dirty image, I^D_{comb}, and the true sky brightness distribution, I. Deconvolving the composite dirty image with the composite beam hence solves for I.

There is no single existing program (task) that employs this method, but a linear combination of maps can be easily obtained in any package for radio data reduction, followed by a favorite choice of a deconvolution algorithm. As an example, Figure 9 shows the 'linear combination' method applied on the ATCA mosaic and Parkes telescope HI observations of the SMC at several stages of data processing. Note that B_{int} in the case of mosaic observations represents a whole

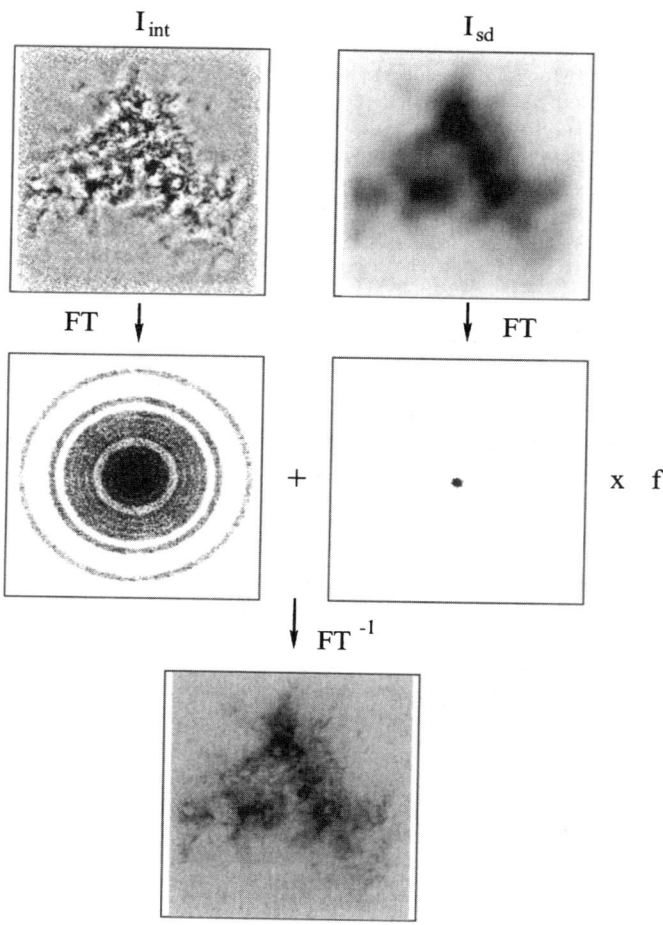

Figure 8. The first row: the ATCA and Parkes images of the SMC at heliocentric velocity 169 km s^{-1}. The second row: amplitudes of the Fourier transforms of the two input images (shown in the first row). Note how the Fourier transform of the clean interferometer image does not have a hole in the center, this is the result of extrapolation before deconvolution. However, the central visibilities are not correctly represented. The third row: the output of IMMERGE obtained by weighting the distributions from the second row, combining them, and Fourier transforming the result back to the image domain.

Short-Spacings Correction

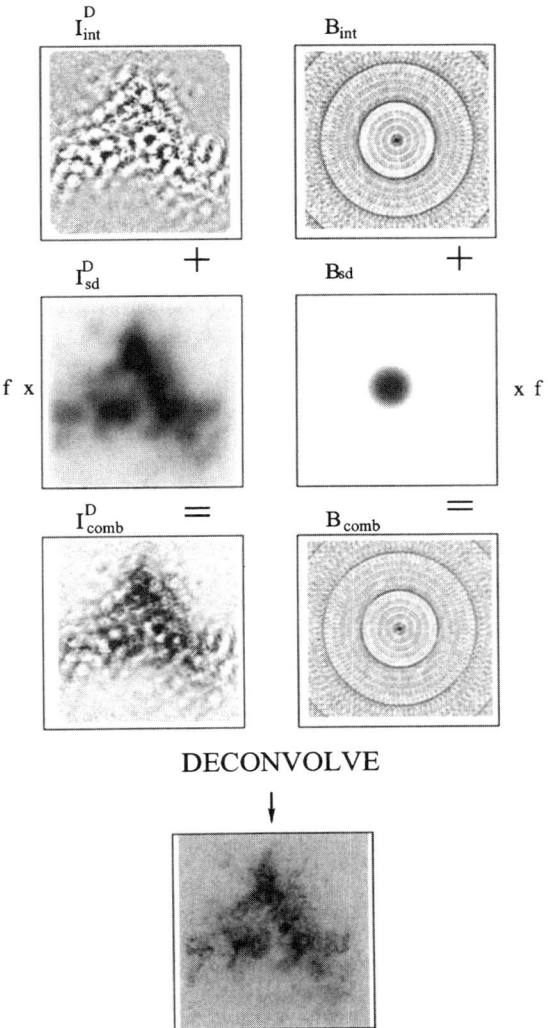

Figure 9. The first row (top): the ATCA 'dirty' image of the SMC at 169 km s^{-1}, and the ATCA 'dirty' beam for a single pointing. The second row: the Parkes image of the SMC at the same heliocentric velocity, and the Parkes beam assumed to be a Gaussian function with the FWHM of 18.8 arcmin. The third row: $I^D_{\text{comb}} = (I^D_{\text{int}} + \alpha f I^D_{\text{sd}})/(1+\alpha)$ and $B_{\text{comb}} = (B_{\text{int}} + \alpha B_{\text{sd}})/(1+\alpha)$, where $\alpha = 7.5 \times 10^{-3}$. The fourth row (bottom): the result of the MEM deconvolution of I^D_{comb} by B_{comb}.

cube of beams, one for each pointing in the mosaic. The combined dirty image was deconvolved using MIRIAD's maximum entropy algorithm (Sault, Staveley-Smith, & Brouw 1996). The model was finally restored with a 98-arcsec Gaussian function.

6.2. Merging during deconvolution

Besides the missing information in the center of the $u - v$ plane, an interferometer $u - v$ coverage suffers from spatial-frequency gaps. Since the missing information can be introduced in an infinite number of ways, the convolution equation ($I^D = I * B$) has a non-unique solution. Hence, the deconvolution has the task of selecting the 'best' image from all those possible (Cornwell 1988). Since deconvolution has to estimate missing information, a non-linear algorithm must be employed (Cornwell 1988; Sault et al. 1996). Cornwell (1988) and Sault et al. (1996) showed that the deconvolution algorithms implementing the so called 'joint' deconvolution scheme, whereby observations of many pointing centers are combined prior to deconvolution, produce superior results in the case of mosaicing, since more information is fed to the deconvolver. We expect that the same argument might apply for the addition of the single-dish data, resulting in the merging before and during deconvolution being more successful than the merging of clean images in the spatial frequency domain.

The maximum entropy method (MEM) is one of the non-linear deconvolution algorithms. It selects the deconvolution solution so it fits the data and, at the same time, has a maximum 'entropy'. Cornwell (1988) explains this entropy as something which when maximized produces a positive image with a compressed range in pixel values. The compression criterion forces the final solution (image) to be smooth, while the positivity criterion forces interpolation of unmeasured Fourier components. One of the commonly used definitions of entropy is:

$$\aleph = - \sum_i I_i \ln \left(\frac{I_i}{M_i e} \right), \qquad (15)$$

where I_i is the brightness of i'th pixel of the MEM image and M_i is the brightness of i'th pixel of a 'default' image incorporated to allow *a priori* knowledge to be used (e is the base of the natural logarithms). The requirement that the final image fits the data is usually incorporated in a constraint such that the fit χ^2_{int} of the predicted visibilities to those observed (Cornwell 1988) is close to the expected value:

$$\chi^2_{\text{int}} \leq N \sigma^2_{\text{int}}, \qquad (16)$$

with N being the number of independent pixels in the map and σ^2_{int} being the noise variance of the interferometer data.

The single-dish data can be incorporated during the maximum-entropy deconvolution process in two ways.

1. **The 'default' image**
   The easiest way is to use the single-dish data as a 'default' image in Equation 15 since, in the absence of any other information or constraints, this forces the deconvolved image to resemble the single-dish image in the spatial frequency domain where the interferometer data contribute no information. Since this method puts more weight on the interferometer data

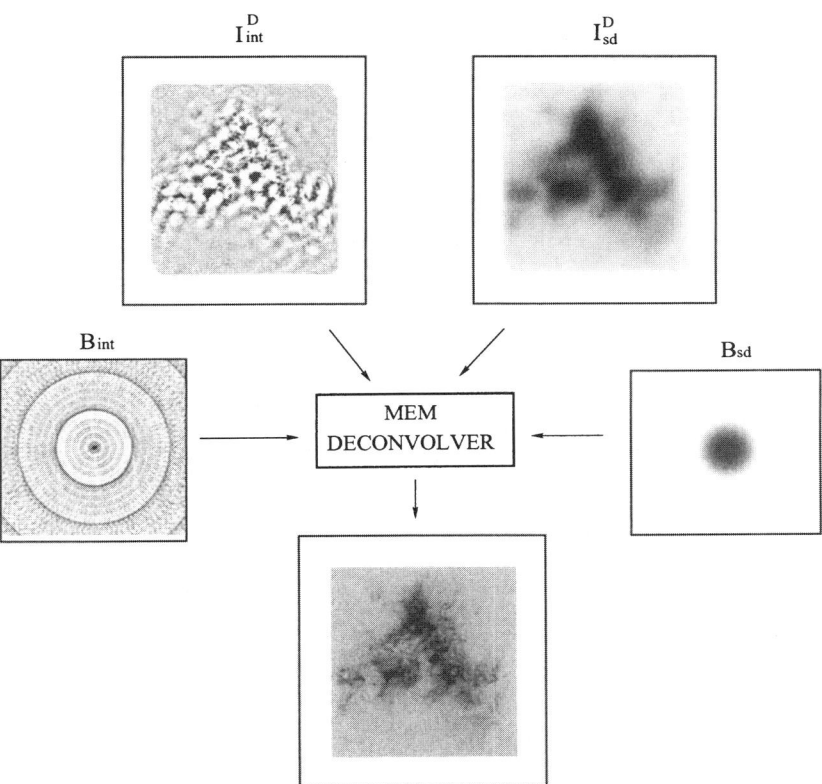

Figure 10. A schematic example of the non-linear approach for adding short-spacings performed by MIRIAD's program MOSMEM. In the **'default' image** method the low-resolution Parkes image of the SMC is used to constrain missing spatial frequencies in the center of the ATCA-image $u-v$ plane. In the **'joint' deconvolution** method, the model is found by fitting both data sets simultaneously while maximizing the entropy.

wherever it exists, the size of the overlapping region plays a very important role (Holdaway 1999). As large an overlap of spatial frequencies as possible is required to provide good quality interferometer and single-dish data within this region, in order to retain the same sensitivity over the image.

2. **The 'joint' deconvolution**
The second way maximizes the entropy while being subject to the constraints of fitting both data sets simultaneously:

$$\aleph = -\sum_i I_i \ln\left(\frac{I_i}{e}\right), \tag{17}$$

$$\sum_i \left\{I_{\text{int}}^D - B_{\text{int}} * I\right\}_i^2 < N\sigma_{\text{int}}^2, \tag{18}$$

$$\sum_i \left\{I_{\text{sd}}^D - \frac{B_{\text{sd}} * I}{f}\right\}_i^2 < M\sigma_{\text{sd}}^2. \tag{19}$$

The 'joint' deconvolution method provides also an alternative way, completely performed in the image domain, for determining the calibration scaling factor. Maximizing the entropy, while fitting both data sets, a 'joint' deconvolution algorithm can iteratively solve for I and f simultaneously.

As a schematic example, Figure 10 shows the non-linear approach for data combination performed by MIRIAD's program MOSMEM.

7. Comparison of the Different Methods

The qualitative comparison of all four methods for the short-spacings correction addressed in Sections 5 and 6, is shown in Figure 11 for the 169 km s^{-1} velocity channel of the SMC data. All four images have the same grey-scale range (-11 to 107 K) and are remarkably similar. They all have the same resolution and show the same small and large scale features with only slightly different flux-density scales. No signs of interferometric artifacts are visible on any of the images. This demonstrates that all four methods for the short-spacings correction give satisfactory results in the first approximation, when an *a priori* determined calibration scaling factor is used. A similar conclusion was reached in Wong & Blitz (2000) where results of data combination using IMMERGE and the 'linear combination' method (Section 6.1.) were compared for the case of BIMA and Kitt Peak 12-m telescope CO observations.

The quantification of the quality of an image depends on the scientific questions we want to address (Cornwell et al. 1993) and is therefore case specific. Something that any short-spacings correction must fulfill, though, is that the resolution of the final image should be the same as for the interferometer data alone, while the integrated flux density of the final image should be the same as measured from the single-dish data alone. Table 1 shows measurements of the total flux density, minimum/maximum values and noise level in the four

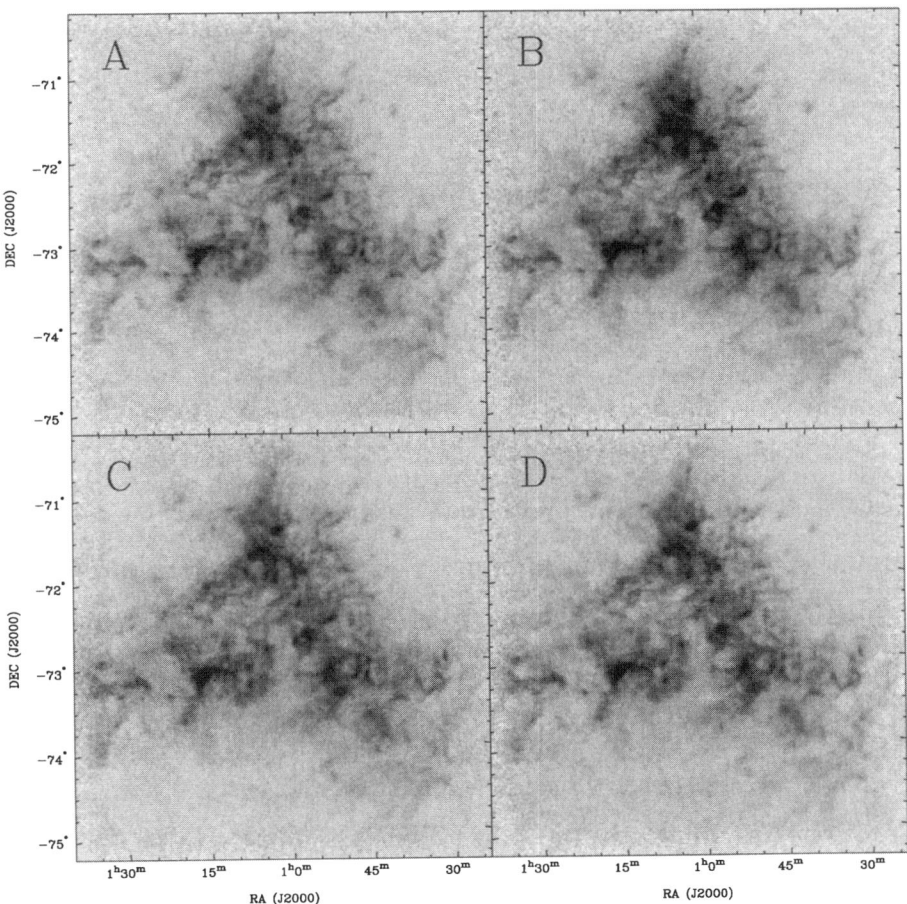

Figure 11. The result of adding Parkes short-spacings into an ATCA HI mosaic of the SMC at heliocentric velocity 169 km s^{-1}. Four different methods are shown from top-left to bottom-right: data combination in the Fourier domain using IMMERGE (panel A), the 'linear combination' (panel B), data combination during deconvolution using Parkes data as a 'default' image (panel C) and the 'joint' deconvolution of both Parkes and ATCA data (panel D). All images have grey-scale range -11 to 107 K with a linear transfer function.

Table 1. Total flux density, minimum and maximum brightness values and noise for the SMC ATCA image at 169 km s^{-1} corrected for Parkes short-spacings. Results of four different methods are shown: data combination in the Fourier domain using IMMERGE (labeled as method A), 'linear combination' (labeled as method B), data combination during deconvolution using Parkes data as a 'default' image (labeled as C) and the 'joint' deconvolution of both Parkes and ATCA data (labeled as D). The total flux density of the Parkes image alone is 6100 Jy.

Method	Total Flux (Jy)	Min (Jy beam^{-1})	Max (Jy beam^{-1})	Noise (mJy beam^{-1})
A	5600	−0.24	1.97	30
B	6500	−0.21	2.16	32
C	6300	−0.21	2.00	28
D	5900	−0.30	2.00	29

resultant images. All four images have very comparable noise levels, minimum values and maximum values. The last two images also have very comparable (within 3%) total flux density, relative to the Parkes value alone, while the first two have lower (by 8%) and higher (by 7%) flux densities, respectively. The differences come, most likely, from the different weighting of the single-dish data employed by the different methods. While IMMERGE slightly over-weights very short interferometric spatial frequencies, the 'linear combination' method slightly over-weights single-dish data in the region of overlap. This results in the total flux density being slightly lower in the first case, and slightly higher in the second case.

A few (general) remarks on all four methods:

- The 'feathering' or Fourier method (A) is the fastest and the least computer intensive way to add short-spacings. It is also the most robust way relative to the other three methods which all require a non-linear deconvolution at the end.

- The great advantage of the 'linear combination' method is that it does not require either Fourier transformation of the single-dish data, which can suffer severely from edge effects, nor deconvolution of the single-dish data which is especially uncertain and leads to amplification of errors.

- The 'default' image method shows surprisingly reliable results when a significantly large single-dish is used.

- Adding during deconvolution when fitting both data sets simultaneously provides, theoretically, the best way to do the short-spacing correction. However, this method depends heavily on a good estimate of the interferometer and single-dish noise variances.

8. Summary

In this article the need for, and methods of combining interferometer and single-dish data have been explained and demonstrated. This combination is an important step when mapping extended objects and it is becoming a standard observing and data processing technique.

After a brief introduction to interferometry in Section 2 the short-spacings problem and general approaches for its solution were discussed in Section 3. To *fully and accurately* fill in the missing short-spacings to interferometric data, a heterogeneous array scheme seems to be preferable. A sufficiently large single-dish, with diameter at least 1.5 greater than the shortest interferometer baseline, provides the simplest option. In order to cross-calibrate single-dish and interferometer data sets, a significant overlap of spatial frequencies is required. Four different combination methods, two linear and two non-linear, for the short-spacings correction have been discussed and the results of applying these methods to the case of HI observations of the SMC presented. Linear methods are data combination in the spatial frequency domain and the 'linear combination' method, while data combination during deconvolution provides two non-linear methods. All four techniques yield satisfactory and comparable results.

Acknowledgments. Many thanks to Darrel Emerson, Chris Salter and John Dickey for reading the article and providing valuable suggestions for improvement. I am also grateful to Matthew Wyndham for his help with figures, as well as assisting with last minute crises in organizing the meeting.

References

Bajaja, E. & van Albada, G. D. 1979, A&A, 75, 251

Braun, R. & Walterbos, R. A. M. 1985, A&A, 143, 307

Burke, B. F. & Smith, F. G. 1997, Radio Astronomy (Cambridge: Cambridge University Press), 76

Cornwell, T. J. 1988, A&A, 202, 316

Cornwell, T. & Braun, R. 1989, in ASP Conf. Ser. Vol. 6, Synthesis Imaging in Radio Astronomy, ed. R. Perley, F. Schwab & A. Bridle (San Francisco: ASP), 167

Cornwell, T., Braun, R., & Briggs, D. S. 1999, in ASP Conf. Ser. Vol. 180, Synthesis Imaging in Radio Astronomy II, ed. G. B. Taylor, C. L. Carilli & R. A. Perley (San Francisco: ASP), 151

Cornwell, T. J., Holdaway, M. A., & Uson, J. M. 1993, A&A, 271, 697

Ekers, R. D. & Rots, A. 1979, in IAU Colloq. 49, Image Formation from Coherence Functions in Astronomy, ed. C. van Schooneveld (Dordrecht: Reidel), 61

Holdaway, M. A. 1999, in ASP Conf. Ser. Vol. 180, Synthesis imaging in radio astronomy II, ed. Taylor G. B., Carilli C. L. & Perley R. A., (San Francisco: ASP), 401

Morita, K-I. 2001, ALMA Memo 374

Roger, R. S., Milne, D. K., Wellinton, K. J., Haynes, R. F., & Kesteven, M. J. 1984, Publ. Astr. Soc. Australia, 5(4), 560
Sault, R. J. & Killeen, N. 1998, Miriad Users Guide, Australia Telescope National Facility
Sault, R. J., Staveley-Smith, L., & Brouw, W. N. 1996, A&AS, 120, 375
Schwarz, U. J. & Wakker, B. P. 1991, in IAU. Colloq. 131, Radio Interferometry: Theory, Techniques and Applications, ed. T. J. Cornwell & R. A. Perley (San Francisco: ASP), 188
Stanimirovic, S. 1999, PhD thesis, University of Western Sydney Nepean
Stanimirovic, S., Staveley-Smith, L., Dickey, J. M., Sault, R. J., & Snowden, S. L. 1999, MNRAS, 302, 417
Stewart, R. T., Caswell, J. L., Haynes, R. F., & Nelson, G. J. 1993, MNRAS, 261, 593
Taylor G. B., Carilli C. L., & Perley R. A. 1999, Synthesis imaging in radio astronomy II, ASP Conf. Ser. Vol. 180 (San Francisco: ASP)
Thompson, A. R. 1994, in ASP Conf. Ser. 6, Synthesis Imaging in Radio Astronomy, ed. R. Perley, F. Schwab & A. Bridle, (San Francisco: ASP), 11
Thompson, A. R., Moran, J. M., & Swenson, G. W. J. 1986, Interferometry and Synthesis in Radio Astronomy (New York: John Wiley & Sons), 265
Vogel, S. N., Wright, M. C. H., Plambeck, R. L., & Welch, W. J. 1984, ApJ, 283, 655
Wilner, D. J. & Welch, W. J. 1994, ApJ, 427, 898
Wong, T. & Blitz, L. 2000, ApJ, 540, 771
Ye, T. & Turtle, A. J. 1991, MNRAS, 249, 722
Zhou, S., Evans, N. J., & Wang, Y. 1996, ApJ, 466, 296

Stray Radiation: Causes, Curses, and Cures

Felix J. Lockman

National Radio Astronomy Observatory, P.O. Box 2, Green Bank, West Virginia 24944, USA

Abstract. Radio telescopes have some response to signals from all directions, not just the direction of the main beam, and thus observations can be contaminated by "stray" radiation which enters the receiver through a sidelobe. This paper considers the origin and characteristics of stray radiation, and discusses some techniques which have been developed to minimize its effects, or to remove it from measurements. The new Green Bank Telescope should be substantially free from stray radiation, and a bootstrap technique using its data might be developed to reduce the stray radiation in Arecibo 21cm HI measurements as well.

1. Introduction

The function of a radio telescope is twofold: to receive emission from a particular direction in the sky, and to reject emission from all other directions. It is an unfortunate fact that all radio telescopes have some response outside of the main beam and thus any extended source of emission will couple to an antenna in complicated ways. The set of radio signals of celestial origin which enters the receiver through *side*lobes rather than through the *main* antenna lobe (the main beam) is referred to as "stray radiation". It can be the limiting factor in the accuracy of many measurements. In this lecture I will concentrate on stray radiation in spectral line observations, but most of the comments apply also to the continuum and to polarization studies. The sidelobes of an antenna also admit radio frequency interference – a devastating problem at some frequencies (e.g. Thompson, Moran, & Swenson 1986, p. 470; J.R. Fisher, this volume).

The papers by Goldsmith and Campbell in this volume contain the fundamental equations which describe the response of a radio telescope. For a discussion of stray radiation it is useful to rewrite Goldsmith's Equation 22 in a somewhat different form:

$$T_a = \eta_r \int_{4\pi} T_b(\theta, \phi) P(\theta, \phi) \, e^{-\tau_{at}} \, d\Omega, \qquad (1)$$

where T_a is the observed antenna temperature (properly calibrated), η_r is the radiation efficiency of the telescope (a constant near unity), $T_b(\theta, \phi)$ is the sky brightness temperature at a given location, τ_{at} is the atmospheric opacity in that direction, and P is the antenna power pattern, where

$$\int_{4\pi} P(\theta,\phi)\, d\Omega \equiv 1. \tag{2}$$

All quantities are functions of frequency. The form of Equation 1 emphasizes that the output of a radio telescope is the all-sky cross-correlation of the true sky brightness with the antenna response, modified slightly by the atmospheric opacity and ohmic losses. Equation 1 applies to an aperture synthesis telescope as well as a single-dish with the modification that the integral in Equation 2 is then zero.

Equation 1 can be written

$$T_a \eta_r^{-1} = \int_{\Omega_{mb}} T_b P_{mb} e^{-\tau_{at}}\, d\Omega + \int_{4\pi-\Omega_{mb}} T_b P_{sl} e^{-\tau_{at}}\, d\Omega, \tag{3}$$

where the power pattern has been divided into a main beam (mb) and sidelobes (sl). The quantity of interest in an observation is the main-beam averaged brightness temperature:

$$\langle T_b \rangle_{mb}\, \eta_{mb} \equiv \int_{\Omega_{mb}} T_b P_{mb}\, d\Omega. \tag{4}$$

where η_{mb} is the main beam efficiency, which is the fraction of the total power pattern that is contained in the main beam.

The desired result is thus related to the observable T_a by

$$\langle T_b \rangle_{mb} = \frac{e^{\tau_{at0}} T_a}{\eta_r \eta_{mb}} - \frac{e^{\tau_{at0}}}{\eta_{mb}} \left[\int_{\Omega_n} T_b\, P_n e^{-\tau_{at}} d\Omega + \int_{\Omega_f} T_b\, P_f e^{-\tau_{at}} d\Omega \right], \tag{5}$$

where $e^{\tau_{at0}}$ is the atmospheric correction in the direction of the main beam, and the power pattern outside of the main beam has been divided into "near" (P_n) and "far" (P_f) sidelobes.

For many astronomical sources the entire second term can be neglected (a happy occurrence which renders the remainder of this paper moot), and a beam-averaged brightness temperature is derived by simply scaling the observed antenna temperature to correct for dilution of the signal by the sidelobes. Radio sources which are larger than the main beam, however, couple to the antenna pattern in ways which may not be simple, bedeviling calibration and contaminating the observations. Molecular and atomic clouds, HII regions and supernova remnants – these are all extended at some frequencies to some telescopes. There are even circumstances in which most of the observed T_a comes through the sidelobes rather than the main beam (e.g. Lockman, Jahoda, & McCammon 1986; hereafter LJM).

There are several reasons for making the distinction between sidelobes that are "near" the main beam and those which are "far" from it. The near sidelobes are caused by the illumination pattern, reflector surface errors and some blockage, while for most telescopes the far sidelobes result from spillover and blockage alone. Near sidelobes are relatively strong (17 to 30 dB below the main beam) and are concentrated near the main beam. They can be measured with reasonable accuracy by mapping a bright isolated compact radio source. Far sidelobes are broad and very weak (-50 dB) and often quite difficult to measure. As important, the stray radiation entering the near sidelobes most often comes from some part of the source under observation, while stray radiation in a far sidelobe may originate in a completely different object may tens of degrees away.

An emission line in a sidelobe will have a different projected velocity with respect to the LSR than emission in the main beam, and will appear at a velocity which varies with the direction of the main beam on the sky, with the time of day, and with the season of the year (as sidelobes rise and set, and possibly rotate with respect to the celestial frame). The Doppler shift of a radio source can act as a filter against stray radiation: the HI line from a galaxy at 1000 km s^{-1} is not likely to be confused by Galactic HI emission in a sidelobe, or by emission from a nearby galaxy at 10,000 km s^{-1}.

Table 1. Errors in Galactic HI Measurements ($60^س$ integration).

Noise	$< 10^{18}$ cm^{-2}
Baseline	$< 10^{18}$ cm^{-2}
Calibration	$< 3\%$
Stray Radiation	$> 10^{19}$ cm^{-2}

Telescope sidelobes limit many measurements. Table 1, adapted from Lockman (1993), shows how stray radiation can dominate the error budget in the total 21cm HI column density, N_H, after a short integration. This has been a problem in Galactic HI studies for many years (e.g. Raimond 1966; Westerhout 1969; Heiles 1975; Heiles, Stark, & Kulkarni 1981; Kalberla, Mebold, & Reich 1980; LJM). Emission in near sidelobes has also compromised studies of molecular lines (Kutner, Mundy, & Howard 1984). The sidelobes on large reflectors which pick up varying amounts of ground radiation make it difficult to calibrate total power continuum measurements, which is why the cosmic microwave blackbody radiation was first detected on an antenna that has extremely low and well-understood sidelobes (Penzias & Wilson 1965). Sidelobes can also be polarized, and confuse polarization measurements of extended sources (e.g. Troland & Heiles 1982; Heiles, this volume).

Antenna systems are often designed to maximize the ratio of gain to system temperature, which is usually equivalent to maximizing the aperture efficiency and minimizing the spillover (see Section 2.3). For observation of extended sources, however, it can be more important to have a high main beam efficiency even at the expense of reduced point-source gain. These considerations are only now being added to discussions of antenna design. The impact of stray radiation is likely to get worse rather than better in the future, for new large antennas with high angular resolution make more and more sources extended to the main

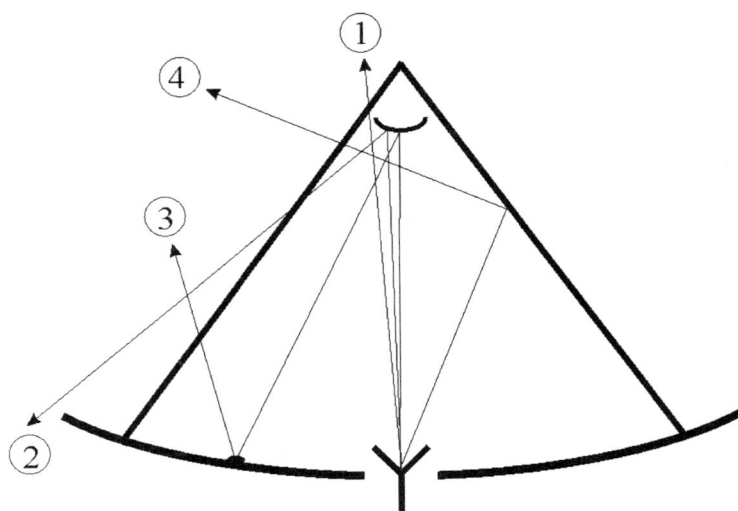

Figure 1. Schematic of a telescope showing spillover from the feed past the secondary (1) and primary (2) reflectors, scattering from a surface irregularity (3) and scattering from the struts (4).

beam, and improved receivers can push down the noise level quickly, revealing a "floor" of stray radiation.

2. The Origin of Sidelobes

Sidelobes arise naturally because of the wave nature of light, but they also result from surface irregularities, blockage, and any defect in the optics of a telescope. It is worth considering these concepts in some detail, if only to remind us of the complexity of real telescopes.

Figure 1 shows a schematic of a radio telescope and some common sources of sidelobes. Radiation can spill past a subreflector or the main dish. It can be scattered and diffracted by feed support legs, secondary or tertiary subreflectors, and guy wires. It can be scattered by surface irregularities. The shadows cast by blockage disrupt aperture currents and modify the main beam. In all these cases antenna response is moved from the main beam to some other direction.

The power pattern of a reflector antenna is determined by the feed illumination pattern, the quality of the surface, and any blockage in the aperture. The pattern near the main beam can be calculated if the aperture field distribution is known (see Goldsmith's paper in this volume), but the aperture field is generally too complicated to measure or calculate for an antenna with significant blockage. However, the basic character of the power pattern can be understood as the superposition of a few simple elements.

2.1. The main beam

The main beam of most radio telescopes is approximately Gaussian with a HPBW ≡ FWHM ≈ 1.2 λ/D where D is the dish diameter. The main beam shape is also a strong function of the location of the feed in the focal plane – lateral or radial defocussing will modify the beam and can produce coma sidelobes (e.g. Rusch et al. 1982; Christiansen & Hogbom 1985). An antenna will have a recognizable main beam even if the surface is distorted considerably or has significant blockage. In general, degradation of the telescope's optics raises sidelobes, reduces the aperture and main beam efficiencies, and may displace the main beam on the sky, but will still leave a recognizable main beam of angular size $\sim \lambda/D$ even if surface errors are a significant fraction of a wavelength.

2.2. The illumination pattern

The beam shape and sidelobe levels are functions of the feed illumination pattern. The sharp edge of the dish, where the illumination must go to zero and often does so abruptly, produces Fraunhoffer diffraction, resulting in sidelobes analogous to the Airy rings of optical systems. Their amplitude can be reduced by tapering the illumination at the expense of an increased beamwidth (because the outer parts of the dish are underweighted relative to the inner parts) and a consequent loss of aperture efficiency (see Fig. 6-106 of Kraus 1986). Examples of illumination taper and the resultant beams for simple linear apertures are given in Figure 2 (see also Olver 1982; Rohlfs & Wilson 1996). In practice, the illumination pattern is usually chosen from considerations of gain and spillover, not sidelobes, as discussed below.

The line feeds at Arecibo are an interesting case because they illuminate the dish in a torus with a central hole. The resultant power pattern has a narrow main beam but elevated near sidelobes, a definite disadvantage for observation of extended sources but of no concern for point sources like pulsars.

2.3. Spillover

Some of the illumination does not hit the dish but "spills over" to make a broad sidelobe. Examples can be seen in Hartsuijker et al. (1972), and Kalberla et al. (1980, hereafter KMR). Spillover can be in the forward direction past a secondary reflector, or backward past the primary. The part of the spillover which reaches the ground increases the noise in the measurements, while spillover which reaches the sky can admit stray radiation. At Arecibo, with its fixed reflector and movable feed, the spillover is variable and can be large at large zenith angles; a ground screen has been installed to divert most of it to the sky reducing the total system noise, but the screen increases the potential for problems from celestial stray radiation.

During design of the NRAO Green Bank Telescope (GBT) the tradeoffs between illumination taper and spillover were explored in some detail and an example for the 5 GHz receiver at the Gregorian secondary focus is given in Table 2 (from Srikanth 1992). The "Edge Taper" in this Table is the ratio of the feed power at the dish edge to the maximum in the aperture. For large values of the taper the aperture efficiency is reduced, but the spillover incident on the ground is also reduced, lowering the total system temperature (this calculation

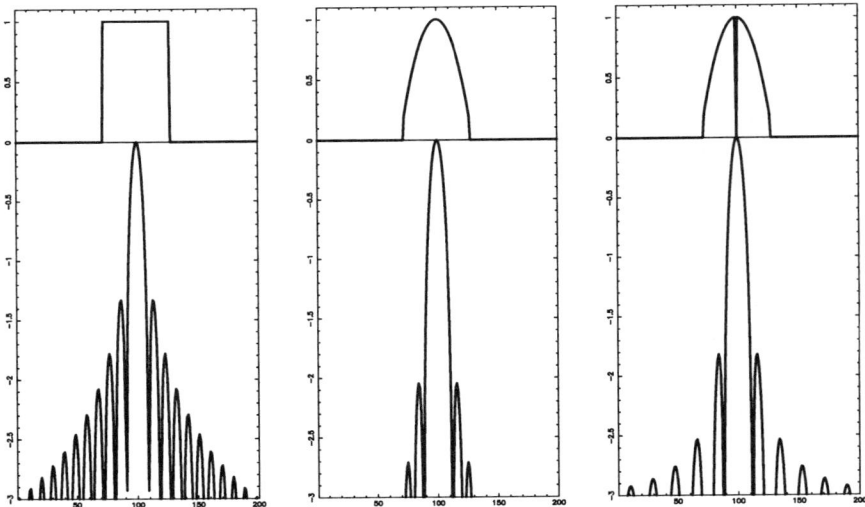

Figure 2. Several schematic illumination patterns and the resultant power patterns for a linear antenna. The illumination pattern at the top of each panel is on a linear vertical scale normalized to unity at peak, and on an arbitrary horizontal scale which is identical for each panel. The power patterns shown below are normalized to the peak and plotted in the log10 from 0 to -3, all on the same horizontal and vertical scale. Note the large decrease in the near sidelobes as the pattern is tapered, accompanied by an increase in the width of the main beam. The third panel shows the effects of a central gap in the illumination pattern caused by subreflector blockage or the illumination pattern of a line feed.

Table 2. Effects of Feed Taper for the GBT at 5 GHz.

Edge Taper (dB)	Aperture Efficiency %	T_{spill} forward K	rear K
-12	70.0	0.4	2.6
-13	69.9	0.4	2.2
-14	69.3	0.4	1.9
-15	68.4	0.3	1.6
-16	67.3	0.2	1.4
-17	66.1	0.2	1.2

was for an elevation of 90°; results at other elevations are similar). A large taper also increases the beamwidth somewhat, implicit in the reduced η_a.

One measure of antenna quality is its signal-to-noise ratio on a point source, which is proportional to the geometric area scaled by $\eta_a / (T_{Rx} + T_{sp} + ...)$. If the total system temperature neglecting spillover is $\approx$ 20 K, the highest point-source sensitivity in Table 2 comes at a taper of -15 dB, and this was adopted for the 5 GHz GBT feed. At frequencies where the atmosphere contributes significant noise, spillover onto the ground is proportionately less important to the total system temperature, and the edge taper is often reduced to increase the telescope gain, resulting in a larger spillover lobe.

2.4. Surface irregularities

Defects in the reflector surface scatter radiation out of the main beam into sidelobes (see Christiansen & Hogbom 1985, Section 3.2). The operating range of a radio telescope is often given as $\lambda \geq 16\sigma$, where σ is the rms deviation of the surface from the desired shape. At the 16σ limit the gain is about 0.54 of that from a perfect surface and thus almost half of the power has been directed out of the main beam. The location on the sky of this scattered radiation depends on the nature of the surface deformations. If the irregularities are small and somewhat random the scattered radiation typically forms a broad plateau around the main beam (e.g. Kutner, Mundy, & Howard 1984). If the surface suffers severe astigmatism the radiation may be scattered into multiple near sidelobes which can be almost as strong as the main beam (e.g. Brown 1980).

A reflector with a single, wildly misaligned surface panel, will have a power pattern which is the combination of one for an antenna which is missing that panel, and the power pattern of the panel itself, pointed in some other direction. True holes in the surface or gaps between panels will also produce sidelobes which are typically weak but can be a significant fraction of the total response in the case of a mesh antenna operated at its short wavelength limit (e.g. Condon, Broderick, & Seielstad 1989).

2.5. Blockage

Anything that blocks the aperture – feed support legs, a Cassegrain house, a secondary reflector, cables – will disrupt the illumination pattern and contribute to sidelobes. For most antennas the blockage is fixed with respect to the illumination so that the sidelobe pattern is fixed with respect to the main beam. At Arecibo, where the feed moves with respect to the blockage, the sidelobes are a function of azimuth and zenith angle.

The structural elements which block an aperture often have a size in one dimension which is comparable to the wavelength of observation, and the effective blockage can be significantly larger than the geometric blockage (see Goldsmith, this volume, Section 4.2 Equation 44). Feed support legs (struts) are a major source of blockage and are typically complex structures which can be difficult to analyze electrically. Their effect is weighted by the illumination pattern across the dish. These complications make it difficult to calculate an accurate power pattern for an antenna with significant blockage, especially the portion of the pattern at large angles from the main beam. The sidelobes must be measured. Examples of such measurements are those of Higgs (1967), Hartsuijker et al. (1972), KMR, and Murphy (1993). Large antennas typically have geometric blockage in the range 4% to 7%, with a somewhat larger *effective* blockage.

It is worthwhile to contrast the effects of blockage and spillover. An antenna with 1% spillover has an effective area reduced by 1% from its potential value, and has a sidelobe which contains 1% of the antenna's response. An antenna with 1% blockage, however, suffers in two ways. About 1% of the power is scattered out of the beam by the blockage, while the aperture itself is also reduced in area by 1%. The effect of a fractional blockage η_b on the gain is thus approximately $(1 - \eta_b)^2 \approx (1 - 2\eta_b)$.

A similar consideration applies to the differing effects of a central hole in the aperture fields caused by subreflector blockage, and a central hole produced by a line feed. The blockage is twice as bad. In this example, the shape of the fields in the aperture can be identical, but the power pattern will be different because power intercepted by the blocking elements will reemerge in a sidelobe at some angle to the main beam. Knowledge of the fields in the aperture plane (in the usual meaning of this term) is generally not sufficient to determine P in all directions for an antenna with blockage.

The feed support legs or "struts" of large symmetric reflectors are a major source of blockage. Each strut produces a ring sidelobe centered on the sky in the direction of the long axis of the strut and passing through the main beam. The rings have a diameter of twice the angle of the strut with the vertex line, typically 70°, and a width $\approx \lambda/L_A$ where L_A is the length of the strut projected onto the aperture (Rusch et al. 1982; Lamb 1990). The amplitude of a ring sidelobe is typically 50dB below the main beam. There is one ring per strut, typically containing $1 - 2\%$ of the antenna's response. Landecker et al. (1991) have made a special study of the shape of sidelobes from feed support struts with respect to ground pickup (see also Anderson et al. 1991).

3. The 140-Foot Telescope at 21cm

In an unpublished MS thesis, E.M. Murphy (1993) analyzed the all-sky response of the 140-Foot Telescope of the NRAO at Green Bank at a wavelength of 21cm. That and subsequent work by Murphy, Sembach, & Lockman (in preparation) have refined the characterization of the telescope. Figure 3 shows the amplitude of the illumination pattern at 21cm and Figure 4 shows the calculated power pattern within a few degrees of the main beam. This calculation agrees with measurements of the location and amplitude of the principal near sidelobes.

Table 3 lists estimates for the various components of the 140-Foot power pattern. Although the amplitude of most sidelobes is many tens of dB below the main beam, their area can be considerable and add up to significant total response. Comparison of the calculations with the measurements indicates that the effective cross section of the 140-Foot feed support legs at 21cm is a factor 1.14 times larger than their geometric cross section.

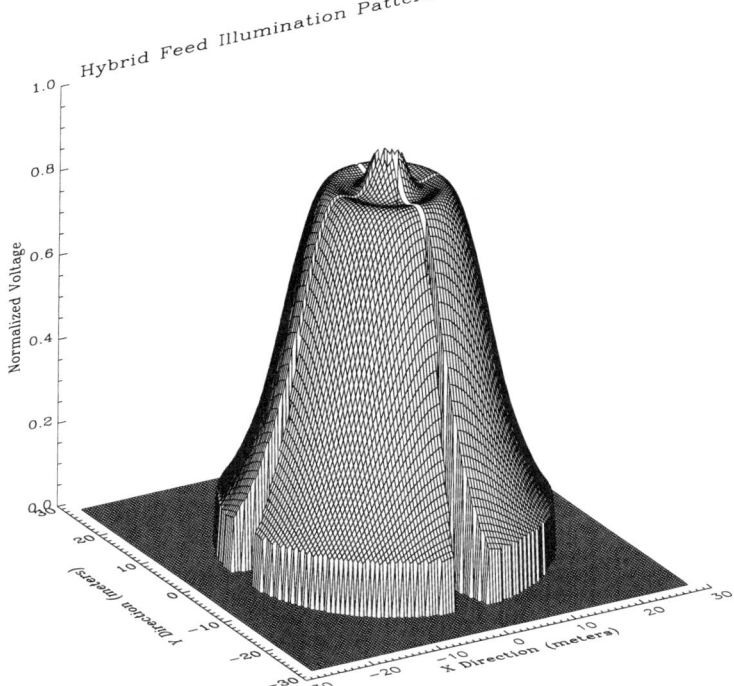

Figure 3. The illumination pattern of the 140-Foot Telescope at 1390 MHz derived from measurements of the feed and other components. Note the shadows cast by the struts, the strong taper of the hybrid-mode feed, the sharp cutoff at the dish edge, and the blockage by the Cassegrain house and subreflector which produces a central hole (from Murphy 1993).

Table 3. Power in the 140-Foot beam pattern at 21 cm.

Lobe	Integral P
Main Beam ($r < 1°$)	83.2%
Near Sidelobes ($1° \leq r \leq 5°$)	2.1
Cassegrain	3.8
Feed Legs	5.6
Spillover	0.8
Panel Gaps	0.4
Other	4.0

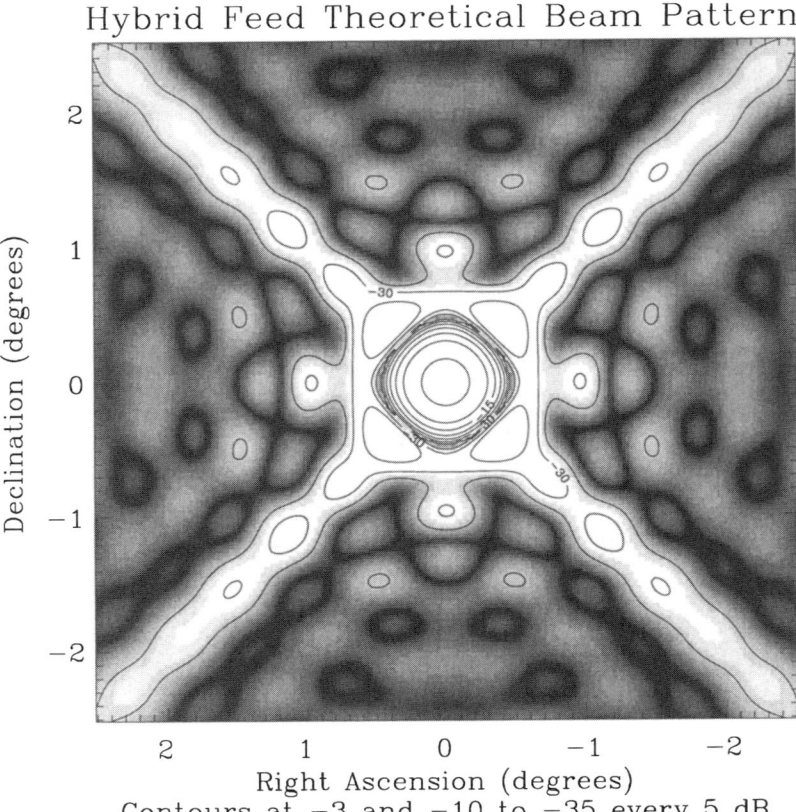

Figure 4. The expected power pattern within 2°.5 of the main beam of the 140-Foot Telescope at 1.4 GHz, as calculated from the illumination pattern of Figure 3. The prominent diagonal sidelobes are caused by the four feed support legs. This calculated pattern is in good agreement with measurements (Murphy 1993).

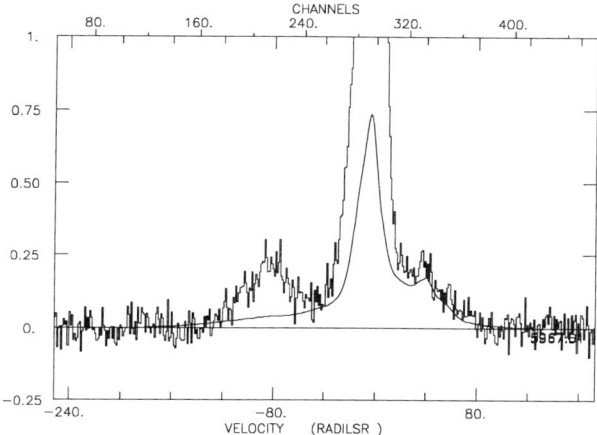

Figure 5. An HI spectrum observed on the 140-Foot Telescope with the stray component (smooth curve) calculated according to Equation 5. The stray component contains 5.3×10^{19} cm^{-2} of HI, and increases the observed signal by about 40% in this direction. Almost all of the line wing at $V > 20$ km s^{-1} is spurious. The spectrum is toward $\ell, b = 203° + 57°$.

4. Stray Radiation in Galactic HI Spectra

Low velocity Galactic HI emission is found in all directions on the sky from a peak T_b of 0.5 K to over 120 K. Every telescope sidelobe that is not on the ground intercepts some of this emission. The stray 21cm signals can sum to a $T_a \sim 1$ K or more, depending on the antenna properties and direction of observation. The stray emission usually peaks near zero velocity (LSR), but in the Galactic Plane the lines can be several hundred km s^{-1} wide and the emission will be Doppler-shifted by the projected motion of the LSR. Stray radiation is conspicuous in Galactic HI spectra because of its large amplitude and because the stray component is a strong function of hour angle as sidelobes rise and set, and of time of the year as the stray component's Doppler shift varies. Near the Galactic plane most of the stray radiation comes via the near sidelobes which lie on the brightest HI; for high latitude observations most of the stray radiation comes through far sidelobes which intersect the Galactic plane.

An example of the stray HI in a 140-Foot spectrum is shown in Figure 5. It can be shocking to the uninitiated to realize that a fairly strong component of an observed HI spectrum may be entirely spurious, but that certainly is the case for most telescopes.

5. Clean Telescopes

Offset paraboloids can be completely unblocked and thus have minimal sidelobes. If their reflecting surface is good, the main beam can be made as clean as desired by choice of taper of the illumination pattern. Examples of offset paraboloids include the horn-reflector and the 7-meter mm-wave telescopes of Bell Labs, and the new 100-meter Green Bank Telescope (GBT). These instruments have sidelobes which are concentrated within a few tens of beam-widths of the main beam, and their amplitude and location can be calculated with high precision because the sidelobes derive entirely from the feed illumination pattern and the surface. The Bell Labs horn-reflector has been used for many HI observations despite its small aperture (Wrixon & Heiles 1972; Heiles et al. 1981; Stark et al. 1992). The GBT, with its 100-meter unblocked aperture and angular resolution of $9'$ in the 21cm line, will surely have an even greater impact if spillover is minimized.

6. Coping with Stray Radiation

Most radio telescopes do not have unblocked apertures and so one must be aware of the possibility of stray radiation. Several strategies have been developed for dealing with stray radiation in data, with varying degrees of success.

6.1. Position switching

If it is possible to "position switch" between the direction of interest and a nearby reference position, most stray radiation may be canceled. Stray radiation from far sidelobes is a slowly changing function of telescope position because far sidelobes are much broader than the main beam, so most of it will be removed in observations made via position switching by a few degrees or less. The near sidelobes change structure more rapidly with position so there may be residual effects from these. Position switching is usually not an option for measurement of low-velocity Galactic HI because of its ubiquity, but does work for many other spectral lines and for HI at high velocities.

In continuum mapping applications, it is often common to chose regions on either side of the source to be studied, and use these to establish a background reference level which can then be subtracted from the map. This will have the effect of removing the slowly-varying component of stray radiation as well as unwanted radio continuum, but its drawbacks are obvious in its suppression of continuum emission with spatial wavelengths greater than the map size.

6.2. Deconvolution

Anyone approaching this topic with a fresh heart will likely study Equation 5 and realize that the problem of stray radiation removal can be viewed as a deconvolution similar in spirit, if not in practice, to the "cleaning" process that is applied to aperture synthesis data. One simply calculates the "expected" stray radiation given P and T_b over the sky, and subtracts it from the observed T_a, properly scaled.

Alas, there are two difficulties with this approach. First, the quantity T_b necessary for the deconvolution is exactly the quantity that one is trying to determine in the direction of the main beam. Second, it is a long and complex task to develop an adequate model for the power pattern, P. It usually cannot be calculated to the required accuracy, it is quite difficult to measure, and it extends over the entire sky. In contrast, the power pattern of an aperture synthesis instrument is determined largely from the configuration of the individual antennas.

Despite these complexities, a brave group led by Peter Kalberla at Bonn University has used this method to make significant corrections to HI spectra from the 100-meter telescope of the MPIfR, by constructing a model for the sidelobes and using antenna temperatures in place of T_b in an iterative process (KMR). Kalberla has also led a similar effort to correct spectra in the Leiden-Dwingeloo HI survey (Hartmann et al. 1996).

6.3. Bootstrapping: the solution for Arecibo?

It would be a formidable task to develop a deconvolution algorithm for the Arecibo telescope because of the fixed reflector and the complexities of the blockage. At Arecibo, the effect of the blockage changes continually during an experiment as the illumination pattern moves across the fixed structures. The power pattern is a function of both main-beam azimuth and zenith angles (though likely with some 3-fold symmetries), and solving for stray radiation via Equation 5 is likely not practical. A "bootstrapping" technique, though, might be simple and appropriate and give significant improvements over raw data. Bootstrapping has been used by LJM for 140-Foot HI spectra, by Jahoda, Lockman, & McCammon (1990) for 300-foot Telescope HI data, and was used in an approximate way by Heiles (1975) for the Hat Creek survey. Techniques of this sort have also been used at other wavelengths, e.g. by Schlegel et al. (1998) in the infrared.

The technique takes advantage of the fact that stray radiation typically varies much more slowly with position on the sky than radiation in the main beam (because the far sidelobes are so broad). If clean data at low resolution are available, a higher resolution map can be spatially convolved to match it, and any difference between the two is then interpreted as the stray component of the higher resolution data. An application to 21cm HI data taken at Arecibo might be as follows: a clean HI spectrum is obtained from the GBT at 9' resolution: $T_b(\text{GBT})$. The Arecibo Observatory telescope maps the area of the GBT main beam to produce a set of $T_b(\text{AO})$, calibrated and corrected for atmospheric absorption, but containing stray radiation at an unknown level. The Arecibo data are then degraded to the GBT resolution to obtain

$$T_{st}(\text{AO}) = T_b(\text{GBT}) - \int T_b(\text{AO}) P(\text{GBT}) \, d\Omega \, . \tag{6}$$

The single spectrum $T_{st}(\text{AO})$ is an estimate of the stray spectrum over the entire Arecibo map, and it can be subtracted from the individual Arecibo spectra. The requirement that the stray radiation not vary too much over the Arecibo map means that the map must be made as quickly as possible.

This technique has some obvious limitations (any noise or systematic problem in the GBT data propagates throughout the corrected Arecibo data – see

LJM), but it may offer a significant advantage in data quality compared to uncorrected Arecibo spectra, and should be tested.

Acknowledgments. Much of what I think I know about this topic was derived from or refined during discussions with J.R. (Rick) Fisher, most recently in consideration of the difficulties in calculating an all-sky power pattern for a telescope with blockage. Thanks, Rick. Thanks also to Tom Bania for a careful reading of the manuscript and Chris Salter for editorial advice.

References

Anderson, M. D., Landecker, T. L., Routledge, D., & Vaneldik, J.F. 1991, Radio Science, 26, 353

Brown, R. L. 1980, NRAO Engineering Memo No. 138

Christiansen, W. N. & Hogbom, J. A. 1985, Radio Telescopes (2nd edition; Cambridge and New York: Cambridge University Press)

Condon, J. J., Broderick, J. J., & Seielstad, G. A. 1989, AJ, 97, 1064

Hartmann, D., Kalberla, P. M. W., Burton, W. B., & Mebold, U. 1996, A&AS, 119, 115

Hartsuijker, A. P., Baars, J. W. M., Drenth, S., & Gelato-Volders, L. 1972, IEEE Trans. Antennas Prop., AP-20, 166

Heiles, C. 1975, A&AS, 20, 37

Heiles, C., Stark, A. A., & Kulkarni, S. 1981, ApJ, 247, L73

Higgs, L. A. 1967, Bull. Astr. Inst. Netherlands Sup., 2, 59

Jahoda, K., Lockman, F. J., & McCammon, D. 1990, ApJ, 354, 184

Kalberla, P. M. W., Mebold, U., & Reich, W. 1980, A&A, 82, 275 (KMR)

Kraus, J. D. 1986, Radio Astronomy (2nd edition; Powell, Ohio: Cygnus-Quasar Books)

Kutner, M., Mundy, L., & Howard, R. J. 1984, ApJ, 283, 890

Lamb, J. 1990, NRAO GBT Memo Series No. 13

Landecker, T. L., Anderson, M. D., Routledge, D., Smegal, R. J., & Vaneldik, J. F. 1991, Radio Science, 26, 363

Lockman, F. J. 1993, in Workshop on Databases for Galactic Structure, ed. A. G. D. Philip, B. Hauck, & A. R. Upgren (Schenectady: L. Davis Press), 181

Lockman, F. J., Jahoda, K., & McCammon, D. 1986, ApJ, 302, 432 (LJM)

Murphy, E. M. 1993, unpublished M.S. Thesis, Univ. Virginia

Olver, A. D. 1982, in Handbook of Antenna Design, 1, ed. A. W. Rudge et al. (London: P. Peregrinus), 1

Penzias, A. A. & Wilson, R. W. 1965, ApJ, 142, 419

Raimond, E. 1966, Bull. Astr. Inst. Netherlands Sup., 1, 33

Rohlfs, K. & Wilson, T. L. 1996, Tools of Radio Astronomy (2nd edition; New York: Springer-Verlag)

Rusch, W. V. T., Chu, T. S., Dion, A. R., Jensen, P. A., & Rudge, A. W. 1982, in Handbook of Antenna Design, 1, ed. A. W. Rudge et al. (London: P. Peregrinus), 128

Schlegel, D. J., Finkbeiner, D. P., & Davis, M. 1998, ApJ, 500, 525

Srikanth, S. 1992, NRAO GBT Memo No. 87 (http://info.gb.nrao.edu/GBT/memos)

Stark, A. A., Gammie, C. F., Wilson, K. W., Bally, J., Linke, R. A., Heiles, C., & Hurwitz, M. 1992, ApJS, 79, 77

Thompson, A. R., Moran, J. M., & Swenson, G. W. Jr. 1986, Interferometry and Synthesis in Radio Astronomy (New York: Wiley-Interscience)

Troland, T. H. & Heiles, C. 1982, ApJ, 252, 179

Westerhout, G. 1969, Maryland-Green Bank Galactic 21-cm Line Survey (2nd edition; College Park, Md.: University Maryland)

Wrixon, G. T. & Heiles, C. 1972, A&A, 18, 444

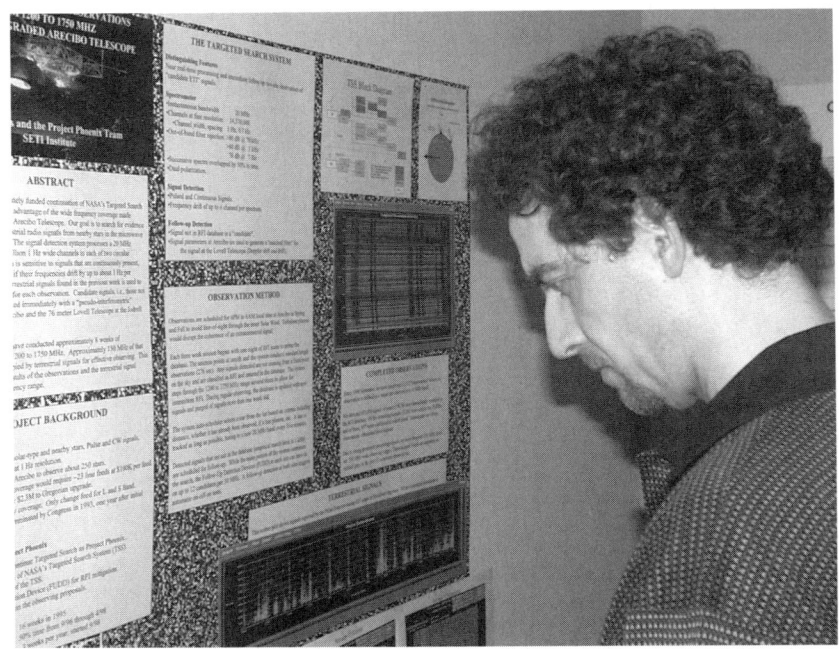

Luca Olmi contemplates the latest on SETI.

The Effects of the Atmosphere

Luca Olmi[1]

LMT/GTM Project, Dept. of Astronomy, 815J Lederle GRT Tower B, University of Massachusetts, 710 N. Pleasant St., Amherst, MA 01003, USA

Abstract. In this lecture I will discuss the effects of the neutral atmosphere and, to a lesser extent of the ionosphere, on radio astronomical observations carried out from single-dish telescopes on the surface of the Earth. We are concerned with three types of effects: large-scale refractive effects, absorption and scattering by the turbulent structure in the media. The phenomenon of scattering results in "seeing", and creates refractive index variations which limit the resolution and sensitivity of observations of astronomical sources. In the troposphere, water vapor plays a fundamental role in radio propagation. The refractivity of water vapor is about 20 times greater in the radio range than in near-infrared or optical regimes. As a consequence, phase fluctuations at frequencies higher than about 1 GHz are predominantly caused by fluctuations in the distribution of water vapor, and I will thus concentrate on troposphericinduced radio seeing. On filled-aperture telescopes radio seeing shows up as an anomalous refraction (AR), i.e. an apparent displacement of a radio source from its true position. The magnitude of this effect, as a fraction of the beam width, is bigger on larger telescopes, and thus its impact on the pointing is likely to become critically important in the next generation of electrically large $(D/\lambda > 10^4)$, filled-aperture radio telescopes. I will thus present the results of recent, systematic AR measurements and discuss a model study of AR effects, obtained producing numerical simulations of two-dimensional phase screens. I will finally discuss the basic concept and requirements of a tip-tilt compensation system at millimeter wavelengths, and will also present a proposed design based on a scanning microwave radiometer as a wave front sensing device.

1. Introduction

Despite the extensive use of orbital and sub-orbital platforms for small submillimeter telescopes, much (sub)millimeter astronomy is, and will still be, conducted from the surface of the Earth. This is because of the need for relatively large ($\gtrsim$ 10m) filled-aperture telescopes to achieve a variety of scientific goals, from the study of small, low surface brigthness objects to observations of CMB

---

[1]Current Address: Department of Physics, P.O. Box 23343, University of Puerto Rico, Rio Piedras, Puerto Rico 00931-3343, USA

anisotropies on small ($\simeq 1'$) angular scale. As a consequence, understanding the effects of the atmosphere on the astronomical observations is still of fundamental importance and, at millimeter and sub-millimeter wavelengths, several techniques have been proposed to compensate for some of the degrading effects of the troposphere.

In describing the effects of the neutral atmosphere and the ionosphere on radio astronomical observations carried out with single-dish telescopes, we are concerned with three types of effects, which also affect laser communications, radar systems and remote sensing: large-scale refractive effects (such as deflection of the wave, propagation velocity and polarization rotation), absorption and scattering by the refractive inhomogeneities generated by the turbulent structure of the medium. The phenomenon of scattering results in "seeing", and creates refractive index variations which limit the resolution and sensitivity of observations of astronomical sources.

In the lowest part of the atmosphere the temperature decreases monotonically from the surface until it reaches an altitude of approximately 11km. In this lowermost layer, called the troposphere, water vapor plays a fundamental role in radio propagation. The refractivity of water vapor is about 20 times greater in the radio range than in the near-infrared or optical regimes. As a consequence, phase fluctuations at frequencies higher than about 1 GHz are predominantly caused by fluctuations in the distribution of water vapor, and I will thus concentrate on tropospheric-induced radio seeing.

Even at frequencies where the ionospheric effects are dominant, the scales of the ionospheric fluctuations are generally much larger than the size of any current or planned long-wavelength single-dish telescope, and they are thus of little importance to the users of filled-aperture antennas. At frequencies $\gtrsim$ 1 GHz, radio seeing on single-dish telescopes shows up as an anomalous refraction (AR), i.e. an apparent displacement of a radio source from its true position. The magnitude of this effect as a fraction of the beam width is greater on larger telescopes, and thus its impact on the pointing is likely to become critically important in the next generation of electrically large, filled-aperture radio telescopes.

2. Turbulence Effects on Propagation

Index of refraction fluctuations originate with turbulent air motion. The source of energy for this air motion is the differential heating and cooling of the surface of the Earth. The resulting large scale inhomogeneities are broken into smaller scale inhomogeneities by air flow (e.g. wind shear and convection), which is nearly always turbulent. Turbulent air motion gives rise to randomly distributed pockets of air, or turbulent eddies, each having a characteristic temperature or water vapor density (or electron density in the ionosphere). The index of refraction of air in the lower atmosphere is quite sensitive to temperature at visible wavelengths, or to water vapor density at radio wavelengths, and thus the index of refraction is random, causing variations in the optical path length and hence phase (see Figure 1).

In a fluid, the passage from a laminar flow (i.e. smooth and regular) to a random fluid motion, i.e. turbulent, is characterized by an increase of the

Reynolds number, Re, a non-dimensional number defined as:

$$Re = \frac{V_{\text{avg}} l}{k_\nu}, \tag{1}$$

where V_{avg} is the average velocity of the viscous fluid, l is a characteristic scale size and k_ν is the kinematic viscosity. The kinematic viscosity of the air is $k_\nu = 1.5 \times 10^{-5}$ m^2 s^{-1}, and if we assume a scale size of $l = 10$ m and a velocity $V_{\text{avg}} = 1$ m s^{-1}, then $Re = 6.7 \times 10^5$, sufficiently high to ensure that atmospheric air flow is turbulent in essentially all situations of practical interest.

The kinetic energy of large-scale motions is introduced at scales larger than $\sim L_\circ$ (known as the *outer scale* of the turbulence) and then transferred to smaller scales, where eventually the turbulent motion stops and kinetic energy is dissipated as heat by viscous friction at a scale $l_\circ$ (or *inner scale*). $L_\circ$ may also be thought of as the largest scale-length over which the refractive index fluctuations are correlated. Kolmogorov's theory is strictly applicable in the so-called *inertial regime* of spatial wave numbers k:

$$\frac{2\pi}{L_\circ} < k < \frac{2\pi}{l_\circ}. \tag{2}$$

The sources of the index of refraction variations are mainly temperature and water vapor density inhomogeneities arising from turbulent air motion in the neutral atmosphere, and electron density variations caused by ionospheric winds and traveling disturbances in the ionosphere. The Kolmogorov model describes the spectral statistics of the turbulence-induced index of refraction fluctuations, i.e. the model can calculate the Power Spectral Density (PSD) of the pockets of air with a uniform refractive index:

$$\Phi_n(k) = \frac{0.033 C_n^2}{(k^2 + k_\circ^2)^{11/6}} \exp(-k^2/k_m^2), \tag{3}$$

where $k_\circ = 2\pi/L_\circ$ and $k_m = 5.92/l_\circ$. The parameter C_n is called the *structure parameter* (see Section 4.1.) and represents a measure of the strength of the fluctuations. Equation 3 is also known as the von Karman spectrum and shows how for large k (i.e. small scale-lengths) the PSD is quite small.

This spectral representation has associated, through the Fourier transform, a *spatial* representation in the form of spatial correlation function and spatial structure function, which is often used to describe the effects of turbulence on a variety of physical quantities. For a detailed discussion of the relationships between these functions see, e.g. Roggeman & Welsh (1996). In the case of the refractive index fluctuations the spatial structure function is defined as

$$\mathcal{S}_n(\boldsymbol{b}) \equiv \left\langle [n(\boldsymbol{r}+\boldsymbol{b},t) - n(\boldsymbol{r},t)]^2 \right\rangle \equiv \left\langle (\Delta n)^2 \right\rangle, \tag{4}$$

where n is the refractive index measured at two positions separated by the distance $\boldsymbol{b}$. The structure function represents a mean squared fluctuation over a span $\boldsymbol{b}$ along the direction $\boldsymbol{r}$, and thus the standard deviation of the fluctuations is simply

$$\sigma_n = \sqrt{\langle (\Delta n)^2 \rangle} = \sqrt{\mathcal{S}_n}. \tag{5}$$

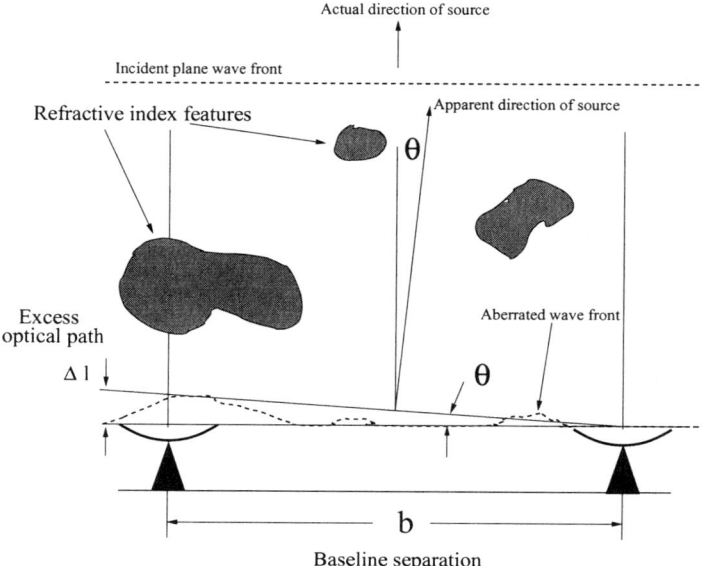

Figure 1. Variation in the refractive index causes a phase delay of electromagnetic radiation.

In many useful cases, the structure function can be approximated with a power law, $S = c^2 b^\alpha$, in which case the slope, α, is a measure of how rapidly the intensity of the fluctuations of the parameter of interest (e.g. n) increases with increasing b. The coefficient c^2, also known as the *structure coefficient* (see Section 4.2.), represents the strength of the turbulence and depends on the site altitude, location, time of the day and wavelength.

3. Ionospheric Effects

At altitudes $\gtrsim 60$ km the ionization of the atmosphere is caused by the ultraviolet radiation from the sun. The electron distribution and the total electron content (TEC) vary with geomagnetic latitude, time of the day, time of the year, and sunspot cycle. There are also ionospheric winds and other disturbances that can cause both small and large scale irregularities in the electron density distribution. The ionosphere is also permeated by the magnetic field of the earth.

Absorption effects by the ionosphere are of some importance only at frequencies less than a few 10's of MHz. On the other hand, the effects of phase path irregularities (see for example Figure 1) are even more dramatic than those due to the troposphere at cm wavelengths. The scales of the irregularities, however, are typically km to hundreds of km, with speed of motion 50-300 m s^{-1} and a height of 300 km upwards. Since the scales of the fluctuations are typically larger or much larger than the size of existing interferometers at these frequencies, their primary effect is to tilt the wave front in a similar fashion to

AR as observed on millimeter single-dish antennas (for a review see Baldwin & Shouguan 1990).

Observations of the ionospheric seeing effects have been carried out, among others, by Warner (1990). He used the Cambridge Low Frequency Synthesis Telescope at 151 MHz to form images of 3C295. The variations of the ionosphere were on scales larger than the maximum baseline of $b = 5$ km, so the incoming wave fronts were tilted but remained nearly plane. Warner, however, was able to correct the seeing by removing a linear phase gradient. The standard deviation of the corrections applied was $18°$ km^{-1}. We can compare this value to the linear phase gradient that would cause a wave front tilt on the current largest filled-aperture telescope, the Arecibo 305-m dish.

In the simple scenario of a wave front tilt there is no degradation of the beam shape. The net tilt of the wave front, $\Delta l/D$, caused by a difference in the electrical path Δl from one side of the telescope to the other (see for example the antenna to the left in Figure 1), is given at the zenith by

$$\Delta\theta = \frac{\Delta l}{D} = \frac{\lambda}{D} \frac{\Delta\phi}{2\pi}, \qquad (6)$$

where $\Delta\phi = 2\pi\Delta l/\lambda$ represents the variation in the wave front phase across the dish, having diameter D. Tilt effects will be negligible if, say, $\Delta\theta < (1/10)(\lambda/D)$, which leads to the condition $\Delta\phi/D < 118°$ km^{-1} for Arecibo, and one can clearly see that the measured phase gradients are typically much smaller than this limiting value. Furthermore, because the ionospheric phase fluctuations vary as $\sim \nu^{-2}$ (see, e.g. Thompson, Moran, & Swenson 1986) the phase gradients will be even smaller at higher frequencies.

As we said earlier, when the phase fluctuations $\Delta\phi \ll 1$ the wave front emerging from the ionospheric screen is crinkled but the amplitude is unchanged. However, if fluctuations in the TEC are high enough the phase variations are large, $\Delta\phi > 1$, and thus significant amplitude fluctuations can occur that are known as *ionospheric scintillations*. The interested reader may find a thorough description and useful references in Thompson et al. (1986) and also in Baldwin & Shouguan (1990).

4. Tropospheric Effects

4.1. Refractive index fluctuations

The dilute gas of molecules in the neutral atmosphere can be modeled, in the classical theory of dispersion, as bound oscillators characterized by a resonance frequency and a damping constant. The resonances of water vapor and oxygen are pressure broadened and cause attenuation far from the resonance frequencies. Although the electric dipole transitions of H_2O, a polar molecule, are much stronger than the magnetic dipole transitions of O_2, a homonuclear molecule, oxygen is much more abundant than water and the two can thus have comparable effects on radiation from a distant source. Their relative strength depends on the specific wavelength range, as shown below. However, two more differences characterize H_2O and O_2: their scale heights are 2 and 8.4 km, respectively, and it is thus possible to mitigate the water vapor effects by selecting high-altitude observing sites; furthermore, water vapor is not at hydrostatic equilibrium, hence there may be moisture variations even on short-size scales ($\lesssim 100$ m).

Let us consider a plane wave propagating along the z direction in a uniform dissipative dielectric medium, as represented by the equation

$$\boldsymbol{E}(z,t) = \boldsymbol{E}_\circ \exp[j(knz - 2\pi\nu t)] \,, \tag{7}$$

where $k = 2\pi\nu/c$ (c is the speed of light in free space), $\boldsymbol{E}_\circ$ is the electric field amplitude and $n = n_R + jn_I$ is the complex index of refraction. Then, the imaginary component of the index of refraction determines the absorption of the wave, whereas the real component will determine its *phase* variations. For visible, infrared and radio-wave regions the real component of the refractivity of air, defined as $N = (n-1)\,10^6$, (suppressing the sub-index R for the sake of simplicity), can be written as

$$N = N_c + D(\nu) \,, \tag{8}$$

where N_c is the continuum value and $D(\nu)$ is the dispersive contribution. The continuum value can be expressed as the sum of a contribution from dry air, N_d, and from water vapor, N_w, that is

$$N_c = N_d + N_w. \tag{9}$$

Hill, Clifford, & Lawrence (1980) have shown that the fluctuations of the real component of the refractive index can be expressed in terms of the fluctuations of temperature (T [K]), water vapor density (Q [molecules cm^{-3}]) and pressure (P [atm]) in the following way (pressure fluctuations can be neglected, as shown below):

$$C_n^2 = A_T^2 \frac{C_T^2}{\langle T\rangle^2} + A_Q^2 \frac{C_Q^2}{\langle Q\rangle^2} + 2A_T A_Q \frac{C_{TQ}}{\langle Q\rangle\langle T\rangle} \,, \tag{10}$$

where the coefficients A_T, A_Q and A_P are derived from writing the differential change in N_c, and depend on frequency and on T, Q and P. The parameters C_n^2, C_T^2, C_Q^2 and C_{TQ} represent the *structure parameters* of refractive index, temperature and humidity, measured between two points at a distance δr, and defined by

$$C_x^2 \equiv \langle(\delta x)^2\rangle/(\delta r)^{2/3} \tag{11}$$

and $C_{TQ} \equiv \langle \delta Q \, \delta T\rangle/(\delta r)^{2/3}$.

It is impossible to be categorical regarding the relative magnitudes of $\delta T/\langle T\rangle$, $\delta P/\langle P\rangle$ and $\delta Q/\langle Q\rangle$. However, one finds that typically:

$$\frac{\langle(\delta P)^2\rangle^{1/2}}{\langle P\rangle} \ll \frac{\langle(\delta T)^2\rangle^{1/2}}{\langle T\rangle} < \text{or} \ll \frac{\langle(\delta Q)^2\rangle^{1/2}}{\langle Q\rangle} \tag{12}$$

and because $(A_Q/A_T)_{\text{radio}} \gg (A_Q/A_T)_{\text{visible}}$ then at radio wavelengths one can approximate Equation 10 as:

$$C_n^2 \simeq A_Q^2 \frac{C_Q^2}{\langle Q\rangle^2}. \tag{13}$$

The A_Q/A_T ratio has its maximum in the microwave region, $(A_Q/A_T) = -0.3$, and then it has a decreasing trend until it reaches a value of 75×10^{-3} at $\lambda = 25\,\mu$m and 4.6×10^{-3} at $\lambda \simeq 7\,\mu$m (Hill et al. 1980) at which point the temperature fluctuations start to become the dominant source of refractive index fluctuations. Although in the radio region (more specifically, between 10 and 500 GHz) C_n^2 varies by no more than 50%, the phase RMS increases as ν^2 (see below). The quantity A_P/A_T is a measure of the importance of pressure fluctuations in producing refractive index fluctuations. At visible frequencies Hill et al. (1980) finds that $(A_P/A_T) = -1$ but because $\langle(\delta P)^2\rangle^{1/2}/\langle P\rangle \ll \langle(\delta T)^2\rangle^{1/2}/\langle T\rangle$ the pressure fluctuations make an insignificant contribution to refractive index fluctuations.

In the radio regime, the sum of all line tails gives a nearly constant value of the dispersive component, $D(\nu)$, of the refractivity therefore providing the water vapor sensitivity of radio waves. In fact, the excess optical path, $c\Delta t$, can be written as:

$$l(\text{cm}) = 10^{-6} \int N \, dl = 0.228 P_\circ(\text{mbar}) + 6.3 w(\text{cm}) , \qquad (14)$$

where $P_\circ$ is the ground pressure and w is the precipitable water vapor (PWV). The second term in Equation 14 is frequency-independent and can be used throughout the radio regime to calculate the excess optical path if w is measured.

4.2. Phase structure function

Variations in atmospheric water vapor column abundance above a dual element interferometer, with a baseline b, produce a phase delay whose structure function can be described using a power law:

$$\mathcal{S}_\phi = c_\phi^2 b^\alpha , \qquad (15)$$

where $\alpha = 5/3$ if b is much less than the scale height of the water vapor, and $\alpha = 2/3$ if b is larger than the scale height of the water vapor. The intermediate region, approximately 5 m $< b <$ 1200 m, is known as the *spectral gap* (Coulman & Vernin 1991) where α undergoes a smooth transition from 5/3 to 2/3. The phase structure coeffcient, c_ϕ^2, is given by:

$$c_\phi^2 = 2.91 \left(\frac{2\pi}{\lambda}\right)^2 C_n^2 L , \qquad (16)$$

where the refractive index structure parameter, C_n^2, is assumed constant up to a height L and zero thereafter. The spatial structure function can also be written as:

$$\mathcal{S}_\phi = 6.973 \left(\frac{b}{r_\circ}\right)^{5/3} , \qquad (17)$$

where the Fried parameter,

$$r_\circ = 0.186 \lambda^{6/5} (C_n^2 L)^{-3/5} , \qquad (18)$$

represents the seeing cell, or the distance between two points for which the phase RMS is 2.2 rad. It is a slightly chromatic parameter and measured values so far indicate that it may vary between ~ 100 m and 5 km at $\lambda = 3.47$ mm (Olmi & Downes 1992).

Because many measurements of seeing effects rely on time series, it is useful to see how the *temporal* phase structure function, $\mathcal{D}_\phi(\tau)$ where τ is the time lag, is defined and how it is related to $\mathcal{S}_\phi$:

$$\mathcal{D}_\phi(\tau) \equiv \left\langle [\phi(\boldsymbol{r}, t+\tau) - \phi(\boldsymbol{r}, t)]^2 \right\rangle = \mathcal{S}_\phi(v\tau) , \qquad (19)$$

where we have used Taylor's hypothesis of frozen turbulence, i.e. (Lawrence & Strohbehn 1970):

$$\phi(\boldsymbol{r}, t+\tau) = \phi(\boldsymbol{r} - \boldsymbol{v}\tau, t) , \qquad (20)$$

where v is the average wind velocity component perpendicular to the line-of-sight of the imaging system.

The temporal structure function, $\mathcal{D}_\theta(\tau)$, for the observable AR-induced pointing error $\Delta\theta(t)$ (i.e. the variation of the angular distance of the target source from the beam center) can be defined as (Olmi 2000a):

$$\mathcal{D}_\theta(\tau) \equiv \left\langle [\Delta\theta(t+\tau) - \Delta\theta(t)]^2 \right\rangle = c_\theta^2 \tau^\alpha , \qquad (21)$$

where the second equality applies when the AR structure function can be fit using a power law. The parameter c_θ^2 is called the AR structure function coefficient and is a direct measure of the strength of the AR effects. It can be shown (Olmi 2001) that $\mathcal{D}_\theta(\tau)$ can also be expressed as a function of $\mathcal{S}_\phi$, which can be written for the case of Kolmogorov turbulence as:

$$\mathcal{D}_\theta(\tau) \simeq \begin{cases} \dfrac{2 c_\phi^2 D^{2/3}}{k} & \text{if } v\tau \gg D \\[1em] \dfrac{2 v^{5/3} c_\phi^2}{kD} \tau^{5/3} & \text{if } v\tau \ll D , \end{cases} \qquad (22)$$

where $k = 2\pi/\lambda$ and we have:

$$c_\theta^2 = \frac{2 v^{5/3} c_\phi^2}{kD} , \qquad (23)$$

where c_ϕ^2 has units of $\text{m}^{-5/3}$ and thus c_θ^2 has units of $\text{s}^{-5/3}$.

5. Description of Seeing Effects on Single-dish Telescopes

5.1. Phenomenological description of AR

As we said earlier, the wave front from a distant source is distorted by the atmospheric variations, causing phase variations across the telescope aperture. In the simplest scenario, as described in Section 3., the wave front encounters a refractive "wedge" of water vapor. On passage through this linear phase screen the wave front is tilted from its original direction because of the phase difference introduced between the opposite extremities of the receiving aperture, as the propagation paths traverse air masses of varying humidity. However, no degradation of the beam-shape is generated, and the resulting pointing errors are also known as angle of arrival fluctuations in the field of clear- or dry-air propagation effects (see, e.g. Fante 1975 and Lawrence & Strohbehn 1970).

The seeing effects on single-dish antennas depend on the telescope diameter to Fried parameter ratio, $D/r_\circ$, as it can be seen combining Equations 5 (for the case of phase fluctuations), 6 and 17 to obtain the standard deviation of the angle of arrival fluctuations, in units of the beam FWHM (assumed here equal to $1.2\lambda/D$):

$$\sigma_\theta[\text{beams}] = 0.504 \left(\frac{D}{r_\circ}\right)^{5/6}. \qquad (24)$$

It is important to remark that while the absolute value of AR is wavelength-independent, from Equation 24 one can see that the magnitude of the pointing error as a fraction of the beam width is bigger on larger telescopes and at shorter wavelengths. Therefore, its impact on the pointing is likely to become critically important in the next generation of electrically large, filled-aperture radio telescopes.

Because $r_\circ$ depends on the wavelength of observation we can define a normalized $D/r_\circ$ parameter which includes both the frequency and spatial dependence, assigning a given value of the Fried parameter at a given wavelength, e.g.:

$$^{50}r_\circ(\nu) = 50 \text{ m} \left(\frac{100 \text{ GHz}}{\nu}\right)^{6/5}, \qquad (25)$$

where we have assumed that, under conditions of strong turbulence, $r_\circ = 50$ m at $\nu = 100$ GHz. Thus we define the non-dimensional parameter

$$m_{50} \equiv \left(\frac{D}{^{50}r_\circ(\nu)}\right)^{5/6} = \frac{\nu}{100 \text{ GHz}} \left(\frac{D}{50 \text{ m}}\right)^{5/6}, \qquad (26)$$

indicating that a telescope would be AR-limited (i.e. $\sigma_\theta > 0.5$ beams) under these specific conditions if $m_{50} > 1$. However, AR-effects will seriously affect the pointing and calibrations of the telescope also for $m_{50} > 0.5$. Values of m_{50} vary from 0.4 for Arecibo at 10 GHz, to 1.8 and 3 for the GBT[2] and LMT/GTM[3], at 100 and 300 GHz, respectively. For somewhat milder turbulence, e.g. $r_\circ = 100$ m at $\nu = 100$ GHz, one finds that $m_{100} = 1$ for the GBT and 1.7 for the LMT/GTM, showing that seeing effects are potentially more critical on the LMT/GTM at higher frequencies.

5.2. Numerical simulations of AR

The phenomenological description of AR presented in the previous section is useful to rapidly and easily determine the (potential) strength of the AR-effects on various telescopes at different wavelengths, but it cannot describe the effects

---

[2] Green Bank Telescope

[3] Large Millimeter Telescope, or Gran Telescopio Milimetrico

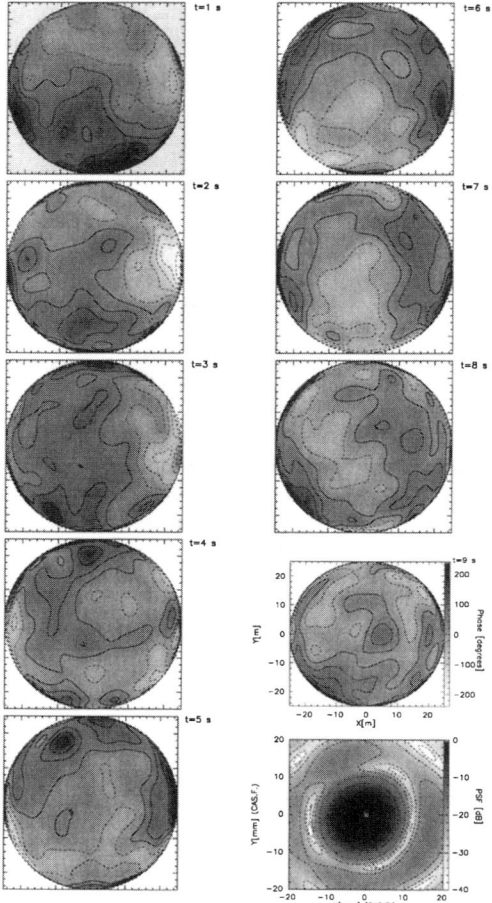

Figure 2. Nine random, time-sequential images of the evolving phase screen over the LMT/GTM aperture plane during conditions of strong turbulence. In the simulation $j_{\max} = 136$ Zernike polynomials were used with $r_\circ = 25$ m, $v_w = 5$ m s^{-1}, and zenith angle $z = 0$. The phase interval between adjacent contour levels is 50°. The image at the bottom right shows the resulting antenna beam, at $\lambda = 1$ mm, after an integration time $t_{\text{int}} = 10$ s. The first contour level is -3 dB, and successive levels are separated by -6 dB. CAS. F., Cassegrain focal plane (from Olmi 2000b).

of the higher-order aberrations present in the wave front and, even more important, it cannot predict the time evolution of the phase screen under specific environmental conditions. Numerical simulations of the phase screens and of their effects on the antenna beam or, equivalently, on the Point Spread Function (PSF), can provide a full description of the aberrated wave front and can also be used to simulate the operation of a *compensating* device.

Unlike reproduceable aberrations induced by the deformation of the telescope structure, turbulence-induced wave front aberrations fluctuate randomly, both spatially and temporally, causing the PSF of the telescope in the focal plane to be random. The dominant effect is a random wave front tilt in the entrance pupil, which, according to Kolmogorov turbulence theory, accounts for most of the degradations in long-exposure images (see, e.g. Roggeman & Welsh 1996). The phase screen generators are computer programs that create random arrays of phase values on a grid of sample points which have the same statistics as the turbulence-induced atmospheric phase, $\phi(x, y, z; t)$. The sample grid points can be spaced in both space, (x, y, z), and time, t.

In the adaptive optics literature there are a number of approaches used to generate random phase screens with the proper point statistics and spatial and temporal correlation properties. One common method consists of expressing the phase perturbations of the wave front, ϕ, as a linear combination of orthonormal basis functions. In many practical cases the Zernike polynomials, $Z_j(\rho, \theta)$, are used as the set of basis functions:

$$\phi(R\rho, \theta) = \sum_{j=1}^{\infty} a_j \, Z_j(\rho, \theta) \; , \tag{27}$$

where $\phi(R\rho, \theta)$ is the turbulence-induced phase in the polar coordinates ρ and θ, normalized for an unobscured aperture of radius $R = D/2$. In all practical simulations the expansion contains polynomials only up to a maximum order $j_{\max}$. Propagation through the turbulent layers is expressed in terms of the effect on the Zernike coefficients a_j. Phase screens can be generated that have the correct (Kolmogorov's) spatial statistics within each phase screen and the correct space-time statistics across the stack of phase screens (see Olmi 2000b for a complete discussion). The most interesting modes for the present discussion are the Zernike modes $Z_2(\rho, \theta)$ and $Z_3(\rho, \theta)$, which are referred to as tilt and correspond to the orthogonal components of the least squares sense best fit plane to $\phi(R\rho, \theta; t)$ in the telescope aperture plane. Tilt does not affect image quality but does cause the PSF to be displaced from the position corresponding to the $a_2 = a_3 = 0$ condition.

An example of a time sequence of images representing an evolving phase screen is shown in Figure 2 at the zenith for a very small Fried parameter $r_o(\lambda = 3\,\mathrm{mm}) = 25$ m and a wind speed $v_w = 5$ m s^{-1}; the time interval between successive frames is 1 s. Figure 3 shows the simulated PSF at the Cassegrain focal plane of the LMT/GTM, with the telescope pointed at the zenith and assuming a ground wind speed of 5 m s^{-1}. The six images presented correspond to three different values of r_o and integration time, and they show three main features. First, all PSFs in the Cassegrain focal plane are shifted from their nominal $(0, 0)$ position, and the amount of AR decreases when a larger Fried parameter is used. Second, the magnitude of AR decreases using

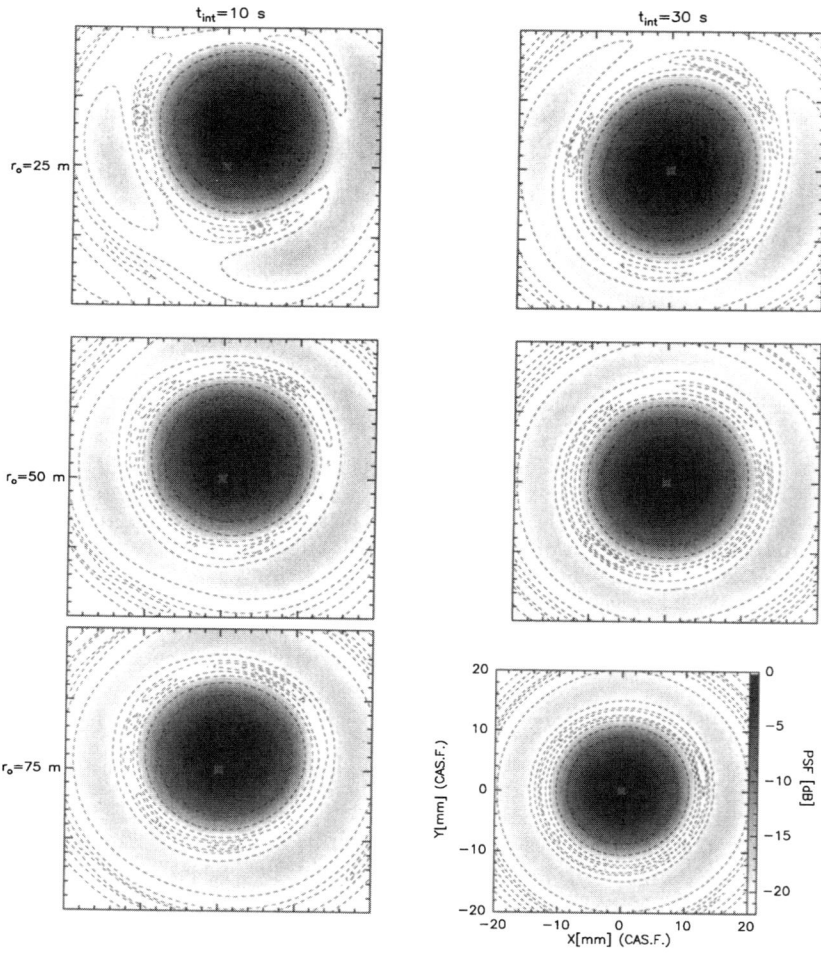

Figure 3. Simulated PSFs in the Cassegrain focal plane of the LMT/GTM at $\lambda = 1$ mm and with the telescope pointed at the zenith. The column on the left corresponds to a 10 s integration time, whereas in the column on the right we used 30 s. The Fried parameters used were $r_\circ(\lambda = 3 \, \text{mm}) = 25, 50,$ and 75 m for the three rows from top to bottom, respectively. A wind speed of 5 m s^{-1} was used (from Olmi 2000b).

The Effects of the Atmosphere   425

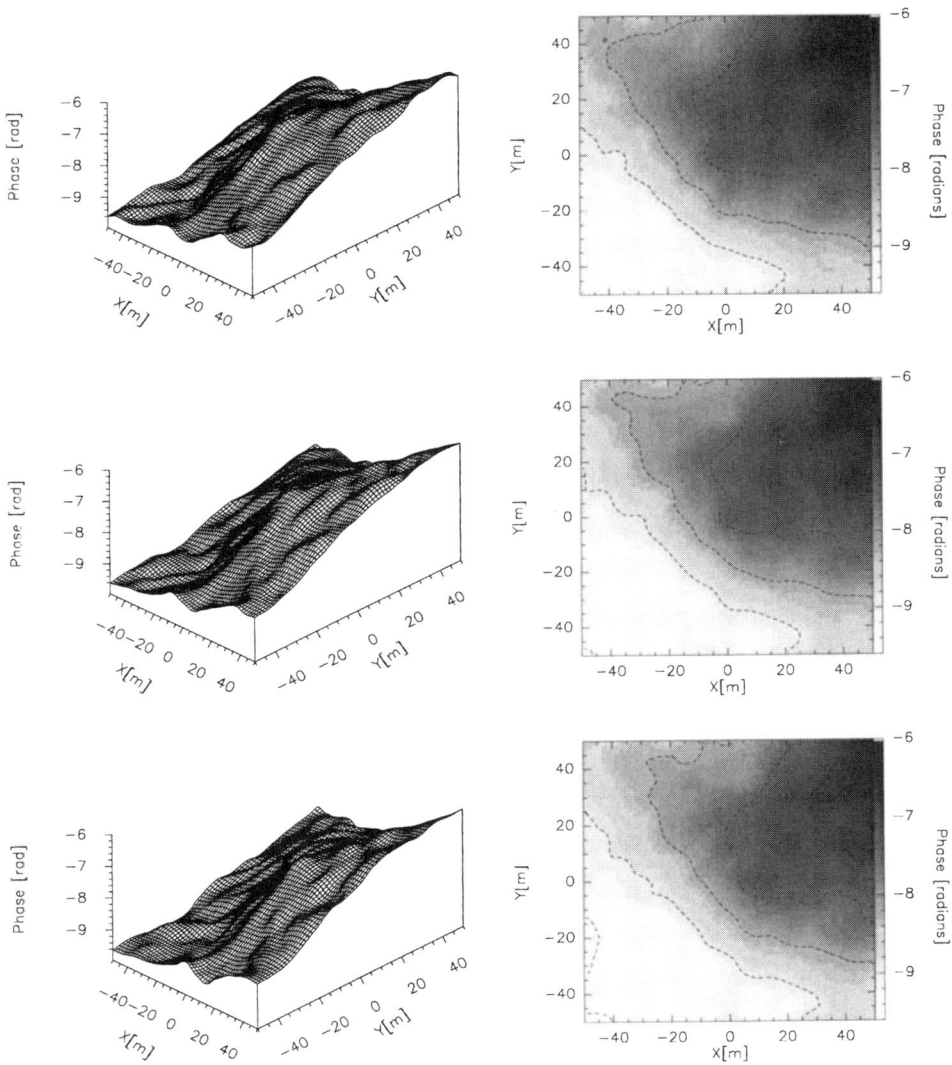

Figure 4. Three random, time-sequential images for the case of atmospheric layers moving in the same direction and speed, and a von Karman spectrum. The time interval between each image is 1s. Surface plots of the phase screen are shown on the left-hand column, whereas grey scale images are shown on the right-hand column.

longer integration times; the averaging process also partly restores the circular symmetry of the PSF. Finally, in the case of strong turbulence, when $r_o(\lambda = 3\text{mm}) = 25$ m for example, the PSF is highly distorted, and the sidelobes rise. In this case the averaging process can reduce the AR level, but the shape of the PSF still remains quite aberrated.

A more complete phase screen generation can be derived for both the von Karman and the Kolmogorov turbulence power spectra, $\Phi_n(k, z)$, where z is the distance from the aperture and k is the spatial wave number of the refractive index fluctuations. The model can include several turbulent layers with different structure parameters, $C_n^2(z)$, and the transverse velocity of turbulence caused by wind can also be characterized by a velocity profile, $v(z)$ (Roggeman & Welsh 1996). The set of basis functions used in this case consist of a collection of delta functions located over a rectangular grid of points in the pupil of the imaging system.

In Figure 4 we show three random, time sequential images for the case of atmospheric layers moving in the same direction and speed ($v_y = 0$ and $v_x = 5$ m s^{-1}), and a von Karman spectrum (see Equation 3). Thus, Taylor's frozen field hypothesis has been applied to the whole atmosphere in this specific case, and the atmosphere is moving as a monolithic block in the x direction (i.e. rightward). Fixed structures in the phase screen should be moving in the x direction. In generating Figure 4 the Hufnagel-Valley turbulence profile (Roggeman & Welsh 1996) was used, with four different atmospheric layers at altitudes varying from 0.2 km to 18 km and thickness varying from 0.4 km to 12.8 km.

This improved model also allows one to enter the atmospheric parameters in terms of measurable quantities at the telescope site. In the GBT example (Figures 4 and 5) the temperature and relative humidity at the site were assumed to be 5 C and 40 %, respectively, corresponding to PWV=5.5 mm and an RMS fluctuation about this value of ΔPWV=0.05 mm. The refractive structure parameter, $C_n^2(z)$, was then calculated according to the formulae presented in Section 4.1. and varied from 2.8×10^{-19} m$^{-2/3}$ to 5.0×10^{-14} m$^{-2/3}$ in the four layers, whereas the Fried parameter varied from about 310 m to 127 km. In Figure 5 the PSF of the GBT at $\lambda = 3$ mm is shown as a result of the phase screen of Figure 4, and is clearly decentered with respect to the nominal position (indicated by a star). The corresponding AR pointing error was $2\rlap{.}''2$ in this specific example.

6. Measurements of AR

The first extensive measurements of AR were carried out by Altenhoff et al. (1987), Downes & Altenhoff (1990), and also Church & Hills (1990) who found that AR events are characterized by angular displacements of the sources from their true positions by a few arc seconds, in both azimuth and elevation, for a few seconds of time, but occasionally showing much larger events that could last for tens of seconds. This is similar to what is observed in near-infrared astronomy, where, for small telescope diameter to Fried parameter ratio, $D/r_o(\lambda) \lesssim 6$, the short-exposure PSF randomly moves in the focal plane (e.g. Close & McCarthy 1994).

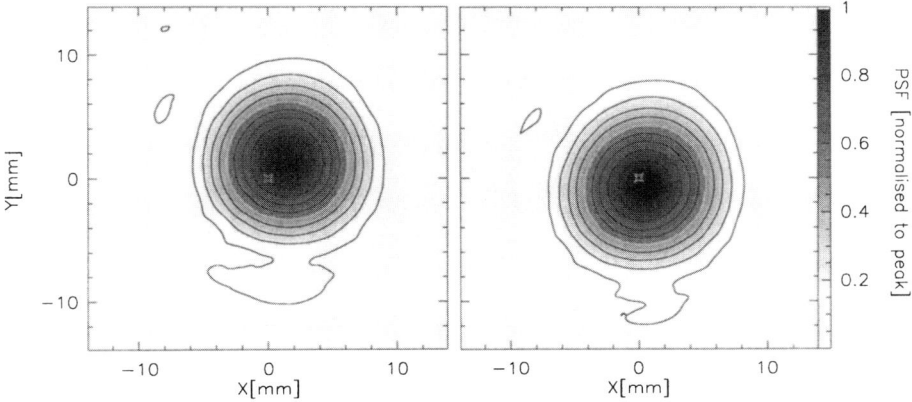

Figure 5. Resulting PSF at the Gregorian focal plane of the GBT at $\nu = 100$ GHz for the phase screen shown in Figure 4. The simulated atmospheric conditions at the GBT site were such that PWV=5.5 mm and its RMS fluctuation was ΔPWV=0.05 mm. *Left*: uncompensated, the angular pointing shift is 2″.2. *Right*: compensated using a wave front sensor (Section 7.); the residual pointing shift is 0″.8.

On the new large radio telescopes that are either under construction (GBT, LMT/GTM) or being designed (ALMA, SRT[4]), for which $D/r_0(\lambda) \sim 1$ at the highest frequencies, phase gradients across the antenna aperture will dominate, but there will be also higher order aberrations that can effectively broaden the primary beam, as we have seen in the previous section. In more recent years these projects have also prompted serious investigations of techniques to compensate AR effects (Holdaway 1997; Butler 1997; Holdaway & Woody 1998; Olmi 2000a, 2000b).

The most extensive survey to date of AR-effects on a single-dish mm-wave telescope has been carried out by Olmi (2001) using the 13.7-m radome-enclosed telescope of the Five College Radio Astronomy Observatory[5] (FCRAO) located in New Salem (U.S.A.) at an elevation of 314 m above sea level. The data are based on the time series of the fluctuations of the angular distance of the source from the beam center, as measured at the -3 dB points. AR-induced pointing errors were detected with the FCRAO $\simeq 60''$ beam and the measured values ranged from $\simeq 2''$ (winter) to $\simeq 20''$ (summer). The widths of the probability density distributions (see Figure 6) corresponding to about 75% of the data were

---

[4] Atacama Large Millimeter Array, Sardinia Radio Telescope

[5] The Five College Radio Astronomy Observatory is operated with support from the National Science Foundation and with permission of the Metropolitan District Commission.

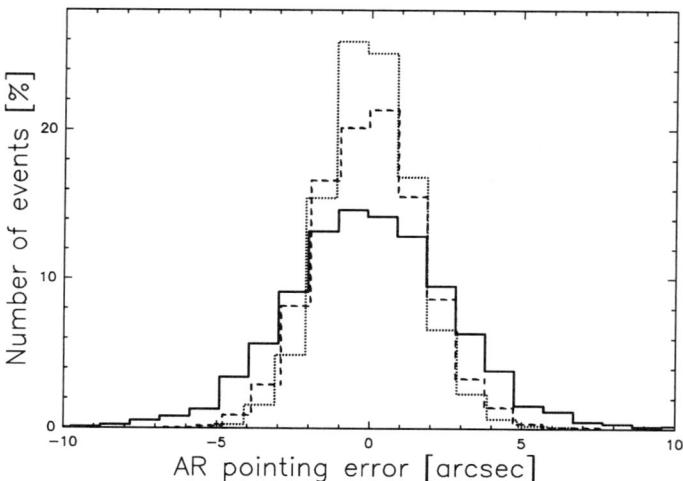

Figure 6. Histograms of AR-induced angular displacements observed on different days: 13-APR-2000 with PWV=2.7 mm (dotted line) corresponding to an optical depth at 225 GHz of $\tau_{225} = 0.27$; 14-APR-2000 with PWV=4.0 mm or $\tau_{225} = 0.33$ (dashed line); and 26-MAY-2000 with PWV$\simeq$ 7.2 mm or $\tau_{225} = 0.59$ (solid line). The bins have a width $\simeq 1''$ (from Olmi 2001).

$\simeq 6''$ for high PWV (PWV> 7 mm, or $\tau_{225} > 0.59$ at the FCRAO site[6]) and $\simeq 4''$ for relatively low PWV (PWV< 4 mm, or $\tau_{225} < 0.33$).

Olmi (2001) used several statistical functions to analyse the data, and found that many structure functions can be fit with a single power law of type $\mathcal{D}_\theta(\tau) = c_\theta^2 \tau^\alpha$, where usually $\alpha < 0.4$ and $\log c_\theta^2 \sim -4$ to -2. The slopes of the AR structure functions, however, were much lower than those of the phase structure functions measured with millimeter-wave interferometers. Power spectra and Allan variance plots could also be fit with single power laws, and the three different power law slopes correlate and are consistent with the standard model of atmospheric turbulence. The magnitude of the AR fluctuations, c_θ^2, correlated well with PWV and ground-level temperature, decreased with increasing elevation angle and also varied during the day. These characteristics indicate that stronger AR fluctuations are associated with increased convective activity near the ground, which is typical of warmer, and more humid, weather when strong thermal gradients create considerable ground-level turbulence (Coulman 1991).

Furthermore, because the largest AR-induced pointing errors will occur at the shortest wavelengths, and because it is likely that large antennas such as the LMT/GTM, GBT and SRT will operate at their highest-frequency regime only during conditions of low ($\lesssim 5-10$ m s^{-1}) wind-speed, we should not

---

[6] τ_{225} is the zenith optical depth at 225 GHz.

expect that the reduced time-scale of AR events resulting during conditions of high wind-speed will contribute to average the AR effects down. Although it is clear that extrapolating these results to sites with different characteristics and at higher frequencies is difficult, it seems reasonable to expect similar AR effects during conditions of similar PWV and ambient temperature. If this is indeed the case then the expected magnitude of the AR-induced pointing errors under certain conditions can be comparable with the beam width of the LMT/GTM, for example, and all antenna measurements (OTF mapping, pointing, focusing, beam switching, etc.) would then be seriously affected.

7. Compensation for AR

A system to remove AR effects should ideally perform two functions: (1) sense the wave front tilt, and (2) compensate in real-time for the phase gradient across the antenna aperture. Olmi (2000a) discussed a conceptual framework for a tip-tilt compensating system at millimeter wavelengths, consisting of two main components: the wave front sensor (WFS) and an adjustable mirror that would be able to remove AR pointing errors with timescales longer than the inherent finite system temporal response. The most likely candidate as the adjustable mirror is the subreflector, which can correct the tilt of the wave front simultaneously for all astronomical receivers.

The WFS must provide the means of measuring the spatial slope of the wave front, i.e. the phase gradient across the antenna aperture. One method to achieve this is by sampling the spatial distribution of the atmospheric brightness temperature T_b over the antenna and along the line of sight. The fluctuating component of the PWV can be calculated from T_b using an atmospheric model, and the excess optical path would then be obtained using Equation 14. Using the conversion equation RMS variations in column abundance then result in corresponding RMS variations in the phase of the wave.

The spatial distribution of T_b over the antenna aperture can be sampled using either an array of radiometers or a single scanning radiometer. Olmi (2000a) compared the two systems and concluded that in a single scanning radiometer the absolute gain calibration is not as nearly as critical as in the array, because a differential measurement is involved. However, the scanning radiometer has its own problems because of timing issues, moving parts and, since the WFS is positioned off-axis to allow the simultaneous use of the astronomical receivers with the WFS, there is potentially a different spillover when the beam is scanned across the antenna aperture. The radiometer(s) would under-illuminate the telescope aperture to measure the emission of the PWV contained in the near field (which is "tube-shaped") of four spatially separated columns, corresponding to the four quadrants, along the line of sight.

The LMT/GTM is currently proposed to be equipped with a WFS based on a single-scanning 183-GHz radiometer, employing a Schottky diode mixer front-end, followed by a broad-band low-noise IF amplifier. The correction sensed by the WFS is fed back into the subreflector positioner or the wobbler, and the gradient would be removed from the wave front. It is important to sample T_b at three different frequencies at least, to enhance the receiver sensitivity under various atmospheric conditions.

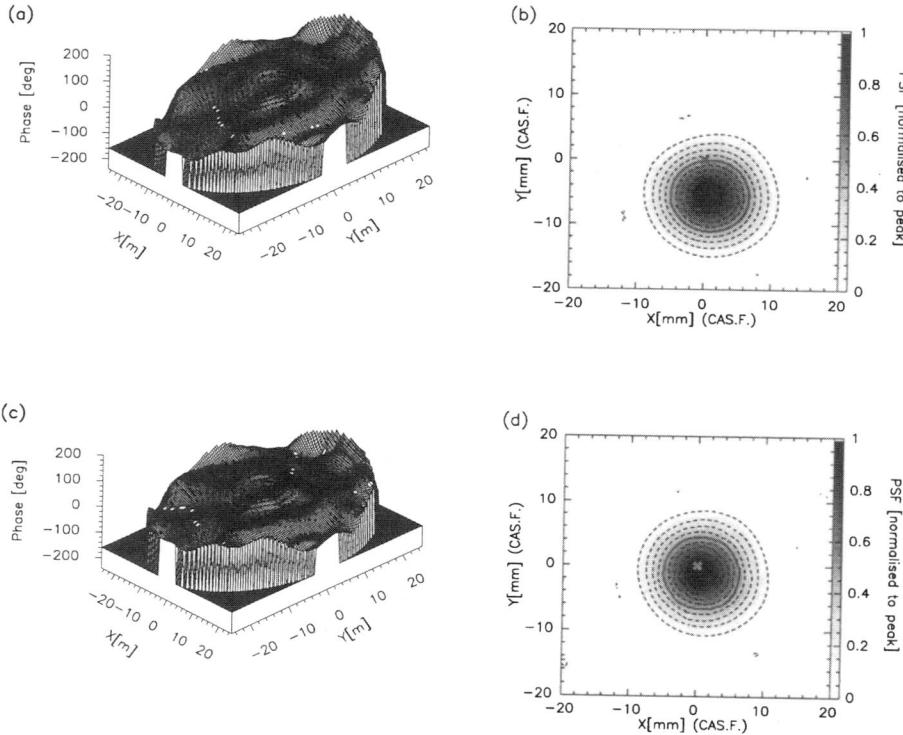

Figure 7. (a) Surface plot of the corrupted phase screen at $t = t_1$ on the LMT/GTM aperture plane. (b) Time-averaged PSF. (c) Phase screen at $t = t_1$ after tip-tilt correction has been applied. (d) Resulting PSF obtained after tip-tilt correction has been applied at each time-instant. Parameters used in the simulation: screen with $v_w = 2$ m s^{-1}, $r_\circ(\lambda = 3$ mm$) = 25$ m, $L_\circ = 25$ m, and $t_{\rm int} = 10$ s.

It is possible to use the numerical method discussed in Section 5.2. to simulate the basic operation of a wave front sensor that would respond to the Zernike modes Z_2 and Z_3 (i.e. tilt). The simulation of the measurement and correction of the wave front tilt consists of the following steps: (1) the average phase is measured over an area of the telescope aperture corresponding to the cross section of the radiometer beam in each quadrant. (2) A least squares fit of a plane to the four phase points over the aperture plane is then performed, and the corresponding tilt of the wave front is calculated. (3) The estimated correction to compensate the wave front slope error is fed back into the simulation, which assumes that the drive system of a flat mirror dedicated for the purpose (e.g. the subreflector) can be adjusted to tilt in two orthogonal directions. An example of the correction that could be achieved on the LMT/GTM is shown in Figure 7. In Figure 5 we showed an example of AR-compensation applied to the GBT; in this case the simulation also included the estimated radiometer noise contribution.

An alternative approach is to measure the PWV in the wavelength range 455-516 cm^{-1} ($\sim$ 20 μm) (Smith, Naylor, & Feldman 2001). This technique has the advantage that the PWV signature is stronger and the instrumental bandpass is much wider ($\sim$ 1.5 THz). Furthermore, photoconductive cells are simple and robust, compared to water vapor radiometers. However, the RF radiometer looks through the same atmosphere as the telescope and its beam can be adjusted to illuminate a finite circular section of the aperture plane. On the other hand the IR radiometer samples a different atmospheric column, adjacent to the telescope, and its aperture is much smaller than that of the telescope. This makes the IR radiometer better suited to measure phase fluctuations on (sub)mm-wave interferometers, rather than to measure phase gradients over filled-aperture telescopes. Emission and scattering from ice crystals may also affect the IR radiometer's readings.

Acknowledgments. This work was sponsored by the Advance Research Project Agency, Sensor Tech. Off. DARPA Order No. C134 Program Code No. 63226E issued by DARPA/CMO under contract No. MDA972-95-C-0004.

References

Altenhoff, W. J., Baars, J. W. M., Downes, D., & Wink, J. E. 1987, A&A, 184, 381

Baldwin, J. E. & Shouguan, W. 1990, in URSI/IAU Symp., Radioastronomical Seeing, ed. J. E. Baldwin & W. Shouguan (New York: Pergamon Press), 1

Butler, B. 1997, MMA Memo 188

Church, S. & Hills, R. 1990, in URSI/IAU Symp., Radioastronomical Seeing, ed. J. E. Baldwin & W. Shouguan (New York: Pergamon Press), 48

Close, L. M. & McCarthy Jr., D. W. 1994, PASP, 106, 77

Coulman, C. E. 1991, A&A, 251, 743

Coulman, C. E. & Vernin, J., 1991, Appl. Opt., 30, 199

Downes, D. & Altenhoff, W. J. 1990, in URSI/IAU Symp., Radioastronomical Seeing, ed. J. E. Baldwin & W. Shouguan (New York: Pergamon Press), 31

Fante, R. L. 1975, Proc. IEEE, 63, 1669

Hill, R. J., Clifford, S. F., & Lawrence, R. S. 1980, J. Opt. Soc. Am., 70, 1192

Holdaway, M. A. 1997, MMA Memo 186

Holdaway, M. A. & Woody, D. 1998, MMA Memo 223

Lawrence, R. S. & Strohbehn, J. W. 1970, Proc. IEEE, 58, 1523

Olmi, L. 2000a in Proc. SPIE, Astronomical telescopes and instrumentation 2000, 4015, ed. H. R. Butcher (Bellingham, Wash.: SPIE), 390

Olmi, L. 2000b, Radio Science, 35, 275

Olmi, L. 2001, A&A, 374, 348

Olmi, L. & Downes, D. 1992, A&A, 262, 634

Roggeman, M. C. & Welsh, B. 1996, in Imaging through Turbulence, ed. M. J. Weber (Boca Raton: CRC Press)

Smith, G. J., Naylor, D. A., & Feldman, P. A. 2001, Int. J. Inf. Mm. W., in press

Thompson, A. R., Moran, J. M., & Swenson, G. W. Jr. 1986, Interferometry and Synthesis in Radio Astronomy (New York: J. Wiley & Sons)

Warner, P. J. 1990 in URSI/IAU Symp., Radioastronomical Seeing, ed. J. E. Baldwin & W. Shouguan (New York: Pergamon Press), 157

RFI and How to Deal with It

J. R. Fisher

National Radio Astronomy Observatory, P. O. Box 2, Green Bank, West Virginia 24944, USA

Abstract. Interference from active users of the radio spectrum will always be a problem for radio astronomy, but careful planning, active data monitoring, and a few data editing techniques allow us to use more spectrum for science than one might expect. A good plan starts with knowing what to expect in the frequency range that you want to use. Then there are a number of receiver options, signal processing configurations, and observing strategies that will help to minimize the effects of interference to your data. The degree to which you can remove interference from data after it is recorded depends considerably on having chosen the best observing parameters before the observing run began. Research into real-time signal processing for interference excision offers some hope that future radio astronomy receivers will better separate cosmic signals from man-made transmissions. Some of these excision techniques under investigation are reviewed.

1. Introduction

Many radio astronomers are accustomed to observing in the protected radio astronomy frequency bands or at high enough frequencies where radio frequency interference (RFI) is, at most, a minor annoyance. To a large extent this will continue to be the case, but the advent of wider band receivers and spectrometers and an increasing desire to pursue science outside of the protected bands present new challenges. A few frequency ranges are so heavily filled with radio transmitters that their use for astronomy is unrealistic. Other frequency bands are sparsely filled or used intermittently on time scales from microseconds to days. If the science is important enough, a careful study of the spectrum and RFI signal characteristics may permit observations that, at first glance, would appear impractical. Even where interference is isolated in frequency or is present only a small fraction of time, a little advanced planning can decrease the fraction of data that must be discarded due to RFI contamination.

To focus this discussion I will concentrate on spectral line observing at Arecibo and Green Bank. Many of the same considerations apply to continuum and pulsar observations and to other observatories. Continuum observers may want to consider the use of a spectrometer for their observations at frequencies where RFI is a problem since this offers the possibility of rejecting interference in the frequency domain. Data rates and data volumes made this impractical in the past, but faster processors and much larger storage capacities have changed the picture considerably.

Much of what I have to say will be self-evident, but it can serve as a check list for setting up an observing run when you know that RFI could be a problem. I encourage you to develop your own techniques and algorithms for interference removal and contribute these to the community.

2. Early Planning

Several resources are available at Arecibo and Green Bank for getting a preview of the interference environment that you will encounter. These are of most help in writing your observing proposal and in communicating with the observatory staff before arriving for your observing run. If appropriate, learn the name of the contact for interference matters at the observatory, and make them aware of your observing plans and the frequencies that you expect to use to give them a chance to offer advice on recent RFI experience and, in a few special cases, to coordinate your observing times with users of the spectrum with whom a coordination agreement exists. Do not rely on a comment on your observing proposal to catch the eye of the local spectrum manager.

Both Arecibo and Green Bank have web pages devoted to interference matters that may be accessed from their home pages at the following addresses:

http://www.naic.edu click [Scientific Users] [Spectrum Management]

http://www.gb.nrao.edu click [Quiet Zone] and [RFI Protection]

Spend a little time browsing these web pages to familiarize yourself with the RFI activities at the respective sites.

2.1. Spectrum surveys

The Arecibo Observatory has an antenna and spectrum analyzer on the hilltop above the telescope near the lab. This antenna provides a recent survey of the spectrum between 0 and 1450 MHz, where the interference is most dense, and between 1550 and 10000 MHz. These measurements may be found at

http://www.naic.edu/~rfiuser/smarg-hplots.html

There are also a number of statistical plots and lists of frequencies of known signals seen in the two 20-cm (L-band) receiver frequency ranges which can be found at

http://www.naic.edu/~rfiuser/smarg-vig1.html

At Green Bank, spectrum surveys are available for the frequency ranges of most of the receivers that were available on the 140-foot telescope between 50 MHz and 3.4 GHz. These surveys were done with the FFT spectrometer at the receiver output with the telescope pointed at the zenith. The spectral resolution was 40 kHz. The spectrum plots and an explanation of the data acquisition and analysis may be found at

http://www.gb.nrao.edu/RFI/

A new spectrum survey of GBT receivers is in the works.

2.2. Coordination

Spectrum sharing is becoming a more frequently used spectrum management tool, particularly in the allocated bands where radio astronomy is a co-primary or secondary "service" in the band as defined by the International Telecommunications Union. At least one sharing agreement is already in place with the Iridium satellite operators and others may be negotiated in the future. These agreements usually require some sort of expected use forecasts from the observatories for scheduling purposes. Also, the Arecibo Observatory has established informal coordination procedures with users of the radio spectrum near the most frequently used observing bands.

To make these coordination procedures work, information about your expected frequency use during your observing run should be provided to the observatory's spectrum management staff well in advance of your observing run. At the present time there are very few observing programs that will be affected by coordination procedures, but it is a good idea to check the latest status of such agreements. The most frequently used coordinated band is at Arecibo below 1400 MHz where radar operators may be asked to limit their transmissions to their lowest frequencies when given sufficient notice. See the web page

http://www.naic.edu/~rfiuser/smarg-coord.html

for a description of current agreements.

2.3. Interference characteristics

The published spectrum survey, where available, will tell you roughly what to expect in the way of RFI in your observing band. If interference looks like it is going to be a problem, you will want more detail about spectrum usage. Requests to the observatory for more detailed measurements of particular parts of the radio spectrum are welcome with the caveats that the appropriate equipment must be available and that the requests can be accommodated within the work load of the staff. This is an area where the dialogue between observers and spectrum managers can be enhanced with better communication. Well specified and limited requests are the most likely to be within the realm of possibility.

Both observatories are working to provide better information on current spectrum use. A useful data set would be something like a time-frequency matrix of the telescope receiver output for the times of day when we expect to make our observations. A time resolution of one second and a frequency resolution of 1 kHz seem like good starting points. Eight hours of floating point data in a 10 MHz wide spectrum would require 1.2 gigabytes of data storage, which is not unrealistic. From this we could determine the fraction of time which is likely to be available for astronomical measurements at each frequency in the spectrum and try summing portions of the data to look for very weak interference. With a little more work we might identify many of the signals and infer a bit more about the usage of this spectrum.

Of course, it is unlikely that we can ask for telescope time for a spectrum sample of this type. A close approximation to the same sensitivity can be obtained with a well placed RFI sampling antenna and a reasonably low noise preamplifier. The typical gain of radio telescope sidelobes to RFI is about -10 dBi so a sampling antenna with 0 dBi gain and ten times the system temperature is roughly equivalent. The other half of the sensitivity equation is integration time so the RFI sampling receiver must be connected to an integrating back-end, such as a Fourier transform spectrometer or autocorrelator, rather than a swept-frequency spectrum analyzer. Only observer experience and feedback can tell whether there is enough benefit to the telescope's science to warrant the allocation to such resources.

Armed with the ideal spectrum information, one can then ask the question, which of the RFI signals are from licensed users of the radio spectrum and which are from incidental emissions from equipment in the vicinity of the observatory. If the receiving antenna is fixed, and a signal does not change much in intensity over time, there is a good chance that it can be tracked down and turned off. This detective work can be time consuming, but it has general benefits to the observatory's environment.

3. Observing Considerations

With a good picture of the interference environment, you can plan your observing strategies accordingly. Some questions to begin answering are: What dynamic range is required of the spectrometer? What total bandwidth is feasible and what frequency resolution is required to isolate the interference? What time resolution is necessary to take advantage of interference-free intervals? Are there any very strong signals that need to be avoided?

3.1. Spectrometer type and configuration

Some spectrometers are more robust to RFI signals than others. The two-bit sampling autocorrelators at Arecibo and Green Bank and the six-bit sampling FFT spectrometer at Green Bank are the extreme examples. An interfering signal whose power is a significant fraction of the total noise power in the spectrum will distort the output of a two-bit sampler and cause problems to the spectrum even well separated in frequency from the offending signal. The six-bit FFT spectrometer can swallow a signal ten times the total noise power with relatively little spectral distortion well away from the strong signal. Both the Arecibo and Green Bank autocorrelators have a three-bit sampling mode, which has an intermediate dynamic range. The penalty for using the spectrometers with larger sampling words is that fewer spectral channels and/or narrower total bandwidths are available. You need to look at the spectrometer specifications to see what trade-off suits your needs best. At frequencies where RFI is a major consideration the widest spectrometer bandwidths are often not necessary, so you can usually find a useful compromise.

Autocorrelators have a reputation for "ringing" of narrowband interference that tends to destroy an entire spectrum that otherwise would be undistorted. By this we mean that the instrumental profile (called a point spread function by opticians) of a single channel of an autocorrelation spectrum is a $\sin(\pi x)/\pi x$ function, where x is the separation from the narrowband signal in number of channels. This looks like a damped sine wave on either side of the strong signal than can cover most of the spectrum. The ringing from weak to moderate signal strengths can be effectively suppressed with a "Hanning" convolution function. Better convolution functions are available to deal with stronger signals, but I am not aware of any investigations on whether sampler output distortions or residual "ringing" is the limiting factor for strong signals. My guess is that the modern 3-bit correlators could benefit from better convolution functions, so some of the spectral "ringing" folklore is probably out of date.

In an FFT spectrometer, the spectrum convolution must be done in real time. You do not have the option of changing your mind about the convolution after the data are recorded as you do with an autocorrelator's spectrum. An FFT spectrometer does not do the convolution in the frequency domain because it is computationally cheaper to do the equivalent by tapering or windowing the voltage samples fed into the FFT in the time domain. The taper function must be selected before your observing begins. Because tapering underweights some of the data entering the FFT, some loss of sensitivity is associated with tapering. This is another trade-off to weigh in your observing setup.

At frequencies where interference is not a problem, spectrum integrations of 60 seconds or more before storing a data unit on disk are commonly used to keep the recorded data volume reasonably small. If, from a high-time-resolution spectrum sample, you find that shorter integrations are necessary to make use of RFI-free intervals, a much shorter integration time may be justified. On-the-fly mapping uses integration times as short as 100 milliseconds or so, so it is not completely unreasonable to do the same for RFI editing. There may not be a great deal to be gained from really short time resolutions so consider the merits carefully before being presented with a truckload of Exabyte tapes or CDs.

At some time scale less than a second or so it makes more sense to throw away contaminated data in real time. Several possibilities currently exist, although they are not in common use or have not been tried at all. At Arecibo, Bill Sisk and colleagues have built a very effective blanking pulse generator that synchronizes itself with the 1330/1350 MHz FAA radar signal at Pico del Este (Hagen 1988). When this pulse is applied to the autocorrelator it turns off the spectrum integrators for a selected time interval, typically 512 microseconds, to allow the radar pulse and echoes to clear out of the correlator delay line before resuming integration. The radar pulses are spaced about 3 milliseconds apart so not too much data are lost. Be aware, however, that in its current configuration the autocorrelator waits until it has integrated for the specified amount of time before recording it on disk so the clock time duration of a scan will be extended proportionately.

Another blanking possibility is that the Iridium satellite system provides a blanking pulse for the duration of 50% of the 90-millisecond on-off period of the transmission cycle. During the off fraction the satellites are not transmitting in the 1612-1625 MHz band so, in principle, this time is available for radio astronomy. The useful interval available is smaller than 45 milliseconds because of the differences in delays to various satellites in the sky. The two correlators in Arecibo and Green Bank and the FFT spectrometer in Green Bank can accept this blanking pulse, but it is not commonly used so ample notification needs to be given to the staff to implement and test it.

Any other blanking pulse that can be generated in synchronism with burst type interference is a candidate for use with one of the spectrometers. At present, no others are routinely provided, mainly because there has been no specific requirement. The FFT spectrometer at Green Bank has a built-in burst detector that proved reasonably effective on lightning interference, but it was so hard to set up that it fell into disuse. It may be resurrected in a more modern form as priorities dictate.

3.2. Receiver filters

A radio astronomy receiver typically has a total amplifier gain of 100 dB or more. Hence, a strong interfering signal will overload the receiver somewhere in the amplifier chain if it is not first attenuated by a frequency-selective filter. A few signals, like the radar signals near 1300 MHz at Arecibo and Green Bank and a few satellite signals, can overload receivers in the telescope far sidelobes after only 40 to 50 dB of receiver gain. These signals require filters immediately following the cooled low-noise amplifier stages. When a receiver overloads it is not useful anywhere in its passband.

Check the published frequency lists of known strong signals and consult with the observatory staff about the optimum filter and frequency conversion setup for the frequencies that you intend to use. A standard setup for the appropriate receiver probably takes into account the existence of known strong signals, but it is a good idea to understand the filter configuration that you will be using. If you do not have a block diagram of the receiver system, ask for one, or ask someone to draw one for you. Many observers take for granted that the receiver works and do not particularly care about the details, so we have fallen out of the habit of providing lots of system information. In demanding situations, such as a high RFI environment, you should not be shy about asking to understand more about your observing setup. Most engineers will be pleased that you are interested.

Keep in mind that all mixers have two sidebands. If the sidebands are not very well separated in frequency, the unwanted (image) sideband may not be suppressed sufficiently to reject a strong signal in its frequency range. Single-sideband mixers are especially vulnerable to unwanted sideband interference because they are limited to about 30 dB of image rejection. The FFT spectrometer uses a single-sideband mixer in its conversion to baseband, so pay special attention to this one. Think about arranging an RF or IF filter to reject signals that would appear in the unwanted sideband. See the section on diagnostics below for a discussion on how to determine whether an interfering signal is leaking through an unwanted sideband.

3.3. Optical fibers

A subtle dynamic range issue has been introduced into the receiver equation by the use of analog optical fibers for the transmission of wideband signals from the telescope receiver room to the control room. Present day optical fiber modems have a very high equivalent noise temperature on the order of 5×10^6 K so they require about 80 dB of gain ahead of them. A strong RFI signal can overload the amplifier stage immediately before the modem or the modem itself. Whenever possible, use the narrowest available filter before the input to the fiber modem that will pass your observing frequency band. As with all receiver stages, do not use any more gain ahead of the fiber modem than is necessary to achieve a good overall noise performance. We tend to concentrate on optimum receiver temperature and use a little more gain throughout the system than is absolutely necessary, so there may be some room for dynamic range optimization in tight RFI situations.

3.4. Diagnostics and reporting

Once an observing run has begun, there is not much time to deal with interference problems because you will want to be concentrating on science issues. Unless the data are completely wiped out, we cannot expect an observer to spend much time diagnosing RFI. Nevertheless, the observatory staff would like to hear about interference during your observing run and to know as much about the observing setup and interference characteristics as possible. If you have time to run a few diagnostics, all the better. Sometimes the RFI is generated locally or even within the receiver itself, and there may be something that can be done about it with prompt notification.

The following information about observed RFI is particularly helpful:

Frequency - What is the exact measured sky frequency of the signal? At what frequency does this same signal appear in the IF(s) and/or baseband? What sideband are you using, if you have a choice? If you move the observing center frequency, does the signal remain fixed in sky frequency, intermediate frequency, or neither?

Time - When is the interference seen? What is its duration or other temporal characteristics? Is it pulsed, steady, modulated, or periodic?

Bandwidth - Is the signal unresolved in your spectrum or does it have frequency structure? Does it drift in frequency with time? Sample spectra of the interference with clearly marked frequency and intensity axes are very helpful.

Direction - Does the signal have any dependence on the pointing direction of the telescope?

Polarization - Is the signal stronger in one polarization than the other? Does this change with time, telescope position, or feed rotation?

Observing conditions - What integration time, switching mode, polarization, LO settings, and filter selections were you using? Are there any other parameters that might be relevant?

The purpose of this information is to isolate the source of the interference. If the RFI source is external to the telescope it will generally vary in intensity with time, telescope orientation, and polarization. If the signal enters the system ahead of the first mixer it will have a constant sky frequency but its intermediate frequency will change when the first LO frequency is changed. If the signal's intermediate frequency changes in the opposite direction from what you expect, it is entering through the unwanted sideband. If the RFI signal's intermediate frequency changes by more than or a strange multiple of the LO frequency change, the interfering signal may be the result of a spurious mixing product in the receiver system.

The signal's bandwidth is a good hint to its origin. Very narrowband signals at telltale frequencies, such as exact integer multiples of 1, 5, 10 or 100

MHz, usually originate somewhere in the observatory. Broadband signals without much frequency structure are often associated with lightning, relay arcing, welders, fluorescent lights, ignition noise, etc. This type of radiation can often be identified by its temporal structure. For example, heater thermostat relays typically cycle on a 3 to 20 minute period, and ignition noise has a pulse frequency associated with gasoline engine speed. Most broadband noise is stronger at lower frequencies, mostly below 1 GHz.

The spectrum of RFI from computers and other digital equipment is a mixture of narrow and broadband emissions with the narrowband spikes being spaced at intervals from kilohertz to many megahertz. There is usually some periodicity to the narrowband frequencies, but there will also be a jumble of spikes that do not fit any well-defined pattern. A common preconception is that computers radiate primarily at their CPU clock frequency, but the intensity at this frequency is often no higher than at a lot of other frequencies in the spectrum. Most digital RFI is strongest below a few hundred megahertz, but strong digital noise can be seen to several gigahertz and above.

The signals from active users of the radio spectrum come in a wide variety of bandwidths and temporal characteristics. These are beyond the scope of this lecture so I suggest that you contact the observatory's spectrum manager for information on the signals expected in your observing frequency range. The experience of local RFI experts can also be very helpful in identifying interfering signals.

4. Data Reduction

I think it is fair to say that, currently, there are very few tools available for RFI mitigation in the standard data analysis packages, mainly because there has been little demand. There are many reasons for this. No single or well-defined solutions to RFI excision exist. Automated excision has tended to be only marginally effective. Data rates and storage volumes have not been adequate for RFI isolation. Finally, observers tend to avoid RFI or be resigned to its existence. If the science warrants a greater effort on RFI mitigation, interested observers can gradually change this situation by being specific about the RFI problems that need to be solved and by contributing any tools for RFI excision at the data analysis level that they find to be effective.

4.1. Convolution/ACF tapering

As mentioned earlier, the native instrumental profile of a channel in an autocorrelation spectrum is $\sin(\pi x)/\pi x$. This function has quite significant amplitude many channels away from a narrowband signal, which means that a strong interfering signal can affect a large portion of the spectrum. This instrumental profile can be made to fall off much more rapidly by either convolving the spectrum with an appropriate function or tapering the autocorrelation function before performing the Fourier transform to produce the power spectrum. The most common convolution function, called a Hanning function, is a [0.25, 0.5, 0.25] weight of three adjacent channels. This is computationally easy to do and is quite sufficient for interference of modest intensity. More sophisticated suppression of the instrumental profile sidelobes is probably better done with tapers to the autocorrelation function. A few useful taper functions may be found in Rabiner and Gold (1975). One or more of these should be added to the standard data analysis packages.

4.2. Manual editing

The most common form of RFI excision in data analysis consists of manually deleting spectra that are obviously contaminated more extensively than average. If there are not too many data records per scan, this can be done by simply examining each record and deleting contaminated spectra from the average of all spectra in each scan. Keep in mind that deleting a data record will change the effective time over which the data are averaged. If you are using total power on-off position switching, it is often a good idea to delete corresponding records in the on and off scan pair, even though only one of the two records is affected. This preserves the intrinsic symmetry of the on-off pair that is often important to accurately subtract baseline distortions due to changing ground or atmospheric noise input to the system.

A more informative and flexible way of editing data is to display the data from one or a pair of scans on a grey scale plot of frequency vs time, where each line in the plot is a spectrum from a single data record. Even if you have recorded a large number of data records per scan by selecting a short integration time, the data will generally fit nicely on a single image. Weak interference that is distributed over adjacent pixels will tend to stand out better in a grey-scale plot as will subtle large-scale changes in the spectra over the scan. RFI can be edited with a few graphical tools, such as drawing a box around data in the time-frequency domain or marking ranges of data records for deletion. If you delete an isolated domain in both time and frequency space, the frequency channels that have less data in the scan average will have a higher rms noise value. This will lead to artifacts in your spectrum that you must be careful not to interpret as real signals. Deleting all records at a given frequency isn't terribly useful since this just leaves a hole in the averaged spectrum.

A useful grey-scale display requires that the receiver gain be normalized across the spectrum. A good normalizing spectrum may be hard to derive if interference signals are present for the whole scan since they will tend to get normalized out of the display. One can play a few tricks, such as averaging or median filtering the normalization spectrum in the frequency domain to accentuate narrowband signals in the grey-scale display.

With or without effective data editing, we would like to have some measure of the quality of data that went into a scan average. Displays of the peak, rms, and statistical asymmetry of each spectral channel's data are a few quality measures that can be useful. Since most astronomical signals are weakly polarized, a comparison of intensities in the two receiver polarization channels can be a very useful flag for the presence of weak interference.

4.3. Algorithms

Automating the process of manually editing data to remove RFI is surprisingly difficult. It is hard to duplicate all of the factors that go into a judgment about which data are valid and which are contaminated. An editing algorithm requires one or more unambiguous properties that distinguish RFI from normal data. The more noise-like an interfering signal looks in its intensity distribution, the more difficult it is to remove from data. An effective algorithm often needs more than a few built-in assumptions about the nature of the interference that it is intended to remove.

If an interfering signal is strong and either on or off, it can be detected and rejected with a threshold algorithm. Either a range of channels around the interference frequency or entire spectra may be discarded when the interfering signal is detected.

A more likely situation is that an RFI signal will vary continuously and randomly in intensity from strong to undetectable. A threshold detection algorithm will not discard weak interference which will usually appear quite prominently in the integrated spectrum for the full scan. If the interference has a fairly low duty cycle, say less than 30%, using a median instead of a mean average for each channel in the scan may be more effective. The noise in a median is not much worse than for a mean as long as the amplitude distribution for each spectral channel is very nearly gaussian. If there is a gain drift which is a significant fraction of the standard deviation of individual channel noise in one record, the median spectrum for a scan will be considerably noisier than a mean spectrum without interference. If the gain variations are uniform across the spectrum, each record's spectrum may be normalized to an average gain before computing the median for each channel.

Simply discarding spectral channels in the scan average that contain RFI is usually not very useful because the astronomical information in those channels is then lost. However, if you can afford to record spectra with much better spectral resolution than is required by the science, then very narrowband interference may be removed by replacing isolated channels with averages of adjacent channels. Since one's eye can ignore isolated channel interference, this type of rejection tends to be of a cosmetic nature.

5. Real-Time RFI Excision

Successful excision of interference with data editing relies on the fact that the interference is present for a reasonably small fraction of the time within the time resolution that one can afford to record for later processing. If the RFI is present most or all of the time at the frequencies of interest, then it must be isolated in the spatial domain or some combination of time and spatial domains that clearly distinguish it from the natural cosmic signals. A simple example would be to place an antenna null in the direction of the interfering signal so that the radio telescope does not see it.

Active nulling of interference has been used by the radar, communications, and acoustics industries for some time. Noise-cancelling headphones are a good example. While some of their basic techniques are applicable to radio astronomy, the degree to which interference must be rejected is quite different. Astronomical signals are almost invariably at the detection limit of our instruments after many minutes of signal integration. A weak interfering signal is nearly as destructive as a strong one and can even be more harmful because of its subtle nature that can lead to false detections. A 10 dB reduction in acoustic noise can be quite useful because the desired signal is well above the detection threshold. The luxury of moderately strong signals in radio astronomy is all too rare.

In the past few years research into signal processing techniques for rejecting interference in radio astronomical receiving systems has become quite active. Initial results are encouraging, but they have also uncovered the complexities that must be understood and accounted for before applications are commonly available to observers. A few of the lines of pursuit are briefly outlined below.

5.1. Addition of information

As a rule, the information available in the radio telescope data stream is insufficient to distinguish interference from astronomical signals. Even the simplest data editing adds the bit of information that the interference is variable in intensity and the astronomical signal is not. If the interfering signal is too weak to detect with short integrations, this added information is insufficient. Any technique or *a priori* knowledge about RFI that enhances our ability to detect it will also increase our ability to reject it.

Similarly, any property of RFI that cleanly isolates it from cosmic signals offers the possibility of building a filter that rejects the RFI but passes the astronomical information. The following techniques use some form of RFI signal enhancement and signal isolation.

5.2. Possible techniques

Blanking On time scales much less than one second it becomes difficult to store data with sufficient time resolution to remove RFI with data editing. In that case data rejection in the time domain must be done in real time. Two examples where this is effective are on pulsed radar and Iridium satellite transmissions as mentioned in Section 3.1. Radars typically transmit pulses of 2 to 200 microseconds in length with the pulses spaced by about 3 milliseconds. After about 500 microseconds most of the detectable echoes from nearby reflections have died away, and the time thereafter, until the next pulse, is available for astronomical measurements. The blanking generator must know the radar's pulse spacing sequence so that it is only required to track slow drifts in the pulse generation clock. Blanking on individually detected pulses is ineffective because the weak pulses are below the detection threshold but add up to a significant feature in the integrated spectrum.

Blanking on time scales much less than one millisecond is problematic with current spectrometers because they require contiguous runs of data samples for intervals greater than the inverse of their frequency resolution. For example, 5 kHz resolution requires a 200 microsecond minimum data window. A spectrometer that works with non-contiguous data samples is conceivable, but none presently exist.

Radar and Iridium signals have frequency structure that is particularly bothersome to astronomical spectral lines, so these interfering signal must be suppressed well below the detection threshold. Some forms of broadband pulsed interference, such as lightning and ignition noise, do not have so much structure on frequency scales less than a few MHz, so some residual interference may be tolerable. For this type of RFI, pulse-by-pulse detection is necessary because of the unpredictable pulse timing, but this may provide enough suppression to be useful, particularly if some antenna gain can be added to the detection receiver.

Cancellation Any signal in a radio receiver can be cancelled by finding a clean copy of the signal and adding it to the receiver signal path with the appropriate amplitude and reversed phase. One way to obtain a copy of an interfering signal is to receive it with a high gain antenna that has little or no response in the direction of the astronomical object being observed (Barnbaum & Bradley 1998). The reference copy of the signal must have a relatively high signal-to-noise ratio to produce adequate suppression and to avoid adding noise to the system from the reference receiver channel. The reference signal must be processed by a fairly complex filter to replicated the delay of the interfering signal received in the main radio telescope receiver. In many cases, the RFI signal in both the main and reference channels will arrive through several or even many propagation paths, each with its own delay and attenuation (Fisher 2001). This complicated delay and attenuation must be duplicated in the reference channel filter. The relative phase and amplitude of the main and reference channel RFI signals will change with time, sometimes quite rapidly, so an adaptive feedback scheme is necessary to maintain cancellation.

Another method, called parametric cancellation, involves deriving a copy of the interfering signal from detailed knowledge of the transmitted signal. Then only a few, slowly varying parameters, such as intensity and delay, need to be solved for and tracked in real time. Ellingson, Bunton, & Bell (2001) showed this technique to be quite effective at suppressing the signal from one of the GLONASS satellites received in the sidelobes of the Australia Telescope Compact Array. The satellite signal had one clearly dominant propagation path which required a solution for delay, phase, Doppler shift, and amplitude roughly every tenth of a second. The GLONASS signal has a complex phase-switched signal structure, but the switching sequence is entirely predictable for 20-millisecond intervals. From the derived parameters, a digitally synthesized version of the interfering signal was produced and added to the sampled telescope output to achieve RFI suppression below the detection level in ten seconds of integration.

Null steering A radio telescope is quite effective at suppressing RFI in the spatial domain by having 50 or 60 dB more gain in the direction of an astronomical source than in the direction of an interfering transmitter. Unfortunately, many interfering signals are considerably stronger than 60 dB above our detection threshold. In principle, one can create nulls in the telescope antenna pattern for selected interfering sources, either by adding the signal from one or more null-producing auxiliary antennas in the case of single-dish radio telescopes or by properly weighting the signal from an array of antennas that make up a radio telescope. In many ways the signal processing to produce spatial nulls is very much like the cancellation technique with a reference antenna described above. Many of the same signal-to-noise ratio and adaptive requirements apply. One extra factor that enters into steering nulls in an array is that the desired astronomical signal may also be suppressed unless extra restrictions on the null-steering algorithm are enforced.

The concept of null-steering is clearest when the interfering signal arrives from a well-defined direction. When significant multi-path propagation is present, as mentioned above, the required nulls are not simply in the spatial domain. They exist in the combined spatial-temporal domain. An interfering signal arriving from many directions is coherent with itself, but it may be cancelled only by the proper combinations of delay and amplitude from array elements or main telescope and auxiliary antennas. These combinations do not produce a directional null in space.

Post-correlation The equivalent to real-time signal cancellation using a reference antenna may be done with post-correlation data since the phase and amplitude of the main radio telescope and reference signals are preserved in the cross-correlation process. Briggs, Bell, & Kesteven (2000) give the details of the RFI subtraction mathematics and show examples on data recorded at the Parkes radio telescope. One advantage of post-correlation interference subtraction is that the data may be processed off-line and iteratively optimized for the best cancellation parameters, if necessary. Aside from the correlation process itself, the data processing requirements, in numbers of arithmetic calculations per second, can be much less for post-correlation interference subtraction than for real-time adaptive cancelling since the correlation products may be integrated for as long as the interfering signal is stable in phase and amplitude, typically on the order of a second.

5.3. Where to from here?

There is no magic bullet for RFI excision. For the foreseeable future, a somewhat different approach may be required for each interfering signal type depending on how it is distributed in the time, frequency, and spatial domains as seen from the radio telescope. A technique that works well for pulsar observations may not be adequate for spectral line measurements. As we gain more experience with signal processing techniques for removing RFI, some common tools will emerge. This is an interesting research topic, and I encourage anyone who is interested to join the effort.

A fairly comprehensive compilation of RFI mitigation work in radio astronomy may be found at the following web site:

http://www.atnf.CSIRO.AU/SKA/intmit/

References

Barnbaum, C. & Bradley, C. F. 1998, AJ, 116, 2598
Briggs, F. H., Bell, J. F., & Kesteven, M. J. 2000, AJ, 120, 3351
Ellingson, S. W., Bunton, J. D., & Bell, J. F. 2001, ApJS, 135, 87
Fisher, J. R. 2001, 'Analysis of Radar Data from February 6',
　　http://www.gb.nrao.edu/~rfisher/Radar/analysis.html
Hagen, J. 1988, Arecibo Observatory Electronics Department Manual No. 8806
Rabiner, L. W. & Gold, B. 1975, Theory and Application of Digital Signal Processing (Englewood Cliffs: Prentice-Hall)

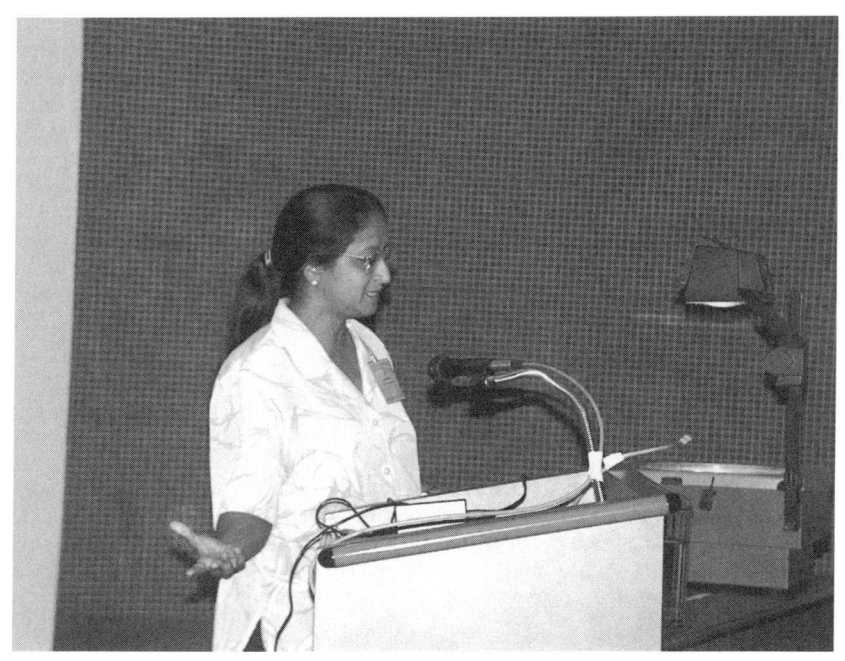

"RFI is a BIG PROBLEM!", Tapasi Ghosh

Spectrum Management

Tapasi Ghosh

National Astronomy and Ionosphere Center, Arecibo Observatory, HC 3 Box 53995, Arecibo, Puerto Rico 00612, USA

Abstract. Present day life without the use of radio frequencies is totally unimaginable. The mode of usage of the different frequency bands is diverse and ever-changing with constant development of newer technology. In the midst of all this, there is a minority group of spectrum users who are labeled "passive". Radio astronomy and remote sensing groups come under this title, having no control over the signal they try to receive, they are often most vulnerable to interference from man-made radio waves. However, through constant need to develop new techniques to detect very weak signals, the research done by these passive users of spectrum often benefits the commercial users. It is therefore of utmost importance that both the active and the passive users of radio spectrum may survive and grow together in a manner of peaceful coexistence. This is the goal of spectrum management. As radio waves "do not know" of any national boundaries, such management issues have to be agreed upon globally. The mechanisms that have been set up for such purposes with global and national counterparts is explained. How, as a minority, the interests of radio astronomy may be best served now and in future is also discussed.

1. Introduction

In our daily lives, we all use radio waves in ever more ingenious ways. With the revolutionary developments in telecommunications of the past 20 years, mobile phones, satellite communications and radio navigation, have all become integral parts of our existence, as are microwave ovens, remotely operated car locks, and garage door openers. These days, national borders are protected using highly sophisticated systems that come under the common heading of "electronic warfare".

However, much of the radio energy that is generated in such utilities is wasted/radiated away forming a general background of radio radiation all around us. Through all this human-generated "radio fog", we radio astronomers try to detect the exceedingly weak celestial radio signals coming to us from across the universe. We have control over neither the nature of these cosmic signals nor their strength. Just how weak is this "weak" cosmic signal compared to the "radio fog"? In the absence of interference, the typical signal-to-noise ratio (SNR) in the RF and IF parts of a receiver is in the range -20 to -60 dB, i.e. the power contributed by the source under study is usually a factor of 10^{-2} to 10^{-6}

lower than the noise power from the atmosphere, the ground, and the electronics of the receiver itself ($T_{signal} = 10^{-6} T_{noise}$). In comparison, the corresponding SNR for most communication systems is of the order of unity or greater. It is, therefore, impossible to detect a cosmic signal in the presence of a man-made source (e.g. a TV transmitter) that is a million times stronger.

This was realized from the very early days of Radio Astronomy. In 1959, via intense efforts of the radio astronomers, the International Telecommunication Union was convinced to recognize the existence of passives users such as Radio Astronomy and Remote Sensing as a type of "radio service". This laid the foundation stones on which ground-based astronomy still stands today. It opened up the possibility of prohibiting active users (services that generate radio radiation, in addition to receiving it) from using certain parts of the spectrum, and thereby keeping these relatively "clean" for exploitation by the passive users.

The important process by which the peaceful coexistence of spectral neighbors can be achieved, has been named "Spectrum Management". It has global, regional, and local aspects, and these will be discussed in the next three sections.

2. The International Arena

At the international level, regulations for spectrum usage are organized under the auspices of the International Telecommunication Union (ITU), an agency of the UNO. The present ITU has three sectors, Telecommunication Development, Telecommunication Standardization, and Radiocommunication (ITU-R). Radio regulations, stating rules for spectrum usages, are treaties between nations agreed upon at World Radiocommunication Conferences (WRC) organized by ITU-R. At a WRC, representatives of national governments have voting rights, and radio astronomers have to work with their respective national governments to assert their points of view within the national agenda.

One very important international organization which represents all passive scientific users of radio spectrum was formed in 1960 and named, the Inter Union-Commission for the Allocation of Frequencies for Radio Astronomy and Space Sciences (IUCAF). The three founding unions of IUCAF were IAU, URSI and COSPAR. IUCAF participates in WRCs as a recognized international organization, but it has no voting rights. A wonderful history of the (often heroic) activities of IUCAF over the past 40 years has been written recently by Brian Robinson (Robinson, 1999). The other avenue through which radio astronomers participate in WRC/ITU-R proceedings is via study groups and working parties, usually active in preparation for a WRC.

For allocating frequency bands to different services, ITU divides the whole world into three regions; Region I being Europe and Africa, Region II the Americas and Region III the Asia/Pacific region. In any given frequency band, the radio services permitted to operate are allocated a primary, co-primary or secondary status, or (at the least) given some protection via footnotes. For example, radio astronomy has a PRIMARY status in the λ 21-cm band where all human generated emission is forbidden between 1400 and 1427 MHz.

Radio Regulations adopted at WRCs are binding rules by which various services are required to co-exist. However, most of the protection criteria needed for the passive radio spectrum users, such as Radio Astronomy, are described in Radio Recommendations. These are rules that the national administrations are *recommended* to follow, though any particular nation is not *required* to do so unless the recommendations are passed into a law within the country. For example, the recommendation, ITU-R RA.769-1 specifies protection criteria for radio astronomical measurements, and administrations in any country are free to convert those into laws (or not) depending upon the situation/public opinion in that particular country.

A few very good articles describing the complexities of spectrum management activities have recently been published, (e.g. Cohen, 2001; Gergely, 2001). In addition, a comprehensive handbook describing many important concepts and definitions was published by the ITU in 1995 (European Science Foundation 1995).

3. National Level

Within the USA, non-Government use of the radio spectrum is regulated by the Federal Communications Commission (FCC) and Governmental use by the National Telecommunications and Information Administration (NTIA). Representatives of the radio-astronomy community work with these agencies to ensure that their interests and rights are preserved. The National Science Board has a subcommittee named the "Committee for Radio Frequencies (CORF)", while the National Science Foundation (NSF) has a Spectrum Management office that oversees the interests of passive scientific users of the radio spectrum. On a non-Governmental basis, the American Astronomical Society (AAS) also has a "Committee on Light Pollution, Radio Interference and Space Debris", that works as a clearing house for information and provides support for various spectrum management activities.

The FCC mandates the rules for national use of the various spectral ranges via public debates, the so called "Negotiated Rule Making" procedures. All interested persons/organizations can participate in writing to such proceedings. Once a set of rules has been adopted, then these become mandatory within the USA. Any company/organization wishing to use a particular piece of the spectrum has to abide by these dictates. Astronomers participate in such rule making procedures whenever their rights need protecting either through individual institutions, CORF, or through the NSF's spectrum management office.

4. Local

In addition to the wider national or international scene, local authorities have helped radio astronomers by forming "Quiet zones" or "Coordination Zones" around specific radio observatories. Examples of these are the National Radio Quiet Zone (NRQZ) around Green Bank, West Virginia, and the Puerto Rico Coordination Zone (PRCZ) in respect of the Arecibo Observatory.

The NRQZ was set up in 1959, and covers 13,000 sq. miles around the NRAO Green Bank Observatory. Within this area, other users of the radio spectrum may operate ONLY on a non-interfering basis as specified by the Radio Astronomers (http://www.gb.nrao.edu/nrqz.html). The services that come under such jurisdiction are, public mobile, wireless communications, maritime,

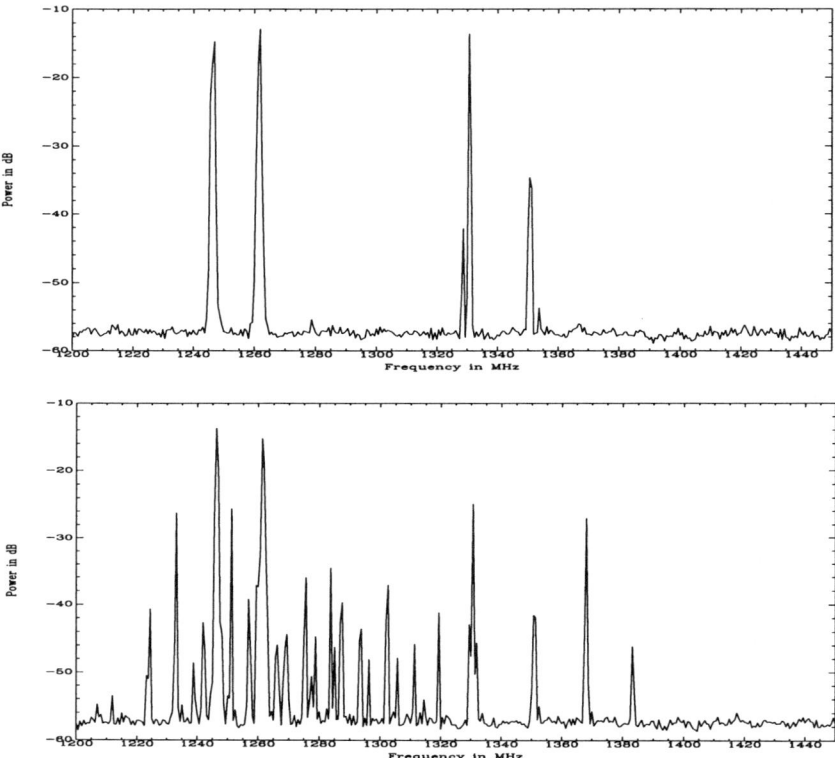

Figure 1. Top: The normal spectrum usage in L-band as of 1998. Spikes at 1330 & 1350 MHz are due to the air traffic control radar at the San Juan Airport, while those at 1241/1261 MHz are due to TARS system of the National Guard. Bottom: The situation after the advent of the "new user."

aviation, private land mobile, personal radio, fixed microwave domestic, public fixed, international fixed, public satellite communications, radio broadcast, special broadcast cable television relay, and amateur radio (repeaters, beacons).

In 1997, after a number of years of negotiation, FCC established a Coordination Zone in Puerto Rico (see http://www.naic.edu/techinfo/prcz/prczinfo.htm). Since then, before applying for a FCC license, an applicant must coordinate with the Arecibo Observatory, and subsequently be able to supply a "no-objection" certification from the interference office of the observatory along with the FCC application.

Earlier, during the days of construction of the Observatory, the Government of Puerto Rico had formed a "zone law" which forbids any radio transmission within a 4-mile zone around the observatory that is harmful to the observations carried out using the 305-m telescope.

All these measures are extremely useful for the functioning of a radio observatory. However, at the same time, observatories do not wish to inhibit "progress" within their communities. To counter any such false impression (that can occur sometimes), constant interaction with other local spectrum users and general public is essential. In Puerto Rico, this has led to the setting up of a Puerto Rico Spectrum Users Group (PRSUG), at whose regular meetings, all spectrum users on the island are invited to meet, discuss, and resolve issues of interest and concern.

A success story for coordination in Puerto Rico has been a recent Memorandum Of Understanding (MOU) signed between the Observatory and the Puerto Rico Air National Guard (PRANG). A large fraction of Arecibo users employ the 305-m telescope to search for neutral hydrogen (HI) in galaxies at high redshifts. Such observations have always been carried out routinely down to the lowest frequency covered by the Arecibo's L-band receivers (e.g. down to 1100 MHz using the L-wide receiver). However, in October 1998, it was discovered via the "hill-top monitoring system" (which samples almost the complete spectrum up to 10 GHz on a round-the-clock basis) that the entire spectrum between 1200 and 1400 MHz had suddenly been taken up by an unknown system (Figures 1a and b).

Military radar units are the prime user of this band within the USA, albeit with footnote protection for Radio Astronomy between 1330 – 1400 MHz (adjacent to the Radio Astronomy band of 1400 – 1427 MHz). When, through the Government agency's "Area Frequency Coordinator", Arecibo had successfully identified the user, and demonstrated that both sets of users could work to their full potential by adopting certain coordinated modes of operations, signing the MOU to that effect became a pleasure. Competitors, potentially in conflict over spectrum usage, had become friends and collaborators. Now, whenever Arecibo observers need to work at some particular part of the L-band, the observatory requests the PRANG to use a part of the spectrum that is away from the astronomer's requirements. Without such coordinations, the presence of another (powerful) user as shown in Figure (1b) would have stopped all 1200 – 1400 MHz L-band observations (hence, most HI studies of galaxies) at Arecibo!

5. The Future

In order to project future directions in radio astronomy, we need to evaluate the situation both in the immediate past and the present. In respect of the spectrum usage within USA, there has been a dramatic change in the way FCC operates. In 1993, it was decided that the right of use of certain parts of the radio spectrum would be sold to the highest bidder, as long as they are a service of a particular kind and have the global/national approval for using the particular piece of spectrum.

Since then, there has been a constant move towards more efficient use of the spectrum. This is partly beneficial for radio astronomy as it has meant technologies involving less radio pollution. However, there is also tremendous pressure from satellite-based communication companies, aimed at attracting a global clientele, to be able to use more and more spectrum. The operations of these companies, involving as it does transmissions down towards the Earth, could undermine the needs of passive spectrum users if the authorities are not watchful. Certainly, when it comes to "highest bidders", they can offer a much higher price per MHz of the electromagnetic spectrum.

The other direction towards which the administrators at the national and ITU level are moving, and which is also potentially detrimental to radio astronomy, is deregulation. Wanting global mobility, people in general (including radio astronomers) welcome such moves. However, as such systems are often satellite based, their uncontrolled growth may lead to a situation where there will be "nowhere to hide" on our globe, UNLESS the needs of the passive users are also considered simultaneously. Even the traditionally "sacrosanct" bands are not necessarily exempt now. New technological development towards Ultra-wide band devices (spanning a frequency range from 50 MHz to 4 GHz) point towards a review of the entire scenario of spectrum usage.

So, from this perspective, the future of ground-based radio astronomy does not look very promising now. However, the best hope is perhaps the intense research that scientists such as Rick Fisher at Green Bank are undertaking to mitigate RFI. (These developments are described in detail in Rick's lecture in this volume). In addition, perhaps it is also time to seek out other avenues for better regulations within the UNO, but not restricted to the ITU alone.

6. Epilogue

Ideally, all frequency bands should be accessible for scientific pursuits for some of the time from some part of the world. This is not an unreasonable goal, and scientists themselves need to believe in it first. Only then, will they be able to explain their goal effectively to the administrators, law makers and general public. If money did not play a major part in this venture, then perhaps we would be closer to such a goal by now ! Sadly, that is another story.

References

Cohen, R. J. 2001, in IAU Symp. 196, Preserving the Astronomical Sky, eds. R.J. Cohen & W. T. Sullivan, (San Francisco: ASP), 220

European Science Foundation 1995, Handbook on Interference (Dwingeloo, The Netherlands: CRAF Secretariat)

Gergely, T. 2001, in IAU Symp. 196, Preserving the Astronomical Sky, eds. R.J. Cohen & W. T. Sullivan, (San Francisco: ASP), 236

Robinson, B. 1999, ARA&A, 37, 65

Focal Plane Arrays

John M. Payne

National Radio Astronomy Observatory, 949 N. Cherry Avenue, Campus Bldg. 65, Tucson, Arizona, USA

Abstract. Over the past few years, the sensitivity of mm/sub-mm receivers has improved dramatically. We are now at the point of the atmosphere and other inevitable sources of noise being equal to, or perhaps greater than, the noise contributed by the receiver itself. Under such conditions, the observing efficiency (for mapping extended regions) may be improved by adding receivers in the focal plane in the manner of the CCDs used in optical astronomy. Details of the principles involved and brief descriptions of existing systems will be given.

1. Introduction

In recent years, the performance of mm/sub-mm receivers for radio astronomy has improved dramatically and recognition has been made of the fact that further improvements in the observing time needed to achieve a given sensitivity-at least for the mapping of extended sources-must rely on the addition of receivers to the focal plane of the telescope much in the manner of CCDs in the optical regime. This subject has assumed great importance in the development of instrumentation for radio astronomy, so much so that a workshop was held on the subject in 1994 and, although a little dated, the record of this workshop provides a valuable source of reference on the subject of multi-feed systems for radio telescopes (Emerson & Payne 1995). References to this workshop will be made throughout this paper. Since 1994, some of the most interesting and exciting results have been obtained with focal plane arrays. Some are continuum systems, others spectral line, but all share the capability of dramatically increasing the output of the telescope on which they are deployed. It is appropriate to mention at the outset that this lecture is intended to be a non-rigorous discussion of focal plane arrays. Over-simplistic in its approach and not complete in many ways, it is intended to stimulate the reader to greater study of the subject. In the spirit of this approach, it is stated at the outset that the lecture will assume that focal plane arrays are divided into two distinct types: 1) separate simple receivers in the focal plane and 2) an array that completely samples the electric field in the focal plane with a series of sub-apertures. This approach is more difficult to realize and to understand than the first approach but gives the potential of great flexibility and is well suited to this age of ever increasing computer power. The purists will argue that this split into two types of array is incorrect and a rigorous treatment will permit the understanding and analysis of both types

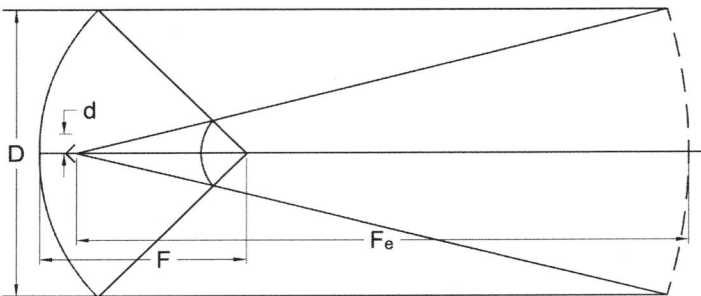

Figure 1. A simple Cassegrain antenna.

using common analysis and tools. This is correct; nevertheless, it is felt that this approach is simpler and more appropriate for this lecture.

2. Simple Feed Arrays

Consider the simple antenna arrangement shown in Figure 1. A convenient way of characterizing the Cassegrain antenna is "the equivalent antenna", a concept that is dealt with in detail in many references. A vital concept here is "equivalent focal length, F_e in the diagram, and in the figure a displacement of d in the focal plane, will displace the beam in the sky by d/F_e radians".

A typical radio telescope consists of a steerable parabolic reflector of high precision, equipped with a sensitive radio receiver usually placed on the optical axis at either the prime focus or at a Cassegrain focus. For several reasons, the Cassegrain focus has become the preferred optical arrangement for radio astronomy applications. A brief summary of these reasons follows.

1. The Cassegrain focus, usually situated behind the main reflector, permits the mounting of the large, cryogenically cooled receivers that are needed for high sensitivity.

2. Several receivers covering different frequency ranges may be mounted in the area behind the main reflector. Receivers may then be selected by a simple reflector arrangement.

3. Beam switching at all frequencies may be accomplished by switching the subreflector.

4. The greatly increased depth of focus permits the use of so-called "quasioptical" devices for performing many of the receiver functions such as polarization diplexing and local oscillator injection.

5. The Cassegrain arrangement results in lower system temperatures due to spillover from the receiver intercepting the cool sky temperature rather than the ground temperature.

6. The long equivalent focal length of the Cassegrain arrangement permits off-axis operation with little degradation in performance.

This last point is a crucial one for multi-beam operation. A lateral displacement of the receiver feed away from the axis of the paraboloid results in a tilt of the telescope beam in the opposite direction (the desired effect), a decrease in the antenna gain, an increase in the level of the side lobe nearest the optical axis (coma lobe), and an increase in the system noise due to increased spillover. A consideration of antenna gain reduction due to phase errors introduced by off-axis operation is sufficient to show that any practical multi-beam receiver must be implemented at the Cassegrain focus. One convenient criterion for defining the field of view of an antenna system is how far off axis the feed must be moved in order to produce a gain reduction of 10%. Using this definition, the 12-m Telescope (the mm-wave telescope previously operated by NRAO at Kitt Peak) operated at prime focus would be restricted to scanning angles of less than four half-power beamwidths, an unacceptable value for any practical multi-beam receiver. Using this same criterion, the 12-m Telescope used in the Cassegrain configuration with an effective focal length of 155 m would have a maximum scanning angle of 15 half-power beamwidths. For conventional feeds in the same focal plane, the minimum spacing between the feeds is restricted by the physical dimensions of the feeds and results in a beam spacing on the sky of approximately three half-power beamwidths. This is virtually independent of frequency and optical arrangement. For the 12-m Telescope, an array of 10x10 feeds (100 separate receivers) would provide a maximum scanning angle of 15 beamwidths and would result in a maximum loss of gain of slightly more than 10% for the outermost feeds. A rigorous approach to calculating the various aberrations along with an excellent complete analysis of array feeds is given by Padman (1995).

Most Cassegrain telescopes are designed so that rays from the periphery of the subreflector are reflected to the periphery of the main reflector. In such cases, there is no spillover on the main reflector when the feed is at the secondary focus. If, however, the feed is displaced laterally in the focal plane, there will be some displacement of the illumination pattern on the main reflector and some associated spillover. This will have two deleterious effects. First, the gain of the antenna will decrease due to incorrect illumination; second, the system noise will increase due to spillover at the main reflector. The second effect will be more serious with very low noise receiver systems. These effects are dealt with fully by Lamb (1983).

2.1. Practical systems

Several multi-beam heterodyne receivers for radio astronomy have been built and have given good results on radio telescopes. The first reported millimeter-wave, multi-beam receiver was an eight-beam, 230-GHz receiver (Payne 1988) using cooled Schottky mixers that was subsequently upgraded to an eight-beam SIS receiver (Payne & Jewell 1995). At this point, it is probably appropriate to

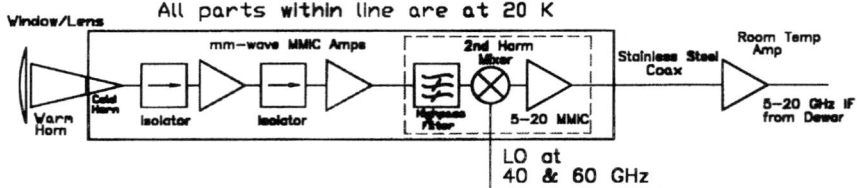

Figure 2. Sequoia overview: Cryogenic focal plane array designed for the 85 - 115.6 GHz range; 16 dual polarized (next year) pixels arranged in a 4 x 4 array; Uses InP MMIC preamplifiers with 35 - 40 dB gain; Noise temperatures 50-80 K over most of the band.

issue a word of caution. The time taken to achieve a certain sensitivity in a radio astronomy observation is proportional to the square of the system temperature and inversely proportional to the number of feeds in a simple multi-beam system. Therefore, it is of crucial importance that the receivers in a multi-beam system are close in noise temperature to the best single-beam system if real gains are to be made.

Another point to make is that the total cost of a multi-beam system for spectral-line work may well be dominated by the cost of the spectrometer. Recent results are shown in Table 1. Various views of the SEQUOIA system and the CSIRO system are shown in Figures 2–5.

Table 1. Two Focal Plane Arrays for Radio Astronomy Spectroscopy.

Telescope/Group	Name	Number of Elements	Frequency Range (GHz)	First Stage	Noise Temp. (K)
FCRAO LBT	Sequoia	32 (16 Dual Pol)	85 - 115.6	MMIC	59 - 80
CSIRO/ATNF		26 (13 Dual Pol)	1.2 - 1.5	HFET	6

3. Full Sampling Feed Arrays

As previously mentioned, this type of array is more difficult to realize and analyze than the very simple arrays described earlier.

As pointed out by Fisher (1995) and several others, with feed arrays that efficiently sample the entire electric field at a surface near the focal region of a reflecting telescope, we can relax a number of constraints on the reflector that affect its cost. The surface does not need to be continuous; hence, it does not need to be as tall as a conventional telescope. The surface does not need to be positioned to a fraction of a wavelength; we just need to know where it is to this accuracy. The surface does not need to conform to a particular shape to a

Focal Plane Arrays 457

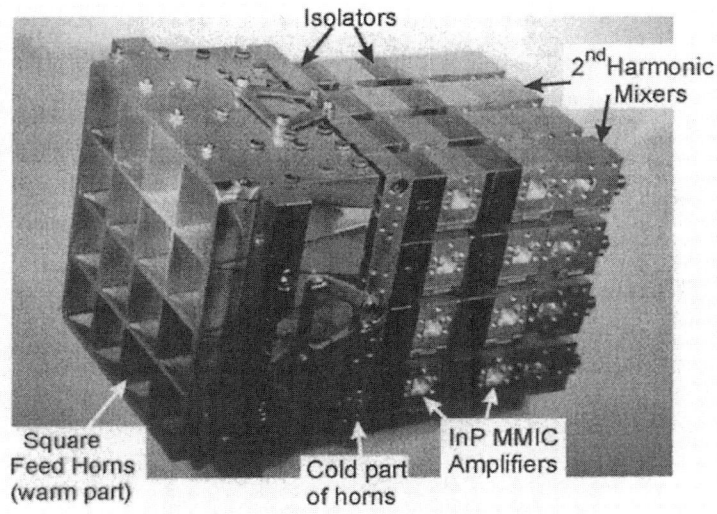

Figure 3. SEQUOIA System - 16 channels.

Figure 4. CSIRO System - overall view.

Figure 5. Dual polarization feed - CSIRO System.

fraction of a wavelength; it only needs to be close enough and smooth enough to direct all of its reflected energy somewhere on the feed array.

The relaxation of these constraints shifts much of the burden of beam formation into the surface measurement and signal processing domains. With the current trend of decreasing costs of signal processing and relatively stable costs of steel and aluminum, the next big increase in the collecting area for radio astronomy will require much greater reliance on signal processing than is done in even our current synthesis arrays. Another potential advantage of the full sampling feed array is the ability to generate simultaneously many beams on the sky, each having a high efficiency and with arbitrary separation between beams.

3.1. General considerations

A phased array feed can be defined by:

- Signals from the array elements combined in the complex voltage domain. Each array element contributes to many beams; and

- Complete sampling of the focal plane fields to the edge of the array, i.e., no grating lobes.

These imply that

- Reflector illumination depends primarily on the array factor rather than on the individual element pattern.

Figure 6. Sinuous feed.

- There is no restriction on the minimum spacing of the reflector antenna beams. Overlapping beams have partially correlated outputs, but no signal or sensitivity loss.

- One has full control of focal plane field matching, restricted only by the array size.

- Off-axis distortion and reflector error correction are now possible.

3.2. A practical system

As can be gathered from the above, the potential of phased array feeds is considerable. Several prototype systems have been fabricated and one system will be described in a little detail. A prototype 19-element feed array has been constructed by Bradley and Fisher and installed on the 140-ft radio telescope in Green Bank, West Virginia. The system uses uncooled components at the present time; the intention is to demonstrate the technique and, when demonstrated, improve the sensitivity by cooling central F_e components. The center frequency of this prototype system was chosen to be 1.5 GHz and the feed elements selected are "sinuous antennas" chosen for their ease of fabrication. A view of this antenna is shown in Figure 6, the array in Figure 7, and a view of one antenna together with its amplifier in Figure 8. A block diagram of the system is shown in Figure 9, in which the outputs of the 19 receivers are combined to produce telescope beams of spacing dictated by the control segments with the switching matrix. A preliminary report on this system is given in Fisher & Bradley (2000).

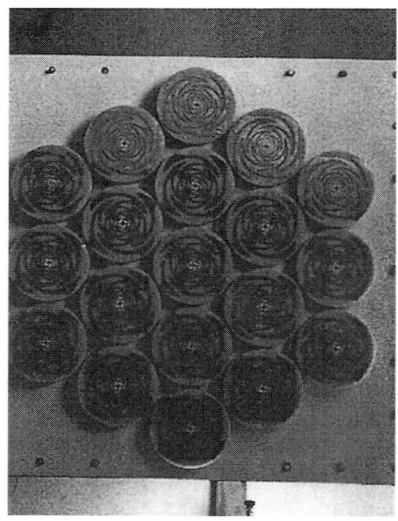

Figure 7. Feed array.

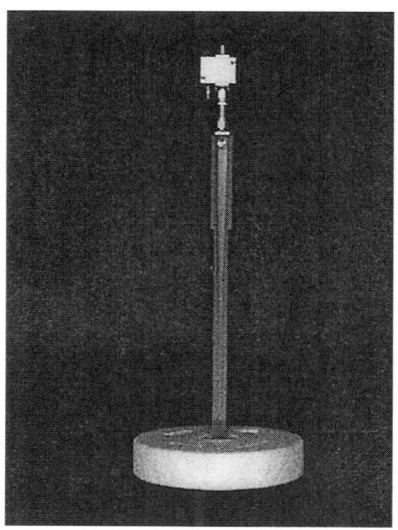

Figure 8. Single feed with amplifier.

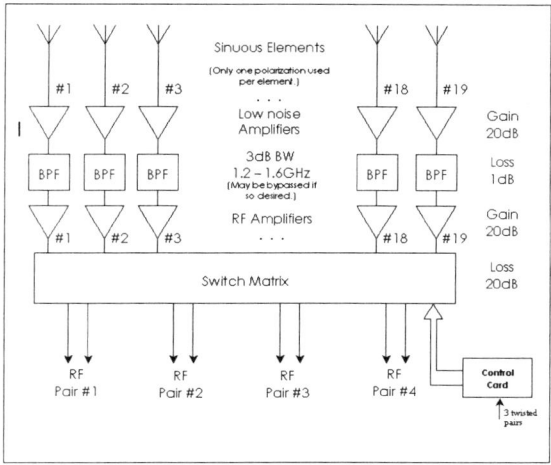

Figure 9. A block diagram of the 19-element feed system installed on the 140-ft radio telescope in Green Bank.

4. Conclusions

The conclusion of this brief lecture is that simple focal plane arrays are here today. These have already resulted in spectacular results and greater utilization of telescope time. Great potential exists for full sampling array feeds. Systems of this kind may well affect how future large aperture antennas are built. The decreasing cost of fabricated steel may well be a significant factor in the future. Such systems overcome the limitations of simple focal plane arrays and permit full sampling of the focal region of an antenna.

References

Emerson, D. T. & Payne, J. M. 1995, Multi-Feed Systems for Radio Telescopes, ASP Conf. Ser. Vol. 75 (San Francisco: ASP)

Fisher, J. R. 1995, in ASP Conf. Ser. Vol. 75, Multi-Feed Systems for Radio Telescopes, ed. D. T. Emerson & J. M. Payne (San Francisco: ASP), 27

Fisher, J. R. & Bradley, R. F. 2000, in SPIE Proc., Radio Telescopes, ed. H. R. Butcher (Bellingham, Wash.: SPIE), 308

Lamb, J. W. 1983, Report of the Helsinki University of Technology Radio Laboratory, S147

Padman, R. 1995, in ASP Conf. Ser. Vol. 75, Multi-Feed Systems for Radio Telescopes, ed. D. T. Emerson & J. M. Payne (San Francisco: ASP), 3

Payne, J. M. 1988, Rev. Sci. Instrum., 59, 9

Payne, J. M. & Jewell, P. R. 1995, in ASP Conf. Ser. Vol. 75, Multi-Feed Systems for Radio Telescopes, ed. D. T. Emerson & J. M. Payne (San Francisco: ASP), 144

Bolometers for Submillimeter and Millimeter Astronomy

Wayne Holland and William Duncan

UK-Astronomy Technology Centre, Royal Observatory, Blackford Hill, Edinburgh EH9 3HJ, UK

Matthew Griffin

Department of Physics and Astronomy, University of Wales, Cardiff, 5 The Parade, Cardiff CF24 3YB, UK

Abstract. The past decade has seen considerable progress in the development of bolometric detectors for submillimeter- and millimeter-wave astronomy. With the introduction of imaging arrays these wavebands have been undergoing a major revolution with many new and exciting discoveries being made in almost all areas of astronomy. In this paper the basic principles of bolometry will be explained, together with the "figures of merit" that characterize performance. Particular examples will focus on using wide-bandwidth devices in conditions where the atmospheric background can change on short timescales.

Practical devices – from single pixels to the current imaging arrays – will be discussed, with particular emphasis on the need for ultra-low temperature operation for sky-background limited performance. Examples will be given of current instrument performance on ground-based telescopes, including the issue of optimum coupling of the small detector element to a large telescope aperture. The development of integrated filled arrays of many thousand pixel cameras bodes well for the future. The new generation of both ground-based and space-borne instruments will also be described, as well as other exciting prospects for the future.

1. Introduction

Bolometric detectors have been used for far-infrared (FIR), submillimeter (submm) and millimeter-wave (mm) astronomy for the past 40 years. Bolometers are the most sensitive detectors of broadband radiation between wavelengths of $200\,\mu$m (the limit for stressed photoconductors) and 2-3 mm (where heterodyne mixers become competitive). However, their versatility has been adopted for a wide-range of applications from X-ray spectrometers to photon counting in the optical. Bolometers are also called *incoherent* detectors in that they only measure the intensity of incoming radiation and do not preserve phase information. Only in the past few years have imaging bolometer arrays started to operate on large submm/mm telescopes. The move away from single-pixel devices to multi-element *cameras* has led to a number of well-publicized discoveries and the emergence of the submm/mm as one of the most important wavelength regimes for astrophysics.

Submm/mm astronomy is most sensitive to *cold gas and dust* - the material at the *origins* of many phenomena, from primaeval galaxies to galactic star and planet formation. In the case of very distant dusty galaxies the emission is redshifted into the submm, allowing us to study galaxy formation and evolution in the early universe. The radiation from deeply embedded star-forming cores is often completely obscured at optical and IR wavelengths. Low opacities in the FIR-mm region mean we can *see* into the centre of such objects and study the very earliest stages of star formation. Physical characteristics (e.g. masses, geometries) for such systems are therefore much less model-dependent than in the infrared. Finally, the spectral energy distribution of the 2.7 K cosmic microwave background, which emanated from the epoch of recombination some 300,000 years after the Big Bang, peaks at a wavelength of 1.3 mm, and allows study of structure in the primordial universe.

Fundamental in the emergence of submm/mm astronomy have been bolometer arrays such as the SCUBA camera (Holland et al. 1999) operating on the James Clerk Maxwell Telescope (JCMT) and the balloon-borne Boomerang instrument (Masi et al. 1999). SCUBA discoveries include resolving the FIR/submm background into discrete (high-z) galaxies (Hughes et al. 1998), studying star formation far from the nucleus of nearby galaxies (Israel, van der Werf, & Tilanus 1999), finding new populations of protostars in molecular clouds (Johnstone & Bally 1999) and the discovery of vast debris disks of cold dust surrounding nearby stars (Holland et al. 1998a). Boomerang mapped over 1800 sq-degrees of the southern sky and provided the first detailed high-resolution image of the early Universe a few hundred thousand years after the Big Bang (de Bernardis et al. 2000).

Such discoveries by SCUBA and Boomerang have continued to stimulate the development of bolometer technology. Crucial developments in recent years include the availability of compact cryogenic systems (necessary to achieve the sensitivities needed for submm astronomy) and advances in quasi-optical instrument design (to control the coupling of the small detector to the large telescope aperture). BoloCAM on the Caltech Submm Observatory (Mauskopf et al. 2000), and SPIRE on the Herschel satellite (Gear & Griffin 2000) are examples of large format arrays currently under development. However, the traditional semiconductor technology, used successfully in SCUBA, will soon be obsolete as a new generation of superconducting detectors and advances in silicon micromachining will allow the construction of large format integrated arrays of many hundreds (or even thousands) of pixels. Such instruments are already in development and include the HAWC camera on SOFIA and, potentially, a 13,000 pixel camera for the JCMT ("SCUBA-2"). Next generation space missions, perhaps employing large deployable reflectors, will have even more exacting requirements, as the low-background environment of space will present sensitivity goals significantly lower than those needed for ground-based astronomy.

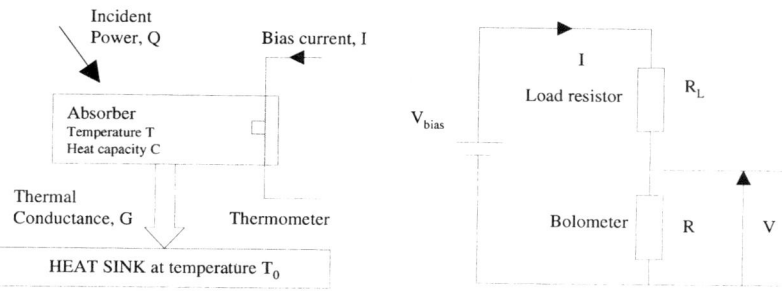

Figure 1. (left) Schematic illustration of bolometer operation; (right) bolometer bias circuit.

2. Basic Principles of Operation

2.1. What is a bolometer and how does it work?

A bolometer measures changes in the heat input from the surroundings and converts this into a measurable quantity such as a voltage or current. Since the mid-1980s most bolometers have adopted a composite design in which the functions of thermometry and radiation absorption are separated. This allows each to be independently optimized to suit a particular application. A bolometer therefore typically consists of an absorber and a thermometer of heat capacity C, connected by a small thermal conductance G, to a heat sink held at a fixed temperature T_0 (Figure 1: left). The energy E of the incident radiation is converted into heat in the absorber, leading to a temperature rise $\Delta T = T - T_0 = E/C$, until the radiation power flowing into the absorber is equal to the power flowing into the heat sink through the weak thermal link. The temperature rise is subsequently measured and is directly proportional to the deposited energy.

The principle of operation of a bolometric detector has been discussed by a number of authors in the past (see for example: Mather 1982; Griffin & Holland 1988; Richards 1994). For a classical germanium bolometer, such as those used in SCUBA (see Section 3), the detector is placed in a bias circuit with a voltage source and load resistor (Figure 1: right). A constant current I, generated from the load resistor R_L and bias voltage V_{bias}, flows through the bolometer. Provided this bias power ($P_{bias} = V_{bias}I$) remains constant, the incoming signal power (P_{signal}), collected by the telescope, will cause the bolometer temperature T to increase according to:

$$T = T_0 + (P_{signal} + P_{bias})/G. \tag{1}$$

The temperature rise causes a change in the resistance of the bolometer and consequently in the voltage across it. This change in voltage is amplified and measured. The thermometer is therefore made of a material that ideally has a large change in resistivity for a small change in temperature (see Section 2.2.2).

2.2. Ideal bolometer theory

2.2.1 Performance parameters The two most important "figures of merit" for a bolometer are the noise equivalent power (NEP) and time constant (τ). The NEP is a measure of the sensitivity of a bolometer, and is defined as the power absorbed that produces a signal-to-noise of unity at the output. In the presence of a varying background the NEP can be written as:

$$NEP^2 = NEP^2_{(detector)} + NEP^2_{(background)}. \tag{2}$$

Ideally, the overall NEP should be limited by the contribution from the thermal background (i.e. the sky and telescope) and not by contributions from the detector i.e. achieve background limited performance. The units of NEP are sometimes quite confusing, as the NEP is most often quoted after post-detection averaging, generally by a filter with a 1 Hz bandwidth. The adopted units are therefore $W/Hz^{1/2}$.

For an ideal bolometer the detector noise contributions should arise from only two sources: Johnson noise and phonon noise. Johnson noise is due to the random motion of electrons in the thermometer and can be written as:

$$NEP^2_J = 4k_B T R/S^2 , \tag{3}$$

where k_B is the Boltzmann constant, and S is called the responsivity of the device and gives the output voltage per input power (units of V/W). Phonon noise is due to the quantization of photons that transport energy between the absorber and the heat sink along the thermal conductance G:

$$NEP^2_P = 4k_B T^2 G. \tag{4}$$

The background contribution to the overall NEP arises from the sky, telescope and instrument. This photon noise arises from the random fluctuations in the rate of absorption of radiation. If the Rayleigh-Jeans approximation is valid ($h\nu < kT$) and the radiation passband is limited to a frequency interval $\Delta\nu$ around a central frequency ν_0, then the photon noise contribution to the NEP can be written as:

$$NEP^2_{PH} = 2Q(h\nu_0 + \eta\epsilon kT) , \tag{5}$$

where Q, the absorbed incident power is given by $Q = A\Omega B(\nu_0, T)\eta\epsilon$, where $A\Omega$ is the telescope throughput (A is the telescope primary area and Ω the beam solid angle), $B(\nu_0, T)$ the Planck function, T the temperature of the background radiation, η is the overall transmission of the system and ϵ the emissivity of the background.

Each of the NEP terms is an uncorrelated noise source and so these can be added together in quadrature. Hence the overall NEP for an ideal bolometer can be written as:

$$NEP^2 = 4k_B T R/S^2 + 4k_B T^2 G + 2Q(h\nu_0 + \eta\epsilon kT). \tag{6}$$

From this expression it can be seen that optimum NEPs (the lower the value the better...) can be obtained by minimizing R, T, G and Q and maximizing S.

Equation 6 readily highlights the critical importance of the operating temperature. Lowering the operating temperature decreases the NEP and results in a more sensitive device.

The thermal time constant is a measure of the response time of the bolometer to incoming radiation and is given by:

$$\tau = C/G. \tag{7}$$

From this expression it can be seen that the larger the value of G the faster the detector response time. Hence, there is a trade-off between NEP and τ in terms of the selection of G with the constant $NEP\tau^{1/2}$ (lower values are better) often being used as a figure of merit.

Whilst the phonon and photon noise contributions to the NEP are frequency independent, the responsivity varies with frequency and hence the Johnson noise component becomes:

$$NEP_J(\omega) = NEP_J(0)(1 + \omega^2\tau^2)^{1/2}, \tag{8}$$

where $NEP_J(0)$ is the DC value of the Johnson noise. This is an important consideration in FIR–mm astronomy where the large sky background necessitates a chopping modulation scheme to extract the weak source signal (see Section 4.5).

2.2.2 Characterizing ideal performance Ideal bolometer performance can be further characterized by assuming that the resistance of the bolometer depends only on temperature and follows a simple power law. For a semiconductor bolometer the following is an R-T relationship commonly measured:

$$R = R^* \exp(T_g/T)^{1/2}, \tag{9}$$

where R^* and T_g are constants that depend on the properties of the material used. This has also been extended to include electrical non-linearities (Holland 1990; Grannan et al. 1997).

The temperature coefficient of resistance, α, describes the change of resistance with temperature according to:

$$\alpha = (T/R)(dR/dT). \tag{10}$$

Hence, it is advantageous to make α as large as possible. Typical values for α for semi-conductor bolometers are 5 – 10.

The ideal thermal conductivity between the bolometer and the heat sink is assumed to follow a power law according to:

$$k(T) = k_0(T/T_0)^\beta, \tag{11}$$

where β is the thermal conductivity index.

These definitions now enable the performance of an ideal bolometric detector to be described by 3 dimensionless parameters:

Bias parameter	$\phi = T/T_0$,
Material parameter	$\delta = T_g/T_0$,
Loading parameter	$\gamma = \eta Q/(GT_0)$.

The bias parameter can be controlled by adjusting the bias current and has typical values of 1.1 – 1.5. The material parameter characterizes the variation of the resistivity of the bolometer material with temperature (typical values for a 300 mK device are 5 – 10). The loading parameter is a measure of the incident power loading on the detector and can have values of 0.06 – 0.2 depending on sky conditions.

2.3. Operation in a variable background loading environment

One of the main attractions of using bolometric detectors as total power detectors is that they respond uniformly over a wide frequency bandwidth. For a ground-based instrument, the limiting factor controlling the bandwidth is often the width of the atmospheric window in which they are being used. However, a problem in using such a large bandwidth, often several tens of GHz wide, is the increased sensitivity to degradation in performance through background power loading. That is, not only does the background radiation contribute photon noise, but it also heats the cooled bolometer. This is particularly a problem for the ultra-low temperature devices now in operation. Hence, the total NEP is modified to:

$$NEP^2 = NEP^2_{(detector)} + NEP^2_{(background)} + NEP^2_{(loading)}. \qquad (12)$$

As an illustration of the effects of power loading, consider an ideal bolometer operating at a low temperature (100 mK) on a large submm telescope. The background power from the sky dominates the overall NEP, and can vary by more than a factor of 3 on short timescales. Figure 2 illustrates the effects that power loading would have on an ideal device as a function of the loading parameter. Figure 2a shows how the voltage across the bolometer varies with applied current for an ideal device. These curves are commonly referred to as $V - I$ or load curves and are an important measurement for characterizing the electrical properties of a bolometer. The dashed line is a load line, i.e. represents a typical bias current passing through the bolometer for a given value of V_{bias} and R_L. As the background power increases ($\gamma > 0$) the bias point moves to a different (lower) curve. In Figure 2b the normalized NEP is plotted as a function of bias parameter. As can be seen there is a broad minimum in the NEP around some optimum value of ϕ. As ϕ is decreased below this optimum value (under-biasing the detector) the NEP rises dramatically, since although the phonon NEP contribution decreases somewhat, the Johnson noise increases significantly. On the other hand ϕ can be made greater than the optimum with little degradation in sensitivity. Figure 2b also shows that the optimum bias point shifts to higher values of ϕ (i.e. higher bias currents) as the power loading increases. Hence, in practice it is often advantageous to over-bias a bolometer if it is to be used in such a varying background situation.

Figure 2c shows how the minimum NEP (from Figure 2b) varies with the loading parameter. It can be shown (Holland 1990) that this takes the form:

$$NEP_{min} \propto T_0 G^{1/2} + Q/G^{1/2} + (photon\ noise\ term). \qquad (13)$$

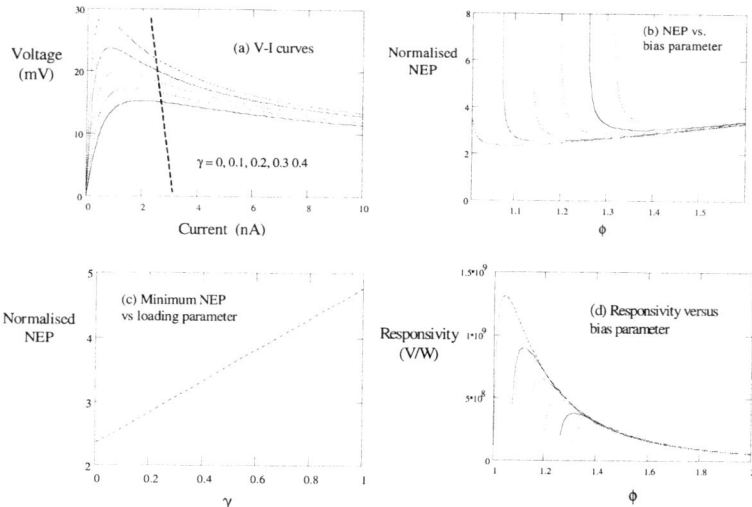

Figure 2. Illustrating the effects of operating a bolometer in a variable background-loading environment.

The first term in the equation represents the inherent NEP of the detector that decreases as G is made smaller. The second term represents the degradation in the NEP by the heating effect of the background power, and increases with decreasing G. Hence there is another trade-off in terms of having a small G to minimize the detector noise contributions to the NEP, and a large G to help reduce the effects of background power loading.

Figure 2d shows how the responsivity varies with bias parameter as a function of background loading. Although the responsivity drops dramatically as the loading parameter increases, it also becomes less sensitive to changes in the background as ϕ is increased. This is another reason for moderately over-biasing a bolometer in such circumstances. It is clear that in the presence of a varying sky background the bolometer becomes a significantly non-linear device. It is essential to calibrate out changes in bolometer responsivity at the time of an astronomical observation.

3. Types of Bolometers

3.1. "Classical" semiconductor bolometers

The vast majority of instruments currently in operation on telescopes use a composite bolometer design that incorporates a semiconductor resistance thermometer and a metal-coated dielectric as the absorber. The theory of operation has already been discussed above. The SCUBA bolometers (Figure 3: left) are made from $0.3\,\text{mm}^3$ chips of Neutron Transmutation Doped (NTD) germanium (Haller 1985) epoxied to the center of hexagonal sapphire substrates. The substrates are coated with a thin film of bismuth to provide an impedance match

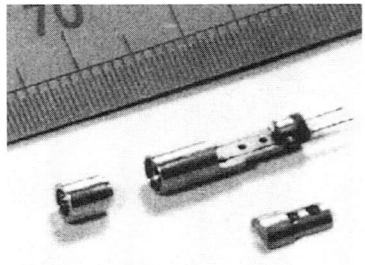 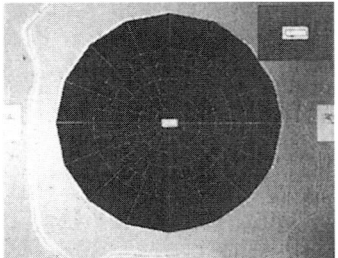

Figure 3. (left) Photograph of an individual SCUBA bolometer module; (right) A "spider-web" bolometer (Bock et al. 1998).

to incoming radiation. The electrical connections to a bolometer are made from 10 μm diameter brass wires, which also dominate the thermal conductance. The assembly is held rigidly by glass rods within a mount that also includes an over-moded cavity.

The current state-of-the-art in this field is represented by the "Spider-Web" composite bolometers. In this design the absorber is a metallized silicon nitride web structure with a small NTD germanium thermometer bump-bonded to the centre (see Figure 3). Since the spacing of the web is much smaller than the wavelength the structure acts as a plane surface to incoming radiation. The main advantage is that the cross-section of the absorber is minimized, thereby reducing the overall heat capacity, and leading to a device with a greatly increased speed of response for a given NEP and operating temperature. These devices are particularly suited to space applications where the low cross-section reduces the susceptibility to ionizing radiation, such as cosmic rays.

The theory and practice of semiconductor bolometers is now well understood. They have a considerable heritage having been used in instruments for a number of years on ground-based, balloon and space-borne telescopes. NTD germanium can be made for a wide range of resistivity values suitable for many applications in astronomy (Haller 1985). Typical low-temperature detector systems have bolometers with impedances of a few MΩ, which is well suited to a low-noise JFET read-out circuit. Although excess low-frequency (1/f) noise is generally not a serious problem, modulating the bias at a frequency higher than that set by the bolometer time constant can be used to greatly reduce the 1/f from all electronic components following the bolometer.

Table 1 gives a few examples of practical bolometer systems in operation or under development. The NEP value is the value needed to achieve background-limited performance and is therefore the goal set for the detector NEP. The time constant requirement is usually dictated by the observing modes (and sometimes by the readout circuit). The detector operating temperature and the performance figure of merit NEP$\tau^{1/2}$ are also shown.

Table 1. NEP, time constant, operating temperature and NEP $\tau^{1/2}$ figure of merit for current and future bolometer instruments.

Instrument	Wavelength range (μm)	NEP$_{det}$ (W/Hz$^{1/2}$)	τ (msec)	T$_{op}$ (mK)	NEP$\tau^{1/2}$ ($\times 10^{-18}$ J)
SCUBA	350–850	1.5×10^{-16}	6	100	12
SCUBA-2	450–850	7×10^{-17}	1.5	100	2.7
BoloCAM	1100–2000	3×10^{-17}	10	300	3
SPIRE	250–500	3×10^{-17}	8	300	2.7
HFI	350–3000	1×10^{-17}	5	100	0.7

3.2. Superconducting TES devices

Superconductors demonstrate very strong dependence of resistivity with temperature in the transition region between the superconducting and normal states. The transition edge sensor (TES) is a superconducting thin film, biased in the transition between the normal and superconducting states (Irwin 1995; Benford et al. 2000). If biased in the transition region, a small change in temperature will lead to a large change in resistance as illustrated in Figure 4a. This allows large values of α to be realized (50 – 2000). TES devices have low impedances (typically a few mΩ) and so to achieve stable biasing a voltage bias is adopted. This is in contrast to the current bias generally used in the typically MΩ semiconductor devices. Hence, the film is held at a constant voltage bias so that a change in resistance results in a change in the current through the film (Figure 4b). This change in current is read out using a SQUID (superconducting quantum interference device) amplifier (Chervenak et al. 1999).

The sharp n-s transition means that there is a fixed power level that a TES can handle. This is the power required to warm the device from the bath (heat sink) to the transition temperature, and is the sum of the bias power and background power i.e. $P_{total} = P_{bias} + P_{back}$. If the device is voltage-biased the electrical bias power is given by $P_{bias} = V_0^2/R$. This leads to *strong electrothermal feedback* (ETF) in which an increase in background power, warms the device and causes an increase in resistance. This in turn causes the bias current to decrease. This negative feedback means that the detector is self-biasing. Changes in the incident power are compensated for by changes in the bias power that can occur on timescales shorter than the thermal time constant of the detector. Thus, ETF has the effect of speeding up the detector and linearizing its response over a wide-range of input power. It also has the effect of suppressing Johnson noise at low frequencies and thus reducing the NEP.

The responsivity (in units of A/W) of a TES device is given by:

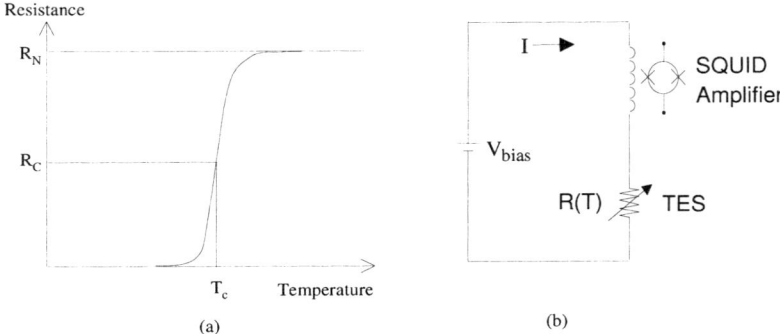

Figure 4. (a) Schematic R-T curve for a TES device. The detector is voltage-biased on the normal-superconducting transition. The resistance has a very steep dependence on temperature in the transition region. (b) TES bias circuit. The TES is held at constant voltage bias – a change in resistance results in a change in current through the film.

$$S = -1/V_{bias}[L/(L+1)], \qquad (14)$$

where L is called the loop gain, given by:

$$L = \alpha/n[1 - (T_0/T)^n]P_{bias}/P_{total}, \qquad (15)$$

where n is the index of the power law which controls thermal conduction to the bath. The transitions are generally so sharp that P_{total} is approximately constant within the transition. If the device is operated in a high background situation (e.g. ground-based FIR–mm astronomy) then P_{bias} may be significantly less than P_{total}, effectively suppressing some of the ETF effect and reducing the overall loop gain. In this case the responsivity will be a strong function of the background power. As discussed in Section 2.3 it then becomes necessary to calibrate out changes in bolometer responsivity at the time of an observation.

The total power for a TES device is fixed. If this power level is exceeded during actual operating conditions the device will go into the normal state and not respond at all. Though this can be remedied in ground-based situations, space-borne instruments will need to design for some power margin to allow for mis-estimation of the actual background power achieved on the spacecraft.

As mentioned above the Johnson noise contribution to the NEP is reduced by the square of the loop gain according to:

$$NEP_J^2 = 4k_B T P_{bias}/L^2, \qquad (16)$$

for low frequencies within the effective bandwidth set by the time constant. The time constant is decreased according to:

$$\tau_{eff} = \tau/(1 + L). \tag{17}$$

A normal-superconducting bilayer is a practical choice for a TES film, as the transition temperature can be adjusted by changing the relative thickness of the normal metal and superconducting layers. In such a bilayer, the very thin normal metal layer is made weakly superconducting by the proximity effect, in which Cooper pairs from the superconducting layer diffuse into the normal metal.

One of the major benefits of this kind of detector is that the signals can be read out by SQUID amplifiers. The SQUIDs operate at the same temperature as the bath so the electrical connections can be very short. The bias to a SQUID may be turned off and the whole device goes into a superconducting state where it adds no noise. Thus, by switching on or off rows or columns of SQUIDs in an array (one SQUID for each pixel) a cold multiplexer may be realized.

Though at relatively early stages of development for FIR–mm use, TES devices hold the potential to produce large format arrays for astronomical use. Crucially, they can be fabricated using lithographic techniques. They have already demonstrated outstanding performance in the optical and X-ray regions where they photon count and energy resolve. Electrical NEPs (i.e. blanked-off to incoming radiation) of 1×10^{-17} W/Hz$^{1/2}$ and time constants of ≈ 10 msec have already been demonstrated at operating temperatures of 300 mK. The most serious potential problems for a TES device is coping with uncertain background levels. The background often turns out to be higher than desired because of imperfect coupling to the telescope, or due to stray light degrading the detector sensitivity. Once the background power drives the operating temperature into the normal operating regime the device will fail to work.

3.3. Silicon "pop-up" bolometers

In addition to the classical germanium semiconductor and TES devices there are a number of other types of bolometer currently in use or under development. Pop-up detectors (PUDs) are being developed (primarily) by NASA Goddard for new generation instruments on the CSO and the SOFIA airborne observatory (Moseley et al. 2000). These use ion-implanted silicon semiconductor thermometers that are folded into a linear array structure. The electrical connections and heat sink are hidden behind the line of pixels which allows the linear structures to be close-packed together to form a 2-D array. These devices have already demonstrated excellent electrical, thermal and optical properties.

4. Practical Instrument Design

The basic components of a practical bolometer instrument design for FIR–mm astronomy are shown in Figure 5. Since the sky background dominates the weak source signal a sky-chopping modulation scheme is applied in which the telescope measures the difference between the sky+source signal and that from the sky alone. Modulating the secondary mirror between 5–10 Hz accomplishes this and also moves the detection signal away from any possible excess low-frequency

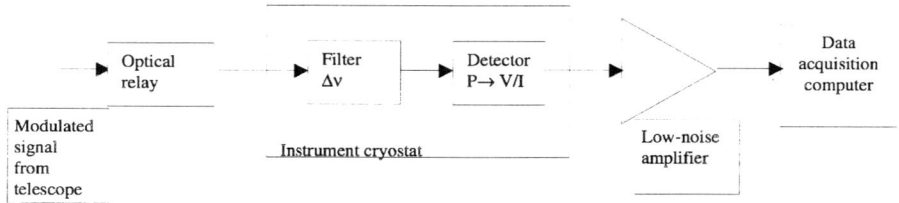

Figure 5. The major components of a practical bolometer instrument.

(1/f) noise (see Section 4.5). The signal from the telescope is invariably re-imaged quasi-optically to the instrument via mirrors and lenses. The signal is then pre-filtered (spatially) by a bandpass filter that selects the waveband of observation. The detector then converts the incoming power into either a voltage or current that is subsequently amplified and fed into a data acquisition system. For a highly sensitive bolometer device the two key areas are the refrigeration system that allows low-temperature operation, and the optical coupling to the telescope.

4.1. Detector operating temperature

As shown in Equation (4) the intrinsic detector NEP (i.e. with no background loading) is proportional to $TG^{1/2}$. For a fixed response time the NEP is therefore proportional to $TC^{1/2}$. Typical bolometers have materials which have heat capacities which are proportional to T (metals) or T^3 (dielectrics) and so the intrinsic NEP can be written as:

$$NEP \propto T^{3/2} - T^{5/2}. \qquad (18)$$

Hence, minimizing the detector operating temperature will result in the lowest achievable NEP. This has presented quite a technical challenge over the years! In order to get the necessary performance in terms of bolometer NEP and time constant for state-of the art astronomical applications it is generally necessary to cool the devices to below 0.5 K. This can be achieved using a variety of cryogenic refrigerators. ^{3}He systems are appropriate in the temperature range 250 to 400 mK, and provide cooling powers from a few μW at 250 mK to $\approx$ 100μW at 400 mK. ^{3}He systems require cooling to 4 K and a point below 2 K (usually obtained by pumping on a ^{4}He bath) to condense the ^{3}He gas into a liquid. High-pressure, sorption pumped, self-contained systems are available which provide an integral ^{4}He system for the ^{3}He condensation. These are bolted directly to the 4 K plate of a cryostat. ^{3}He systems require high cooling powers from the 4 K system during condensation and pumping down to base temperature.

If lower temperatures (250 mK to a few mK) are required, dilution or adiabatic demagnetization (ADR) refrigerators are most commonly used. ADRs are cryogenically efficient, compact, and may be operated liquid cryogen-free using a closed-cycle cooler to provide the 4 K stage. They are generally used in situations where the cooling power required is no more than a few μW. An ADR consists of a magnet surrounding one or more paramagnetic salt pills. During normal operation the paramagnetic ions in the salt are aligned when the magnet is ramped up to maximum field (entropy is reduced). The heat generated in this process is transferred to a liquid helium bath that is connected to the salt pill via a heat switch. Next the heat switch is opened and the magnetic field is ramped down to zero. The salt pill starts to cool adiabatically to its base temperature. To stabilize the salt pill above base temperature, the ramp down of the field is stopped at a specific temperature and the magnetic field is reduced slowly to compensate for the residual heat leak. In this way the temperature can be stabilized anywhere between about 30 mK and 4.2 K. ADRs also have high magnetic fields (a few Tesla during magnetization of the salt pill) which generally require some shielding to keep the detector and electronics in a low field region. Practical ADRs reach about 30 mK as their lowest temperature. ADRs and ^{3}He systems do not provide continuous operation and so every 12 – 24 hours they require re-cycling to maintain the base temperature.

Dilution refrigerators (DR) are used whenever higher cooling powers are required (tens of μW) and continuous operation at base temperature is needed. The price paid for this is system complexity and size. The dilution cycle works by utilizing the entropy increase that occurs when ^{3}He atoms are transferred from a concentrated solution of ^{4}He. When a mixture of ^{3}He and ^{4}He is cooled below 0.9 K it separates out gravitationally. The layers are not the pure isotopes but a layer rich in ^{3}He, the concentrated phase, floating on a layer poor in ^{3}He, the dilute phase. A phase boundary, similar to that between a liquid and a gas, exists and energy is needed to move ^{3}He across the boundary. If the ^{3}He is forced across the boundary artificially then energy is drawn in from the surroundings and cooling is produced. In a dilution refrigerator this is done by reducing the concentration of ^{3}He in the dilute phase. The dilute phase is connected by a capillary to the still at $\approx$ 0.6 K, where the vapour pressure of ^{3}He is about a thousand times that of ^{4}He. The ^{3}He boils off preferentially and is drawn off by a pump. As the concentration of ^{3}He in the still is reduced, osmotic pressure causes ^{3}He to migrate from the mixing chamber (where the phase boundary is) to the still. This draws ^{3}He across the phase boundary producing cooling. The pump is part of a circulating system and the ^{3}He is returned to the concentrated phase via a series of heat exchangers, which ensure the returning liquid is as close as possible to the temperature of the mixing chamber. The incoming ^{3}He is condensed at a temperature of 1.2 K, produced by pumping on ^{4}He in a small pot. The ^{4}He is vented to the atmosphere and replenished from the main liquid He dewar through a needle valve which is adjustable externally. The circulating ^{3}He is kept clean by passing through a cold trap (usually at 77 K). Practical instruments that incorporate DRs include the SCUBA bolometer array on the JCMT. Some manufacturers have successfully automated their systems, and within a few years it is likely that liquid cryogen-free DRs will become commercially available.

4.2. Optical coupling

At submm wavelengths the size of a bolometric detector (typically a few mm in diameter) is usually small compared to the telescope diffraction spot. To overcome this problem the traditional approach has been to place the bolometer in an integrating cavity and couple it to the incident beam using either a Winston concentrator or a conical horn with a section of cylindrical waveguide (Holland 1990). Conical feedhorns have been adopted more commonly in recent years (e.g. with SCUBA and in development for the SPIRE instrument on the Herschel satellite). Feedhorns can be close packed together in the telescope focal plane, fitting as many as possible into the available area. The feedhorn defines the detector field-of-view and gives a tapered (approximately Gaussian) illumination of the telescope (Figure 6: left). Maximum efficiency for the detection of a point source is achieved for a horn of diameter close to $2F\lambda$ where F is the focal ratio of the final optics and λ the wavelength. This corresponds to a beam spacing on the sky of $\approx 2\lambda/D$ where D is the telescope diameter. To fully-sample the image plane requires the horn diameter and spacing to be $0.5F\lambda$. If the maximum aperture efficiency spacing of $2F\lambda$ is adopted 16 separate telescope pointings are needed to create a fully sampled image. This makes the observing modes quite complex. In addition to achieving high aperture efficiencies feedhorns allow the bolometer field-of-view to be restricted to that of the telescope which results in good stray light rejection (see Section 4.6.4). Instruments using feedhorns have considerable heritage – they are well understood in terms of their control of the beam coupling, are relatively easy to manufacture, are reproducible in moderately large numbers and offer good rejection of electro-magnetic interference (the feedhorn and detector cavity effectively act as a Faraday enclosure).

An alternative approach to coupling to the telescope is to dispense with feedhorns and simply have bare pixels of size $0.5F\lambda$ in the focal plane. In this case the bolometers have a large $\approx \pi$ steradians field-of-view, and so it is necessary to use a cold stop within the optical system to define the coupling to the telescope. As shown in Figure 6 (right) the telescope illumination is flatter than the Gaussian profile of a feedhorn and results in an almost "top-hat" pattern at the primary. This is advantageous if the application calls for the best point source sensitivity, but since the illumination pattern is not tapered at the telescope, it will often result in higher side-lobe levels. By close packing $0.5F\lambda$ pixels together in an array configuration an instantaneous sampling of the image plane is possible, which, in principle, simplifies observing procedures. The main disadvantage of bare pixels is the increased susceptibility to stray light (see Section 4.6.4).

Some applications (e.g. sensitive CMB measurements using the HFI on the Planck satellite) require suppression of the sidelobe levels to a very high order. This is being achieved using a triple feed system (Griffin 2000) in which the first horn defines the illumination pattern on the telescope with no other optical components in front (that might scatter or diffract the beam). A second horn (back-to-back with the first) has a lens that collimates the beam through a bandpass filter. A final horn/lens combination feeds the incoming radiation onto the detector element. This system is expected to produce extremely clean beam profiles on the sky with minimum sidelobe levels.

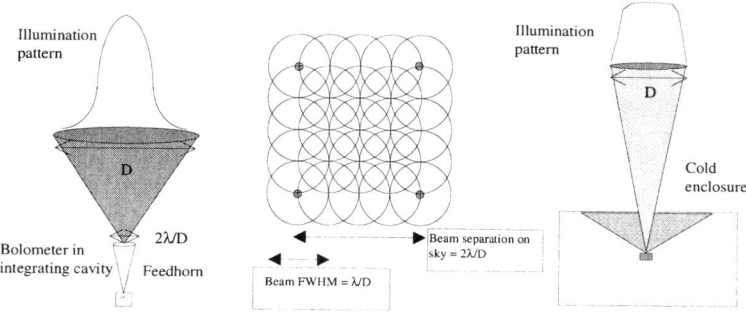

Figure 6. (left) Feedhorn-coupled bolometer and illumination pattern, (middle) 16-point jiggle-pattern needed to achieve a fully-sampled map with 2Fλ-spaced feedhorns, (right) bare pixel bolometer and illumination pattern.

4.3. Wavelength selection

Ground-based astronomy in the FIR–mm waveband is restricted to observing through particular transmission windows in the atmosphere. Throughout this region atmospheric water vapor is the main absorber of radiation from astronomical sources. Excellent observing conditions are therefore also referred to as dry conditions, meaning that the water vapor content is low above the observing site. A bandpass filter carefully designed to match the transmission window makes the selection of observing wavelength (see Figure 7). These filters are invariably multi-layer, metal-mesh interference filters (Hazell 1991). They have excellent transmission (typically over 80%) and also less than 0.1% out-of-band power leakage. The latter design characteristic is particularly important, as it ensures that there is minimum contribution to the source signal from extraneous sky emission.

4.4. Low-noise electronics

For traditional semiconductor devices the first stage amplification is typically a cooled JFET amplifier. Ideally, the JFET should not add significantly to the overall NEP. JFETs configured as voltage followers are well matched in their noise characteristics to typical low-temperature bolometers which have resistances in the range 5–20 MΩ. Hence the first stage amplification is invariably a JFET voltage follower cooled to 100-200 K to minimize gate current noise. The output from the FETs then usually passes through RF filters (see Section 4.6.2) to an external high-gain AC-coupled amplifier. A critical component in the design of the electronics is the wiring to bring the signals from the high impedance bolometers to the low-impedance JFET. Ribbon cables constructed from superconducting wires (to minimize heat leaks) have been successfully employed in bolometer instruments, although care has to be taken to avoid excess noise contributions from microphonic pick-up (see Section 4.6.1).

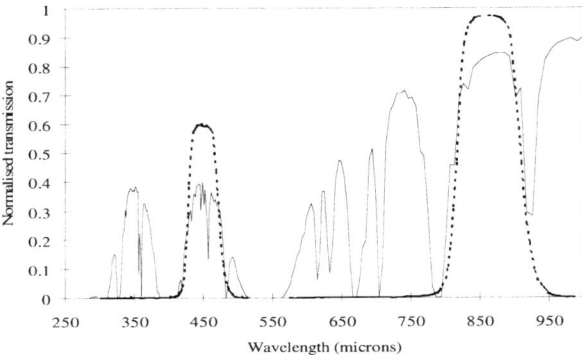

Figure 7. Atmospheric transmission curve for 1 mm precipitable water vapor on Mauna Kea showing the semi-transparent windows in the submm region. Theoretical filter profiles are shown overlaid on the 450 and 850 μm windows.

4.5. Data acquisition techniques

Sources in the FIR–mm waveband are generally very faint compared to the background. The background can be astronomical in nature – zodiacal light for instance, terrestrial – atmospheric and telescope emission for ground based observations, or electronic – offsets which may drift with time. Thus the background has to be subtracted in order to reveal the source. The technique adopted depends on the time and position stability of the whole system – background power, bolometer and electronics. In an ideal scenario the astronomer determines the background from the data and subtracts it. For instance, the astronomer determines which regions are source free and fits a plane or higher order polynomial to these regions and uses that as an estimate of the background. Alternatively, the sources are modeled based on the data and subtracted. These techniques preserve information on the largest spatial scales, which is generally desirable to determine the total flux. In the worst of circumstances the subtraction happens very early on in the data acquisition chain and spatial frequencies may be lost. It should also be noted that any finite aperture system (e.g. telescope) is a low-pass filter with a hard cut-off in terms of spatial frequency.

The modulation frequency is usually determined by the 1/f performance of the system. The modulation technique serves to move the information about the source to frequencies above the 1/f knee of the whole system. In ground-based submm astronomy the sky can have a high 1/f component (emission due to the poor mixing of water vapor) and it has been traditional to chop with the telescope secondary mirror or a focal plane chopping system. This produces alternating samples of the atmosphere emission (background) and source plus atmosphere. Phase sensitive detection at the chop frequency (set to be above the system 1/f knee) reveals the source signal. This technique works well, especially for point source measurements but tends to be limited in large area mapping as the spatial frequency response of a two-position chop is a sinusoid. This eliminates information at and near the zeroes of the sinusoid. It is thus an AC technique and the electronics may be AC coupled throughout. Alternatives to chopping include fast scanning the telescope to move astronomical spatial frequencies to electrical frequencies above the 1/f knee of the system. This is a broadband technique requiring a low noise floor over a range of frequencies.

If the detector has low 1/f then the electronics 1/f may be eliminated or reduced by AC biasing the bolometer at a frequency above the inverse of the time constant of the bolometer, and well above any 1/f frequency in the electronics. Phase sensitive demodulation then reproduces the background signal (and its 1/f) at DC. A further modulation technique is then needed to discriminate sources (fast or slow scanning the telescope or using a map if the system is an array). In ground-based submm astronomy the 1/f in the background is generally well correlated across the focal plane, so arrays of detectors make background subtraction much easier (particularly if the arrays instantaneously sample all of the flux in the focal plane area).

Data acquisition techniques are a complex area of system design, and modeling of the complete system is required to determine the best modulation scheme for a given experiment.

4.6. Environmental issues

4.6.1 Microphonics and pickup Microphonics arise from the motion of the bolometer or its wiring/amplifiers relative to the ambient electric and magnetic fields. If a wire moves, its capacitance to ground changes. A voltage-biased time varying capacitor gives a current $V_b \, dC/dt$ at the microphonic frequency. Pickup arises if the ambient fields change, inducing currents and voltages in the wiring and the bolometer. They are reduced by rigidly mounting the bolometer and wiring, shielding them from ambient fields and by keeping loop areas low and wire lengths short. This applies particularly between the bolometer and the first stage amplifier. The use of differential amplification is recommended to suppress common-mode signals. It should be remembered that modern bolometers detect changes in signal levels at the 10^{-17} W level or less, and a very careful consideration of mounting, screening, grounding and wire routing must be done at the design stage of any experiment.

4.6.2 Radio frequency interference Semiconductor bolometers tend to have resistances in excess of $1\,\mathrm{M}\Omega$ and are sensitive to RF interference that can be a source of heating and 1/f noise. RF filtering is essential on all lines entering the cryostat and care must taken to see that the cryostat is effectively a "Faraday cage". For instance, RF can enter via O-ring joints that are not made metal-to-metal when the O-ring is compressed. Hence, they can be very sensitive to electrical pick-up. Superconducting devices are low impedance and not very sensitive to RF but are considerably more sensitive to magnetic fields. The fields must be heavily attenuated using high permittivity materials and superconducting screens in order that SQUID amplifiers and TES devices will work properly.

4.6.3 Grounding and shielding Care must be taken in the grounding of the bolometer and electronics to avoid multiple ground points. Current can flow between ground points and consequently inject mains frequencies (50/60 Hz and harmonics) into the electronics and signals. A single-point ground at the cryostat is recommended. SQUID amplifiers have to be carefully shielded from magnetic fields. Cold enclosures that provide stray light rejection (see Section 4.6.4) made from a high permittivity material also provide magnetic shielding for the SQUIDs.

4.6.4 Stray light Bolometers respond to any form of energy that can heat the device. As discussed in Section 2.4 this can change the performance by increasing the background loading and, if the energy is electro-magnetic, by adding photon noise. Typically, the main source of unwanted energy is *stray light*. This can come from the environment through the bandpass filters if they "leak" at unwanted frequencies, or by paths through wiring channels, holes in radiation shields etc. The problem is particularly acute when the photons being measured are much less numerous and energetic than those present in the surroundings. In the case of submm astronomy, visible/near-IR photons need to be suppressed to the level of 10^{-9} or better to avoid stray light problems. A great deal of care is required to make a cryostat truly dark (as close as possible to zero background) when the optical path is blanked-off with a cold shutter.

The other contribution to the stray light problem comes from photons within the passband but outside the optical path. Controlling these is a question of controlling the area-solid angle product of the beam as it progresses through the optical system (it's a good idea to think in time reverse – what the bolometer "sees" – for such problems). Use of beam controlling devices (e.g. feedhorns, antenna-coupled bolometers), cold stops and proper diffraction analysis is required to control and understand the in-band stray light problem. With the move to fully sampled arrays with bare pixels (not feedhorn coupled), where the solid angle of the pixels will be $\approx \pi$ steradians, this problem will be especially difficult, and require a low temperature black box around the array.

5. Imaging Arrays

With a few exceptions until the mid-1990s the only instruments available for continuum astronomy in the FIR–mm waveband were single-pixel broadband photometers. Not only was mapping extended regions of sky painstakingly slow, but also instrument sensitivity was invariably detector-noise limited at all wavelengths of operation. Improving the sensitivity to achieve background-limited performance was desirable, but the next logical step was to increase the data collection rate by having more than one detector in the focal plane. Such a multiplex advantage would potentially give substantial gains in observing time. Compared with IR arrays (even those in operation 20 years ago) the first generation of submm cameras have modest numbers of pixels (up to about a hundred or so).

On ground-based telescopes arrays currently in operation include SCUBA on the JCMT (at 350-850 μm) – see Figure 8 (left), SHARC (Wang et al. 1996) on the 10.4 m Caltech Submillimeter Observatory telescope (350 μm),

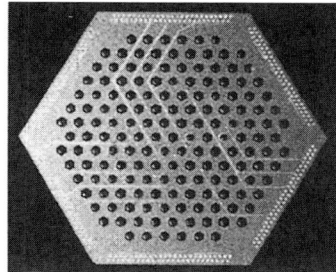

Figure 8. (left) Photograph of the SCUBA long-wave feedhorn array; (right) Photograph of the BoloCAM array. (Figure courtesy of Jamie Bock.)

MAMBO (Kreysa et al. 1998) on the 30 m Institut de Radio Astronomie Millimetrique (1.3 mm) and SIMBA on the 15 m Swedish-ESO Submillimeter Telescope (1.3 mm). More specialized projects include the Sunyaev-Zeldovich radiometer SuZie on CSO, the bolometer-spectrometer SPIFI on JCM, and the polarization imager Hertz on CSO. Bolometer arrays have been successfully operated from space (on satellites such as COBE and IRTS), and balloons (e.g. Boomerang, Maxima). Instrument under development include BoloCAM – a 151-pixel camera for the CSO (see Figure 8: right), and the SPIRE tri-band imaging camera for the Herschel satellite.

5.1. Basic design: Feedhorn versus filled arrays

The choice of pixel architecture is crucial to the design of an array instrument. The two basic options are to have each pixel coupled by a feedhorn or to have a filled array of bare pixels. Each option has advantages and disadvantages that are summarized below.

(1) Mapping speed. Feedhorns provide the best mapping speed per pixel. For extended sources the mapping speed of a filled array is approximately 3.5 and 1.5 times faster than a $2F\lambda$ and $F\lambda$ feedhorn array.

(2) Jiggle positions. Feedhorn coupled arrays with 1 and $2F\lambda$ spaced pixels require some jiggling of the secondary mirror to produce a fully-sampled image. There may also be a reduction in data quality and image fidelity because of the finite time needed to complete the jiggle pattern (during which the atmosphere may change).

(3) Number of detectors. Filled arrays require up to 16 times more pixels to cover the same field-of-view as feedhorn arrays. This introduces additional complexities in terms of reading out the number of pixels and a higher data rate.

(4) Background limited NEP. The background power on filled $0.5F\lambda$ pixels is typically a factor of ≈ 5 less than that for $2F\lambda$ feeds because the necessity to

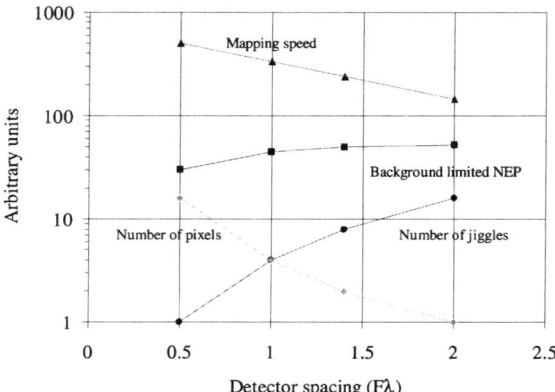

Figure 9. Relative trade-offs in pixel architecture (based on Glenn et al. 1998).

control the beam with a cold stop results in less energy (background power) throughput.

(5) Field-of-view control. It is generally considered to be easier to control the field-of-view of a feedhorn-coupled bolometer than a bare pixel with cold stop. However, even with a feedhorn-coupled array (e.g. SCUBA or BoloCAM) it is often very advantageous to use a cold stop to define the final telescope illumination.

(6) Observing methodology. Filled (bare) arrays allow instantaneous "snap-shot" imaging of a field. In principle, observations are inherently simpler with a fully-sampled array.

The above parameters are summarized graphically for various detector spacings in Figure 9.

5.2. Telescope performance

On the telescope the sensitivity is represented by the Noise Equivalent Flux Density. This is the flux density that produces a signal-to-noise of unity in a second of integration, and is given by:

$$NEFD = NEP/(\eta_c \eta_t A_e e^{-\tau A} \Delta\nu) , \qquad (19)$$

where η_c is a chopping efficiency factor, η_t is the overall optical efficiency, A_e is the effective area of the telescope primary, $e^{-\tau A}$ is the sky transmission in which τ is the zenith optical depth and A the source airmass, and $\Delta\nu$ is the filter passband.

The NEFD is very much dependent on the weather and varies with sky transmission particularly in the submm wavebands (Holland et al. 1999). On many occasions the fundamental limitation is set by sky-noise. Sky-noise manifests itself in a DC offset and in spatial and temporal variations in the emissivity of the atmosphere above the telescope on short timescales. This can degrade the NEFD by more than an order of magnitude in unstable conditions. The standard techniques of sky chopping and telescope nodding remove the DC offset, but since the chopped beams travel through slightly different atmospheric paths, the effects of sky variability are often not completely removed. However, if the sky-noise arises from features that are larger than the array size it is possible to remove the effects to high order. The scale height of the screen layer in the atmosphere, that is moving over the telescope and causing the sky variations, is of order of 2 km or less above a good submm site (such as Mauna Kea). Combining this with typical wind speeds, and the observed frequency of the fluctuations, implies that the scale size is of order 1000 arcsec or more. Hence, the sky-noise component is caused by features larger than the array, and so in principle can be subtracted as a time-varying DC offset. Sky noise removal algorithms have been successfully applied to SCUBA data with the result that NEFDs close to theoretical values are measured (see Jenness, Lightfoot, & Holland 1998 for more details of sky-noise removal using large-format arrays).

5.3. Current bolometer arrays

Submillimetre Common-User Bolometer Array (SCUBA)   SCUBA consists of two arrays of bolometers. The Long-Wave (LW) array has 37 pixels operating in the 750 and 850 μm atmospheric windows, while the Short-Wave (SW) array has 91 pixels for observations at 350 and 450 μm. Each of the pixels has diffraction-limited resolution on the telescope, and are arranged in close-packed hexagons (as shown in Figure 10). Both arrays have approximately the same field-of-view on the sky (diameter of 2.3 arcmin), and can be used simultaneously by means of a dichroic beamsplitter. The sensitivity is limited primarily from the photon noise for the sky and telescope background at all wavelengths. This is achieved by cooling the bolometers to 100 mK using a dilution refrigerator, while limiting the background power by a combination of single-moded conical feedhorns and narrow-band filters. The multiplex advantage means that SCUBA acquires data thousands of times faster than the previous (single-pixel) instrument to the same noise level.

SCUBA is mounted on the Nasymth platform of the JCMT. A complex relay of mirrors forming two back-to-back Gaussian-beam telescopes re-image the f/16 beam to f/4 at the detector arrays. This preserves a focal plane which is frequency-invariant and minimizes the size of the feedhorns on the low-temperature stage. Wavelength selection is made using a rotatable drum that contains pairs of filters to select the observing frequency. When blanked-off to incoming radiation the bolometers achieve a NEP of 6×10^{-17} W/Hz$^{1/2}$. Under low background levels from the zenith sky, with a water vapor level of < 1 mm, optical NEPs are typically 3.5 and 1.8×10^{-16} W/Hz$^{1/2}$ at 450 and 850 μm respectively. The NEP is also very uniform across the arrays with variations of no more than 20% (Holland et al. 1998b). Under good observing conditions the NEFD on the sky is typically 500 and 80 mJy/Hz$^{1/2}$ per pixel at 450 and 850 μm respectively.

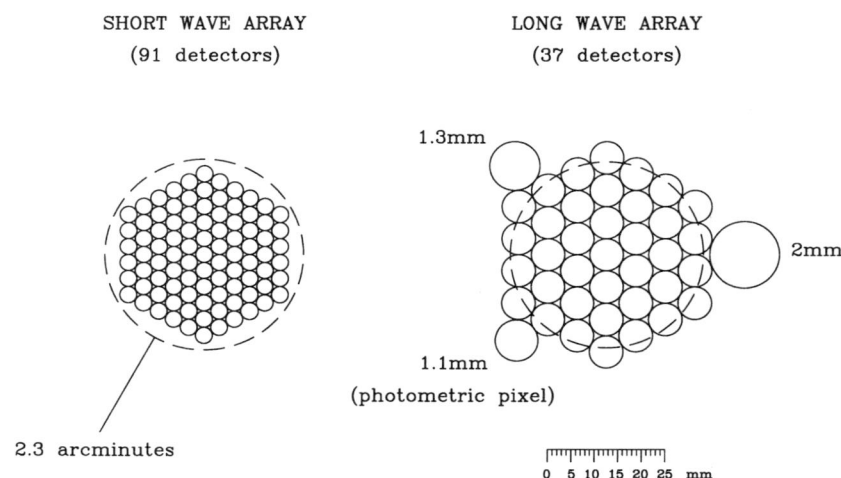

Figure 10. Layout of the SCUBA arrays. The photometric pixels extend the wavelength range into the millimetre region.

SCUBA is both a camera and a photometer. Point-source photometry is carried out with the center pixel of each array simultaneously at two wavelengths or with any of the photometric pixels independently. The conventional techniques of secondary mirror chopping and telescope nodding remove the dominant background. Extensive tests have shown that better S/N is obtained by performing a small 3 × 3 grid (of spacing 2 arcsec) around the source. This compensates for the slight offset between the arrays and presumably helps cancel submm scintillation (seeing) effects and/or slight pointing or tracking uncertainties.

Jiggle mapping is the adopted mode for sources that are smaller than the array field-of-view. Since the bolometers are spaced in the focal plane by 2 × FWHM beam on the sky, the secondary mirror is jiggled to fill in the gaps and produce a fully-sampled (Nyquist) map. It requires 16 jiggles to fully-sample a source at one wavelength. When using both arrays simultaneously (the default) a 64-point jiggle pattern is needed. Limited mosaicing of jiggle-maps is also possible. For regions that are extended compared to the field-of-view the technique of scan mapping is used. This is an extension of the raster (on-the-fly) technique used for single-pixel photometers, where the telescope is scanned across a region, whilst chopping, to produce a differential map of the source. Because of the spacing of the SCUBA pixels the arrays have to be scanned at one of 6 angles to produce a fully sampled map. Data is acquired using a method first described by Emerson (1995) where maps are taken at several different chop throws and directions. They are coadded together in Fourier space and the resultant image has significant improvements in S/N over the traditional raster technique (chopping along the scan direction).

Submillimetre High-Angular Resolution camera (SHARC) SHARC is a 24-pixel camera operating at wavelengths of 350 and 450 μm on the 10.4 m CSO telescope on Mauna Kea (Wang et al. 1996). It uses a linear monolithic array of silicon bolometers cooled to 300 mK by a ^{3}He refrigerator. The telescope illumination is not defined by feedhorns but by a cold aperture and field stop inside the cryostat. The optics consists of a number of mirrors that re-focus the telescope beam onto the array. The pixel size is 5 arcsec in the direction of the array and 10 arcsec in the cross direction, which gives a Nyquist sampled map in the array direction. The beamsize on the sky is approximately 9 arcsec and is close to the diffraction limit.

The observing modes consist of an ON-OFF chopping mode used for point source measurements, and an on-the-fly (OTF) dual-beam mapping mode for extended sources. Since the array is quite small, SHARC does not employ any sophisticated sky-noise removal algorithms, and so spatial and temporal changes in the sky emission in the high frequency 350/450 μm windows limits the sensitivity of the instrument under most conditions. The optical NEP for each pixel is $\approx 5 \times 10^{-15}$ W/Hz$^{1/2}$ and under good weather conditions the instrument can achieve an NEFD of 1 Jy/Hz$^{1/2}$. An upgrade to SHARC (with a much larger array) is planned for 2002/3.

MPIfR bolometer arrays (MAMBO) The Max-Planck Institut fur Radioastronomie have produced a series of bolometer arrays for the IRAM 30m telescope on Pico Veleta near Granada in Spain (Kreysa et al. 1998). All the bolometers and their interconnections are micro-machined on a single silicon wafer and then matched to a separate $2F\lambda$ feedhorn array. The one manual step in the array manufacture is the gluing of the NTD germanium thermometer onto the membrane substrate of each pixel. The current instrument, which has been used since early 1998, is a compact ^{3}He-cooled 37-channel array operating at 1.3 mm. Observing is largely carried out using an OTF method. The performance of the 37-channel array on the telescope routinely gives NEFDs of 50mJy/Hz$^{1/2}$ per pixel in good weather conditions. A next generation 119-channel instrument has already undergone some limited tests on the IRAM telescope, and this coming winter 2001/2 should see this instrument in routine operation. The MPIfR have also been developing similar arrays for the SEST telescope in Chile (SIMBA) and Heinrich-Hertz telescope on Mount Graham in Arizona (the latter optimized for higher frequencies 350/450 μm).

Bolometer CAMera (BoloCAM) BoloCAM is a 151-element silicon nitride bolometer array designed to operate in the 1.1 – 2 mm atmospheric windows (Mauskopf et al. 2000). The instrument has already undergone engineering tests at the CSO and the current plans are to be operating routinely by the end of 2001. The instrument uses a compact ^{3}He refrigerator to achieve background-limited performance with the arrays at a temperature of 300 mK. The electronics uses a stable DC readout scheme based on that developed for Boomerang. This allows the array to be used in both a drift scan and OTF mode for mapping large objects, without the need of a chopping secondary mirror. This has exacting requirements on the low-frequency noise stability of each bolometer. The detectors are coupled to the telescope using 1.2 Fλ feedhorns, and hence, according to Figure 9, would only need about 4 pointings to fully-sample the field-of-view. This is a compromise between having a high efficiency 2 Fλ horn, but needing 16-pointings per map, and a less efficient filled array of 0.5 Fλ detectors that instantaneously samples the field.

The per-pixel NEP under low backgrounds is estimated at 3×10^{-17} W/Hz$^{1/2}$ and the projected NEFD on the CSO telescope is between 30 – 40 mJy/Hz$^{1/2}$. A BoloCAM instrument is also planned for the 50 m Large Millimetre-wave Telescope, currently under construction in Mexico. In this case the projected performance at 1.1 mm is for a per-pixel NEFD of 1 mJy/Hz$^{1/2}$ and diffraction limited beam of 6 arcsec.

6. Current and Future Developments

Bolometer technology has advanced considerably since the design of SCUBA. In the near future a new generation of large-format, wide-field instruments will become available for the first time in FIR–mm astronomy. Eventually these may contain many thousands of pixels and be more akin to IR arrays or optical CCD cameras than the instruments currently in use today. Conventional bolometer technology, such as that used in SCUBA and Boomerang, limits the number of pixels to just a few hundred. This is primarily due to the wire count (there are no multiplexed read-outs schemes) and the high heat dissipation in using large numbers of cold JFET amplifiers.

6.1. Semiconductor bolometer arrays for space applications

Both the Herschel and Planck satellites will have bolometer arrays for their respective missions. They will use traditional semiconductor arrays of bolometers (spider-webs), coupled with feedhorns, rather than the superconducting TES devices. The major drivers behind this decision were the timescale involved for the completion of the instruments (TES devices are still unproven in large formats), and the uncertain levels of background power once in orbit (it will be low, but it is hard to predict to the high accuracy needed). The bolometer instrument on Herschel, SPIRE, will have three imaging arrays operating at 250, 350 and 500 μm, as well as a broad-band Fourier transform spectrometer covering the 200-670 μm range. The field-of-view of the photometer will be approximately 4×8 arcmin at each wavelength (see Figure 11 for the array footprints), with each pixel having diffraction-limited resolution. The wire count and the volume of cold electronics limit the number of pixels. Data collection is also limited by the telemetry rates to the spacecraft.

The SPIRE arrays will undersample the image plane of the telescope, and so observations will require the use of a novel "beam steering mechanism", which is a mirror placed at an image of the telescope secondary. This mirror performs the functions of chopping and jiggling, the latter of which compensates for the undersampled array (just as in SCUBA). SPIRE has projected per-pixel NEFDs of between 30 - 40 mJy/Hz$^{1/2}$ in the low-background environment of space.

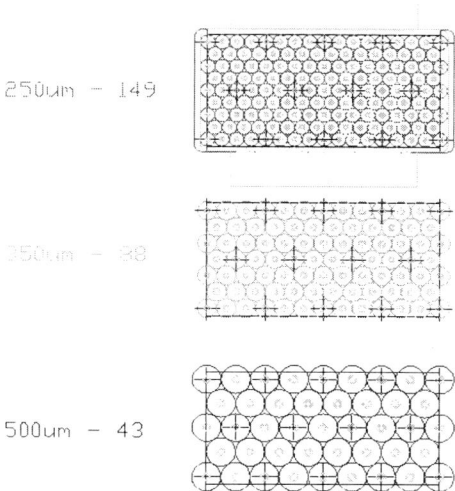

Figure 11. Array footprints for the three SPIRE detector arrays.

6.2. Large-format TES arrays

A number of groups (e.g. NIST, Berkeley and LLNL in USA, ISAS in Japan and SRON in Europe) are developing arrays of TES devices and multiplexed SQUID amplifiers (as described in Section 3). The use of silicon micro-machining and thin film depository techniques will soon make it feasible to construct large format arrays of many thousands of pixels. These arrays can be either feedhorn coupled or bare filled arrays, but once the pixel count increases above a few hundred mass production of feedhorns is likely to become too complex and expensive. Hence, in the future most of the large-format cameras will likely adopt a pixel architecture consisting of filled arrays of bare pixels.

SCUBA-2 (Robson, Holland, & Duncan 2000) is an example of such a new development for the JCMT. It will incorporate two arrays of 6,400 pixels covering a field-of-view of 8 × 8 arcmin. As in the current SCUBA design the new instrument will operate at two wavelengths simultaneously. The arrays will have a hybrid structure consisting of a micro-machined silicon wafer that forms a lattice structure for the pixels with the TES devices photo-lithographically deposited on the underside of the wafer (see Figure 12). A second silicon wafer is bump-bonded to the first and has first-stage SQUID amplifiers for multiplexing the array signals. The arrays will be cooled to 120 mK to give a per-pixel sensitivity limited by the sky and telescope background. With a much larger field-of-view and sky-background limited sensitivity, SCUBA-2 will be able to map large areas of sky up to 1000 times faster than the current SCUBA camera to the same noise level.

Table 2. Bolometer array projects for the coming decade. Notes: (1) Most arrays do not instantaneously sample the image plane - this has been taken into account in the mapping speed calculations, (2) Mapping speeds for both point source and extended region are calculated with respect to SCUBA-2 at $850\,\mu m$, (3) Mapping speed calculations for extended regions are based on the fraction of a $degree^2$ mapped in 1 hour to a 1-σ noise level of 10 mJy at $850\,\mu m$ (this is equal to $\approx$ 7 deg^2/hr for a theoretical complete SCUBA-2 operating at $850\,\mu m$). For completeness, a ν^3 dust-spectrum is assumed and noise levels are calculated at other wavelengths with respect to this (for both point source and extended regions).

Instrument Name	Year	Wavelength (μm)	Field of view ($arcmin^2$)	FWHM beamsize (arcsec)	Per-pixel NEFD ($mJy/Hz^{1/2}$)	Point source map speed	Extended region map speed
SCUBA (JCMT)	1996	450 850	4.2 4.5	7.5 14.5	500 80	0.21 0.08	0.0012 0.0018
BoloCAM (CSO)	2001	1400	50	30	35	0.02	0.01
MAMBO-II (IRAM)	2001	1250	6	8	50	0.08	0.003
MIPS (SIRTF)	2002	160	2.6	45	150	13	34
SHARC-II (CSO)	2003	350	2.7	10	500	0.3	0.03
HAWC (SOFIA)	2004	200	9.3	45	560	6.9	2.1
BoloCAM (LMT)	2004	1100	2.1	5.5	2	8.5	0.7
SCUBA-2 (JCMT)	2006	450 850	64 64	7.5 14.5	100 30	2.5 1.0	2.5 1.0
SPIRE (Herschel)	2007	250 350 500	16 16 16	18 25 36	34 35 41	490 62 10	293 37 6

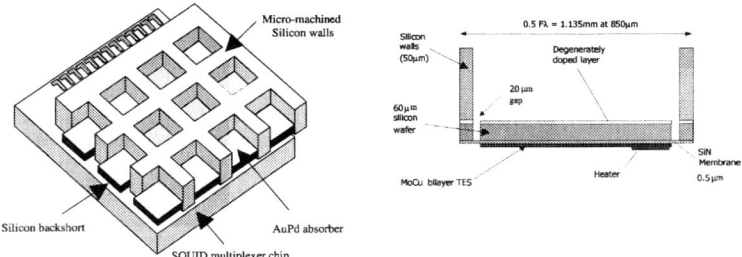

Figure 12. (left) A section of a SCUBA-2 array. The TES is on the underside of a silicon backshort with a AuPd absorber providing an impedance match to incoming radiation. The multiplexer chip underneath the TES/backshort is bump-bonded to the detector chip; (right) Cross-section of a single SCUBA-2 pixel. The silicon walls provide mechanical strength. The detector is thermally isolated on a Si_3N_4 membrane. The heater is used to compensate for the expected variations in background power due to changes in sky opacity.

6.3. Antenna coupled devices

Detector-feedhorn assemblies, such as the Planck HFI design (Griffin 2000), are very complex and expensive to build. Because of this they do not lend themselves to mass production for large-format arrays. Several groups are now investigating a much more streamlined approach of coupling the telescope beam to the detector element. Planar lithographed antennas, first developed for SIS mixers, are an example of such a development which is now being adapted for large arrays of bolometric detectors. The designs include microstrip patch antennas, planar antennas on photonic bandgap crystals and integrated lens antennas (van der Vorst 1999). The advantage of these approaches is that entire assemblies of detectors, antennas and filters could be fabricated onto the same chip, making array focal planes much easier to build and mass produce.

6.4. Array developments over the next decade

Table 2 summarizes the current status of known bolometer array development over the next 10 years for ground- and space-based platforms. This is not necessarily a definitive list!

7. Conclusions

The development of bolometric detectors for the FIR–mm waveband continues to gather pace. Bolometers are intrinsically very simple devices – they detect the total power across a wide range of wavelengths. Optimizing their performance for a given application is considerably more difficult, and needs careful consideration of a whole range of issues from minimizing extraneous background power that could degrade sensitivity, to the practicalities of the design of the readout electronics and the low-temperature refrigeration system. The "modest" array instruments currently in operation today have certainly given tantalizing glimpses of what is still to come.

Rapid developments in detector technology, the ability to build large single-dish telescopes, and construct complementary giant interferometer arrays is a testament to the fact that the FIR–mm wavelength regime is regarded as critical to the understanding of a wide-range of astrophysical phenomena. The ability to construct much larger format imaging arrays will bring about the same kind of revolution in FIR–mm astronomy that occurred at infrared wavelengths in the 1980s. They will not only provide a wide-field complementarity to the new generation of submm/mm interferometers (such as ALMA), but will allow the entire FIR–mm sky to be studied in great detail for the first time.

Acknowledgments. We would like to thank Kent Irwin, Jamie Bock and Tim Jenness for helpful comments and contributions to this paper.

References

Benford, D. J., et al. 2000, in ASP Conf. Ser. Vol. 217, Imaging at Radio Through Submillimeter Wavelengths, ed. J. G. Mangum & S. J. E. Radford (San Francisco: ASP), 134

de Bernardis, P., et al. 2000, Nature, 404, 955

Bock, J., et al. 1998, in Proc. SPIE, 3357, Advanced Technology MMW, Radio and Terahertz Telescopes, ed. T. G. Phillips (Bellingham, Wash., USA : SPIE), 297

Chervenak, J. A., et al. 1999, Appl. Phys. Lett., 74, 4043

Emerson, D. T. 1995, in ASP Conf. Ser. Vol. 75, Multi-feed Systems for Radio Telescopes, ed. D. T. Emerson & J. M. Payne (San Francisco: ASP), 309

Gear, W. K. & Griffin, M. G. 2000, in ASP Conf. Ser. Vol. 217, Imaging at Radio through Submillimeter Wavelengths, ed. J. G. Mangum & S. J. E. Radford (San Francisco: ASP), 126

Glenn, J., et al. 1998, in Proc. SPIE, 3357, Advanced Technology MMW, Radio and Terahertz Telescopes, ed. T. G. Phillips (Bellingham, Wash., USA : SPIE), 326

Grannan, S. M., et al. 1997, Int. J. Infrared & Millimeter Waves, 18, 319

Griffin, M. J. 2000, in AIP Conf. Proc. 605, 8th International Workshop on Low Temperature Detectors, eds. P. de Korte & T. Peacock (New York: Springer-Verlag), 397

Griffin, M. J. & Holland, W. S., 1988, Int. J. Infrared & Millimeter Waves, 17, 669

Haller, E. E. 1985, Infrared Phys., 25, 257

Hazell, A. S. 1991, Ph.D Thesis, Queen Mary & Westfield College, University of London

Holland, W. S. 1990, Ph.D Thesis, Queen Mary & Westfield College, University of London
Holland, W. S., et al. 1998a, Nature, 392, 788
Holland W. S., et al. 1998b, in Proc. SPIE, 3357, Advanced Technology MMW, Radio and Terahertz Telescopes, ed. T. G. Phillips (Bellingham, Wash., USA : SPIE), 305
Holland, W. S., et al. 1999, MNRAS, 303, 659
Hughes, D. H., et al. 1998, Nature, 394, 241
Irwin, K. D. 1995, Appl. Phys. Lett, 66, 1998
Israel, F. P., van der Werf, P. P., & Tilanus, R. P. J. 1999, A&A, 344, L83
Jenness, T., Lightfoot, J. F., & Holland, W. S. 1998, in Proc. SPIE, 3357, Advanced Technology MMW, Radio and Terahertz Telescopes, ed. T. G. Phillips (Bellingham, Wash., USA : SPIE), 548
Johnstone, D. & Bally, J. 1999, ApJ, 510, L49, 1999
Kreysa, E., et al. 1998, in Proc. SPIE, 3357, Advanced Technology MMW, Radio and Terahertz Telescopes, ed. T. G. Phillips (Bellingham, Wash., USA : SPIE), 319
Masi, S., et al. 1999, in AIP Conf. Proc. Vol. 476, 3K Cosmology, ed. L. Maiani, F. Melchiorri & N. Vittorio (Woodbury, N.Y.: American Institute of Physics), 237
Mather, J. 1982, Appl. Opt., 21, 1125
Mauskopf, P. D., et al. 2000, in ASP Conf. Ser. Vol. 217, Imaging at Radio Through Submillimeter Wavelengths, ed. J. G. Mangum & S. J. E. Radford (San Francisco: ASP), 115
Moseley, H., et al. 2000, in ASP Conf. Ser. Vol. 217, Imaging at Radio through Submillimeter Wavelengths, ed. J. G. Mangum & S. J. E. Radford (San Francisco: ASP), 140
Richards, P. L. 1994, J. Appl. Phys, 76, 1
Robson, E. I., Holland, W. S., & Duncan, W. D., 2000, in ASP Conf. Ser., Deep Millimeter Surveys: Implications for Galaxy Formation and Evolution, in press
Van der Vorst, M. J. M. 1999, Ph.D Thesis, University of Eindhoven, Holland
Wang, N., et al. 1996, Appl. Opt., 35, 6629

Carl Heiles investigating the polarization properties of the Arecibo Observatory pool during a lunch break.

Single-Dish Radio Telescopes of the World

C. J. Salter

National Astronomy and Ionosphere Center, Arecibo Observatory, HC 3 Box 53995, Arecibo, Puerto Rico 00612, USA

Abstract. A compendium of over fifty of the world's radio telescopes is presented. Tables are given containing the location, size, operational parameters, operating institution, public availability, and web address for each. It is hoped that this will help astronomers in selecting "instruments of choice" for their investigations. The reader is cautioned concerning completeness of the compendium, accuracy of some entries, the ephemeral nature of much of the information listed, and is encouraged to approach institutions for collaboration even if their telescopes are not marked here as regularly available to outside users.

1. Introduction

During the Single-Dish School, one session was dedicated to "The GBT, the Arecibo Telescope and Other Single-Dish Telescopes". For detailed information concerning the first two instruments discussed there, potential users of Arecibo and the GBT are referred to the full descriptions of these telescopes on the relevant web pages, (http://www.naic.edu and http://www.gb.nrao.edu). Expanded from the third presentation of that session, a compendium containing information on many of the single-dish telescopes of the world is presented here.

It is hoped that the tables appearing below will help prospective single-dish users to select the instruments which potentially fit their needs in terms of location, frequency coverage, angular resolution, sensitivity and backend availability. A web reference is given for each telescope so that, having selected possible instruments of interest, an astronomer can visit the relevant web pages to find out about them in detail, and how to apply for observing time. We stress the importance of visiting the appropriate web pages, as the lists here doubtless omit important details (or contain errors), while policies, procedures and instrumental provisions also change with time at every observatory, and many of the details in these tables will soon be out-dated.

The present compiler does not claim that this compendium is all-inclusive, but hopes that most of the world's highest profile single dishes are included. The omission of any telescope should be viewed only as the compiler's oversight or ignorance, and not any adjudication by him as to the importance of a particular observing facility. Another caveat is that telescopes used predominantly, or totally, as elements of VLBI or synthesis arrays have mostly not been included unless it is clear that they are also used regularly for single-dish radio astronomy *per se*.

A few telescopes that are presently (June 2002) under construction have been included in the tables, and are marked as such in the appropriate places. As witness a "lively discussion" at the school itself, the inclusion of ALMA was "a moot point" (in the British sense of the phrase). However, this new facility appears in the tables below on the principle that it is better to err on the side of inclusion, than be damned for omission!

2. The Tables

The information contained in the four tables presented below attempts to summarize the details of individual single-dish telescopes. Much of it has been drawn from the web pages of the individual host observatories. As such, there is no substitute for a full perusal of the web pages themselves by any interested reader.

Table 1 contains information on the geographical locations of the telescopes included in the compendium. The columns are as follows;

1. A "Tag" Name that will identify the telescope is given in the first column of each of the tables. This name was chosen to be as descriptive of the instrument as possible, often being the officially adopted name of the telescope (e.g. Arecibo, GBT, LMT/GTM, RATAN-600).

2. An approximate description of the location.

3. The telescope longitude, with eastern longitudes positive.

4. The telescope latitude.

Table 2 lists a few of the more useful parameters for each telescope. The columns are as follows;

1. The "Tag" Name.

2. The physical size of the telescope in meters. A single diameter is given for circular dishes, while two dimensions are listed for elliptical or rectangular telescopes. Note that the resolution (HPBW) and sensitivity of an instrument can be roughly estimated from its size, D (in m), via HPBW$\sim 1.2\lambda/D$ radians, and gain$\sim 2.8 \times 10^{-4}\eta_A D^2$ K/Jy, where η_A is the aperture efficiency.

3. The frequency range over which receivers are available. Note that this does not imply continuous coverage. To find typical system temperatures, the interested astronomer should consult the appropriate web pages.

4. The rough sky coverage of the instrument. In many cases this was estimated from the geographic latitude of the observatory, and the elevation drive limits of the telescope. For some sites that do not list the drive limits on their web pages, coverage of $0° <$ elevation $< 90°$ has been assumed, though this may well be in error.

Table 3 lists a few of the operational details for each telescope. The columns are as follows;

1. The "Tag" Name.

2. The operating organization.

3. If the web page of relevance clearly invites proposals for telescope time from "outside users", then 'Yes' is entered in this column. The interested user should consult the web pages for themselves as there may be additional requirements for some facilities, such as the finding of an "inside collaborator". However, it is cautioned that most institutions will entertain exciting proposals to use their facilities productively, and the lack of an entry indicating a regular proposal system should not be taken to mean that you should not approach the appropriate management concerning use of their telescope, if it is your instrument of choice. Appropriate contact information is usually provided on the relevant web page (see Table 4).

4. A comment entry listing the observing disciplines supported (often as gleaned from web pages). The entries here may not be complete, or necessarily totally accurate. The abbreviations are; C = Continuum observations possible, SL = Spectral Line observations (i.e. a spectrometer is available), P = Pulsar research has been carried out with the instrument, PR = Planetary Radar work is done (see Arecibo and the DSN Goldstone antenna), and IPS = Inter-Planetary Scintillation measurements. Other comments, including completion dates for telescopes under construction or upgrade, should be self-explanatory.

Table 4, perhaps the most important to the prospective user, gives the URL for a web page to which an interested party can proceed in order to learn more about a given telescope, its equipment, its operations and observatory policies. The two columns are as follows;

1. The "Tag" Name.

2. The appropriate URL.

Table 1. Telescope Location

Tag Name	Location	Long (°) (′)	Lat (°) (′)
ALMA	Llano de Chajnantor, Chile	−67 45.3	−23 01.4
AST/RO	South Pole	−45 53.0	−89 59.7
Arecibo	Arecibo, Puerto Rico	−66 45.2	+18 20.6
Bonn 100-m	Effelsberg, Germany	+6 53.0	+50 31.5
Caltech 40-m	Owens Valley, CA, USA	−118 16.9	+37 14.0
Caltech 5.5-m	Owens Valley, CA, USA	−118 16.9	+37 14.0
Ceduna 30-m	Ceduna, S. Aus., Australia	+133 48.6	−31 52.2
CSO	Mauna Kea, HI, USA	−155 28.5	+19 49.3
DRAO	Penticton, Canada	−119 37.2	+49 19.2
DSN-C	Tidbinbilla, Australia	+148 58.9	−35 24.1
DSN-G	Goldstone, CA, USA	−116 53.4	+35 25.6
DSN-M	near Madrid, Spain	−4 14.9	+40 25.8
Dwingeloo	Dwingeloo, Netherlands	+6 23.8	+52 48.8
FCRAO 14-m	Quabbin, MA, USA	−72 20.7	+42 23.5
GBT	Green Bank, WV, USA	−79 50.4	+38 26.0
Hartebeesthoek	near Krugersdorp, S. Africa	+27 41.1	−25 53.2
Haystack 37-m	Westford, MA, USA	−71 29.3	+42 37.4
Itapetinga	Itapetinga, Atibaia, SP, Brazil	−46 33.8	−23 11.0
IRAM 30-m	Pico Veleta, Spain	−3 23.5	+37 04.0
JCMT	Mauna Kea, HI, USA	−155 28.2	+19 49.3
Jodrell Mk-II	Jodrell Bank, UK	−2 18.2	+53 14.1
Kitt Peak 12-m	Kitt Peak, AZ, USA	−111 36.9	+31 57.2
KOSMA	Gornergrat, Switzerland	+7 47.5	+45 59.0
LMT/GTM	Sierra Negra, Puebla, Mexico	−97 18.8	+18 59.1
Lovell	Jodrell Bank, UK	−2 18.5	+53 14.2
Medicina 32-m	Medicina, Italy	+11 38.7	+44 31.2
Metsähovi	Metsähovi, Kylmälä, Finland	+24 23.6	+60 13.1
Mopra 22-m	near Siding Spring, Australia	+149 06.0	−31 16.1
Nancay	Nancay, France	+2 11.8	+47 22.8
Nobeyama 45-m	Nobeyama, Japan	+138 28.5	+35 56.5
Noto 32-m	Noto, Sicily, Italy	+14 59.3	+36 52.6

Table 1. Telescope Location (continued)

Tag Name	Location	Long (°) (′)	Lat (°) (′)
Onsala 20-m	Onsala, Sweden	+11 55.6	+57 23.7
Onsala 25-m	Onsala, Sweden	+11 55.0	+57 23.6
Ooty	Ootacamund, India	+76 40.0	+11 22.9
Parkes 64-m	Parkes, Australia	+148 15.7	−33 00.0
Pisgah 26-m's	near Rosman, NC, USA	−82 52.3	+35 12.0
		−82 52.5	+35 11.9
Pisgah 12.2-m	near Rosman, NC, USA	−82 52.6	+35 11.7
Purple Mtn.	Delinha, Qinhai, China	+97 44.0	+37 22.0
Puschino RT-22	Puschino, Russia	+37 55.0	+56 00.0
RATAN-600	Zelenchukskaya, Russia	+41 35.5	+43 49.9
RRI 10.4-m	Bangalore, India	+77 34.8	+13 00.6
Simeiz RT-22	Simeiz, near Yalta, Ukraine	+34 01.0	+44 32.1
SEST	La Silla, Chile	−70 44.1	−29 15.6
SMTO/HHT	Mt. Graham, AZ, USA	−109 53.5	+32 42.1
SRT	Pranu Sanguni, Sardinia, Italy	+9 14.7	+39 29.8
Taeduk 14-m	Taeduk, Taejon City, S. Korea	+127 24	+36 24
Tasmania 26-m	Mt. Pleasant, Tasmania, Australia	+147 26.3	−42 48.3
Tasmania 14-m	Mt. Pleasant, Tasmania, Australia	+147 26.3	−42 48.3
Torun 32-m	Piwnice, Poland	+18 33.8	+53 05.7
UMRAO 26-m	near Dexter, MI, USA	−83 56.2	+42 23.9
Urumqi 25-m	Nan Mtn, Urumqi, Xingjiang, China	+81 10.7	+43 28.3
Ventspils 32-m	Ventspils, Latvia	+21 51.3	+57 33.2
Villa Elisa	Villa Elisa, Argentina	−58 08.2	−34 52.1
Yebes	Yebes, Guadalajara, Spain	−3 05.4	+40 31.4

Table 2. Telescope Parameters

Tag Name	Size	Present Freq. Range	Sky Coverage
ALMA	64 × 12-m	31 – 950 GHz	−90 < Dec < +53
AST/RO	1.7-m	230 – 809 GHz	−90 < Dec < 0
Arecibo	305-m	47 – 6000 MHz	−01 < Dec < +37, ZA < 19.7
Bonn 100-m	100-m	408 MHz – 86 GHz	−30 < Dec < +90
Caltech 40-m	40-m	10 – 24 GHz	−48 < Dec < +90
Caltech 5.5-m	5.5-m	96 GHz	−50 < Dec < +90
Ceduna 30-m	30-m	2.2 – 23 GHz	−90 < Dec < +48
CSO	10.4-m	200 – 950 GHz	−70 < Dec < +90
DRAO	25.6-m	408 MHz – 8.4 GHz	−34 < Dec < +90
DSN-C	70-m	1.6 – 26 GHz	−90 < Dec < +49
DSN-G	70-m	1.6 – 26 GHz	−49 < Dec < +90
DSN-M	70-m	1.6 – 26 GHz	−45 < Dec < +90
Dwingeloo	25-m	1.38 – 1.72 GHz	−37 < Dec < +90
FCRAO 14-m	14-m	85 – 115 GHz	−32 < Dec < +90
GBT	100-m × 110-m	290 MHz – 52 GHz	−46 < Dec < +90
Hartebeesthoek	26-m	1.6 - 12.2 GHz	−90 < Dec < +45
Haystack 37-m	37-m	22 – 115 GHz	−42 < Dec < +90
Itapetinga	13.7-m	22 – 90 GHz	−90 < Dec < +62
IRAM 30-m	30-m	80 – 280 GHz	−44 < Dec < +90
JCMT	15-m	215 – 880 GHz	−50 < Dec < +90
Jodrell Mk-II	37-m × 25-m	150 MHz – 24 GHz	−36 < Dec < +90
Kitt Peak 12-m	12-m	68 – 300 GHz	−43 < Dec < +90
KOSMA	3-m	210 – 820 GHz	−45 < Dec < +90
LMT/GTM	50-m	75 – 345 GHz	–
Lovell	76.2-m	150 MHz – 5 GHz	−34 < Dec < +90
Medicina 32-m	32-m	1.35 – 45 GHz	−40 < Dec < +90
Metsähovi	14-m	5 – 120 GHz	−29 < Dec < +90
Mopra 22-m	22-m	1 – 115 GHz	−90 < Dec < +47
Nancay	200-m × 35-m	1.1 – 3.5 GHz	−39 < Dec < +90
Nobeyama 45-m	45-m	20 – 230 GHz	−42 < Dec < +90
Noto 32-m	32-m	1.4 – 22.5 GHz	−49 < Dec < +90

Table 2. Telescope Parameters (continued)

Tag Name	Size	Present Freq. Range	Sky Coverage
Onsala 20-m	20-m	21 – 116 GHz	$-30 <$ Dec $< +90$
Onsala 25-m	25.6-m	?	$-30 <$ Dec $< +90$
Ooty	530-m × 30-m	326.5 ± 7.5 MHz	$-60 <$ Dec $< +60$
			$-4 <$ HA $< +5.5$
Parkes 64-m	64-m	440 MHz – 24 GHz	$-90 <$ Dec $< +27$
Pisgah 26-m's	Two × 26-m	340 MHz – 12 GHz	$-55 <$ Dec $< +90$
Pisgah 12.2-m	12.2-m	3.3 – 12.75 GHz	$-55 <$ Dec $< +90$
Purple Mtn.	13.7-m	85 – 115 GHz	$-43 <$ Dec $< +90$
Puschino RT-22	22-m	To Mm-wavelengths	$-29 <$ Dec $< +90$
RATAN-600	576-m circle	935 MHz – 30 GHz	$-42 <$ Dec $< +90$
RRI 10.4-m	10.4-m	6 – 115 GHz	$-77 <$ Dec $< +90$
Simeiz RT-22	22-m	327 MHz – 116 GHz	$-42 <$ Dec $< +90$
SEST	15-m	78 – 363 GHz	$-90 <$ Dec $< +45$
SMTO/HHT	10-m	230 – 860 GHz	$-55 <$ Dec $< +90$
SRT	64-m	300 MHz – 100 GHz	$-50 <$ Dec $< +90$
Taeduk 14-m	13.7-m	40 – 150 GHz	$-54 <$ Dec $< +90$
Tasmania 26-m	26-m	0.66 – 22 GHz	$-90 <$ Dec $< +31$
Tasmania 14-m	14-m	630 – 1400 MHz	?
Torun 32-m	32-m	1.4 – 6.8 GHz	$-35 <$ Dec $< +90$
UMRAO 26-m	26-m	4.8 – 14.5 GHz	$-40 <$ Dec $< +90$
Urumqi 25-m	25-m	327 MHz – 23 GHz	$-42 <$ Dec $< +90$
Ventspils 32-m	32-m	327 MHz – 12.2 GHz	$-33 <$ Dec $< +90$
Villa Elisa	Two × 30-m	1.4 – 3.3 GHz	$-90 <$ Dec < -09
Yebes	13.7-m	2.3 – 49 GHz	$-46 <$ Dec $< +90$

Table 3. Telescope Operations

Tag Name	Operating Inst.	Publ Access	Comments
ALMA	NRAO	Yes	C, SL: Projected for 2006-11
AST/RO	CARA	Yes	SL
Arecibo	NAIC/Cornell	Yes	C, SL, P, PR: Fixed reflector
Bonn 100-m	MPIfR	Yes	C, SL, P
Caltech 40-m	Caltech	–	C
Caltech 5.5-m	Caltech	–	C (bolometer)
Ceduna 30-m	UTasmania	–	C
CSO	Caltech	Yes	C (bolometer array), SL
DRAO	DRAO	Yes?	C, SL
DSN-C	NASA	Yes	C, SL, P
DSN-G	NASA	Yes	C, SL, P, PR
DSN-M	NASA	Yes	C, SL, P
Dwingeloo	NFRA	Yes?	SL
FCRAO 14-m	FCRAO	Yes	C, SL: λ3-cm multibeam
GBT	NRAO	Yes	C, SL, P Unblocked aperture
Hartebeesthoek	NRF	Yes	C, SL, P
Haystack 37-m	NEROC	Yes	C, SL: Education projects
Itapetinga	CRAAM	–	C, SL
IRAM 30-m	IRAM	Yes	C (bolometers), SL
JCMT	JAC	Yes	C (bolometer array), SL
Jodrell Mk-II	U. Manchester	–	Mostly dedicated to MERLIN
Kitt Peak 12-m	UASO	Yes	C, SL: Subject to funding
KOSMA	U. zu Köln	Yes	C, SL
LMT/GTM	UMass/INAOE	–	To be completed ∼2004
Lovell	U. Manchester	Yes	C, SL, P: Upgrade completed 2002
Medicina 32-m	IRA/CNR	–	C, SL
Metsähovi	Helsinki UT	–	C, SL
Mopra 22-m	UNSW/ATNF	Yes	C, SL
Nancay	Paris Obs/CNRS	Yes	C, SL, P: Upgraded 2000
Nobeyama 45-m	NRO	Yes	C, SL
Noto 32-m	IRA/CNR	–	C, SL

Table 3. Telescope Operations (continued)

Tag Name	Operating Inst.	Publ Access	Comments
Onsala 20-m	Chalmers UT	Yes	C, SL
Onsala 25-m	Chalmers UT	Yes ?	C, SL
Ooty	TIFR/NCRA	Yes	C, SL, P, IPS: Unblocked aperture
Parkes 64-m	ATNF/CSIRO	Yes	C, SL, P: L-Band multibeam
Pisgah 26-m's	PARI	Yes	C, SL, P
Pisgah 12.2 m	PARI	Yes	C, SL
Purple Mtn.	PMO/CAS	–	C, SL
Puschino RT-22	Lebedev/ASC	–	C, SL
RATAN-600	SAO	Yes	C, SL
RRI 10.4-m	RRI	–	C, SL
Simeiz RT-22	CrAO	–	C, SL
SEST	Chalmers UT/ESO	Yes	C (bolometers), SL
SMTO/HHT	UASO	Yes	C (bolometers), SL
SRT	IRA/CNR	–	C, SL, P: Under construction
Taeduk 14-m	TRAO/KAO	Yes	C, SL
Tasmania 26-m	UTasmania	Yes	C, SL
Tasmania 14-m	UTasmania	–	P: Vela PSR monitoring
Torun 32-m	N. Copernicus U	–	C, SL, P
UMRAO 26-m	UMRAO	–	C (Full Stokes)
Urumqi 25-m	UAS/CAS	–	C, SL, P
Ventspils 32-m	VIRAC/LAS	–	C
Villa Elisa	IAR	Yes	C (Dish 1), SL (Dish 2)
Yebes	OAN	–	SL

Note: In "Comments", the available types of observing are listed, where; C = Continuum, SL = Spectral Line, P = Pulsar, PR = Planetary Radar, IPS = Interplanetary Scintillation.

Table 4. Web Page

Tag Name	WWW Page
ALMA	www.alma.nrao.edu
AST/RO	cfa-www.harvard.edu/~adair/AST_RO
Arecibo	www.naic.edu
Bonn	www.mpifr-bonn.mpg.de/index_e.html
Caltech 40-m	www.ovro.caltech.edu
Caltech 5.5-m	www.ovro.caltech.edu
Ceduna 30-m	www-ra.phys.utas.edu.au/observatories/ceduna.html
CSO	www.submm.caltech.edu/cso
DRAO	www.drao.nrc.ca/facilities/telescopes/26m
DSN-C	dsnra.jpl.nasa.gov
DSN-G	dsnra.jpl.nasa.gov
DSN-M	dsnra.jpl.nasa.gov
Dwingeloo	www.astron.nl/astron/dwingeloo25m.htm
FCRAO 14-m	www.astro.umass.edu/~fcrao
GBT	www.gb.nrao.edu/GBT/GBT.html
Hartebeesthoek	www.hartrao.ac.za
Haystack 37-m	fourier.haystack.edu/37m
Itapetinga	www.craam.mackenzie.br/roi2.htm
IRAM 30-m	www.iram.es
JCMT	www.jach.hawaii.edu/JACpublic/JCMT
Jodrell Mk-II	www.merlin.ac.uk
Kitt Peak 12-m	kp12m.as.arizona.edu/index.html
KOSMA	www.ph1.uni-koeln.de/kosma.html
LMT/GTM	www-lmt.phast.umass.edu
Lovell	www.jb.man.ac.uk/tech/lovell
Medicina 32-m	medvlbi.ira.bo.cnr.it
Metsähovi	kurp-www.hut.fi
Mopra 22-m	newt.phys.unsw.edu.au/~ramesh (UNSW)
	www.narrabri.atnf.csiro.au/mopra (ATNF)
Nancay	www.obs-nancay.fr/html_an/a_rt.htm
Nobeyama 45-m	www.nro.nao.ac.jp/ nro45mrt/index-e.html
Noto 32-m	www.ira.noto.cnr.it

Table 4. Web Page (continued)

Tag Name	WWW Page
Onsala 20-m	www.oso.chalmers.se/20m/index.html
Onsala 25-m	www.oso.chalmers.se/25m/index.html
Ooty	www.ncra.tifr.res.in/ncra_hpage/ort/ort.html
Parkes 64-m	www.parkes.atnf.csiro.au
Pisgah 26-m's	www.pari.edu
Pisgah 12.2-m	www.pari.edu
Purple Mtn.	www.pmo.ac.cn/pmo-eng.htm
Puschino RT-22	www.prao.psn.ru/english/3.html
RATAN-600	brown.nord.nw.ru
RRI 10.4-m	www.rri.res.in/htmls/aa/tenmt_tel.html
Simeiz RT-22	giub.geod.uni-bonn.de/vlbi/stations/rt22.html
SEST	www.ls.eso.org/lasilla/Telescopes/SEST
SMTO/HHT	maisel.as.arizona.edu:8080/smt.html
SRT	www.ca.astro.it/srt/benvenuto.htm
Taeduk 14-m	www.trao.re.kr/trao/index_en.html
Tasmania 26-m	www-ra.phys.utas.edu.au/observatories/26m-intro.html
Tasmania 14-m	www-ra.phys.utas.edu.au/observatories/14m-intro.html
Torun 32-m	www.astro.uni.torun.pl
UMRAO 26-m	www.astro.lsa.umich.edu:80/obs/radiotel/radiotel.html
Urumqi 25-m	From www.evlbi.org/evn.html
Ventspils 32-m	www.astr.lu.lv/virac/virac.htm
Villa Elisa	www.iar.unlp.edu.ar/ES/iar-es.htm (Spanish)
Yebes	www.oan.es/cay/14m/antena

The audience managed mostly to stay awake...

Part 5
Poster Papers

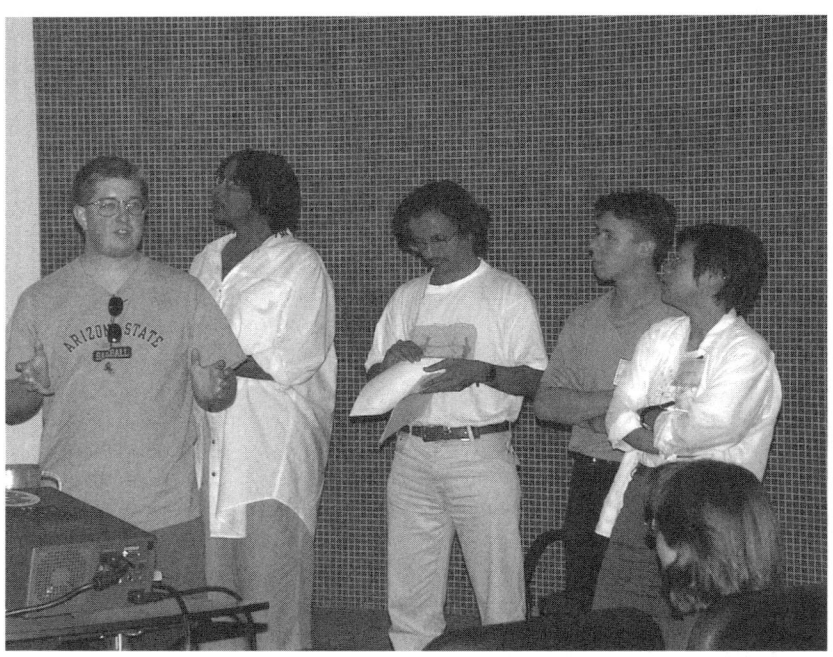

Presentation of hands-on projects at the end of the School. From left to right: Kevin Healy, Rolaine Young Owl, Ashish Asgekar, Esteban Araya and Shih-Ping Lai.

First *VLBI* Observations with Arecibo in an International S2 Ad-hoc Array

Igor Molotov, Andrey Chuprikov, Sergey Likhachev

Astro Space Center, P. N. Lebedev Physical Institute, Profsoyuznaya 84/32, Moscow 117810, Russia

Chris Salter, Tapasi Ghosh

National Astronomy and Ionosphere Center, Arecibo Observatory, HC 3 Box 53995, Arecibo, Puerto Rico 00612, USA

Frank Ghigo

National Radio Astronomy Observatory, P.O. Box 2, Green Bank, West Virginia 24944, USA

Sean Dougherty

Dominion Radio Astrophysical Observatory, Herzberg Institute of Astrophysics, NRC, PO Box 248, Penticton, British Columbia V2A 6K3, Canada

Abstract. The international ad-hoc *S2* array that is being developed as a part of the *LFVN* project is described. The results of the first *S2* ground-based *VLBI* observations with Arecibo are presented. The possibility of future cooperation is discussed.

1. International Ad-hoc *S2* Array

Of all astronomical measurements, Very Long Baseline Interferometry (*VLBI*) achieves the highest angular resolution, equivalent to that which could be obtained by a hypothetical single telescope of continental or intercontinental size. The loan of an *S2 VLBI* recording terminal to NAIC by the Space Geodynamics Laboratory at CRESTech in Toronto, Canada for support of the *VSOP* mission opened the possibility of Arecibo participating in ground-based *VLBI* observations using an ad-hoc international *S2* array. Such an array has begun to operate under the Low Frequency *VLBI* Network (*LFVN*) project (Molotov et al. 1999) which arranges *VLBI* experiments using telescopes equipped with *S2* recorders in 11 countries. It is based on the informal collaboration of two Russian antennas, Bear Lakes *RT-64* and Puschino *RT-22* (both near Moscow), with the Noto *RT-32* (Italy), Shanghai *RT-25* (China), and the *DRAO S2* correlator in Canada. The ad-hoc *S2* array has made two observing runs per year at $\lambda 18$ cm (a survey of compact extragalactic sources, monitoring of OH-masers in evolved stars, studies of turbulent structure in the solar wind, and *SETI*) and at $\lambda 6$

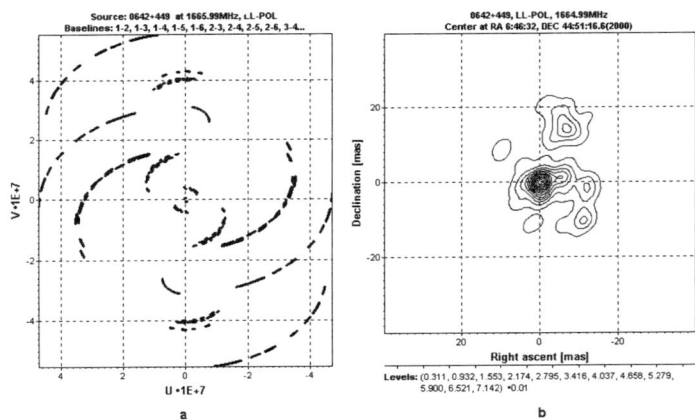

Figure 1. *INTAS99.4*: (u,v)-plane coverage (a), and an image (b), of quasar *0642+449*. The Bear Lakes, Puschino, Noto, Shanghai, Hartebeesthoek and Svetloe antennas were used.

cm (*VLBI* radar research of Earth-group planets, asteroids crossing the Earth's orbit and Space Debris using the planetary-radar equipment at the Evpatoria *RT-70*, Crimea, Ukraine). Seven observing sessions have been organized over the past 3 yr: *INTAS98.2* and *INTAS98.5* in 1998, *VLBR99.1* and *INTAS99.4* in 1999, *VLBR00.2* and *INTAS00.3* in 2000, and *VLBR01.1* in 2001. Figure 1 shows results for the quasar *0642+449* obtained during *INTAS99.4*. A wide range of baselines up to the largest possible on Earth may be achieved with an ad-hoc *S2* array, while the participation of Arecibo ensures high sensitivity.

2. First *S2* Array Observations Including Arecibo

The first *LFVN S2* observations to include Arecibo were made at λ18-cm on 2 Dec 98 as part of *INTAS98.5*. This experiment involved an array of 3 Russian antennas (Bear Lakes, Puschino and Svetloe), the Green Bank *RT-43* and the Hartebeesthoek *RT-26*, and included 7 hr of observation of weak sources using the Arecibo-Green Bank baseline. The data were successfully correlated with the *DRAO S2* correlator.

Figure 2a shows an example of Arecibo-Green Bank fringes. Data reduction was completed at the Astro Space Center, Moscow, Russia, using the *ASL for Windows* imaging software (Likhachev et al. 2000). These test observations had several scientific goals, including a Radioastron Space-*VLBI* pre-launch survey, *SETI*, OH-masers and pulsars. Figure 2b shows the image of quasar *1156+295* produced from these observations. Another interesting result was obtained for

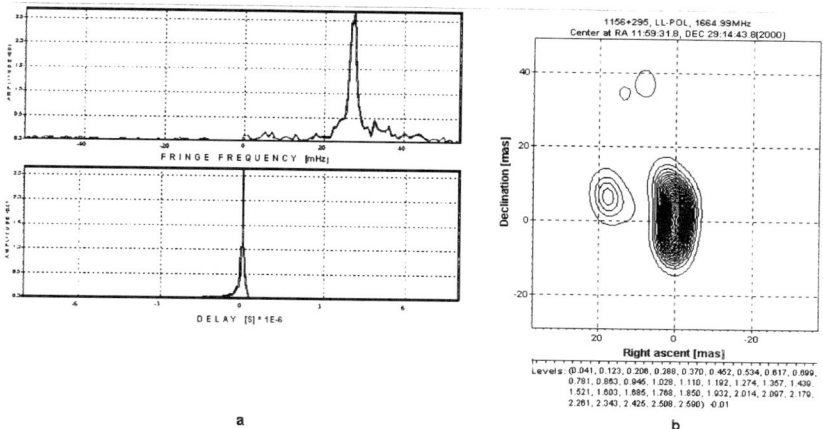

Figure 2. *INTAS98.5*: fringes for the Green Bank-Arecibo baseline (a), and an image (b), for quasar *1156+295*.

SETI source *37 Gem*, which seems to be resolved. As a result, *37 Gem* has been included into the list of objects for the next Cosmic Call transmission in Sept 2001.

3. Future Plans

The excellent results that we have obtained encourage us to plan further collaboration with Arecibo. We are pursuing observations with other radio telescopes having mutual visibility with Arecibo; for example, such an array could include the Green Bank *RT-100*, the *DRAO RT-26* and the Madrid *RT-70* (where the *S2* system will be installed at the end of 2001). The recent arrival of a λ92-cm receiver at Arecibo also opens new opportunities for *LFVN* cooperation, especially as there is a plan to also supply the Algonquin Park *RT-46* and *DRAO RT-26* in Canada with λ92-cm capability. Another field of interest for cooperation could be the study of Solar system bodies with the *VLBI* radar method, where the echo-signals are received in differential *VLBI* mode using the *S2* system to record both the transmitted and reflected signals. *LFVN* carries out investigations directed at the measurement of short-periodic variations of the Earth-group planets proper rotation and the precise definition of near-Earth asteroid orbits in the Radio Reference Frame.

References

Likhachev, S., Guirin, L., Molotov, I., & Chuprikov, A. 2000, in Proc. of the 5th European *VLBI* Network Symposium, eds. J. E. Conway, A. G. Polatidis, R. S. Booth & Y. Pihlstrom (Onsala Space Observatory), 251

Molotov, I. E. et al. 1999, to appear in IAU Symp., The Universe at Low Radio Frequencies, eds. G. Swarup & A. P. Rao

Exploring the Sensitivity Limit: 21-cm Observations of Low Surface Brightness Galaxies with the Arecibo Telescope

K. O'Neil, J. Eder

National Astronomy and Ionosphere Center, Arecibo Observatory, HC 3 Box 53995, Arecibo, Puerto Rico 00612, USA

G. Bothun, J. Schombert

University of Oregon, Physics Department, Eugene, Oregon 97403, USA

Abstract. The wide bandwidth, high sensitivity, and flat baselines of the Arecibo Gregorian telescope make it an excellent instrument to probe the HI content of LSB systems. In this paper we briefly describe a few of the results to come out of Arecibo observations of these enigmatic systems.

1. Introduction

Despite the difficulty in detecting low surface brightness (LSB) galaxies, it has been well established in recent years that these diffuse systems play a significant role in both defining the baryon content of the Universe and in our understanding of the distribution of local galaxy types. Additionally, LSB galaxies afford a unique view into star formation and evolution on galactic scales.

The wide bandwidth, high sensitivity, and flat baselines of the Arecibo Gregorian telescope makes it an excellent instrument to probe the HI content of LSB systems. Obtaining our source list from a variety of (optically determined) galaxy catalogs, ranging from the UGC through new LSB catalogs based on both the Digital Palomar Sky Survey (DPOSS) and other galaxy searches, we have spent the last three years determining the HI properties of LSB systems. Our studies have greatly expanded our knowledge of these diffuse systems, and include - the detection of possibly the highest M_{HI}/L_B galaxies known; discovery of red, gas rich LSB galaxies; extending the known local galaxy surface brightness distribution function by one magnitude; and the discovery a large number of massive LSB galaxies, including a significant number of galaxies with LSB disks surrounding active galactic nuclei.

2. High M_{HI}/L_B Galaxies

LSB galaxies are often thought of as gas rich, unevolved systems. With that in mind, it is perhaps not surprising that LSB galaxies appear to be some of the most gas rich galaxies known. What may be more surprising, though, is that

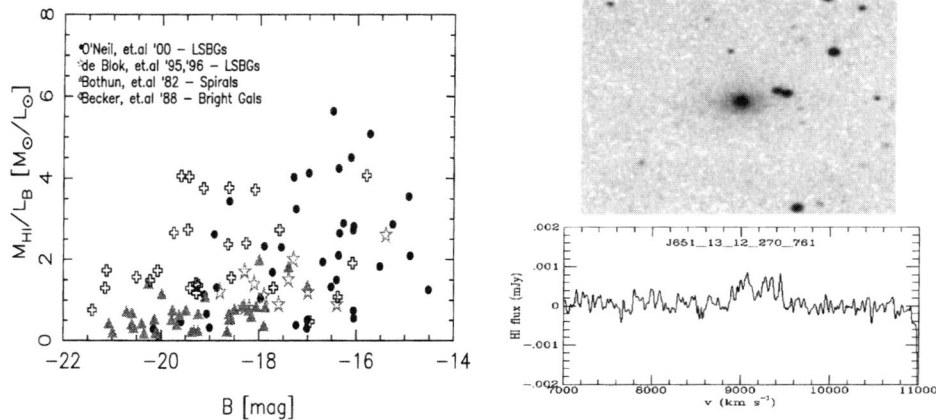

Figure 1. (left) Absolute B magnitude versus HI Mass-to-Luminosity ratio for a variety of both LSB and HSB Galaxies (from O'Neil, Bothun, & Schombert 2000). (right) DPOSS and HI profile of a 'typical' LSB AGN galaxy.

there is no clear correlation between the galaxies' central surface brightness and their M_{HI}/L_B ratio. Additionally, no correlation is seen between the galaxies' gas mass ratio and their colors, with the surveys discovering a number of fairly gas rich LSB galaxies with $(B-V) > 1.0$. This supports the idea of LSB galaxies having a low overall star formation rate, punctuated by localized bursts of star formation.

3. LSB AGN Galaxies

A preliminary search was undertaken last winter using eight plates from DPOSS that contain some fraction of 2MASS coverage (about 3 deg^2). Forty candidates were produced, of which ten were searched with the Arecibo Gregorian system. Eight of the ten candidates were detected at 21-cm, one of which is shown in Figure 1 (right). All detected galaxies have the characteristic AGN nucleus surrounded by an LSB disk. All eight also have HI widths in excess of 350 km s^{-1}. Assuming these galaxies follow the baryonic Tully-Fisher relation (McGaugh et al. 2000), their masses will exceed 10^{12} M$_\odot$.

4. Local Galaxy Surface Brightness Distribution

To obtain a surface brightness distribution through $\mu_B(0)$=25.0 mag/arcsec^2, a bi-variate volume correction was applied to the data in the O'Neil, Bothun, & Schombert (2000) catalog of LSB galaxies, velocities, etc. This extension by one magnitude of the surface brightness distribution function still shows a flat (slope=0) function from the Freeman value of $\mu_B(0)$=21.65 mag/arcsec^2 through the survey limits at $\mu_B(0)$=25.0 mag/arcsec^2, see Figure 2. This indicates that

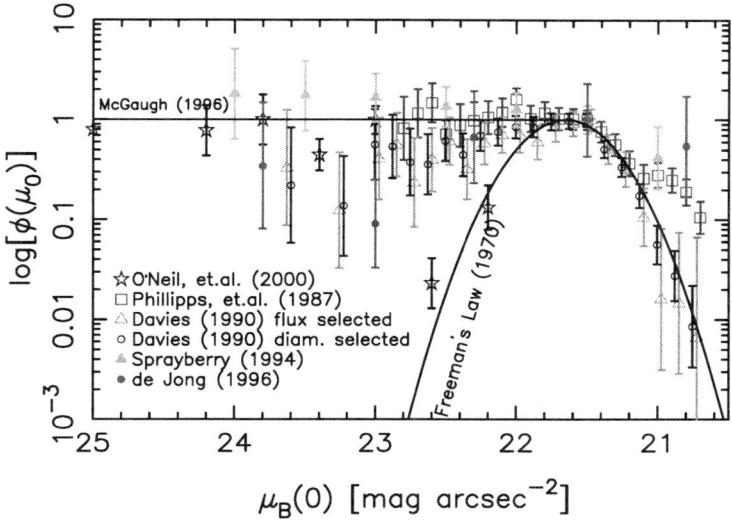

Figure 2. The volume corrected surface brightness distribution as given in O'Neil & Bothun (2000).

a significant percentage of the galaxy number density, and possibly the baryonic density, of the local Universe lies in LSB galaxies.

References

Becker, R., Mebold, U., Reif, K., & van Woerden, H. 1988, A&A, 203, 21
Bothun, G., Sullivan, W. T. III, & Schommer, R. A. 1982, AJ, 87, 725
Davies, J. I. 1990, MNRAS, 244, 8
de Blok, W. J. G., van der Hulst, J. M., & Bothun, G. D. 1995, MNRAS, 274, 235
de Blok, E., van der Hulst, T., & McGaugh, S. 1996, BAAS, 189, 8402
de Jong, R. 1996, A&A, 313, 46
Freeman, K. 1970, ApJ, 160, 811
McGaugh, S. S. 1996, MNRAS, 280, 337
McGaugh, S. S., Schombert, J. M., Bothun, G. D., & de Blok, W. J. G. 2000, ApJ, 533L, 99
O'Neil, K. & Bothun, G. 2000, ApJ, 529, 811
O'Neil, K., Bothun, G., & Schombert, J. 2000, AJ, 119, 136
Phillipps, S., Disney, M. J., Kibblewhite, E. J., Cawson, M. G. M. 1987, MNRAS, 229, 505
Sprayberry, D. 1994, PhD thesis, University of Arizona, Tucson

Presentation of hands-on projects at the end of the School. From left to right: Kristin Kearns, Omar Lopez Cruz, Carmen Pantoja and Carlos Vargas.

Single-Dish Radio Astronomy: Techniques and Applications
ASP Conference Series, Vol. 278, 2002
S. Stanimirović, D. R. Altschuler, P. F. Goldsmith, and C. J. Salter

Simulating the Performance of Large Format Sub-mm Focal-plane Arrays

Edward Chapin, David H. Hughes

Instituto Nacional de Astrofísica, Óptica y Electrónica, Apartado Postal 51 y 216, 72000, Puebla, Pue. México

Abstract. A robust measurement of the clustering amplitude of the sub-mm population of starburst galaxies requires large-area surveys ($\gg 1$ deg^2). Since even the largest-format arrays subtend only 10 arcmin^2 on the sky then scan-mapping is a necessary observing mode. Providing realistic representations of the extragalactic sky and atmosphere, as the input to a detailed simulator of the telescope and instrument performance, allows important decisions to be made about the design of large-area fully-sampled surveys and observing strategies. Our simulations include detector noise, time-constants and array geometry, telescope pointing errors, scan speeds and scanning angles, sky noise and sky rotation.

1. Generating Realistic Synthetic Time-Series

A mapping simulator has been developed to generate synthetic bolometer time series data and associated astrometric information which are then run through a realistic reduction pipeline (*http://www.inaoep.mx/~echapin/scansim.html*). This process allows one to develop and test the observing strategy and analysis software in advance of instrument delivery. Furthermore it is possible to assess the impact of instrumental design aspects on the science objectives.

The simulated sky, typically generated by convolving a synthetic catalogue of extra-galactic sources with the telescope beam (Hughes & Gaztañaga 2000), is given in equatorial coordinates on the celestial sphere. The ground-based telescope is then assigned an arbitrary location and local-time. The geometry of the array is defined and the beam positions on the sky are determined relative to the telescope bore-sight. The beams are scanned across the simulated sky model (in some pattern on an Alt-Az coordinate system) and, using a rigorous calculation of the astrometry, the positional information and *noise-less* flux density time-series is determined for each bolometer in the array.

More realistic time-series are produced by incorporating features that are characteristic of real instruments, the pointing performance of the telescope, and the non-negligible contribution of the atmosphere (sky noise and attenuation).

Sources of instrumental noise in the time-series include: 1) addition of an uncorrelated Gaussian system noise component; 2) multiplication by gain factors (or different responsivities) for each detector; 3) addition of independent low frequency $1/f$ noise for each bolometer time-series to model the slowly-changing

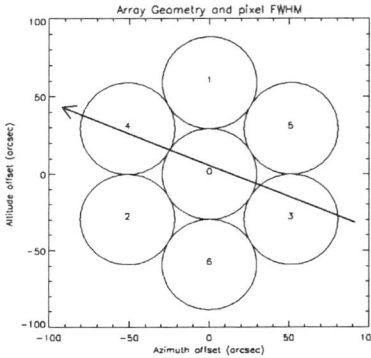 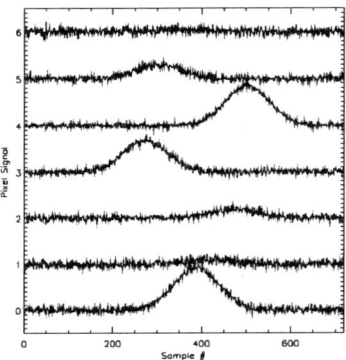

Figure 1. 7-element, 500μm 1-$F\lambda$ spaced hexagonal close-packed array geometry with 59″ FWHM beams, and bolometer signals for a scan across a point source (path of source indicated by arrow).

detector baselines in long duration scans; 4) convolution with an $e^{-t/\tau}$ impulse function to mimic the instantaneous response of the detectors.

The *perfect* astrometry may also be *corrupted* by adding random and systematic pointing errors. Figure 1 shows the time-series for a 7-pixel feed-horn coupled array scanned across a point source.

1.1. Sky noise

Sky noise at sub-mm wavelengths is due to variable emission from cells of water vapour moving across the field-of-view of the telescope. This sky signal is significantly greater than the astronomical signal, and errors introduced through its imperfect subtraction lead to increased noise and artifacts in the reduced data. Thus this important component of noise must be considered in all realistic simulations of ground-based observations.

Data from the SCUBA camera on the James Clerk Maxwell Telescope on Mauna Kea suggests that the sky generates $1/f$-like noise that is correlated across the array (W. Holland & D. Kelly, private communication). An empirical model that reproduces the observed gradients across the SCUBA array at 850μm assumes the sky consists of clouds, generated by convolving Gaussian noise with a symmetrical top-hat $(1/f)^{11/6}$ function, distributed in a flat-plane at some fiducial altitude, moving at the wind-speed above the telescope aperture. This planar sky-model is passed across the simulated detector beams to produce the strong, variable sky signal.

While the total power of the sky may be removed by subtracting the median signal of the off-source sky bolometers, strong gradients still remain and can dominate the astronomical signals.

In the case of short integrations (less than the characteristic timescale on which the detector sensitivities drift - the $1/f$ knee frequency, f_{knee}) residual errors on small spatial scales (of order the beam-size) remain in the data.

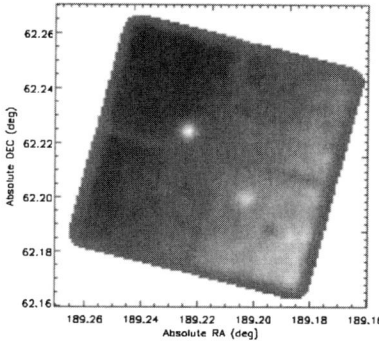 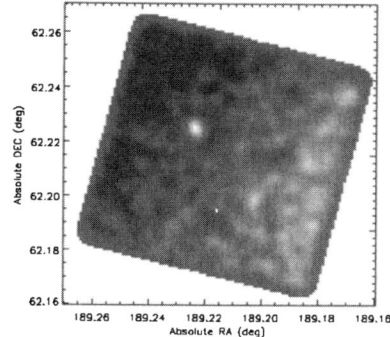

Figure 2. Simulations of one-second point-and-stare total power observations with a fully-sampled (0.5-$F\lambda$) 40 × 40 pixel monolithic array at 850μm with 14″ FWHM beams. In the case of perfect flat-fielding (left panel), two significant sources with a S/N of 25 and 12σ are clearly observed against a typical sky-background gradient. In contrast, the fainter of the sources blends into the background noise when the gains of the bolometers are only known to a precision of $\sim$ 1% (right panel).

The illumination of the array by a uniform source (astronomical or instrumental) provide a measure of the gain corrections, which normalize the bolometer responsivities. Errors in measuring these detector gains result in imperfect flat-fielding. These flat-fielding errors are amplified by the strong sky signal and can dominate the system noise in the sky-subtracted data.

Additional complications arise when the time required to flat-field the array to a specified level of precision, or the length of time between flat-fields during total power scanned observations (for example), exceeds the time-scale on which the detector responsivities are stable ($f > f_{knee}$). Furthermore, an additional source of noise is introduced when the flat-fielding is performed on an astronomical source in presence of variable sky emission.

Figure 2 demonstrates, for pointed total-power observations, two examples of the effects described above. In conclusion, the requirement to flat-field at sub-mm wavelengths with a precision of $\ll$ 1% presents a major obstacle to the efficient use of large-format monolithic arrays operated in total power mode.

Acknowledgments. We thank Wayne Holland and Dennis Kelly (UK ATC, Edinburgh) for their valuable contributions.

References

Hughes, D. H. & Gaztañaga, E. 2000 in ESLAB Symp. 33, Star Formation from the Small to the Large Scale, ed. F. Favata, A. Kaas & S. Haasnoot (Noordwijk: ESA), 29 (available as astro-ph/0004002)

Peter Hofner, Tomas Greve and Jennifer Neakrase.

On the Life and Death of OH/IR Stars

B. M. Lewis

National Astronomy and Ionosphere Center, Arecibo Observatory, HC 3 Box 53995, Arecibo, Puerto Rico 00612, USA

Abstract. The study of emission from OH/IR stars currently provides several topics where we are interested simultaneously in both the strong and weak features. Several are discussed here.

1. Introduction

Visitors often comment on the high signal-to-noise-ratio spectra of my OH/IR stars when I am observing. Yet many research topics are concerned with subtle details of emission in the presence of strong signals, as occurs for example in studying high velocity clouds in proximity to galactic HI, or in searching for the HI signature of a Stromgren sphere about a massive star. OH/IR star studies currently offer several topics of this kind that are made practical by the unprecedented gain of the Arecibo telescope.

2. Deriving Light Curves

Deriving light curves from features with large intensity differences can be a problem. IRAS 22402+1045 has several 1.0-0.1 Jy 1612 and 1667 MHz features, but only one, ~20 mJy, 1665 MHz feature. While the light curves from the stronger features are very similar, those from the weakest are notably different near minimum light. This suggests that the weakest masers are either unsaturated or are influenced by competitive gain between sub-levels of the OH molecule. This star requires good signal-to-noise measurements at intensities differing by factors >20.

3. Uncovering Weak Emission

Weak 1667 MHz emission was discovered in IRAS 22402+1045 when the 1667 MHz spectra collected for its light curve were combined into a single spectrum. The star is then seen to have a 10 mJy plateau extending 2 km s^{-1} beyond the velocity range of its 1612 MHz features: one interpretation, though moot, would cite this as evidence for a recent rapid decrease in its expansion velocity. Similar plateaux are present in Arecibo observations of the proto planetary nebula (PPN) IRAS 18095+2704 and the hypergiant star IRAS 19566+3423. Moreover 22 GHz studies by Bowers (1992) with the VLA have shown that the water maser emission of several Miras exhibit striking low-intensity plateaux.

4. Uncovering and Validating the Death of OH/IR Stars

The "headroom" provided by the sensitivity of the Arecibo telescope is crucial for this study, so let me expand. The total duration of the 1612 MHz emission phase for low progenitor mass OH/IR stars is ~ 1700 yr (Lewis 2000), a result derived from their frequency relative to associated PPN, and the shell-expansion age of one PPN. But a 1700 yr net duration suggests that it is possible to witness the death of one such OH/IR star on average every 10 years from a sample of 170, where the death of a star is set by the disappearance of detectable 1612 MHz masers. The prototype is IRAS 18455+0448, which had a 2.1 Jy maser in 1988 that had faded to a solitary 0.1 Jy maser by 1998. Its final exponential decline was followed by Lewis, Oppenheimer, & Daubar (2001) for two years, until its intensity $I_{1612} < 2$ mJy, an I_{1612} more than 1000 times weaker than in 1988.

The first difficulty in identifying more examples of dead OH/IR stars lies in the sampling statistics. While there are 328 OH/IR stars in the Arecibo sky with a first epoch $I_{1612} > 100$ mJy, which are therefore above the survey limit for other telescopes, just 87 of them have $I_{1612} > 1$ Jy. Moreover most of this last set have progenitor masses $> 1\,M_\odot$, which disqualifies them from being regarded as low mass stars. The second difficulty lies in the inherent variability of OH/IR stars, which commonly vary by a factor of 2 to 3 around a pulsation cycle, though exceptional factor of 10 changes have been seen in the intensity of some 1612 MHz features in the hypergiant IRC+10420. In consequence to verify a "death" we would like to see a factor of > 20 change in the 1612 MHz intensity. So a practical sample of stars for checking on deaths at Arecibo is restricted to those with $0.1 < I_{1612} < 1$ Jy: this would be difficult to work with anywhere else.

In practice four of the Arecibo OH/IR stars detected between 1985 and 1989 with $I_{1612} > 0.1$ Jy have since faded from view, which gives a tally of four "deaths". There is in addition one object that has faded three times by a factor of >30, only to recover by a factor of five. While these last changes can still be attributed to the pulsation cycle, its mainline masers have evolved morphologically in intensity and velocity too, so we appear to be witnessing another slow death. Nevertheless the number of dead stars clearly exceeds prediction if the 1612 MHz emission phase generally occurs as a single solitary phase in the evolution of a star: these numbers are therefore evidence that the 1612 MHz emission phase of most oxygen-rich stars is cyclical.

Acknowledgments. This work is supported by the National Astronomy and Ionospheric Center, which is operated by Cornell University under a management agreement with the National Science Foundation.

References

Bowers, P. F. 1992, ApJ, 390, L27
Lewis, B. M. 2000, ApJ, 533, 959
Lewis, B. M., Oppenheimer, B. D., & Daubar, I. J. 2001, ApJ, 548, L77

Single-Dish Radio Astronomy: Techniques and Applications
ASP Conference Series, Vol. 278, 2002
S. Stanimirović, D. R. Altschuler, P. F. Goldsmith, and C. J. Salter

Arecibo Spectral Baselines in the Presence of Continuum Emission

Tapasi Ghosh, Chris Salter

National Astronomy and Ionosphere Center, Arecibo Observatory, HC 3 Box 53995, Arecibo, Puerto Rico 00612, USA

Abstract. We explore the "double position-switching" approach to improving spectral baselines when observing sources with significant continuum emission.

1. Introduction

Apart from the familiar reflections between the horn and the apex of the dish (Rohlfs & Wilson 1996), the suspended feed platform of the Arecibo telescope can cause multi-path scattering to the horn of radiation from the surrounding hills and other sources of emission (Briggs et al. 1997). The observing technique usually employed to deal with this is ON/OFF position switching, with the OFF traversing the same track on the dish as the ON. This method runs into trouble when emission/absorption-line spectra of sources emitting significant continuum radiation are observed as the component of the standing-wave pattern due to the target's continuum emission is not cancelled at all by subtracting the source-free OFF data from the ON. The residual standing waves degrade the spectral baseline, and have an amplitude proportional to the source intensity. To combat this when observing the spectrum of a continuum source, a "reference" continuum source (at different redshift to avoid it possessing a spectral line near the line frequency of the target) is also observed in the same ON/OFF position-switched mode (e.g. Briggs, Sorar, & Taramopoulos 1993). The azimuth-zenith angle track followed during this "reference ON/OFF" should be as near as possible to that for the target source. Division of the target spectrum by that of the reference source then cancels the residual standing wave, leaving a spectrum whose magnitude across the observed band is proportional to the ratio of the flux densities of the two sources. We have investigated this "double position switching" (DPS) mode for the post-upgrade Arecibo telescope.

2. The Observations

In early 2000, the damped Ly-α (DLyα) absorption system at $z = 0.221$ towards the radio-loud quasar OI363 was observed in the DPS mode using the nearby radio-loud source OI371 as reference. Each double ON/OFF cycle on this pair lasted 4×4.6 min, a total of 112 such cycles being observed for a total ON-source integration time of 8.6 hr. Here we consider data of 6.25-MHz total bandwidth, with 2048 spectral channels per polarization. The HI absorption spectrum from the DLyα absorber was measured with very high signal-to-noise ratio, (Kanekar, Ghosh, & Chengalur 2001). However, here we have used this data to examine the DPS approach, quantifying the expected signal-to-noise ratio, and exploring the best approach to data reduction.

3. Approaches to Dealing with Spectral Baseline Ripples

The OI363/OI371 data were reduced in three different ways; a) forming (ON − OFF)/OFF for just the OI363 data, b) using the DPS approach taking the ratio of (ON − OFF)/OFF for the two continuum sources and, c) similar to b, but taking the ratio of (ON − OFF) for the two sources. Detailed consideration of the factors involved, including the expected signal-to-noise ratios, favored method c). Thus, concentrating on approach c), the derived ratio, R, of (ON − OFF) for the target and reference sources (designated to be S_T and S_R respectively) is;

$$R = \frac{(ON(\nu) - OFF(\nu))_{S_T}}{(ON(\nu) - OFF(\nu))_{S_R}} = \frac{(T_{source}(\nu) - T_{conf}(\nu))_{S_T}}{(T_{source}(\nu) - T_{conf}(\nu))_{S_R}}, \quad (1)$$

where, $T_{source}(\nu)$ is the antenna temperature on either of the continuum sources, and $T_{conf}(\nu)$ is the antenna temperature (including confusion) at the corresponding OFF positions. All terms containing standing wave ripples cancel, assuming these to be of similar form for both target and reference.

In considering the noise levels expected for the different approaches, we employed the concept of System Equivalent Flux Density (SEFD), defined to be the point-source flux density that will double the system temperature. Suppose the SEFD of the observing system is $SEFD_{off}$ when pointing at "blank sky", then when observing a source of flux density S, the $SEFD_{on}$ will be ($SEFD_{off}$ + S).

For quantifying noise levels for DPS, we assume that both of the OFF positions have the same SEFD of $SEFD_{off}$, while for the target and reference source, the SEFDs are $SEFD_{S_T} = (SEFD_{off} + S_{S_T})$ and $SEFD_{S_R} = (SEFD_{off} + S_{S_R})$. Further, let τ be the total ON + OFF integration time for an ON/OFF cycle on the target source *alone*, i.e. 50% of the total integration time for the full DPS cycle, while β is the frequency resolution. Then, the fractional rms noise per polarization is;

$$\frac{\sigma(R)}{R} = \frac{\sqrt{2}}{\sqrt{\beta\tau}} \times \left[\frac{SEFD_{S_T}^2 + SEFD_{off}^2}{(SEFD_{S_T} - SEFD_{off})^2} + \frac{SEFD_{S_R}^2 + SEFD_{off}^2}{(SEFD_{S_R} - SEFD_{off})^2} \right]^{\frac{1}{2}}. \quad (2)$$

Comparing the noise on the actual measurements with the predictions from Equation 2, the agreement is excellent. Note that the declinations of OI363 and OI371 differ by about 30′.

3.1. Integration over many cycles

In the presence of continuum emission, the simplest comparison of the two DPS approaches and the basic ON/OFF technique is to see how the measured noise integrates down with time. This is shown in Figure 1, where the measured noise is plotted against the integrated number of cycles. In this, only data taken at

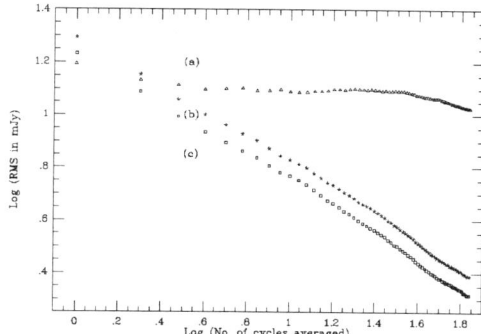

Figure 1. A log-log plot of the measured rms noises (in mJy) plotted versus the integrated number of observing cycles at zenith angle < 15°; (a) is for the basic position-switched approach, while (b) and (c) are the DPS approaches (see Section 3.).

zenith angle <15° was used, as the SEFDs there are practically independent of pointing position, permitting simple unweighted averaging to be used. The plot represents a single polarization. Figure 1 shows that the basic ON/OFF approach ceases to integrate down as $\sqrt{t}$ after only 3 cycles have been averaged! In contrast, even after averaging all 68 cycles, the noises for both DPS cases continue to decrease as $\sqrt{t}$. The explanation is clearly illustrated by Figure 2, which shows the averaged spectra for three typical observing cycles. It is seen that while both DPS spectra look noise-like (apart from the narrow, deep absorption feature from the DLyα system), a standing-wave pattern corrupts the single-source (ON − OFF)/OFF spectrum. This pattern is found to change only slowly with pointing, and presumably epoch, explaining the saturation of the rms noise for approach a). The peak-to-peak amplitude of the standing wave is ∼2.3% of the flux density of the target source for both polarizations.

4. Concluding Remarks

The DPS approach has been shown to be an effective way of obtaining high-quality spectra for objects emitting significant continuum radiation, and for which the basic position-switched technique breaks down. For DPS observations, we have demonstrated that noise levels integrating down as $\sqrt{t}$ can be obtained for total integration times of many hours. For data reduction, taking the ratio of (ON − OFF) for the target and reference sources is to be preferred. In terms of selecting a continuum reference source for such observations, the highest signal-to-noise results are obtained using the strongest available reference source, subject to variation of the standing-wave pattern with the direction of pointing.

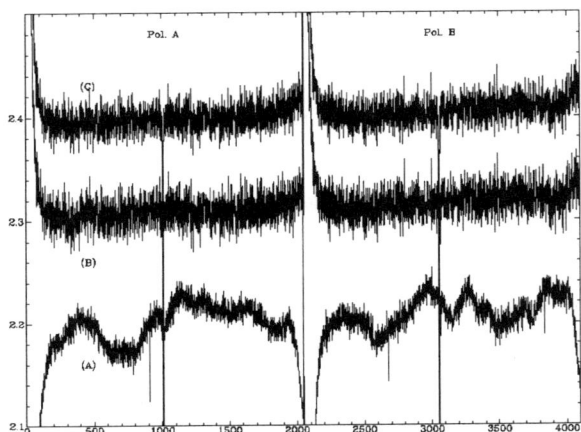

Figure 2. The average of three observing cycles; (A) is for the basic position-switched approach, while (B) and (C) represent cases b) and c) of Section 3. respectively. The vertical axis is in Jy (with the zero level shifted for clarity), with the horizontal axis in spectrometer channel numbers; the first and second 2048 channels represent the two polarizations.

References

Briggs, F.H., Sorar, E., Kraan-Kortweg, R.C. & van Driel, W. 1997, PASA, 14, 37
Briggs, F.H., Sorar, E., & Taramopoulos, A. 1993, ApJ, 415, L99
Kanekar, N., Ghosh, T., & Chengalur, J.N. 2001, A&A, 373, 394
Rohlfs, K. & Wilson, T.L. 1996, Tools of Radio Astronomy (2nd edition; Berlin: Springer-Verlag), 200

Project Phoenix: SETI Observations from 1200 to 1750 MHz with the Upgraded Arecibo Telescope

P. R. Backus and the Project Phoenix Team

SETI Institute, 2035 Landings Drive, Mountain View, California 94043

Abstract. Project Phoenix, the privately funded continuation of NASA's Targeted Search SETI Program, has taken advantage of the wide frequency coverage made possible by the upgraded Arecibo Telescope. Our goal is to search for evidence of narrowband extraterrestrial radio signals from nearby stars in the microwave portion of the spectrum. The signal detection system processes a 20-MHz bandwidth with 1-Hz wide channels in each of two circular polarizations. The system is sensitive to signals that are continuously present, or pulsed regularly, even if their frequencies drift by up to about 1 Hz per second. A database of terrestrial signals found in the previous week is used to match against detections for each observation. Candidate signals, i.e., those not in the database, are checked immediately with a "pseudo-interferometric" observation between Arecibo and the 76-m Lovell Telescope at the Jodrell Bank Observatory.

Since 1998 October, we have conducted approximately 8 weeks of observations at L-Band, 1200 to 1750 MHz. Approximately 180 MHz of that range is too heavily occupied by terrestrial signals for effective observing.

5. Introduction

The NASA SETI Program was composed of two search strategies: an All Sky Survey and a Targeted Search. The Targeted Search was to scrutinize approximately 1000 nearby stars with the largest available telescopes to achieve high sensitivity. The search was to cover the frequency range from 1200 to 3000 MHz, searching for both pulsed and continuous signals with a channel resolution of 1 Hz. The Targeted Search was awarded a total of 2600 hours to observe about 250 stars with the Arecibo telescope. In order to provide the necessary continuous frequency coverage, the NASA SETI Program was prepared to issue a contract to NAIC for the construction of 23 line feeds at a total cost of $2.3 million. Instead, the money was contributed toward the Gregorian Upgrade Project to benefit the entire community. In 1993, one year after the initial 200 hours of observations at Arecibo, Congress terminated the NASA SETI Program.

Since SETI Institute scientists, engineers, and subcontractors formed the core of the Targeted Search team, the Institute raised private funds to continue the search as "Project Phoenix." NASA provided the Targeted Search System (TSS) on long-term loan to the Institute. During 1994, the Phoenix team doubled the bandwidth of the TSS and added a Follow-Up Detection Device (FUDD) for interference mitigation.

The TSS traveled to Parkes Observatory in 1995 where it began a 16-week observing run only one month later than the original NASA schedule. In mid-1995 the TSS returned to the Institute for upgrades to improve reliability and increase automation. The system operated at the NRAO 140-ft Telescope from 1996 September through 1998 April, using the telescope about 50% of the time. In 1998 September, the TSS returned to Arecibo to continue the observing program begun under NASA six years before. Project Phoenix now uses a total of three weeks of telescope time per year, scheduled in two sessions.

6. The Targeted Search System

The Targeted Search System uses near real-time signal processing and immediate follow-up two-site observations of "candidate ETI" signals. The TSS processes an instantaneous bandwidth of 20 MHz into 28,740,096 channels in each of two polarizations. The channels have an equivalent noise bandwidth of 1 Hz with a spacing of 0.7 Hz. A spectrum is generated through two layers of bandpass filters, each with more than 80 dB of out-of-band rejection. Time samples from the resulting 643-Hz resolution spectra are Fourier transformed into the 1-Hz resolution spectra. A Hanning weighting is applied at this resolution to suppress strong, narrow signals by 70 dB at 7 Hz from the edge of the signal. Successive spectra are overlapped in time by 50%.

The TSS is sensitive to both pulsed and continuous signals that may drift by up to plus or minus one channel per spectrum. Any signals not found in a database of known interference are considered "candidates". While the main part of the TSS continues the search, Follow-Up Detection Devices (FUDDs) at Arecibo and Jodrell Bank Observatory observe a narrow frequency band around each candidate signal. The rotation of the Earth causes a different Doppler shift and drift (velocity and acceleration) for the signal at each site. Using the measured parameters of the signal at Arecibo, the FUDD at the Lovell Telescope looks for the signal with the appropriate frequency offset and drift. We call this process pseudo-interferometry since it effectively narrows the beamwidth of the observation.

7. Observation Method

Observations are scheduled for 6 pm to 6 am local time at Arecibo in spring and fall to avoid lines-of-sight through the inner Solar Wind. Turbulent plasma would disrupt the coherence of an extraterrestrial signal. Each three-week session begins with one night of RFI scans to prime the database. The antenna points at the zenith and the system conducts standard length observations (276 seconds). Any signals detected are not coming from a fixed point on the sky and are classified as RFI and entered into the database. The system steps through the 1200 to 1750 MHz range several times to allow for intermittent RFI. During regular observing, the database is updated with new signals and purged of signals more than one week old.

Auto-scheduling software selects a star from a catalog based on weighted criteria including distance, whether it has already been observed, if it has planets, etc. A star is tracked as long as possible, tuning to a new 20-MHz band every five minutes. Detected signals that are not in the database (empirical match criterion is 1 kHz) are scheduled for follow-up. The FUDD can zero-in on up to 12 candidates per 10 MHz. In order to make up for the sensitivity difference between Arecibo (305 m) and the Lovell Telescope (76 m), the FUDD at Jodrell Bank performs a 100 second coherent integration. A FUDD detection at both sites triggers automatic on-off-on tests.

4. Completed Observations

Since 1998 September we have completed 12,712 "observations", where an observation is defined as a target star observed in a 10-MHz band. We detected 635,038 signals of which 576,794 were immediately matched in the RFI database. Of the remaining signals, 55,292 were sent to the FUDDs and of those, 857 were confirmed at Jodrell Bank. Subsequent on-off-on observations eliminated those signals.

Due to strong, persistent interfering signals, some 10-MHz bands cannot be observed. The number of such bands has increased over time. We currently skip 180 MHz of the 550-MHz wide observing range at L-Band. The frequency ranges skipped are: 1225-1265, 1325-1335, 1345-1355, 1375-1385, 1525-1555, 1565-1585, 1605-1615, 1685-1695, and 1715-1755 MHz.

A touch of local flavor. From left to right: Soledad del Rio, Neftalis Rivera-Castillo, Carmen Pantoja, Mayra Lebron, Omar Lopez-Cruz and Marco Grossi. Musicians: Dimas Alvarez, Toño and Pepe Vives.

Epilogue

Larry Solanch

Department of Physics and Astronomy, University of Georgia, Athens, Georgia 30602, USA

Don't Cry For Me Arecibo*

It won't be easy, you'll think it is strange
When I try to explain how I feel.
That I still need you!
After all that I've done,
You won't believe me!

All you will see is someone you once knew,
Although I seem so very different
At such significant distances from you.
I had to go back, even though
I could stay all my life at AO!
Looking out at the sky, staying out of the sun.
But I chose radio astronomy,
Running around trying everything new!
My data never ended my questions:
I never expected them to!

Don't cry for me Arecibo.
The truth is I'll never leave you!
All through my wild days
My mad existence.
I'll keep my promise.
To not keep my distance!

And as for fortune, and as for fame
I never pursued them at all,
Though it seemed to the world they were all I desired
They remain illusions
Not the solutions they promise to be!
The answer was here all the time:
I love you and I will come back! So,

Don't cry for me Arecibo.
The truth is I'll never leave you!
All through my wild days

My mad existence.
I'll keep my promise.
To not keep my distance!

Have I said too much?
There's nothing more I can think of to say to you.
But all you have to do is look at me,
To know that my every word is true!

Don't cry for me Arecibo.
The truth is I'll never leave you!
All through my wild days
My mad existence.
I'll keep my promise.
To not keep my distance!

* To the tune of "Don't Cry For Me Argentina" from the musical "Evita".

Subject Index

absorption coefficient, 194, 211
accumulator, 114
AIPS++, 353
ALMA, 42
amplifier, 29, 93
angular broadening (seeing), 229
antenna, 55
    Cassegrain type, 55, 454
    effective aperture, 158
    effective area, 49
    gain, 319
    Gregorian type, 55
    loss, 318
        blockage, 319
        diffraction loss, 318
        ohmic loss, 318
    power pattern, 50
    prime focus type, 55
    sensitivity, 83
    sky coupling, 81
    solid angle, 50
    temperature, 52, 82, 294, 397
    theorem, 50, 77
aperture,
    antenna, 53
    efficiency, 50
    illumination, 57, 70
    plane, 57
aperture synthesis technique, 376
Arecibo Observatory, 1
    education, 21
    outreach, 21
array,
    aperture synthesis, 28
    focal plane, 453, 515
        140-Foot Telescope, 459
        Parkes Multi-beam, 456
        Sequoia, 456
    phased, 28
asteroids, 272
astigmatism, 310
astrometry, 286
atmosphere, 45, 99, 414

    absorption effects, 324, 416
    anomalous refraction, 420
    ionospheric effects, 416
    refractive index fluctuations, 417
    transmission, 100, 477
    turbulence effects, 414
atmospheric window, 100
autocorrelation function, 118, 124
autocorrelator, 118, 127

back-end, 113
backscatter, 278
basketweaving, 168
beam,
    efficiency, 74
    pattern, 88, 380
    polarized, 148
    solid angle, 158
    squash, 148
    squint, 150
    switching, 177
    switching in software, 179
    width, 67
blackbody brightness, 49
blanking pulse generator, 437
blockage, 59, 403
BoloCAM, 485
bolometer, 101, 463, 515
    and RFI, 479
    array, 480
    performance, 466
    semiconductor type, 469
    silicon pop-up type, 473
    superconducting type, 471
bootstrapping, 306, 409
braking index, 266
brightness temperature, 212

calibration, 293
    absolute, 327
    cm wavelength, 293
    mm wavelength, 313
calibrator, 114
central limit theorem, 126

Chebyshev polynomial coefficients, 266
chopper wheels, 317
circular polarization ratio, 274
COBRA, 255
cold sky, 313
coma, 310
Committee for Radio Frequencies (CORF), 449
communication system, 448
confusion, 163
continuum,
    mapping, 168, 179
    observing, 155
    observing and RFI, 173
    radio sources, 155
convolution, 84
corrugated horn, 92
cosmic microwave background, 85
coupling efficiency, 73

damping, 191
data reduction, 329
    beam-switched observations, 332
    continuum grid mapping, 339
    continuum on-the-fly mapping, 339
    on-off continuum observations, 331
    spectral-line beam switching, 343
    spectral-line frequency switching, 343
    spectral-line position switching, 341
deconvolution, 180, 390
defocus, 70
delay-Doppler mapping, 283
detector, 29, 115
    bolometric, 463
    incoherent, 463
dielectric constant, 271
diffraction, 46
diffractive intensity scintillations (DISS), 229
dilution refrigerator, 475
DISH, 358
dispersion constant, 262
dispersion measure, 228, 252
dispersive time delay, 228
Doppler,
    bandwidth, 280
    dispersion, 280
    effect, 189
    equator, 281
    shift, 278
    temperature, 192
double position switching, 521
double sideband, 107
dual-beam function, 180
dynamic range, 438

edge phase error, 63
Einstein coefficients, 188
Einstein delay term, 262
electron density model, 248
equivalent noise temperature, 93
equivalent paraboloid, 56
error beam response, 319
excitation temperature, 194
experiment,
    bistatic, 272
    continuous wave, 278
    monostatic, 272

Faraday rotation, 228
Federal Communications Commission (FCC), 449
feed, 91
    line, 10
    sinuous, 459
    system, 54
field pattern, 67
filter, 94
    bandpass, 125
    low-pass, 115, 125
flux density, 48, 82, 294, 305
    peak, 157
Fourier transform, 118
frequency conversion, 94
front-end, 91
front-end switches, 113

gain curve, 307
gain-elevation curves, 159
Glish, 354
Gregorian dome, 17
gridding,
    top-hat, 339
group velocity, 252

Hagfors scattering law, 277
Hanning function, 440

heterodyne principle, 102
HI mass of galaxies, 212
hybrid, 143

illumination loss, 320
illumination pattern, 401
image domain, 377
image plane, 215
image rejection, 105, 438
index of refraction fluctuations, 414
integrator, 85
interferometer, 28, 376
intermediate frequency, 84
international atomic time, 261
International Telecommunications Union (ITU), 435, 448
Inter Union-Commission for the Allocation of Frequencies (IUCAF), 448
ionosphere, 5
ionospheric scintillations, 417

Jansky, 48, 83
Jones matrix, 140
Jones vector, 134

Kelvin per Jansky, 75, 83
kinetic temperature, 191
Kirchoff's law, 49
klystron, 272
Kolmogorov model, 415

level populations, 210
load,
    blackbody, 317
    hot & cold, 299
    matched, 51
    thermal, 75
local oscillator, 84, 94, 108
Local Standard of Rest (LSR), 190
loss tangent, 275

main beam, 67, 401
    efficiency, 74
main lobe, 67
MAMBO, 485
mapping speeds, 218
matrix inversion, 181
maximum entropy algorithm, 181, 390
Maxwellian distribution, 215

median filtering, 441
microphonics, 479
microturbulence, 191
mixer, 84, 94, 438
mosaicing technique, 379
Mueller matrix, 140, 260
multibeam surveys, 218
multipath propogation, 229
multiplier, 30

noise,
    $1/f$, 167, 174, 515
    equivalent flux density, 482
    equivalent power, 466
    Johnson, 93, 466
    phonon, 466
    thermal, 93
noise-diode,
    correlated, 145
    switched, 297
NRAO VLA Sky Survey (NVSS), 160
Nyquist sampling, 217

on-the-fly mapping, 437
optical depth, 193, 213
optical fiber, 438
orthomode transducer (OMT), 92

parallactic angle, 144
particle interactions, 197
phase efficiency, 63
phase errors, 63
phase structure function, 419
phase transformer, 55
Planck law, 81, 294
plasma frequency, 252
pointing, 159, 309
polarimeter, 257
polarization,
    astronomical, 131
    calibration, 139
    circular, 132, 137
    elliptical, 133
    linear, 135
polarizer, 54, 92
position switching, 301
power spectrum, 118
project Phoenix, 525
pulsar,

534 Subject Index

age, 267
arrival-time precision, 244
binary, 262
coherent dedispersion, 121, 251
distances, 245
flux calibration, 256
mass, 264
millisecond, 261
observing,
    dedispersion techniques, 233
    post-dectection technique, 237
    post-detection technique, 227
parallax, 247, 262
polarimetry, 255
processor, 121
RFI excision, 255
search, 237
spin-down model, 262
time of arrival, 261
timing, 261
volume, 240
pulse broadening, 229

radar, 40
    albedo, 273
    astronomy, 271
    cross section, 273
    echo, 273
    equation, 286
    military, 451
    north-south ambiguity, 283
radiative processes, 187
radiative transfer, 193, 212
radio fog, 447
radio frequency interference (RFI), 433, 526
    cancellation, 444
    diagnostics, 439
    environment, 436
    excision, 440
    lightning, 438
    mitigation, 440
    nulling, 442
    null steering, 444
    post-correlation, 445
    real-time excision, 442, 443
    suppression, 444
radiometer equation, 85, 218
radio service, 448

radio spectrum,
    active users, 440
    coordination zones, 449
    licensed users, 436
    management, 435, 447
    manager, 440
    negotiated rule making, 449
    passive users, 452
    pollution, 451
    quiet zones, 449
    recommendations, 449
    regulations, 448
    surveys, 434
radio window, 45
ranging, 281
Rayleigh-Jeans approximation, 49, 81, 212, 294
receiver,
    coherent type, 101
    correlation type, 178
    Dicke-switched, 174
    filter, 438
    high frequency, 107
    incoherent type, 101
    linearity, 96
    mm/sub-mm, 453
    multi-beam, 453
    noise, 315
    stability, 95
    super-heterodyne type, 84, 91
receiving systems, 84
reciprocity theorem, 53
recombination lines, 201, 209
redshift,
    optical, 190
    radio, 190
reflection,
    quasi-specular, 276
    specular, 276
reflector,
    efficiency, 159
    secondary, 17
    spherical, 12
    tertiary, 17
refractive intensity scintillation (RISS), 230
refrigeration methods, 110
relays, 440
Reynolds number, 415

ringing, 436
rms phase error, 64
rms surface error, 65
Roemer delay, 264
rotating vector model (RVM), 256
rotation, 278
rotation measure, 228
Ruze formula, 65
Rydberg constant, 202, 210

sampler, 115
satellite,
    Global Positioning System (GPS), 261
    GLONASS, 444
    Iridium, 435
scattering measure, 233
SCUBA-2, 487
SCUBA, 483
semi-transparent vane, 327
SETI, 525
Shapiro delay, 262
SHARC, 485
short-spacings correction, 220, 375
    cross calibration, 383
    in image domain, 387
    in spatial frequency domain, 386
sidelobe, 68, 309, 397
signal,
    Gaussian nature, 126
signal-to-noise ratio, 447
single-dish telescopes, 493
single sideband, 107
sky noise, 324
solar motion, 190
source,
    coupling efficiency, 322
    extended, 52, 76
    flat-spectrum, 159
    point–like, 52, 75
    resolved, 52
    spectral index, 163
    steep-spectrum, 159
    temperature, 296
    thermal, 48
    unresolved, 52
spatial frequency, 33, 88, 379
    domain, 377
specific intensity, 48

spectral broadening, 230
spectral density, 123
spectral-line, 187, 209
    absorption, 211
    baseline fitting, 300, 345
    baseline ripples, 522
    baselines, 521
    broadening, 189, 192
    component fitting, 346
    cube, 222
    emission, 211
    frequency switching, 300
    gridding, 217
    grid mapping, 347
    HI, 3, 200
    mapping, 215
    molecular, 203
    moments, 346
    on-the-fly mapping, 349
    RFI excision, 347
    smoothing, 346
    velocity moments, 223
spectrometer, 118, 436
    autocorrelation type, 120, 436
    FFT type, 119, 436
spillover, 70, 321, 401
spillover efficiency, 59
SPIRE, 486
spontaneous emission, 188
square-law detector, 85, 115
standing-wave pattern, 521
statistical equilibrium, 196
stimulated emission, 188
Stokes parameters, 135, 257
Stokes vector, 134, 260
stray radiation, 73, 397
    and deconvolution, 408
    in position switching, 408
superconductor-insulator-
    superconductor (SIS), 102
superresolution phenomena, 230
surface,
    porosity, 275
    scattering, 275
system,
    equivalent flux density, 88, 522
    noise, 85, 313
    temperature, 85, 295, 316

taper efficiency, 59
telecommunications, 447
TEMPO, 266
terrestrial time, 261
thermal motions, 191
thermodynamic equilibrium, 195
time constant, 466
time of arrival fluctuations, 229
tip-tilt compensating system, 429
transition edge sensor (TES), 471
transmission frequency, 273
transmitter,
    S-band, 14, 272
    X-band, 272

Universal Coordinated Time (UTC), 261
u-v plane, 215

velocity,
    optical, 190
    radio, 190
Very Long Baseline Interferometry, 507
voltage sampling, 116

Wideband Arecibo Pulsar Processor (WAPP), 258
World Radiocommunication Conferences (WRC), 448

Y-factor, 299, 315

Zernike polynomials, 423

A LIST OF THE VOLUMES

Published
by

THE ASTRONOMICAL SOCIETY OF THE PACIFIC
(ASP)

An international, nonprofit, scientific and educational organization
founded in 1889

All book orders or inquiries concerning

THE ASTRONOMICAL SOCIETY OF THE PACIFIC CONFERENCE SERIES (ASP - CS)

and

INTERNATIONAL ASTRONOMICAL UNION VOLUMES (IAU)

should be directed to the:

The Astronomical Society of the Pacific Conference Series
390 Ashton Avenue
San Francisco   CA   94112-1722   USA

    Phone: 800-335-2624   (Within USA)
    Phone: 415-337-2126
    Fax: 415-337-5205

  E-mail: service@astrosociety.org
  Web Site: http://www.astrosociety.org

Complete lists of proceedings of past IAU Meetings are maintained at the IAU Web site at the URL: http://www.iau.org/publicat.html

Volumes 32 - 189 in the IAU Symposia Series may be ordered from:

Kluwer Academic Publishers
P. O. Box 117
NL 3300 AA Dordrecht
The Netherlands

Kluwer@wKap.com

ASP CONFERENCE SERIES VOLUMES
Published by the Astronomical Society of the Pacific

PUBLISHED: 1988 (* asterisk means OUT OF STOCK)

Vol. CS -1 PROGRESS AND OPPORTUNITIES IN SOUTHERN HEMISPHERE
OPTICAL ASTRONOMY: CTIO 25TH Anniversary Symposium
eds. V. M. Blanco and M. M. Phillips
ISBN 0-937707-18-X

Vol. CS-2 PROCEEDINGS OF A WORKSHOP ON OPTICAL SURVEYS FOR QUASARS
eds. Patrick S. Osmer, Alain C. Porter, Richard F. Green, and Craig B. Foltz
ISBN 0-937707-19-8

Vol. CS-3 FIBER OPTICS IN ASTRONOMY
ed. Samuel C. Barden
ISBN 0-937707-20-1

Vol. CS-4 THE EXTRAGALACTIC DISTANCE SCALE:
Proceedings of the ASP 100th Anniversary Symposium
eds. Sidney van den Bergh and Christopher J. Pritchet
ISBN 0-937707-21-X

Vol. CS-5 THE MINNESOTA LECTURES ON CLUSTERS OF GALAXIES
AND LARGE-SCALE STRUCTURE
ed. John M. Dickey
ISBN 0-937707-22-8

PUBLISHED: 1989

Vol. CS-6 * SYNTHESIS IMAGING IN RADIO ASTRONOMY: A Collection of Lectures
from the Third NRAO Synthesis Imaging Summer School
eds. Richard A. Perley, Frederic R. Schwab, and Alan H. Bridle
ISBN 0-937707-23-6

PUBLISHED: 1990

Vol. CS-7 PROPERTIES OF HOT LUMINOUS STARS: Boulder-Munich Workshop
ed. Catharine D. Garmany
ISBN 0-937707-24-4

Vol. CS-8 * CCDs IN ASTRONOMY
ed. George H. Jacoby
ISBN 0-937707-25-2

Vol. CS-9 COOL STARS, STELLAR SYSTEMS, AND THE SUN: Sixth Cambridge Workshop
ed. George Wallerstein
ISBN 0-937707-27-9

Vol. CS-10 * EVOLUTION OF THE UNIVERSE OF GALAXIES:
Edwin Hubble Centennial Symposium
ed. Richard G. Kron
ISBN 0-937707-28-7

Vol. CS-11 CONFRONTATION BETWEEN STELLAR PULSATION AND EVOLUTION
eds. Carla Cacciari and Gisella Clementini
ISBN 0-937707-30-9

Vol. CS-12 THE EVOLUTION OF THE INTERSTELLAR MEDIUM
ed. Leo Blitz
ISBN 0-937707-31-7

PUBLISHED: 1991

Vol. CS-13 THE FORMATION AND EVOLUTION OF STAR CLUSTERS
ed. Kenneth Janes
ISBN 0-937707-32-5

ASP CONFERENCE SERIES VOLUMES
Published by the Astronomical Society of the Pacific

PUBLISHED: 1991 (* asterisk means OUT OF STOCK)

Vol. CS-14   ASTROPHYSICS WITH INFRARED ARRAYS
              ed. Richard Elston
              ISBN 0-937707-33-3

Vol. CS-15   LARGE-SCALE STRUCTURES AND PECULIAR MOTIONS IN THE UNIVERSE
              eds. David W. Latham and L. A. Nicolaci da Costa
              ISBN 0-937707-34-1

Vol. CS-16   Proceedings of the 3rd Haystack Observatory Conference on ATOMS, IONS,
              AND MOLECULES: NEW RESULTS IN SPECTRAL LINE ASTROPHYSICS
              eds. Aubrey D. Haschick and Paul T. P. Ho
              ISBN 0-937707-35-X

Vol. CS-17   LIGHT POLLUTION, RADIO INTERFERENCE, AND SPACE DEBRIS
              ed. David L. Crawford
              ISBN 0-937707-36-8

Vol. CS-18   THE INTERPRETATION OF MODERN SYNTHESIS OBSERVATIONS
              OF SPIRAL GALAXIES
              eds. Nebojsa Duric and Patrick C. Crane
              ISBN 0-937707-37-6

Vol. CS-19   RADIO INTERFEROMETRY: THEORY, TECHNIQUES, AND APPLICATIONS,
              IAU Colloquium 131
              eds. T. J. Cornwell and R. A. Perley
              ISBN 0-937707-38-4

Vol. CS-20   FRONTIERS OF STELLAR EVOLUTION:
              50th Anniversary McDonald Observatory (1939-1989)
              ed. David L. Lambert
              ISBN 0-937707-39-2

Vol. CS-21   THE SPACE DISTRIBUTION OF QUASARS
              ed . David Crampton
              ISBN 0-937707-40-6

PUBLISHED: 1992

Vol. CS-22   NONISOTROPIC AND VARIABLE OUTFLOWS FROM STARS
              eds. Laurent Drissen, Claus Leitherer, and Antonella Nota
              ISBN 0-937707-41-4

Vol CS-23 *   ASTRONOMICAL CCD OBSERVING AND REDUCTION TECHNIQUES
              ed. Steve B. Howell
              ISBN 0-937707-42-4

Vol. CS-24   COSMOLOGY AND LARGE-SCALE STRUCTURE IN THE UNIVERSE
              ed. Reinaldo R. de Carvalho
              ISBN 0-937707-43-0

Vol. CS-25   ASTRONOMICAL DATA ANALYSIS, SOFTWARE AND SYSTEMS I - (ADASS I)
              eds. Diana M. Worrall, Chris Biemesderfer, and Jeannette Barnes
              ISBN 0-937707-44-9

Vol. CS-26   COOL STARS, STELLAR SYSTEMS, AND THE SUN:
              Seventh Cambridge Workshop
              eds. Mark S. Giampapa and Jay A. Bookbinder
              ISBN 0-937707-45-7

Vol. CS-27   THE SOLAR CYCLE: Proceedings of the
              National Solar Observatory/Sacramento Peak 12th Summer Workshop
              ed. Karen L. Harvey
              ISBN 0-937707-46-5

ASP CONFERENCE SERIES VOLUMES
Published by the Astronomical Society of the Pacific

PUBLISHED: 1992 (asterisk means OUT OF STOCK)

Vol. CS-28　AUTOMATED TELESCOPES FOR PHOTOMETRY AND IMAGING
eds. Saul J. Adelman, Robert J. Dukes, Jr., and Carol J. Adelman
ISBN 0-937707-47-3

Vol. CS-29　Viña del Mar Workshop on CATACLYSMIC VARIABLE STARS
ed. Nikolaus Vogt
ISBN 0-937707-48-1

Vol. CS-30　VARIABLE STARS AND GALAXIES
ed. Brian Warner
ISBN 0-937707-49-X

Vol. CS-31　RELATIONSHIPS BETWEEN ACTIVE GALACTIC NUCLEI
AND STARBURST GALAXIES
ed. Alexei V. Filippenko
ISBN 0-937707-50-3

Vol. CS-32　COMPLEMENTARY APPROACHES TO DOUBLE
AND MULTIPLE STAR RESEARCH, IAU Colloquium 135
eds. Harold A. McAlister and William I. Hartkopf
ISBN 0-937707-51-1

Vol. CS-33 *　RESEARCH AMATEUR ASTRONOMY
ed. Stephen J. Edberg
ISBN 0-937707-52-X

Vol. CS-34　ROBOTIC TELESCOPES IN THE 1990's
ed. Alexei V. Filippenko
ISBN 0-937707-53-8

PUBLISHED: 1993

Vol. CS-35 *　MASSIVE STARS: THEIR LIVES IN THE INTERSTELLAR MEDIUM
eds. Joseph P. Cassinelli and Edward B. Churchwell
ISBN 0-937707-54-6

Vol. CS-36　PLANETS AROUND PULSARS
ed. J. A. Phillips, S. E. Thorsett, and S. R. Kulkarni
ISBN 0-937707-55-4

Vol. CS-37　FIBER OPTICS IN ASTRONOMY II
ed. Peter M. Gray
ISBN 0-937707-56-2

Vol. CS-38　NEW FRONTIERS IN BINARY STAR RESEARCH: Pacific Rim Colloquium
eds. K. C. Leung and I.-S. Nha
ISBN 0-937707-57-0

Vol. CS-39　THE MINNESOTA LECTURES ON THE STRUCTURE
AND DYNAMICS OF THE MILKY WAY
ed. Roberta M. Humphreys
ISBN 0-937707-58-9

Vol. CS-40　INSIDE THE STARS, IAU Colloquium 137
eds. Werner W. Weiss and Annie Baglin
ISBN 0-937707-59-7

Vol. CS-41　ASTRONOMICAL INFRARED SPECTROSCOPY:
FUTURE OBSERVATIONAL DIRECTIONS
ed. Sun Kwok
ISBN 0-937707-60-0

ASP CONFERENCE SERIES VOLUMES
Published by the Astronomical Society of the Pacific

PUBLISHED: 1993 (* asterisk means OUT OF STOCK)

Vol. CS-42	GONG 1992: SEISMIC INVESTIGATION OF THE SUN AND STARS ed. Timothy M. Brown ISBN 0-937707-61-9
Vol. CS-43	SKY SURVEYS: PROTOSTARS TO PROTOGALAXIES ed. B. T. Soifer ISBN 0-937707-62-7
Vol. CS-44	PECULIAR VERSUS NORMAL PHENOMENA IN A-TYPE AND RELATED STARS, IAU Colloquium 138 eds. M. M. Dworetsky, F. Castelli, and R. Faraggiana ISBN 0-937707-63-5
Vol. CS-45	LUMINOUS HIGH-LATITUDE STARS ed. Dimitar D. Sasselov ISBN 0-937707-64-3
Vol. CS-46	THE MAGNETIC AND VELOCITY FIELDS OF SOLAR ACTIVE REGIONS, IAU Colloquium 141 eds. Harold Zirin, Guoxiang Ai, and Haimin Wang ISBN 0-937707-65-1
Vol. CS-47	THIRD DECENNIAL US-USSR CONFERENCE ON SETI -- Santa Cruz, California, USA ed. G. Seth Shostak ISBN 0-937707-66-X
Vol. CS-48	THE GLOBULAR CLUSTER-GALAXY CONNECTION eds. Graeme H. Smith and Jean P. Brodie ISBN 0-937707-67-8
Vol. CS-49	GALAXY EVOLUTION: THE MILKY WAY PERSPECTIVE ed. Steven R. Majewski ISBN 0-937707-68-6
Vol. CS-50	STRUCTURE AND DYNAMICS OF GLOBULAR CLUSTERS eds. S. G. Djorgovski and G. Meylan ISBN 0-937707-69-4
Vol. CS-51	OBSERVATIONAL COSMOLOGY eds. Guido Chincarini, Angela Iovino, Tommaso Maccacaro, and Dario Maccagni ISBN 0-937707-70-8
Vol. CS-52	ASTRONOMICAL DATA ANALYSIS SOFTWARE AND SYSTEMS II - (ADASS II) eds. R. J. Hanisch, R. J. V. Brissenden, and Jeannette Barnes ISBN 0-937707-71-6
Vol. CS-53	BLUE STRAGGLERS ed. Rex A. Saffer ISBN 0-937707-72-4

PUBLISHED: 1994

Vol. CS-54 *	THE FIRST STROMLO SYMPOSIUM: THE PHYSICS OF ACTIVE GALAXIES eds. Geoffrey V. Bicknell, Michael A. Dopita, and Peter J. Quinn ISBN 0-937707-73-2
Vol. CS-55	OPTICAL ASTRONOMY FROM THE EARTH AND MOON eds. Diane M. Pyper and Ronald J. Angione ISBN 0-937707-74-0
Vol. CS-56	INTERACTING BINARY STARS ed. Allen W. Shafter ISBN 0-937707-75-9

ASP CONFERENCE SERIES VOLUMES
Published by the Astronomical Society of the Pacific

PUBLISHED: 1994 (* asterisk means OUT OF STOCK)

Vol. CS-57 STELLAR AND CIRCUMSTELLAR ASTROPHYSICS
eds. George Wallerstein and Alberto Noriega-Crespo
ISBN 0-937707-76-7

Vol. CS-58 * THE FIRST SYMPOSIUM ON THE INFRARED CIRRUS
AND DIFFUSE INTERSTELLAR CLOUDS
eds. Roc M. Cutri and William B. Latter
ISBN 0-937707-77-5

Vol. CS-59 ASTRONOMY WITH MILLIMETER AND SUBMILLIMETER WAVE
INTERFEROMETRY,
IAU Colloquium 140
eds. M. Ishiguro and Wm. J. Welch
ISBN 0-937707-78-3

Vol. CS-60 THE MK PROCESS AT 50 YEARS: A POWERFUL TOOL FOR ASTROPHYSICAL
INSIGHT, A Workshop of the Vatican Observatory --Tucson, Arizona, USA
eds. C. J. Corbally, R. O. Gray, and R. F. Garrison
ISBN 0-937707-79-1

Vol. CS-61 ASTRONOMICAL DATA ANALYSIS SOFTWARE AND SYSTEMS III - (ADASS III)
eds. Dennis R. Crabtree, R. J. Hanisch, and Jeannette Barnes
ISBN 0-937707-80-5

Vol. CS-62 THE NATURE AND EVOLUTIONARY STATUS OF HERBIG Ae/Be STARS
eds. Pik Sin Thé, Mario R. Pérez, and Ed P. J. van den Heuvel
ISBN 0-9837707-81-3

Vol. CS-63 SEVENTY-FIVE YEARS OF HIRAYAMA ASTEROID FAMILIES:
THE ROLE OF COLLISIONS IN THE SOLAR SYSTEM HISTORY
eds. Yoshihide Kozai, Richard P. Binzel, and Tomohiro Hirayama
ISBN 0-937707-82-1

Vol. CS-64 * COOL STARS, STELLAR SYSTEMS, AND THE SUN:
Eighth Cambridge Workshop
ed. Jean-Pierre Caillault
ISBN 0-937707-83-X

Vol. CS-65 * CLOUDS, CORES, AND LOW MASS STARS:
The Fourth Haystack Observatory Conference
eds. Dan P. Clemens and Richard Barvainis
ISBN 0-937707-84-8

Vol. CS-66 * PHYSICS OF THE GASEOUS AND STELLAR DISKS OF THE GALAXY
ed. Ivan R. King
ISBN 0-937707-85-6

Vol. CS-67 UNVEILING LARGE-SCALE STRUCTURES BEHIND THE MILKY WAY
eds. C. Balkowski and R. C. Kraan-Korteweg
ISBN 0-937707-86-4

Vol. CS-68 * SOLAR ACTIVE REGION EVOLUTION:
COMPARING MODELS WITH OBSERVATIONS
eds. K. S. Balasubramaniam and George W. Simon
ISBN 0-937707-87-2

Vol. CS-69 REVERBERATION MAPPING OF THE BROAD-LINE REGION
IN ACTIVE GALACTIC NUCLEI
eds. P. M. Gondhalekar, K. Horne, and B. M. Peterson
ISBN 0-937707-88-0

Vol. CS-70 * GROUPS OF GALAXIES
eds. Otto-G. Richter and Kirk Borne
ISBN 0-937707-89-9

ASP CONFERENCE SERIES VOLUMES
Published by the Astronomical Society of the Pacific

PUBLISHED: 1995 (* asterisk means OUT OF STOCK)

Vol. CS-71 TRIDIMENSIONAL OPTICAL SPECTROSCOPIC METHODS IN ASTROPHYSICS,
IAU Colloquium 149
eds. Georges Comte and Michel Marcelin
ISBN 0-937707-90-2

Vol. CS-72 MILLISECOND PULSARS: A DECADE OF SURPRISE
eds. A. S Fruchter, M. Tavani, and D. C. Backer
ISBN 0-937707-91-0

Vol. CS-73 AIRBORNE ASTRONOMY SYMPOSIUM ON THE GALACTIC ECOSYSTEM:
FROM GAS TO STARS TO DUST
eds. Michael R. Haas, Jacqueline A. Davidson, and Edwin F. Erickson
ISBN 0-937707-92-9

Vol. CS-74 PROGRESS IN THE SEARCH FOR EXTRATERRESTRIAL LIFE:
1993 Bioastronomy Symposium
ed. G. Seth Shostak
ISBN 0-937707-93-7

Vol. CS-75 MULTI-FEED SYSTEMS FOR RADIO TELESCOPES
eds. Darrel T. Emerson and John M. Payne
ISBN 0-937707-94-5

Vol. CS-76 GONG '94: HELIO- AND ASTERO-SEISMOLOGY FROM THE EARTH
AND SPACE
eds. Roger K. Ulrich, Edward J. Rhodes, Jr., and Werner Däppen
ISBN 0-937707-95-3

Vol. CS-77 ASTRONOMICAL DATA ANALYSIS SOFTWARE AND SYSTEMS IV - (ADASS IV)
eds. R. A. Shaw, H. E. Payne, and J. J. E. Hayes
ISBN 0-937707-96-1

Vol. CS-78 ASTROPHYSICAL APPLICATIONS OF POWERFUL NEW DATABASES:
Joint Discussion No. 16 of the 22nd General Assembly of the IAU
eds. S. J. Adelman and W. L. Wiese
ISBN 0-937707-97-X

Vol. CS-79 * ROBOTIC TELESCOPES: CURRENT CAPABILITIES, PRESENT
DEVELOPMENTS, AND FUTURE PROSPECTS
FOR AUTOMATED ASTRONOMY
eds. Gregory W. Henry and Joel A. Eaton
ISBN 0-937707-98-8

Vol. CS-80 * THE PHYSICS OF THE INTERSTELLAR MEDIUM
AND INTERGALACTIC MEDIUM
eds. A. Ferrara, C. F. McKee, C. Heiles, and P. R. Shapiro
ISBN 0-937707-99-6

Vol. CS-81 LABORATORY AND ASTRONOMICAL HIGH RESOLUTION SPECTRA
eds. A. J. Sauval, R. Blomme, and N. Grevesse
ISBN 1-886733-01-5

Vol. CS-82 * VERY LONG BASELINE INTERFEROMETRY AND THE VLBA
eds. J. A. Zensus, P. J. Diamond, and P. J. Napier
ISBN 1-886733-02-3

Vol. CS-83 * ASTROPHYSICAL APPLICATIONS OF STELLAR PULSATION,
IAU Colloquium 155
eds. R. S. Stobie and P. A. Whitelock
ISBN 1-886733-03-1

ATLAS INFRARED ATLAS OF THE ARCTURUS SPECTRUM, 0.9 - 5.3 µm
eds. Kenneth Hinkle, Lloyd Wallace, and William Livingston
ISBN: 1-886733-04-X

ASP CONFERENCE SERIES VOLUMES
Published by the Astronomical Society of the Pacific

PUBLISHED: 1995 (* asterisk means OUT OF STOCK)

Vol. CS-84   THE FUTURE UTILIZATION OF SCHMIDT TELESCOPES, IAU Colloquium 148
eds. Jessica Chapman, Russell Cannon, Sandra Harrison, and Bambang Hidayat
ISBN 1-886733-05-8

Vol. CS-85 *   CAPE WORKSHOP ON MAGNETIC CATACLYSMIC VARIABLES
eds. D. A. H. Buckley and B. Warner
ISBN 1-886733-06-6

Vol. CS-86   FRESH VIEWS OF ELLIPTICAL GALAXIES
eds. Alberto Buzzoni, Alvio Renzini, and Alfonso Serrano
ISBN 1-886733-07-4

PUBLISHED: 1996

Vol. CS-87   NEW OBSERVING MODES FOR THE NEXT CENTURY
eds. Todd Boroson, John Davies, and Ian Robson
ISBN 1-886733-08-2

Vol. CS-88 *   CLUSTERS, LENSING, AND THE FUTURE OF THE UNIVERSE
eds. Virginia Trimble and Andreas Reisenegger
ISBN 1-886733-09-0

Vol. CS-89   ASTRONOMY EDUCATION: CURRENT DEVELOPMENTS, FUTURE COORDINATION
ed. John R. Percy
ISBN 1-886733-10-4

Vol. CS-90   THE ORIGINS, EVOLUTION, AND DESTINIES OF BINARY STARS IN CLUSTERS
eds. E. F. Milone and J. -C. Mermilliod
ISBN 1-886733-11-2

Vol. CS-91   BARRED GALAXIES, IAU Colloquium 157
eds. R. Buta, D. A. Crocker, and B. G. Elmegreen
ISBN 1-886733-12-0

Vol. CS-92 *   FORMATION OF THE GALACTIC HALO .... INSIDE AND OUT
eds. Heather L. Morrison and Ata Sarajedini
ISBN 1-886733-13-9

Vol. CS-93   RADIO EMISSION FROM THE STARS AND THE SUN
eds. A. R. Taylor and J. M. Paredes
ISBN 1-886733-14-7

Vol. CS-94   MAPPING, MEASURING, AND MODELING THE UNIVERSE
eds. Peter Coles, Vicent J. Martinez, and Maria-Jesus Pons-Borderia
ISBN 1-886733-15-5

Vol. CS-95   SOLAR DRIVERS OF INTERPLANETARY AND TERRESTRIAL DISTURBANCES:
Proceedings of 16th International Workshop National Solar Observatory/Sacramento Peak
eds. K. S. Balasubramaniam, Stephen L. Keil, and Raymond N. Smartt
ISBN 1-886733-16-3

Vol. CS-96   HYDROGEN-DEFICIENT STARS
eds. C. S. Jeffery and U. Heber
ISBN 1-886733-17-1

Vol. CS-97   POLARIMETRY OF THE INTERSTELLAR MEDIUM
eds. W. G. Roberge and D. C. B. Whittet
ISBN 1-886733-18-X

ASP CONFERENCE SERIES VOLUMES
Published by the Astronomical Society of the Pacific

PUBLISHED: 1996 (* asterisk means OUT OF STOCK)

Vol. CS-98   FROM STARS TO GALAXIES: THE IMPACT OF STELLAR PHYSICS ON GALAXY EVOLUTION
eds. Claus Leitherer, Uta Fritze-von Alvensleben, and John Huchra
ISBN 1-886733-19-8

Vol. CS-99   COSMIC ABUNDANCES:
Proceedings of the 6th Annual October Astrophysics Conference
eds. Stephen S. Holt and George Sonneborn
ISBN 1-886733-20-1

Vol. CS-100   ENERGY TRANSPORT IN RADIO GALAXIES AND QUASARS
eds. P. E. Hardee, A. H. Bridle, and J. A. Zensus
ISBN 1-886733-21-X

Vol. CS-101   ASTRONOMICAL DATA ANALYSIS SOFTWARE AND SYSTEMS V – (ADASS V)
eds. George H. Jacoby and Jeannette Barnes
ISBN 1080-7926

Vol. CS-102   THE GALACTIC CENTER, 4th ESO/CTIO Workshop
ed. Roland Gredel
ISBN 1-886733-22-8

Vol. CS-103   THE PHYSICS OF LINERS IN VIEW OF RECENT OBSERVATIONS
eds. M. Eracleous, A. Koratkar, C. Leitherer, and L. Ho
ISBN 1-886733-23-6

Vol. CS-104   PHYSICS, CHEMISTRY, AND DYNAMICS OF INTERPLANETARY DUST,
IAU Colloquium 150
eds. Bo Å. S. Gustafson and Martha S. Hanner
ISBN 1-886733-24-4

Vol. CS-105   PULSARS: PROBLEMS AND PROGRESS, IAU Colloquium 160
ed. S. Johnston, M. A. Walker, and M. Bailes
ISBN 1-886733-25-2

Vol. CS-106   THE MINNESOTA LECTURES ON EXTRAGALACTIC NEUTRAL HYDROGEN
ed. Evan D. Skillman
ISBN 1-886733-26-0

Vol. CS-107   COMPLETING THE INVENTORY OF THE SOLAR SYSTEM:
A Symposium held in conjunction with the 106th Annual Meeting of the ASP
eds. Terrence W. Rettig and Joseph M. Hahn
ISBN 1-886733-27-9

Vol. CS-108   M.A.S.S. -- MODEL ATMOSPHERES AND SPECTRUM SYNTHESIS:
5th Vienna - Workshop
eds. Saul J. Adelman, Friedrich Kupka, and Werner W. Weiss
ISBN 1-886733-28-7

Vol. CS-109   COOL STARS, STELLAR SYSTEMS, AND THE SUN: Ninth Cambridge Workshop
eds. Roberto Pallavicini and Andrea K. Dupree
ISBN 1-886733-29-5

Vol. CS-110   BLAZAR CONTINUUM VARIABILITY
eds. H. R. Miller, J. R. Webb, and J. C. Noble
ISBN 1-886733-30-9

Vol. CS-111   MAGNETIC RECONNECTION IN THE SOLAR ATMOSPHERE:
Proceedings of a Yohkoh Conference
eds. R. D. Bentley and J. T. Mariska
ISBN 1-886733-31-7

ASP CONFERENCE SERIES VOLUMES
Published by the Astronomical Society of the Pacific

PUBLISHED: 1996 (* asterisk means OUT OF STOCK)

Vol. CS-112　THE HISTORY OF THE MILKY WAY AND ITS SATELLITE SYSTEM
eds. Andreas Burkert, Dieter H. Hartmann, and Steven R. Majewski
ISBN 1-886733-32-5

PUBLISHED: 1997

Vol. CS-113　EMISSION LINES IN ACTIVE GALAXIES: NEW METHODS AND TECHNIQUES,
IAU Colloquium 159
eds. B. M. Peterson, F.-Z. Cheng, and A. S. Wilson
ISBN 1-886733-33-3

Vol. CS-114　YOUNG GALAXIES AND QSO ABSORPTION-LINE SYSTEMS
eds. Sueli M. Viegas, Ruth Gruenwald, and Reinaldo R. de Carvalho
ISBN 1-886733-34-1

Vol. CS-115　GALACTIC CLUSTER COOLING FLOWS
ed. Noam Soker
ISBN 1-886733-35-X

Vol. CS-116　THE SECOND STROMLO SYMPOSIUM:
THE NATURE OF ELLIPTICAL GALAXIES
eds. M. Arnaboldi, G. S. Da Costa, and P. Saha
ISBN 1-886733-36-8

Vol. CS-117　DARK AND VISIBLE MATTER IN GALAXIES
eds. Massimo Persic and Paolo Salucci
ISBN-1-886733-37-6

Vol. CS-118　FIRST ADVANCES IN SOLAR PHYSICS EUROCONFERENCE:
ADVANCES IN THE PHYSICS OF SUNSPOTS
eds. B. Schmieder. J. C. del Toro Iniesta, and M. Vázquez
ISBN 1-886733-38-4

Vol. CS-119　PLANETS BEYOND THE SOLAR SYSTEM
AND THE NEXT GENERATION OF SPACE MISSIONS
ed. David R. Soderblom
ISBN 1-886733-39-2

Vol. CS-120　LUMINOUS BLUE VARIABLES: MASSIVE STARS IN TRANSITION
eds. Antonella Nota and Henny J. G. L. M. Lamers
ISBN 1-886733-40-6

Vol. CS-121　ACCRETION PHENOMENA AND RELATED OUTFLOWS, IAU Colloquium 163
eds. D. T. Wickramasinghe, G. V. Bicknell, and L. Ferrario
ISBN 1-886733-41-4

Vol. CS-122　FROM STARDUST TO PLANETESIMALS:
Symposium held as part of the 108th Annual Meeting of the ASP
eds. Yvonne J. Pendleton and A. G. G. M. Tielens
ISBN 1-886733-42-2

Vol. CS-123　THE 12th 'KINGSTON MEETING': COMPUTATIONAL ASTROPHYSICS
eds. David A. Clarke and Michael J. West
ISBN 1-886733-43-0

Vol. CS-124　DIFFUSE INFRARED RADIATION AND THE IRTS
eds. Haruyuki Okuda, Toshio Matsumoto, and Thomas Roellig
ISBN 1-886733-44-9

Vol. CS-125　ASTRONOMICAL DATA ANALYSIS SOFTWARE AND SYSTEMS VI
eds. Gareth Hunt and H. E. Payne
ISBN 1-886733-45-7

ASP CONFERENCE SERIES VOLUMES
Published by the Astronomical Society of the Pacific

PUBLISHED: 1997 (* asterisk means OUT OF STOCK)

Vol. CS-126   FROM QUANTUM FLUCTUATIONS TO COSMOLOGICAL STRUCTURES
eds. David Valls-Gabaud, Martin A. Hendry, Paolo Molaro, and Khalil Chamcham
ISBN 1-886733-46-5

Vol. CS-127   PROPER MOTIONS AND GALACTIC ASTRONOMY
ed. Roberta M. Humphreys
ISBN 1-886733-47-3

Vol. CS-128   MASS EJECTION FROM AGN (Active Galactic Nuclei)
eds. N. Arav, I. Shlosman, and R. J. Weymann
ISBN 1-886733-48-1

Vol. CS-129   THE GEORGE GAMOW SYMPOSIUM
eds. E. Harper, W. C. Parke, and G. D. Anderson
ISBN 1-886733-49-X

Vol. CS-130   THE THIRD PACIFIC RIM CONFERENCE ON
RECENT DEVELOPMENT ON BINARY STAR RESEARCH
eds. Kam-Ching Leung
ISBN 1-886733-50-3

PUBLISHED: 1998

Vol. CS-131   BOULDER-MUNICH II: PROPERTIES OF HOT, LUMINOUS STARS
ed. Ian D. Howarth
ISBN 1-886733-51-1

Vol. CS-132   STAR FORMATION WITH THE INFRARED SPACE OBSERVATORY (ISO)
eds. João L. Yun and René Liseau
ISBN 1-886733-52-X

Vol. CS-133   SCIENCE WITH THE NGST (Next Generation Space Telescope)
eds. Eric P. Smith and Anuradha Koratkar
ISBN 1-886733-53-8

Vol. CS-134   BROWN DWARFS AND EXTRASOLAR PLANETS
eds. Rafael Rebolo, Eduardo L. Martin, and Maria Rosa Zapatero Osorio
ISBN 1-886733-54-6

Vol. CS-135   A HALF CENTURY OF STELLAR PULSATION INTERPRETATIONS:
A TRIBUTE TO ARTHUR N. COX
eds. P. A. Bradley and J. A. Guzik
ISBN 1-886733-55-4

Vol. CS-136   GALACTIC HALOS: A UC SANTA CRUZ WORKSHOP
ed. Dennis Zaritsky
ISBN 1-886733-56-2

Vol. CS-137   WILD STARS IN THE OLD WEST: PROCEEDINGS OF THE 13th NORTH
AMERICAN WORKSHOP ON CATACLYSMIC VARIABLES
AND RELATED OBJECTS
eds. S. Howell, E. Kuulkers, and C. Woodward
ISBN 1-886733-57-0

Vol. CS-138   1997 PACIFIC RIM CONFERENCE ON STELLAR ASTROPHYSICS
eds. Kwing Lam Chan, K. S. Cheng, and H. P. Singh
ISBN 1-886733-58-9

Vol. CS-139   PRESERVING THE ASTRONOMICAL WINDOWS:
Proceedings of Joint Discussion No. 5 of the 23rd General Assembly of the IAU
eds. Syuzo Isobe and Tomohiro Hirayama
ISBN 1-886733-59-7

ASP CONFERENCE SERIES VOLUMES
Published by the Astronomical Society of the Pacific

PUBLISHED: 1998 (* asterisk means OUT OF STOCK)

Vol. CS-140   SYNOPTIC SOLAR PHYSICS --18th NSO/Sacramento Peak Summer Workshop
eds. K. S. Balasubramaniam, J. W. Harvey, and D. M. Rabin
ISBN 1-886733-60-0

Vol. CS-141   ASTROPHYSICS FROM ANTARCTICA:
A Symposium held as a part of the 109th Annual Meeting of the ASP
eds. Giles Novak and Randall H. Landsberg
ISBN 1-886733-61-9

Vol. CS-142   THE STELLAR INITIAL MASS FUNCTION: 38th Herstmonceux Conference
eds. Gerry Gilmore and Debbie Howell
ISBN 1-886733-62-7

Vol. CS-143 *   THE SCIENTIFIC IMPACT OF THE GODDARD HIGH RESOLUTION SPECTROGRAPH (GHRS)
eds. John C. Brandt, Thomas B. Ake III, and Carolyn Collins Petersen
ISBN 1-886733-63-5

Vol. CS-144   RADIO EMISSION FROM GALACTIC AND EXTRAGALACTIC COMPACT SOURCES, IAU Colloquium 164
eds. J. Anton Zensus, G. B. Taylor, and J. M. Wrobel
ISBN 1-886733-64-3

Vol. CS-145   ASTRONOMICAL DATA ANALYSIS SOFTWARE AND SYSTEMS VII – (ADASS VII)
eds. Rudolf Albrecht, Richard N. Hook, and Howard A. Bushouse
ISBN 1-886733-65-1

Vol. CS-146   THE YOUNG UNIVERSE GALAXY FORMATION
AND EVOLUTION AT INTERMEDIATE AND HIGH REDSHIFT
eds. S. D'Odorico, A. Fontana, and E. Giallongo
ISBN 1-886733-66-X

Vol. CS-147   ABUNDANCE PROFILES: DIAGNOSTIC TOOLS FOR GALAXY HISTORY
eds. Daniel Friedli, Mike Edmunds, Carmelle Robert, and Laurent Drissen
ISBN 1-886733-67-8

Vol. CS-148   ORIGINS
eds. Charles E. Woodward, J. Michael Shull, and Harley A. Thronson, Jr.
ISBN 1-886733-68-6

Vol. CS-149   SOLAR SYSTEM FORMATION AND EVOLUTION
eds. D. Lazzaro, R. Vieira Martins, S. Ferraz-Mello, J. Fernández, and C. Beaugé
ISBN 1-886733-69-4

Vol. CS-150   NEW PERSPECTIVES ON SOLAR PROMINENCES, IAU Colloquium 167
eds. David Webb, David Rust, and Brigitte Schmieder
ISBN 1-886733-70-8

Vol. CS-151   COSMIC MICROWAVE BACKGROUND
AND LARGE SCALE STRUCTURES OF THE UNIVERSE
eds. Yong-Ik Byun and Kin-Wang Ng
ISBN 1-886733-71-6

Vol. CS-152   FIBER OPTICS IN ASTRONOMY III
eds. S. Arribas, E. Mediavilla, and F. Watson
ISBN 1-886733-72-4

Vol. CS-153   LIBRARY AND INFORMATION SERVICES IN ASTRONOMY III -- (LISA III)
eds. Uta Grothkopf, Heinz Andernach, Sarah Stevens-Rayburn, and Monique Gomez
ISBN 1-886733-73-2

ASP CONFERENCE SERIES VOLUMES
Published by the Astronomical Society of the Pacific

PUBLISHED: 1998 (* asterisk means OUT OF STOCK)

Vol. CS-154 COOL STARS, STELLAR SYSTEMS AND THE SUN: Tenth Cambridge Workshop
eds. Robert A. Donahue and Jay A. Bookbinder
ISBN 1-886733-74-0

Vol. CS-155 SECOND ADVANCES IN SOLAR PHYSICS EUROCONFERENCE:
THREE-DIMENSIONAL STRUCTURE OF SOLAR ACTIVE REGIONS
eds. Costas E. Alissandrakis and Brigitte Schmieder
ISBN 1-886733-75-9

PUBLISHED: 1999

Vol. CS-156 HIGHLY REDSHIFTED RADIO LINES
eds. C. L. Carilli, S. J. E. Radford, K. M. Menten, and G. I. Langston
ISBN 1-886733-76-7

Vol. CS-157 ANNAPOLIS WORKSHOP ON MAGNETIC CATACLYSMIC VARIABLES
eds. Coel Hellier and Koji Mukai
ISBN 1-886733-77-5

Vol. CS-158 SOLAR AND STELLAR ACTIVITY: SIMILARITIES AND DIFFERENCES
eds. C. J. Butler and J. G. Doyle
ISBN 1-886733-78-3

Vol. CS-159 BL LAC PHENOMENON
eds. Leo O. Takalo and Aimo Sillanpää
ISBN 1-886733-79-1

Vol. CS-160 ASTROPHYSICAL DISCS: An EC Summer School
eds. J. A. Sellwood and Jeremy Goodman
ISBN 1-886733-80-5

Vol. CS-161 HIGH ENERGY PROCESSES IN ACCRETING BLACK HOLES
eds. Juri Poutanen and Roland Svensson
ISBN 1-886733-81-3

Vol. CS-162 QUASARS AND COSMOLOGY
eds. Gary Ferland and Jack Baldwin
ISBN 1-886733-83-X

Vol. CS-163 STAR FORMATION IN EARLY-TYPE GALAXIES
eds. Jordi Cepa and Patricia Carral
ISBN 1-886733-84-8

Vol. CS-164 ULTRAVIOLET–OPTICAL SPACE ASTRONOMY BEYOND HST
eds. Jon A. Morse, J. Michael Shull, and Anne L. Kinney
ISBN 1-886733-85-6

Vol. CS-165 THE THIRD STROMLO SYMPOSIUM: THE GALACTIC HALO
eds. Brad K. Gibson, Tim S. Axelrod, and Mary E. Putman
ISBN 1-886733-86-4

Vol. CS-166 STROMLO WORKSHOP ON HIGH-VELOCITY CLOUDS
eds. Brad K. Gibson and Mary E. Putman
ISBN 1-886733-87-2

Vol. CS-167 HARMONIZING COSMIC DISTANCE SCALES IN A POST-HIPPARCOS ERA
eds. Daniel Egret and André Heck
ISBN 1-886733-88-0

Vol. CS-168 NEW PERSPECTIVES ON THE INTERSTELLAR MEDIUM
eds. A. R. Taylor, T. L. Landecker, and G. Joncas
ISBN 1-886733-89-9

ASP CONFERENCE SERIES VOLUMES
Published by the Astronomical Society of the Pacific

PUBLISHED: 1999 (* asterisk means OUT OF STOCK)

Vol. CS-169   11th EUROPEAN WORKSHOP ON WHITE DWARFS
 eds. J.-E. Solheim and E. G. Meištas
 ISBN 1-886733-91-0

Vol. CS-170   THE LOW SURFACE BRIGHTNESS UNIVERSE, IAU Colloquium 171
 eds. J. I. Davies, C. Impey, and S. Phillipps
 ISBN 1-886733-92-9

Vol. CS-171   LiBeB, COSMIC RAYS, AND RELATED X- AND GAMMA-RAYS
 eds. Reuven Ramaty, Elisabeth Vangioni-Flam, Michel Cassé, and Keith Olive
 ISBN 1-886733-93-7

Vol. CS-172   ASTRONOMICAL DATA ANALYSIS SOFTWARE AND SYSTEMS VIII
 eds. David M. Mehringer, Raymond L. Plante, and Douglas A. Roberts
 ISBN 1-886733-94-5

Vol. CS-173   THEORY AND TESTS OF CONVECTION IN STELLAR STRUCTURE:
 First Granada Workshop
 ed. Álvaro Giménez, Edward F. Guinan, and Benjamín Montesinos
 ISBN 1-886733-95-3

Vol. CS-174   CATCHING THE PERFECT WAVE: ADAPTIVE OPTICS AND
 INTERFEROMETRY IN THE 21st CENTURY,
 A Symposium held as a part of the 110th Annual Meeting of the ASP
 eds. Sergio R. Restaino, William Junor, and Nebojsa Duric
 ISBN 1-886733-96-1

Vol. CS-175   STRUCTURE AND KINEMATICS OF QUASAR BROAD LINE REGIONS
 eds. C. M. Gaskell, W. N. Brandt, M. Dietrich, D. Dultzin-Hacyan,
 and M. Eracleous
 ISBN 1-886733-97-X

Vol. CS-176   OBSERVATIONAL COSMOLOGY: THE DEVELOPMENT OF GALAXY SYSTEMS
 eds. Giuliano Giuricin, Marino Mezzetti, and Paolo Salucci
 ISBN 1-58381-000-5

Vol. CS-177   ASTROPHYSICS WITH INFRARED SURVEYS: A Prelude to SIRTF
 eds. Michael D. Bicay, Chas A. Beichman, Roc M. Cutri, and Barry F. Madore
 ISBN 1-58381-001-3

Vol. CS-178   STELLAR DYNAMOS: NONLINEARITY AND CHAOTIC FLOWS
 eds. Manuel Núñez and Antonio Ferriz-Mas
 ISBN 1-58381-002-1

Vol. CS-179   ETA CARINAE AT THE MILLENNIUM
 eds. Jon A. Morse, Roberta M. Humphreys, and Augusto Damineli
 ISBN 1-58381-003-X

Vol. CS-180   SYNTHESIS IMAGING IN RADIO ASTRONOMY II
 eds. G. B. Taylor, C. L. Carilli, and R. A. Perley
 ISBN 1-58381-005-6

Vol. CS-181   MICROWAVE FOREGROUNDS
 eds. Angelica de Oliveira-Costa and Max Tegmark
 ISBN 1-58381-006-4

Vol. CS-182   GALAXY DYNAMICS: A Rutgers Symposium
 eds. David Merritt, J. A. Sellwood, and Monica Valluri
 ISBN 1-58381-007-2

Vol. CS-183   HIGH RESOLUTION SOLAR PHYSICS: THEORY, OBSERVATIONS,
 AND TECHNIQUES
 eds. T. R. Rimmele, K. S. Balasubramaniam, and R. R. Radick
 ISBN 1-58381-009-9

ASP CONFERENCE SERIES VOLUMES
Published by the Astronomical Society of the Pacific

PUBLISHED: 1999 (* asterisk means OUT OF STOCK)

Vol. CS-184 THIRD ADVANCES IN SOLAR PHYSICS EUROCONFERENCE:
MAGNETIC FIELDS AND OSCILLATIONS
eds. B. Schmieder, A. Hofmann, and J. Staude
ISBN 1-58381-010-2

Vol. CS-185 PRECISE STELLAR RADIAL VELOCITIES, IAU Colloquium 170
eds. J. B. Hearnshaw and C. D. Scarfe
ISBN 1-58381-011-0

Vol. CS-186 THE CENTRAL PARSECS OF THE GALAXY
eds. Heino Falcke, Angela Cotera, Wolfgang J. Duschl, Fulvio Melia, and Marcia J. Rieke
ISBN 1-58381-012-9

Vol. CS-187 THE EVOLUTION OF GALAXIES ON COSMOLOGICAL TIMESCALES
eds. J. E. Beckman and T. J. Mahoney
ISBN 1-58381-013-7

Vol. CS-188 OPTICAL AND INFRARED SPECTROSCOPY OF CIRCUMSTELLAR MATTER
eds. Eike W. Guenther, Bringfried Stecklum, and Sylvio Klose
ISBN 1-58381-014-5

Vol. CS-189 CCD PRECISION PHOTOMETRY WORKSHOP
eds. Eric R. Craine, Roy A. Tucker, and Jeannette Barnes
ISBN 1-58381-015-3

Vol. CS-190 GAMMA-RAY BURSTS: THE FIRST THREE MINUTES
eds. Juri Poutanen and Roland Svensson
ISBN 1-58381-016-1

Vol. CS-191 PHOTOMETRIC REDSHIFTS AND HIGH REDSHIFT GALAXIES
eds. Ray J. Weymann, Lisa J. Storrie-Lombardi, Marcin Sawicki, and Robert J. Brunner
ISBN 1-58381-017-X

Vol. CS-192 SPECTROPHOTOMETRIC DATING OF STARS AND GALAXIES
ed. I. Hubeny, S. R. Heap, and R. H. Cornett
ISBN 1-58381-018-8

Vol. CS-193 THE HY-REDSHIFT UNIVERSE:
GALAXY FORMATION AND EVOLUTION AT HIGH REDSHIFT
eds. Andrew J. Bunker and Wil J. M. van Breugel
ISBN 1-58381-019-6

Vol. CS-194 WORKING ON THE FRINGE:
OPTICAL AND IR INTERFEROMETRY FROM GROUND AND SPACE
eds. Stephen Unwin and Robert Stachnik
ISBN 1-58381-020-X

PUBLISHED: 2000

Vol. CS-195 IMAGING THE UNIVERSE IN THREE DIMENSIONS:
Astrophysics with Advanced Multi-Wavelength Imaging Devices
eds. W. van Breugel and J. Bland-Hawthorn
ISBN 1-58381-022-6

Vol. CS-196 THERMAL EMISSION SPECTROSCOPY AND ANALYSIS OF DUST, DISKS, AND REGOLITHS
eds. Michael L. Sitko, Ann L. Sprague, and David K. Lynch
ISBN: 1-58381-023-4

Vol. CS-197 XVth IAP MEETING DYNAMICS OF GALAXIES:
FROM THE EARLY UNIVERSE TO THE PRESENT
eds. F. Combes, G. A. Mamon, and V. Charmandaris
ISBN: 1-58381-24-2

ASP CONFERENCE SERIES VOLUMES
Published by the Astronomical Society of the Pacific

PUBLISHED: 2000 (* asterisk means OUT OF STOCK)

Vol. CS-198   EUROCONFERENCE ON "STELLAR CLUSTERS AND ASSOCIATIONS: CONVECTION, ROTATION, AND DYNAMOS"
eds. R. Pallavicini, G. Micela, and S. Sciortino
ISBN: 1-58381-25-0

Vol. CS-199   ASYMMETRICAL PLANETARY NEBULAE II: FROM ORIGINS TO MICROSTRUCTURES
eds. J. H. Kastner, N. Soker, and S. Rappaport
ISBN: 1-58381-026-9

Vol. CS-200   CLUSTERING AT HIGH REDSHIFT
eds. A. Mazure, O. Le Fèvre, and V. Le Brun
ISBN: 1-58381-027-7

Vol. CS-201   COSMIC FLOWS 1999: TOWARDS AN UNDERSTANDING OF LARGE-SCALE STRUCTURES
eds. Stéphane Courteau, Michael A. Strauss, and Jeffrey A. Willick
ISBN: 1-58381-028-5

Vol. CS-202 *   PULSAR ASTRONOMY – 2000 AND BEYOND, IAU Colloquium 177
eds. M. Kramer, N. Wex, and R. Wielebinski
ISBN: 1-58381-029-3

Vol. CS-203   THE IMPACT OF LARGE-SCALE SURVEYS ON PULSATING STAR RESEARCH, IAU Colloquium 176
eds. L. Szabados and D. W. Kurtz
ISBN: 1-58381-030-7

Vol. CS-204   THERMAL AND IONIZATION ASPECTS OF FLOWS FROM HOT STARS: OBSERVATIONS AND THEORY
eds. Henny J. G. L. M. Lamers and Arved Sapar
ISBN: 1-58381-031-5

Vol. CS-205   THE LAST TOTAL SOLAR ECLIPSE OF THE MILLENNIUM IN TURKEY
eds. W. C. Livingston and A. Özgüç
ISBN: 1-58381-032-3

Vol. CS-206   HIGH ENERGY SOLAR PHYSICS – *ANTICIPATING HESSI*
eds. Reuven Ramaty and Natalie Mandzhavidze
ISBN: 1-58381-033-1

Vol. CS-207   NGST SCIENCE AND TECHNOLOGY EXPOSITION
eds. Eric P. Smith and Knox S. Long
ISBN: 1-58381-036-6

ATLAS   VISIBLE AND NEAR INFRARED ATLAS OF THE ARCTURUS SPECTRUM 3727-9300 Å
eds. Kenneth Hinkle, Lloyd Wallace, Jeff Valenti, and Dianne Harmer
ISBN: 1-58381-037-4

Vol. CS-208   POLAR MOTION: HISTORICAL AND SCIENTIFIC PROBLEMS, IAU Colloquium 178
eds. Steven Dick, Dennis McCarthy, and Brian Luzum
ISBN: 1-58381-039-0

Vol. CS-209   SMALL GALAXY GROUPS, IAU Colloquium 174
eds. Mauri J. Valtonen and Chris Flynn
ISBN: 1-58381-040-4

Vol. CS-210   DELTA SCUTI AND RELATED STARS: Reference Handbook and Proceedings of the 6th Vienna Workshop in Astrophysics
eds. Michel Breger and Michael Houston Montgomery
ISBN: 1-58381-043-9

ASP CONFERENCE SERIES VOLUMES
Published by the Astronomical Society of the Pacific

PUBLISHED: 2000 (* asterisk means OUT OF STOCK)

Vol. CS-211　　MASSIVE STELLAR CLUSTERS
eds. Ariane Lançon and Christian M. Boily
ISBN: 1-58381-042-0

Vol. CS-212　　FROM GIANT PLANETS TO COOL STARS
eds. Caitlin A. Griffith and Mark S. Marley
ISBN: 1-58381-041-2

Vol. CS-213　　BIOASTRONOMY `99: A NEW ERA IN BIOASTRONOMY
eds. Guillermo A. Lemarchand and Karen J. Meech
ISBN: 1-58381-044-7

Vol. CS-214　　THE Be PHENOMENON IN EARLY-TYPE STARS, IAU Colloquium 175
eds. Myron A. Smith, Huib F. Henrichs and Juan Fabregat
ISBN: 1-58381-045-5

Vol. CS-215　　COSMIC EVOLUTION AND GALAXY FORMATION:
STRUCTURE, INTERACTIONS AND FEEDBACK
The 3rd Guillermo Haro Astrophysics Conference
eds. José Franco, Elena Terlevich, Omar López-Cruz, and Itziar Aretxaga
ISBN: 1-58381-046-3

Vol. CS-216　　ASTRONOMICAL DATA ANALYSIS SOFTWARE AND SYSTEMS IX
eds. Nadine Manset, Christian Veillet, and Dennis Crabtree
ISBN: 1-58381-047-1　　　　ISSN: 1080-7926

Vol. CS-217　　IMAGING AT RADIO THROUGH SUBMILLIMETER WAVELENGTHS
eds. Jeffrey G. Mangum and Simon J. E. Radford
ISBN: 1-58381-049-8

Vol. CS-218　　MAPPING THE HIDDEN UNIVERSE: THE UNIVERSE BEHIND THE MILKY WAY
THE UNIVERSE IN HI
eds. Renée C. Kraan-Korteweg, Patricia A. Henning, and Heinz Andernach
ISBN: 1-58381-050-1

Vol. CS-219　　DISKS, PLANETESIMALS, AND PLANETS
eds. F. Garzón, C. Eiroa, D. de Winter, and T. J. Mahoney
ISBN: 1-58381-051-X

Vol. CS-220　　AMATEUR - PROFESSIONAL PARTNERSHIPS IN ASTRONOMY:
The 111th Annual Meeting of the ASP
eds. John R. Percy and Joseph B. Wilson
ISBN: 1-58381-052-8

Vol. CS-221　　STARS, GAS AND DUST IN GALAXIES: EXPLORING THE LINKS
eds. Danielle Alloin, Knut Olsen, and Gaspar Galaz
ISBN: 1-58381-053-6

PUBLISHED: 2001

Vol. CS-222　　THE PHYSICS OF GALAXY FORMATION
eds. M. Umemura and H. Susa
ISBN: 1-58381-054-4

Vol. CS-223　　COOL STARS, STELLAR SYSTEMS AND THE SUN:
Eleventh Cambridge Workshop
eds. Ramón J. García López, Rafael Rebolo, and María Zapatero Osorio
ISBN: 1-58381-056-0

Vol. CS-224　　PROBING THE PHYSICS OF ACTIVE GALACTIC NUCLEI
BY MULTIWAVELENGTH MONITORING
eds. Bradley M. Peterson, Ronald S. Polidan, and Richard W. Pogge
ISBN: 1-58381-055-2

ASP CONFERENCE SERIES VOLUMES
Published by the Astronomical Society of the Pacific

PUBLISHED: 2001 (* asterisk means OUT OF STOCK)

Vol. CS-225   VIRTUAL OBSERVATORIES OF THE FUTURE
eds. Robert J. Brunner, S. George Djorgovski, and Alex S. Szalay
ISBN: 1-58381-057-9

Vol. CS-226   12th EUROPEAN CONFERENCE ON WHITE DWARFS
eds. J. L. Provencal, H. L. Shipman, J. MacDonald, and S. Goodchild
ISBN: 1-58381-058-7

Vol. CS-227   BLAZAR DEMOGRAPHICS AND PHYSICS
eds. Paolo Padovani and C. Megan Urry
ISBN: 1-58381-059-5

Vol. CS-228   DYNAMICS OF STAR CLUSTERS AND THE MILKY WAY
eds. S. Deiters, B. Fuchs, A. Just, R. Spurzem, and R. Wielen
ISBN: 1-58381-060-9

Vol. CS-229   EVOLUTION OF BINARY AND MULTIPLE STAR SYSTEMS
A Meeting in Celebration of Peter Eggleton's 60th Birthday
eds. Ph. Podsiadlowski, S. Rappaport, A. R. King, F. D'Antona, and L. Burderi
IBSN: 1-58381-061-7

Vol. CS-230   GALAXY DISKS AND DISK GALAXIES
eds. Jose G. Funes, S. J. and Enrico Maria Corsini
ISBN: 1-58381-063-3

Vol. CS-231   TETONS 4: GALACTIC STRUCTURE, STARS, AND
THE INTERSTELLAR MEDIUM
eds. Charles E. Woodward, Michael D. Bicay, and J. Michael Shull
ISBN: 1-58381-064-1

Vol. CS-232   THE NEW ERA OF WIDE FIELD ASTRONOMY
eds. Roger Clowes, Andrew Adamson, and Gordon Bromage
ISBN: 1-58381-065-X

Vol. CS-233   P CYGNI 2000: 400 YEARS OF PROGRESS
eds. Mart de Groot and Christiaan Sterken
ISBN: 1-58381-070-6

Vol. CS-234   X-RAY ASTRONOMY 2000
eds. R. Giacconi, S. Serio, and L. Stella
ISBN: 1-58381-071-4

Vol. CS-235   SCIENCE WITH THE ATACAMA LARGE MILLIMETER ARRAY (ALMA)
ed. Alwyn Wootten
ISBN: 1-58381-072-2

Vol. CS-236   ADVANCED SOLAR POLARIMETRY: THEORY, OBSERVATION, AND
INSTRUMENTATION, The 20th Sacramento Peak Summer Workshop
ed. M. Sigwarth
ISBN: 1-58381-073-0

Vol. CS-237   GRAVITATIONAL LENSING: RECENT PROGRESS AND FUTURE GOALS
eds. Tereasa G. Brainerd and Christopher S. Kochanek
ISBN: 1-58381-074-9

Vol. CS-238   ASTRONOMICAL DATA ANALYSIS SOFTWARE AND SYSTEMS X
eds. F. R. Harnden, Jr., Francis A. Primini, and Harry E. Payne
ISBN: 1-58381-075-7

Vol. CS-239   MICROLENSING 2000: A NEW ERA OF MICROLENSING ASTROPHYSICS
ed. John Menzies and Penny D. Sackett
ISBN: 1-58381-076-5

ASP CONFERENCE SERIES VOLUMES
Published by the Astronomical Society of the Pacific

PUBLISHED: 2001 (* asterisk means OUT OF STOCK)

Vol. CS-240   GAS AND GALAXY EVOLUTION,
A Conference in Honor of the 20th Anniversary of the VLA
eds. J. E. Hibbard, M. P. Rupen, and J. H. van Gorkom
ISBN: 1-58381-077-3

Vol. CS-241   CS-241 THE 7TH TAIPEI ASTROPHYSICS WORKSHOP ON
COSMIC RAYS IN THE UNIVERSE
ed. Chung-Ming Ko
ISBN: 1-58381-079-X

Vol. CS-242   ETA CARINAE AND OTHER MYSTERIOUS STARS:
THE HIDDEN OPPORTUNITIES OF EMISSION SPECTROSCOPY
eds. Theodore R. Gull, Sveneric Johannson, and Kris Davidson
ISBN: 1-58381-080-3

Vol. CS-243   FROM DARKNESS TO LIGHT:
ORIGIN AND EVOLUTION OF YOUNG STELLAR CLUSTERS
eds. Thierry Montmerle and Philippe André
ISBN: 1-58381-081-1

Vol. CS-244   YOUNG STARS NEAR EARTH: PROGRESS AND PROSPECTS
eds. Ray Jayawardhana and Thomas P. Greene
ISBN: 1-58381-082-X

Vol. CS-245   ASTROPHYSICAL AGES AND TIME SCALES
eds. Ted von Hippel, Chris Simpson, and Nadine Manset
ISBN: 1-58381-083-8

Vol. CS-246   SMALL TELESCOPE ASTRONOMY ON GLOBAL SCALES, IAU Colloquium 183
eds. Wen-Ping Chen, Claudia Lemme, and Bohdan Paczyński
ISBN: 1-58381-084-6

Vol. CS-247   SPECTROSCOPIC CHALLENGES OF PHOTOIONIZED PLASMAS
eds. Gary Ferland and Daniel Wolf Savin
ISBN: 1-58381-085-4

Vol. CS-248   MAGNETIC FIELDS ACROSS THE HERTZSPRUNG-RUSSELL DIAGRAM
eds. G. Mathys, S. K. Solanki, and D. T. Wickramasinghe
ISBN: 1-58381-088-9

Vol. CS-249   THE CENTRAL KILOPARSEC OF STARBURSTS AND AGN:
THE LA PALMA CONNECTION
eds. J. H. Knapen, J. E. Beckman, I. Shlosman, and T. J. Mahoney
ISBN: 1-58381-089-7

Vol. CS-250   PARTICLES AND FIELDS IN RADIO GALAXIES CONFERENCE
eds. Robert A. Laing and Katherine M. Blundell
ISBN: 1-58381-090-0

Vol. CS-251   NEW CENTURY OF X-RAY ASTRONOMY
eds. H. Inoue and H. Kunieda
ISBN: 1-58381-091-9

Vol. CS-252   HISTORICAL DEVELOPMENT OF MODERN COSMOLOGY
eds. Vicent J. Martínez, Virginia Trimble, and María Jesús Pons-Bordería
ISBN: 1-58381-092-7

PUBLISHED: 2002

Vol. CS-253   CHEMICAL ENRICHMENT OF INTRACLUSTER AND INTERGALACTIC MEDIUM
eds. Roberto Fusco-Femiano and Francesca Matteucci
ISBN: 1-58381-093-5

ASP CONFERENCE SERIES VOLUMES
Published by the Astronomical Society of the Pacific

PUBLISHED: 2002 (* asterisk means OUT OF STOCK)

Vol. CS-254   EXTRAGALACTIC GAS AT LOW REDSHIFT
eds. John S. Mulchaey and John T. Stocke
ISBN: 1-58381-094-3

Vol. CS-255   MASS OUTFLOW IN ACTIVE GALACTIC NUCLEI: NEW PERSPECTIVES
eds. D. M. Crenshaw, S. B. Kraemer, and I. M. George
ISBN: 1-58381-095-1

Vol. CS-256   OBSERVATIONAL ASPECTS OF PULSATING B AND A STARS
eds. Christiaan Sterken and Donald W. Kurtz
ISBN: 1-58381-096-X

Vol. CS-257   AMiBA 2001: HIGH-Z CLUSTERS, MISSING BARYONS, AND CMB POLARIZATION
eds. Lin-Wen Chen, Chung-Pei Ma, Kin-Wang Ng, and Ue-Li Pen
ISBN: 1-58381-097-8

Vol. CS-258   ISSUES IN UNIFICATION OF ACTIVE GALACTIC NUCLEI
eds. Roberto Maiolino, Alessandro Marconi, and Neil Nagar
ISBN: 1-58381-098-6

Vol. CS-259   RADIAL AND NONRADIAL PULSATIONS AS PROBES OF STELLAR PHYSICS
eds. Conny Aerts, Timothy R. Bedding, and Jørgen Christensen-Dalsgaard
ISBN: 1-58381-099-4

Vol. CS-260   INTERACTING WINDS FROM MASSIVE STARS
eds. Anthony F. J. Moffat and Nicole St-Louis
ISBN: 1-58381-100-1

Vol. CS-261   THE PHYSICS OF CATACLYSMIC VARIABLES AND RELATED OBJECTS
eds. B. T. Gänsicke, K. Beuermann, and K. Reinsch
ISBN: 1-58381-101-X

Vol. CS-262   THE HIGH ENERGY UNIVERSE AT SHARP FOCUS: CHANDRA SCIENCE
The 113th Annual Meeting of the ASP
eds. Eric M. Schlegel and Saeqa Dil Vrtilek
ISBN: 1-58381-102-8

Vol. CS-263   STELLAR COLLISIONS, MERGERS AND THEIR CONSEQUENCES
ed. Michael M. Shara
ISBN: 1-58381-103-6

Vol. CS-264   CONTINUING THE CHALLENGE OF EUV ASTRONOMY: CURRENT ANALYSIS AND PROSPECTS FOR THE FUTURE
eds. Steve B. Howell, Jean Dupuis, Daniel Golombek, Frederick M. Walter, and Jennifer Cullison
ISBN: 1-58381-104-4

Vol. CS-265   ω CENTAURI, A UNIQUE WINDOW INTO ASTROPHYSICS
eds. Floor van Leeuwen, Joanne D. Hughes, and Giampaolo Piotto
ISBN: 1-58381-105-2

Vol. CS-266   ASTRONOMICAL SITE EVALUATION IN THE VISIBLE AND RADIO RANGE
eds. J. Vernin, Z. Benkhaldoun, and C. Muñoz-Tuñón
ISBN: 1-58381-106-0

Vol. CS-267   HOT STAR WORKSHOP III: THE EARLIEST STAGES OF MASSIVE STAR BIRTH
ed. Paul A. Crowther
ISBN: 1-58381-107-9

Vol. CS-268   TRACING COSMIC EVOLUTION WITH GALAXY CLUSTERS
eds. Stefano Borgani, Marino Mezzetti, and Riccardo Valdarnini
ISBN: 1-58381-108-7

ASP CONFERENCE SERIES VOLUMES
Published by the Astronomical Society of the Pacific

PUBLISHED: 2002 (* asterisk means OUT OF STOCK)

Vol. CS-269 THE EVOLVING SUN AND ITS INFLUENCE ON PLANETARY ENVIRONMENTS
eds. Benjamín Montesinos, Álvaro Giménez, and Edward F. Guinan
ISBN: 1-58381-109-5

Vol. CS-270 ASTRONOMICAL INSTRUMENTATION AND THE BIRTH AND GROWTH OF ASTROPHYSICS: A Symposium held in honor of Robert G. Tull
eds. Frank N. Bash and Christopher Sneden
ISBN: 1-58381-110-9

Vol. CS-271 NEUTRON STARS IN SUPERNOVA REMNANTS
eds. Patrick O. Slane and Bryan M. Gaensler
ISBN: 1-58381-111-7

Vol. CS-272 THE FUTURE OF SOLAR SYSTEM EXPLORATION, 2003-2013
Community Contributions to the NRC Solar System Exploration Decadal Survey
ed. Mark V. Sykes
ISBN: 1-58381-113-3

Vol. CS-273 THE DYNAMICS, STRUCTURE AND HISTORY OF GALAXIES
eds. G. S. Da Costa and H. Jerjen
ISBN: 1-58381-114-1

Vol. CS-274 OBSERVED HR DIAGRAMS AND STELLAR EVOLUTION
eds. Thibault Lejeune and João Fernandes
ISBN: 1-58381-116-8

Vol. CS-275 DISKS OF GALAXIES: KINEMATICS, DYNAMICS AND PERTURBATIONS
eds. E. Athanassoula, A. Bosma, and R. Mujica
ISBN: 1-58381-117-6

Vol. CS-276 SEEING THROUGH THE DUST:
THE DETECTION OF HI AND THE EXPLORATION OF THE ISM IN GALAXIES
eds. A. R. Taylor, T. L. Landecker, and A. G. Willis
ISBN: 1-58381-118-4

Vol. CS 277 STELLAR CORONAE IN THE CHANDRA AND XMM-NEWTON ERA
eds. Fabio Favata and Jeremy J. Drake
ISBN: 1-58381-119-2

Vol. CS 278 NAIC–NRAO SCHOOL ON SINGLE-DISH ASTRONOMY:
TECHNIQUES AND APPLICATIONS
eds. Snezana Stanimirovic, Daniel Altschuler, Paul Goldsmith, and Chris Salter
ISBN: 1-58381-120-6

A LISTING OF IAU VOLUMES MAY BE FOUND ON THE NEXT PAGE

INTERNATIONAL ASTRONOMICAL UNION (IAU) VOLUMES
Published by the Astronomical Society of the Pacific

PUBLISHED: 1999 (* asterisk means OUT OF STOCK)

Vol. No. 190 NEW VIEWS OF THE MAGELLANIC CLOUDS
eds. You-Hua Chu, Nicholas B. Suntzeff, James E. Hesser, and David A. Bohlender
ISBN: 1-58381-021-8

Vol. No. 191 ASYMPTOTIC GIANT BRANCH STARS
eds. T. Le Bertre, A. Lèbre, and C. Waelkens
ISBN: 1-886733-90-2

Vol. No. 192 THE STELLAR CONTENT OF LOCAL GROUP GALAXIES
eds. Patricia Whitelock and Russell Cannon
ISBN: 1-886733-82-1

Vol. No. 193 WOLF-RAYET PHENOMENA IN MASSIVE STARS AND STARBURST GALAXIES
eds. Karel A. van der Hucht, Gloria Koenigsberger, and Philippe R. J. Eenens
ISBN: 1-58381-004-8

Vol. No. 194 ACTIVE GALACTIC NUCLEI AND RELATED PHENOMENA
eds. Yervant Terzian, Daniel Weedman, and Edward Khachikian
ISBN: 1-58381-008-0

PUBLISHED: 2000

Vol. XXIVA TRANSACTIONS OF THE INTERNATIONAL ASTRONOMICAL UNION REPORTS ON ASTRONOMY 1996-1999
ed. Johannes Andersen
ISBN: 1-58381-035-8

Vol. No. 195 HIGHLY ENERGETIC PHYSICAL PROCESSES AND MECHANISMS FOR EMISSION FROM ASTROPHYSICAL PLASMAS
eds. P. C. H. Martens, S. Tsuruta, and M. A. Weber
ISBN: 1-58381-038-2

Vol. No. 197 ASTROCHEMISTRY: FROM MOLECULAR CLOUDS TO PLANETARY SYSTEMS
eds. Y. C. Minh and E. F. van Dishoeck
ISBN: 1-58381-034-X

Vol. No. 198 THE LIGHT ELEMENTS AND THEIR EVOLUTION
eds. L. da Silva, M. Spite, and J. R. de Medeiros
ISBN: 1-58381-048-X

PUBLISHED: 2001

IAU SPS ASTRONOMY FOR DEVELOPING COUNTRIES
Special Session of the XXIV General Assembly of the IAU
ed. Alan H. Batten
ISBN: 1-58381-067-6

Vol. No. 196 PRESERVING THE ASTRONOMICAL SKY
eds. R. J. Cohen and W. T. Sullivan, III
ISBN: 1-58381-078-1

Vol. No. 200 THE FORMATION OF BINARY STARS
eds. Hans Zinnecker and Robert D. Mathieu
ISBN: 1-58381-068-4

Vol. No. 203 RECENT INSIGHTS INTO THE PHYSICS 0F THE SUN AND HELIOSPHERE: HIGHLIGHTS FROM SOHO AND OTHER SPACE MISSIONS
eds. Pål Brekke, Bernhard Fleck, and Joseph B. Gurman
ISBN: 1-58381-069-2

Vol. No. 204 THE EXTRAGALACTIC INFRARED BACKGROUND AND ITS COSMOLOGICAL IMPLICATIONS
eds. Martin Harwit and Michael G. Hauser
ISBN: 1-58381-062-5

INTERNATIONAL ASTRONOMICAL UNION (IAU) VOLUMES
Published by the Astronomical Society of the Pacific

PUBLISHED: 2001 (* asterisk means OUT OF STOCK)

Vol. No. 205   GALAXIES AND THEIR CONSTITUENTS
AT THE HIGHEST ANGULAR RESOLUTIONS
eds. Richard T. Schilizzi, Stuart N. Vogel, Francesco Paresce, and Martin S. Elvis
ISBN: 1-58381-066-8

Vol. XXIVB   TRANSACTIONS OF THE INTERNATIONAL ASTRONOMICAL UNION
REPORTS ON ASTRONOMY
ed. Hans Rickman
ISBN: 1-58381-087-0

PUBLISHED: 2002

Vol. No. 12   HIGHLIGHTS OF ASTRONOMY
ed. Hans Rickman
ISBN: 1-58381-086-2

Vol. No. 206   COSMIC MASERS: FROM PROTOSTARS TO BLACKHOLES
eds. Victor Migenes, Mark J. Reid, and Everton Ludke
ISBN: 1-58381-112-5

Ordering information is available at the beginning of the listing